THE TIMETABLES™
OF TECHNOLOGY

A Chronology of the Most Important People and Events in the History of Technology

Bryan Bunch and Alexander Hellemans

A TOUCHSTONE BOOK

Published by Simon & Schuster

New York London Toronto Sydney Tokyo Singapore

Research and writing assistance: Marianne Bunch; Sally Bunch
Copyediting and indexing: Felice Levy of AEIOU, Inc.
Additional copyediting: Leontina Hellemans and Marianne Bunch
Design and layout: Gerry Burstein and Jeanne Borczuk, G & H SOHO, Inc.

TOUCHSTONE
Rockefeller Center
1230 Avenue of the Americas
New York, New York 10020

The Timetables of Technology is a trademark of Simon & Schuster Inc.
The Timetables of Technology is part of a best-selling series of books,
including *The Timetables of Science*, *The Timetables of History*,
The Timetables of Jewish History, and *The Timetables of American History*.

First Touchstone Edition 1994

Manufactured in the United States of America

1   3   5   7   9   10   8   6   4   2

Library of Congress Cataloging-in-Publication Data

Bunch, Bryan H.
The timetables of technology : a chronology of the most important
people and events in the history of technology / Bryan Bunch and Alexander Hellemans.
p.   cm.
Includes index
1. Technology—History—Chronology—Tables
I. Hellemans, Alexander. II. Title
T15.B73   1993
609—dc20   93-27734   CIP
ISBN 0-671-76918-9
ISBN 0-671-88767-X (pbk)

CONTENTS

Notes on conventions and abbreviations

Cross-references

Many entries in the timetable section have cross-references, at most one to an entry from an earlier date and one to an entry that follows. Often these can be put together to form a string, so that one can use them to trace the history of bridge building or television. In other cases they are used primarily to refer to additional information on a specific topic (such as the Eddystone Lighthouse) or to follow up an entry that occurs some years later. All cross-references use the year as it is listed in the side columns (except "c" is omitted from early dates) and a three-letter abbreviation:

GEN   GENERAL
ARC   ARCHITECTURE AND CONSTRUCTION
COM   COMMUNICATIONS or
      COMMUNICATIONS/TRANSPORTATION
ELE   ELECTRONICS AND COMPUTERS
ENE   ENERGY
FOO   FOOD AND AGRICULTURE or
      FOOD AND SHELTER
MAT   MATERIALS or
      MATERIALS/MEDICAL TECHNOLOGY
MED   MEDICAL TECHNOLOGY
TOO   TOOLS AND DEVICES or TOOLS/MATERIALS
TRA   TRANSPORTATION

The same abbreviations are used to identify entries in the Index.

Date labels and note on early dates

Dates in *The Timetables of Technology* are labeled as BP for "before present"; BC, for a compromise between BCE for "before common era" and BC for "before Christ," the common designation in most Christian countries; and CE for "common era," a designation developed largely in Israel. Dates labeled CE are exactly the same as those generally labeled AD in most Christian countries.

It is assumed that for periods longer ago than about 100,000 years, possible variation in dates is so great that there is no discernible difference between BP and BC. For more recent dates, those between about 5000 years ago and 100,000 years ago, the difference between BP and BC is taken to be 2000 years. From 5000 years ago to the present, BP dates are not used.

Biographical dates and places

Where possible, a person who is mentioned in the book has birth and death information listed with the *first* entry concerning that person in the timetable section. Thus, at the first mention of the inventor of a common type of thermometer, he is listed as Gabriel *Daniel* Fahrenheit [b Danzig (Gdansk, Poland), May 24 1693, d The Hague, Holland, Sep 16 1736]. Note that names commonly used other than the first are shown in italics, birth and death are listed as b and d, the biographical entry—which may also include information on knighthood—is in square brackets, and modern versions of place names are in parentheses. All months are abbreviated with their first three letters and no periods are used with abbreviations.

If partial information is all that is available, that information is provided. Information is not dropped because it is not consistent with the information in other entries. Thus, if a date or birthplace is omitted, it is usually because we could not locate a source for it. The exception of course is that living people do not have death dates.

An effort has been made to use original place names where possible and identify them further with modern names in parentheses. Cities can be recognized from nations or states by the use of *in*, as for example Antioch (in Turkey) or Samarkand (in Uzbekistan). States that have different names do not use *in*, as for example Hanover (Germany), Holland (the Netherlands), or Persia (Iran), even when boundaries are considerably changed from earlier ones. Some states, such as England and France, are presumed to have existed since the fall of the Roman Empire.

OVERVIEW

The Stone Ages

Today we are so accustomed to the idea of a time called the Stone Age that it is easy to forget that the expression was coined less than two centuries ago by Christian Jürgensen Thomsen for a project started in 1816. Thomsen divided early artifacts for a museum collection into stone, bronze, and iron. The museum catalog, published in 1836, enshrined the Stone, Bronze, and Iron ages. In 1865 Sir John Lubbock further subdivided the Stone Age into the Old Stone Age and the New Stone Age. After these simple names were unnecessarily turned into the Greek-derived technical terms *Paleolithic* and *Neolithic,* a middle stone age, the *Mesolithic*, was added. Actually, some paleoarchaeologists make a distinction between, say, the Old Stone Age and the Paleolithic, but we shall not.

Before the stone ages

Stone tools are the earliest artifacts known because stone does not decay. However, there is good reason to believe that tool use, and therefore technology, began among our ancestors well before any tools were made from stone.

Many animals use naturally occurring objects for specific purposes—perhaps the most famous example is the use of rocks by sea otters to crack open abalone shells. Although this is akin to tool use, it is more helpful to reject that use because the natural object, the rock, is not modified by the user for its specific purpose. In other words, the rock is an object used as a tool, not a tool.

Still, tools are not limited to human manufacture and use. Since Oct 1960 science has known that chimpanzees make and use tools in the wild. At that time Jane Goodall observed the chimpanzee she had named David Graybeard trim a wide blade of grass into a tool that could be poked into a termite mound, from which it would emerge as a termite "lollipop." Angry termites clinging to the tool could be easily removed by putting the tool in the chimp's mouth and pulling the tool out between closed lips.

Over more than 30 years of studying chimpanzees, Goodall noted many instances when the apes made similar termite-snaring tools during "termite season." Often the tools were manufactured hundreds of yards from, and well out of sight of, the termite nests, showing that the toolmaking was not just a nonce activity, even though the tools themselves were nonce tools (intended to be used once and discarded). Goodall also observed chimpanzees making "sponges" from leaves for use in drinking water from a stream, another example of toolmaking. In the only example of chimpanzee toolmaking reported before 1960, laboratory studies showed one instance in which a bright chimpanzee named Sultan put two sticks together as a nonce tool to obtain bananas that were otherwise beyond reach.

Beyond the chimpanzee, our closest relative, the orangutan has been observed to make upside-down nests as shelters from the rain. Going beyond primates and even beyond mammals, finches in the Galapagos Islands also make tools. They remove and trim thorns to use in obtaining insect grubs from holes in trees. Perhaps even more dramatically, striated herons break off a small piece of twig to use as a lure, tossing the piece into the water and watching until it attracts attention from an unfortunate fish, who is quickly eaten.

Technology and other hominids

Knowledge of chimpanzee toolmaking strongly suggests that the chimpanzees' closest relatives of about 5 to 7 million years ago—a time in which part of the ape line branched into chimpanzees and the gorilla on the one hand and our ancestors on the other—also were toolmakers. Most likely, any early tools were also of the nonce variety, although we have no way of knowing this. The very earliest manufactured tools, almost surely made from biological materials such as wood, have long since vanished. Perhaps someday, however, we will find some of these tools, nonce or otherwise, in a bog or some other place with unusual powers of preservation.

Not all of our early ancestors are known yet. Paleoanthropologists classify most of our earliest ones as australopithecines, a different genus from *Homo*, but in the same family. It is thought that australopithecines had too little brain power to make tools, but their brains were about the same size as those of modern chimpanzees.

Anthropologists say that the average australopithecine brain was about 430 cm³ (although that of *Australopithecus robustus* was about a hundred cubic centimeters more), while the brain of the first member of the genus *Homo, Homo habilis,* took a giant leap forward to 750 cm³. A look at the actual data shows a more gradual progression from *A. robustus*—who probably is not a direct ancestor, but more like a cousin—with a maximum brain of about 550 cm³ to *H. habilis*. Most *habilis* specimens had a brain capacity closer to 600 cm³ than to 750 cm³, which is the maximum size found in *habilis* fossils.

Brain size is relevant because the first recognizable manufactured tools, the so-called pebble or Oldowan tools, are found at sites that include both *A. robustus* and *H. habilis*. While we assume that our somewhat larger brained direct ancestor chipped and used those first recognizable stone tools, we cannot be certain that our cousin was not involved.

A million years or so after the first tools were made, the situation grew much clearer. By that time all the australopithecines were gone and the only hominid present was our ancestor *Homo erectus* and (depending on how you classify them) the two species of chimpanzee and the gorilla. Tools are clearly associated with *H. erectus*. Furthermore, the tools are much more sophisticated than the simple pebble tools. The main tool was a sort of generalized stone oblong or teardrop shape that anthropologists like to call by the neutral name of *biface,* although the traditional name of *hand axe* is commonly employed by the rest of the world.

H. erectus's advances include more than just better stone tools. These hominids were apparently the first creatures on Earth to control fire. No other organism has accomplished this, so fire is considered a defining characteristic of humanity, perhaps even more so than toolmaking or language, which form something of a continuum from other animals to humans. The early evidence for fire is not unequivocal; some remains thought to be hearths could be the results of lightning strikes. During the million and a half years that *H. erectus* was present, however, the evidence for control of fire gradually improves. Eventually there is definite evidence that *erectus* used fire, although it is not known whether *erectus* could start fires as needed.

All the hominids before *H. erectus* arose and stayed in Africa. *H. erectus* also first appeared in Africa, but soon ventured out to adjoining Europe and Asia. Any land that could be reached by walking or crossing streams was soon populated with the wandering hominids. It was left for the next species to populate the rest of Earth.

One group of hominids succeeding *H. erectus* was the mysterious Neandertal, the characteristic human of most of the Ice Age. Neandertals bring new elements to the human equation—ornaments and burial of the dead.

Anthropologists call the typical set of tools made by a given people a toolkit or a tool industry. The Neandertal tool industry brings new techniques and more complex tools, but like the toolkit of *H. erectus*, the Neandertal toolkit was virtually created once and then left alone. Changes over long periods of time consist of small improvements, not major innovations.

Unexpected pleasures

At some not clearly identified point in the past, people much like ourselves began to appear, probably in

Most people know that stone "arrowheads" were made from a kind of rock called *flint,* but otherwise have no idea about the relationship of stone as a material to tool manufacture and use. Our ancestors were more discerning, as they had to be to survive.

Flint really is the best material for making a great variety of stone tools. Flint is closely related to the semiprecious stones called carnelian, chrysoprase, and jasper, which are uniform, red, green, or yellow forms of the mineral chalcedony. Large deposits of gray or black chalcedony are called chert, while small pieces, called nodules, of gray or black chalcedony found in limestone or chalk are called flint. Thus, the harder and more chemically stable flint can easily be picked from its limestone or chalk background. A nodule broken in two by one process or another could be the first manufactured stone tool, indistinguishable as a tool today except for microscopic wear patterns that indicate use.

No one knows for sure what causes flint nodules to appear where they do. Chalcedony is a form of quartz that has tiny crystals and is very dense; hence it is silicon dioxide, also known to mineralologists as silica, just as quartz is. Limestone and chalk are both calcium carbonate, a different mineral entirely. Veins of silicon dioxide often form as a result of solutions of water containing the silica, especially solutions in superheated water. The solutions travel through a different kind of rock and leave the silica behind when they cool.

Because of its crystal structure, flint breaks in a pattern geologists call a conchoidal fracture; this produces sharp edges but does not propagate throughout a stone, splitting or shattering it. Furthermore, there are no preferential fracture planes, so small pieces of almost any shape can be removed. The beautifully scalloped surfaces of many later stone tools are one result.

In the absence of the fine-grained flints, our ancestors used the best approximations they could find—quartz (silicon dioxide with a larger crystal structure) and materials infused with silica, including petrified wood.

Anyone who has picked up the remains of a shattered glass dish or bottle knows that glass also breaks in a pattern that causes very sharp edges. The break is also a conchoidal fracture, but because glass is more brittle than flint, the glass not only fractures but also easily shatters. A fragment of broken glass can have a very sharp cutting edge that can be used as a tool—who hasn't seen the hero or heroine of an adventure movie cut the ropes binding him or her with a piece of broken glass? Hugh Lofting's Dr. Doolittle often vowed on his adventures to shave with broken glass to save the space that would be taken by packing a razor.

The utility of broken glass was not lost on our ancestors. Although manufacture of glass from silicon dioxide (sand that is formed from small particles of quartz) did not start as far as we know until about 1400 BC, implements made from natural glass are among the earliest stone tools. Natural glass is a rock produced when granite or rhyolite, quartz combined with feldspar and mica, are melted by a volcano and then cooled very quickly. The differences between granite (large crystals), rhyolite (small crystals), and obsidian (no crystals to speak of, although a few tiny ones may be present) are entirely the result of differences in the time of cooling. Slow cooling is needed for crystals to form; the slower the cooling, the larger the crystals. A glass with no crystal structure is called amorphous. Thus, veins of rock that cool slowly deep in Earth become granite, while extruded lavas from volcanos often contain obsidian. Opal, an amorphous form of quartz, was also used for stone tools, but it is less common than obsidian.

As with substitutes for flint, volcanic rocks such as lavas or rocks formed from hot ash flows were sometimes used instead of volcanic glasses.

A third type of stone-tool material includes various rocks, such as quartzites or hardened shales, that had been hardened (metamorphosed) by great heat and pressure in the interior of Earth. Quartzites that are metamorphosed sandstones became particularly useful for axe and adz heads after the practice of grinding edges was introduced as part of the Neolithic Revolution. Quartzite axe heads have the advantage of a structure in which cracks do not propagate far, so the tools maintain their integrity even when their edge is chipped.

Very early stone bifaces (hand axes) were also made from quartzite, which easily can be flaked to produce a useful but not very attractive tool. As our ancestors became more experienced, they first shifted to the volcanic lavas that could yield smaller, better looking flakes with sharper edges. Eventually flint came to predominate, even in regions where flint is not a common rock.

Flint and obsidian were mined and traded in the later Stone Age. As population increased and there was less flint to go around, new techniques, such as manufacture of microliths, were developed. Microliths involve flint or obsidian cutting edges embedded in wood, with the bulk of the tool being the wooden portion.

Africa, the original home of all previous hominids. These people advanced on all the technological fronts pioneered by earlier hominids. But they also introduced something new, which we eventually came to think of as an esthetic sensibility; this was shared only with the Neandertals, who may have been the same species as modern humans in any case.

Although there is evidence that the Neandertals made some ornaments and that they painted something—probably their own bodies, the most common form of artistic expression around the world—it fell to the first humans like ourselves, fully modern *Homo sapiens,* to introduce a wide range of decorative jewelry. Furthermore, our kind made music of some sort and truly great paintings and sculpture. The paintings are considered as a part of the communications strand, and are discussed below.

Major Advances

Architecture and construction.

Whether or not *Homo habilis* had semipermanent homes is far from clear. Certainly there are places where leftover tools and chips from toolmaking are much more common than others. Perhaps, however, these are the sites of major kills, either by *habilis* or by predators (in which case *habilis* used the tools to scavenge the site). Or maybe they were favorite places for water or what anthropologists call "living floors," in which case they might have held some sort of permanent dwelling—the beginning of architecture.

It is not until over a million years later that we today can be sure that people were building recognizable huts, although ambiguous evidence for earlier postholes suggests that the practice may have started with *Homo erectus.* The best evidence for houses built by *erectus* comes from the site Terra Amata, dated at 400,000 BP, near what is now Nice, France. There the postholes form ovals 15 m (50 ft) long and 6 m (20 ft) wide. Furthermore, there is literally rock-hard evidence of housing, since rocks that surround the postholes seem to have been used for bracing or perhaps for holding down the edges of hides. However, some paleoanthropologists remain unconvinced that these are the remains of the earliest known houses.

The biodegradability of wood is a major problem for paleoarchaeologists studying housing. Undisputed identification first comes for dwellings made from bone and brick, although no doubt preceded by houses of straw and sticks.

Bricks began to emerge in the Near East about 12,000 BP in regions where both wood and stone were in short supply. The first bricks were a kind of sun-dried adobe held together without mortar. Mortar and plaster were discovered by at least 8500 BC; they greatly improved construction with sunbaked brick. Although people in Europe learned to fire clay to make it harder (and to mix it with ground-up bone to make it harder yet), populations of the Near East were slow to learn that fire makes clay into ceramic. However, in Europe, where the expertise to make ceramics existed, people did not need the material for construction, having plenty of wood and stone. Thus, ceramics were first used for ornaments and small statues.

By the end of the Stone Age, houses were fairly complex, with some even built of stone. For most of the period, however, stone was not used in construction. Temples, tombs, towers, and mighty stone works of all sorts were yet to come. Indeed, some of the main works of the next period consist solely of giant stones.

Communications/Transportation.

The question of how much and what kind of communication existed before written records is extremely difficult to answer. The 1980s and 1990s saw a continuing battle over whether Neandertals could speak—and Neandertals were probably of our own species. Understanding communication among earlier hominids is more difficult still.

As with toolmaking, perhaps the first place to look for guidance is to the chimpanzee and gorilla, accepted by many on the basis of DNA studies as the only extant hominids other than *Homo sapiens.* Jane Goodall, who has studied chimpanzees in the wild for over 30 years, summarizes the situation in her 1990 book *Through a Window*: "Sometimes when watching the chimpanzees, I have felt that, because they have no human-like language, they are trapped within themselves. Their calls, postures and gestures, together, add up to a rich repertoire, a complex and sophisticated method of communication. But it is non-verbal. How much more might they accomplish if they could *talk* to each other. It is true that they can be *taught* to use the signs or symbols of a human-type language [italics hers]." There is every possibility that other nonhuman hominids were similar in abilities to chimpanzees, although some may have been even more sophisticated in their use of "calls, postures and gestures."

On the other hand, *Homo erectus* at least appears to have more complex hunting behavior than chimpanzees. Perhaps *erectus* invented speech along with learning to control fire. It is impossible to know one way or another.

Looking at the prehistoric archaeological record, however, one can discern a variety of modes of communication that have been preserved. Even ignoring music and ornament, visual art appears to have been used for more purposes than casual copying of nature. Although there are a few carvings that may have been produced by Neandertals, recognizable images all date from the time after completely modern *H. sapiens* replaced all other members of our genus.

No one knows what very early art meant, but there are excellent reasons to suspect that it meant *something*. A prevalence of images that suggest vulvas or women with exaggerated sexual characteristics and some early phalluses provide a theme running through much of art from about 30,000 BP. A different theme emerges about 10,000 years later with many paintings of prey animals, some shown with mysterious "darts" painted at various locations on the body. Many theories

Although some taxonomists want to call *Homo habilis* a composite of several species, while others want to lump *Homo erectus* into *H. sapiens,* most authorities continue to class *habilis* and *erectus* as two and only two species. In that light, we probably owe the first key technology, the use of stone tools, to *H. habilis* and certainly the second key tool, fire, to *H. erectus.*

Stone tools, especially early ones, are simply a better technology for doing the same things as wooden tools or even bare hands. Fire is totally different. No other animal comes even close to controlling fire, so its use sets humans apart from the rest of the natural world. Fire allowed humans of the Ice Age to extend their range from Africa deep into Europe and Asia. Furthermore, fire led directly to the creation of new technology, from the simple fire-hardened wood and stone tools of the Old Stone Age to the pottery of the New Stone Age and the smelting of the Metal Ages. By itself, fire provides light and warmth and keeps away undesirable predators.

Some animals and plants were probably cooked as early as the time of *H. erectus,* although the archaeological evidence is spotty. In the Choukoutien (Zhoukoudian) cave site near Beijing, China, some layers associated with *H. erectus* have many charred bones of prey creatures, which many have interpreted to mean barbecue. On the other hand, evidence from farther north shows that mammoth bones were burned as fuel on occasion.

Cooking does more than make some foods taste better. It kills parasites in animal food and often detoxifies plant food. Some foods, including all the starchy staples (grains and potatoes, primarily), are not digestible in an uncooked state. Fire is involved in the production of food in other ways as well, ranging from its use by early hunters to drive game to its use by farmers up to the present day to clear land (and provide useful fertilizer for the cleared land by way of ashes).

Proving the existence of very early fire is not easy. Charcoal and ashes can arise from natural sources, as can burned bone. At Choukoutien itself, the layers of ash are so extensive that some scientists have suggested that the ash is the result of natural fires caused by spontaneous combustion of bat guano and other organic debris; these fires then charred any bones that were also lying in the cave.

A pattern of stones that looks like a hearth might also be a natural formation. The combination of localized deposits of charcoal inside what appears to be a hearth, especially in the presence of large amounts of heated or charred bone, is generally needed before all experts can agree that fire was present. In tropical regions, where ash does not survive well, there is magnetic evidence that soil was baked in localized patches (although such baking could be caused by burning tree stumps after a canopy fire). While all of the early sites are problematical, the later the site the better the evidence. Certainly, control of fire appeared at some point in the Old Stone Age, but it is difficult to be certain when.

Even less certain is the progression from control to manufacture. Paleoarchaeologists assume without proof that early *H. erectus,* used to handling sticks, picked up the nonburning end of a branch from a natural fire and found that it could catch other wood on fire with its burning end. Thus, fire could be transported from place to place. Australian aborigines favored this method of handling fire when first encountered by Europeans. Another essential discovery is thought to have been the realization that since fire does not propagate on bare earth or stone, small fires can be confined. Finally, if such small fires are fed wood a little at a time, they will continue indefinitely. None of this can be determined from the archaeological record. It is difficult to imagine, however, that use of fire could have started in any other way.

The next step, starting a fire from scratch, also is inferred, not observed. If you know that some materials (wood dust, dry leaves, small sticks, or kindling—generically called tinder) catch fire easily, and if you notice that friction produces considerable heat, you can put the two ideas together to assume that producing enough heat from friction will start the tinder smoldering. It also helps to know that blowing on smoldering materials can cause them to burst into flame. These ideas are the basis of the most common form of fire-making known among technologically simple peoples even today. The friction is produced by a wooden rod rotated as a drill, often powered by a device similar to a bow, although other devices also exist to speed up rotation of the drill.

A faster method for lighting tinder is by producing a spark by striking flint against steel or even against an iron-rich rock, such as pyrite. Pyrites apparently used for this purpose date from the Late Old Stone Age. Matches are a much later invention, and were for a long time too expensive for common use. The combination of flint and steel was used for starting fires well into the 19th century.

have been posed and disposed of with regard to the meaning of the cave paintings. Popular accounts connect them with religion of some sort or perhaps with education. From their inaccessible locations, it is apparent that the cave paintings were not mere decoration.

Cave paintings are clearly intentional, but there is another group of enigmatic markings that could simply be the result of accidents or of doodling. These are bone and antler fragments that contain scratches or notches that some paleoanthropologists have identified as tally marks or even as early calendars. Similar but more clearly intentional scratches on bone have also been identified as maps. All of these scratched or notched bone and antler fragments come from the same cultures that made recognizable portraits of many animals and a few humans. Although it is easy to dispute any single instance of tallies or calendar marks, when all the examples are put together, it appears that early humans were beginning to keep records.

There was increased need to keep records in New Stone Age (Neolithic) times, for trade over long distances became common. At this point, about 8000 BC, people in the Near East began to use a system of fired clay tokens to record numbers of sheep or measures of grain. For example, one token of a particular shape would be equal to a measure of grain. The token system persisted in one form or another for at least 5000 years. During the Metal Ages, it progressed through a sequence of steps that culminated in the invention of numeration and writing.

Although there are some who think that Paleolithic peoples domesticated the reindeer and horse, not only do most scholars doubt this, but even advocates of the domestication scenario think that the purpose was food, not transportation. Thus, most transportation before at least 7000 BC, when cattle were domesticated and could have been used to pull sledges, was on foot. Not all, but most. The remainder was by raft or dugout canoe.

Although the first remains of wooden canoes and paddles date from about 7500 BC, there is circumstantial evidence from much earlier times that humans were crossing bodies of water too wide to swim and too deep to wade. Humans reached Australia probably 60,000 years ago, and certainly 40,000 years ago. They did not walk their way there. If humans reached South America more than 15,000 years ago, as many believe, they probably did not walk there either. Certainly, by 13,000 years ago, obsidian from the island of Melos in the Aegean Sea was turning up in tools made on the mainland. Melos could only be reached by boat or raft.

Where there was snow, sledges (sleds) were used for transporting goods, although the sledges probably were pulled by humans. During this period there is no real evidence of the use of animals in transportation, even though that became theoretically possible as soon as cattle were domesticated. That important next step in transportation began just as the Stone Age was ending and the Metal Ages were beginning.

Food and agriculture. Under this heading we will deal with virtually all of the relationships between people, food, animals, and plants, even when the animals and plants are not directly connected to people as food.

These days nutritionists often remind us that people were hunter-gatherers for a million years, farmers for less than 10,000 years, and sedentary factory and office works for a couple of hundred years. Their point is that since evolution does not proceed quickly, we still have the physical needs of hunter-gatherers, but we eat like farmers and live like kings. As a result we get fat, develop Type II diabetes and high blood pressure, have rotten teeth, and generally are physical wrecks compared to our glorious ancestors. There is some reason to think that this may be true when we compare ourselves to early hunter-gatherers, whose very bones often seem to radiate good health, but it is also clear that today's factory and office workers are generally in better health than farmers have been at any time since the Agricultural Revolution.

The nutritionists' version may be an oversimplification, but it has veracity in describing the main ways of obtaining sustenance. Perhaps the proper progression for human agricultural transition should be gatherers (occasional hunters and scavengers); scavenger-gatherers; hunter-gatherers; farmers; farmer-workers.

Many paleoanthropologists think that our ancestors and early relatives were gatherers, but only incidental hunters. Once again, the best comparison we can make is to the common and pygmy chimpanzees and gorilla. Both species of chimpanzee live primarily on gathered food, especially fruit, nuts, leaves, and insects. Nevertheless, chimpanzees like a little red meat occasionally, and they hunt it when the occasion arises. This seems to happen every 10 days or so. In general, the males do the hunting, and they share the kill with other males and with females. Australopithecines may have lived much like the chimpanzees, although there is some evidence from tooth size and wear that *A. robustus* dug and ate roots a lot. Recent studies of the chemistry of fossils suggest that australopithecines did include meat as a part of their diet, even big-toothed *robustus*.

The gathering way of life, even with a little hunting thrown in, does not require much technology. Gathering roots is much easier if done with a digging stick, or *dibble*, a method still used by modern-day hunter-gatherers, and probably employed by very early hominids as well. But digging sticks are very hard to recover from the fossil record.

There is a fair amount of evidence that the next major revolution in sustenance did not involve hunting with weapons, but scavenging with stone tools. Cuts and scratches on fossilized bones of antelopes and other food species can be separated under the microscope into those caused by tooth or claw and those caused by pebble tools. Some claim that a preponderance of such bones have most of the tool scratches on top of the teeth marks, evidence that the lion or tiger or bear got to the food before the hominid did. This could suggest that while *Homo habilis*, the putative first hunter, may have picked off some small game, such as monkeys, interactions with larger game were limited to scavenging kills. If that is the case, the first tools were primitive butcher knives, not early weapons.

Although there is a good case to be made for *H. habilis* as a scavenger, the evidence strongly backs up the claim of *Homo erectus* as a powerful hunter.

Individual sites may be disputed. One authority says that a site containing 800,000-year-old baboon remains mixed with many bifaces (hand axes) at Olorgesailie in southwestern Kenya resulted from a successful planned hunting expedition that involved a nighttime ambush of nearly a hundred baboons that were sleeping in trees. The baboons were knocked from the trees by thrown bifaces and butchered with stone cleavers. A different authority says that the baboon bones and bifaces present at the site were swept there from many individual locations and concentrated by the action of a river.

Despite such differences, the overall trend is inescapable. Caves with *erectus* remains also contain bones of butchered animals, some quite large. Evidence supports occurrences of trapping the ancestors of cattle in a bog in Africa, driving elephants over a cliff in Spain, and similar major kills elsewhere. Certainly, any creature that can control fire can also use it as a tool in hunting. Thus, nutritionists are right to say that our ancestors were hunter-gatherers for over a million years, since that is how long *erectus* hunted.

By the time of the Neandertals and modern humans, hominids were not just hunters, they were extremely successful hunters, the greatest hunters the planet had known. The evidence of this is largely in the huge piles of butchered bones of antelopes, horses, and mammoths piled up by hunters of the ice ages. We probably hunted the larger species to extinction. Certainly, the hunting pressure was so great that humans eliminated the larger examples of each species, reducing the overall size of horses, antelopes, and beavers by ruthlessly hunting and killing the biggest prey available.

What happened next is analyzed in the essay "What caused the Agricultural Revolution?" on pp 75, 77, and 79. Then people started farming.

A chronological history relies on dates, and there are only a few dates for weapon improvements during the hunting period. The Agricultural Revolution brings with it a vast array of dates as one species after another became domesticated. We know these dates better for arid or temperate regions and for mammals and plants with hard parts. In the tropics, the evidence vanishes from the archaeological record too fast; and the hollow bones of birds or plants with only soft parts, such as squash or peas, also do not last long.

Domestication appears in the archaeological record because domesticated animals and plants have different characteristics from their wild ancestors. Domesticated animals are bred to be less dangerous than wild ones;

many domesticated animals are smaller, for example, than their wild counterparts (dogs and cattle come to mind). Pigs have their tusks bred away. Frequently, breeders aim for neonaty, the evolution of appearance and behavior of a newborn that persists into adulthood. In other words, dogs are bred to be more like puppies, cats to be more like kittens. Often neonaty results in a foreshortening of the face that is observable in the archaeological record. The presence of traits such as increased woolliness in sheep, when observed, is also a good sign of domestication. In addition, archaeologists look at sex ratios and butchered body parts found in the record. Breeders eat most males and keep females for further breeding or for milk. They consume all of a pig butchered in a village or at home, while they bring home only the best bacon from one killed far off in the forest.

Similar rules hold for plants. In wild grasses, most seeds fall to the ground as soon they are ripe (they "shatter"), but domesticated grasses, such as domesticated wheat, are bred from those plants that shatter less. Many wild plants, including wheat and the progenitor of maize, have their seeds covered with thick and hard-to-remove husks; the domesticated varieties have much thinner coverings. Often the number of edible parts increases, as evidenced by the improvement from wild two-row barley to tame six-row barley.

From this kind of evidence and also from drawings and engravings showing people tending animals or plants, archaeologists can trace the step-by-step process of domestication, one that is not complete today. We continue to find new uses for plants or even use genetic engineering to produce new forms of domesticated plants and animals.

Some time after domestication of plants, the digging stick was converted into a simple plow by pulling it along the ground. Probably people did the pulling at first. There is better evidence for implements used to process food, such as grindstones, than there is for equipment used for planting or cultivation, most of which was done by hand for the first several thousand years of agriculture. Stone sickles are known early, however; indeed, they predate domestication, being first used for wild grasses, as were the first grindstones.

Tools/materials. During the early eras of our ancestors' and relatives' development of technology, materials and tools were even more closely blended than they are today. For this period we treat them together; in later periods, we attempt to separate like true Platonists the actual material tool from the concept of the tool.

Although this period is labeled The Stone Ages, it begins before stone was used and continues through the introduction of bone and antler, ceramics, fibers, and even limited use of native metals. Just before the end of this period, a few people had even learned how to smelt at least one ore into metal.

As noted several times above, most biological materials are biodegradable, and for that reason they form only a tiny part of the early archaeological record. Nevertheless, no one seriously doubts that such materials were used as the basis of the first technology. Biological materials have continued to be an important part of technology up to the present. Although a computer may be almost entirely synthetic, it usually rests on a wooden table in a wood-framed house, for example. It would be possible today to replace all the wood in an office or house with other materials, but they would be more expensive, less attractive, and in many cases not as good as wood. Whether or not archaeologists can find its traces, wood has remained the most used material for most of human history. Worldwide, it probably still is. Wood is strong for its weight, can be flexible or stiff as required, and is still cheap.

Our earliest ancestors probably used both wood and stone objects as tools, even before they learned to make wood and stone into tools. Again, about the only window into such early behavior is to consider tool use among the chimpanzees. Although baboons are stronger and more dangerous than chimpanzees, chimpanzees tend to dominate baboons where the two species frequent the same territory. Jane Goodall thinks that this is because chimpanzees can and do throw branches and stones at baboons when the two species quarrel. Baboons do not know how to throw things, and so they come to believe that chimpanzees are more powerful than they really are. It seems likely that our weak ancestors and early cousins also threw whatever came to hand as protection against stronger beasts.

When it comes to throwing an object as a weapon, stone is clearly superior to wood for most purposes. Some paleoarchaeologists think that such early tools as the bifaces we usually call hand axes were essentially developed to be thrown at prey. Certainly, being hit by a well-thrown object over a kilogram (2 lbs) in weight that has sharpened edges would inflict a fairly serious wound.

The use of stone as a primary material for tools in the Stone Age displaced wood for a few purposes but also opened vast avenues for progress that could not be found in wood. A sharpened flint edge is immensely superior to a broken branch for cutting meat, for example. Stone's greater density (weight per unit of volume) not only improves thrown weapons in various ways, but also is of great help when pounding something, living or inanimate—you would not want a hammer with its business end made from wood.

The essay "Stone technology of the Old Stone Age" on p 9 tells of the progress of stone tools through six stages, often labeled Modes I through VI, from over 2 million years ago until the start of the New Stone Age (Neolithic), about 6000 BC. The Neolithic was coincident with Mode VI technology, a technology that we now know did not rely as much on stone tools as the name Neolithic would suggest. It should be noted that different cultures entered and left the various Modes at different points in time. In parts of the world, some cultures have remained at or close to the Mode V technology until the present, while slightly more advanced Mode VI cultures until quite recently (the past 200 years) made up a large segment of the human population.

Throughout Modes IV, V, and VI there was an increasing tendency to rely on materials other than stone in tool manufacture. For one thing, suitable stone became not so easy to find as it once was, despite a tendency from Mode II onward to reduce the size of stone tools. Also, inventive humans were finding that other materials were better than stone for specific purposes. There may also be a slight tendency of archaeologists to overlook somewhat biodegradable materials from early in the Old Stone Age, such as bone and horn, simply because they do not last as well as stone does.

Somewhere during the development of the Mode IV or Mode V technology, the bow and arrow became an important technological innovation, so both stone and bone points were specifically adapted as arrowheads. The term *point* continues to be used as it is fairly neutral as to how the stone object was employed, although barbed harpoon points from Modes V and VI are generally referred to as harpoons, not points. But otherwise similar unbarbed points could have been used for arrowheads or for teeth of a saw, for example.

Spears, harpoons, and arrows are not the only tools used for striking at a distance. The boomerang is known from artifacts that date from well before the first spear thrower, harpoon, and bow. Very likely the sling preceded the boomerang, and may date from 100,000 years before the bow and arrow. It is difficult to tell, however, because the sling itself is biodegradable and the missiles used might be any smooth stones.

An innovation of the Mesolithic was regular and widespread use of the ceramic material called pottery. Although some ceramics are known from earlier times, the earliest known pottery is from Japan. Pottery became common in the Near East toward the end of the Mesolithic as well. Weaving is also known from the later Mesolithic.

Ironically, the other innovation of the New Stone Age was the use of metal. Although the Copper Age follows the Neolithic, finds of copper implements date from a couple of thousand years earlier and smelting of copper ores begins in Egypt as early as 4500 BC.

c
2,400,000
bc

Homo habilis, the first member of the genus *Homo*, of which we are a member, makes an appearance in East Africa; this small, bipedal, large-brained creature is not the first of our family, the hominids; depending on whether you include the chimpanzees and gorilla, our family began some 2 to 5 million years earlier; chimpanzees and gorillas date to as early as 7 million years BP, while a bipedal group known as australopithecines began at least 4 million years BP; DNA evidence indicates that the chimpanzee and the pygmy chimpanzee (the bonobo) are our closest living relatives, with all others ultimately replaced by our own species; *Homo habilis* is nicknamed "Handy Man"

c
2,000,000
BC

The second member of our genus, *Homo erectus*, is believed to have existed this early; all technological innovations for at least the next million years are owed to this ancestor

c
1,900,000
BC

Some archaeological sites in the Olduvai Gorge in Tanzania from around this time until about 1,600,000 BP are thought to be living floors—that is, home bases or campsites; at these sites, tools and hominid remains seem to concentrate; some paleoanthropologists believe that such concentrations are caused by agencies other than use as home bases

c
1,500,000
BC

Although still unverified, patches of baked earth near *Homo erectus* sites at Koobi Fora and Chesowanja in Kenya suggest that hominids are using fire; *See also* 1,000,000 BC GEN

c
1,000,000
BC

A few fossils suggest that some form of archaic *Homo sapiens* developed this early, although fully modern humans are far in the future

Ancient hearths found in the Swartkrans cave in South Africa indicate that *Homo erectus*, the immediate predecessor of modern humans, uses fire; *See also* 1,500,000 BC GEN; 600,000 BC GEN

c
600,000
BC

A thick ash layer forms between this time and 400,000 BP in L'Escale Cave in southeastern France; this layer has sometimes been viewed as the earliest known evidence of fire made by hominids in Europe, although it cannot be definitely established that it is not the result of naturally caused fires; *See also* 40,000 BC GEN

Stone technology of the Old Stone Age

Many different stone industries and toolkits have been labeled. The first, the Oldowan industry, produced pebble tools by simply breaking off bits of rock, called *flakes*, with virtually no retouching of the material left behind. The larger piece of rock from which the flakes were broken is known as the *core*. The core is the main Oldowan tool, although some flakes might also have been used as tools.

Some paleoanthropologists call this toolkit Mode I, according to a system of nomenclature developed by Desmond Clark in 1968 as a generalized replacement for names based on specific sites.

The second industry is a toolkit based on bifaces (hand axes or cleavers) that is called Acheulean or Mode II. Although the tools are still primarily the cores, produced by hammering with another stone or, in later versions, by hammering with something softer than stone, such as wood or bone, the cores and some of the larger flakes are retouched into tools that anyone can recognize immediately as artificial. In Europe, the Mode II toolkit is often called Abbevillian for the early "hard-hammer" phase and Chellan for the later "soft-hammer" stage. This tool industry was virtually coextensive with the rise and domination of *Homo erectus*. It also persisted for at least a million years. In some places, the Acheulean remained the basic industry for half a million years more. When it was replaced, the new industry was coincident with early *Homo sapiens,* including Neandertals.

The next major development, associated in Europe with the Neandertal, is often called Mousterian. In the modal system it is Mode III. The Mousterian kit is associated with side scrapers and projectile points. The later version of the industry is largely based on a new technique, called prepared core or Levallois, in which the core is preshaped to produce a particular type of flake. The flake is the primary tool, not the core, as in Modes I and II.

The next stage is divided into several parts in the European nomenclature, with the early part called the Perigordian industry and the next, and more famous, part labeled Aurignacian. Both are grouped together in Mode IV in the modal system. The main difference is that the tools of the Aurignacian are noticeably better made than those of the Perigordian. These industries used blade knives and the pointed stone tools called *burins*, similar to modern metal burins used in engraving metal and stone. Tools of other materials are also known from these and later times.

Also included in Mode IV are three later industries, which complete the Old Stone Age. The Gravettian used smaller blades than the Aurignacian, often with one edge, called the back, blunted, as if to protect the user's fingers. Together, the Perigordian, Aurignacian, and Gravettian industries in Europe lasted from about 32,000 BP to about 18,000 BP. After 18,000 BP there were two other well-known European tool industries. The Solutrean is primarily known for its laurel-leaf blades, so symmetrical and thin that they are sometimes thought to have been too beautiful to have a practical purpose. The last of the Old Stone Age, or Mode IV, industries made no innovations in stoneworking; they are primarily differentiated by the use of many tools of other materials. This industry is labeled the Magdalenian.

Hominids in Africa, generally thought to be *Homo habilis*, begin to manufacture simple stone tools known as pebble tools; the tool assembly, or toolkit, is called Oldowan after the Olduvai Gorge in Tanzania, although the oldest sites are from the Omo Valley in Ethiopia	c 2,400,000 BC
	c 2,000,000 BC
The oldest known stone tool "industry" is characterized by remains that start about this time at the Olduvai Gorge in northern Tanzania, accounting for the name Oldowan; the choppers and scrapers are simple chips of stone and are sometimes classed as pebble tools	c 1,900,000 BC
Acheulean artifact assemblies, which continue to be produced until 200,000 BP, are first left behind in Africa, probably by *Homo erectus*; this toolkit is characterized by bifaces (hand axes, cleavers, and picks), typically large, ovoid stones from which flakes have been removed by hammerstones and whose exact purpose is unknown	c 1,500,000 BC
	c 1,000,000 BC
	c 600,000 BC

c 400,000 BC

Evidence at Terra Amata, a site near Nice, France, suggests that *Homo erectus* occupies oval huts that are 15 m (50 ft) by 6 m (20 ft); this is the first evidence of housing construction

c 300,000 BC

Very early Neandertals are living in what is now Spain; *See also* 200,000 BC GEN

c 200,000 BC

Neandertals are becoming common in Europe; these puzzling hominids are clearly in the genus *Homo,* but authorities differ on whether they can be classed as a subspecies of *Homo sapiens* or whether they are a different species that arose separately from *Homo erectus* or possibly from archaic *Homo sapiens*; Neandertals have a technology all their own, one that is more diverse and advanced than that of *Homo erectus,* but much more stable than that of fully modern *Homo sapiens*; Neandertal technology changes little over periods as long as 100,000 years

c 150,000 BC

c 130,000 BC

The Apollo-11 site in the Orange River Valley of Namibia, excavated by W.E. Wendt from 1969 to 1972, is first occupied; it remains so until 4000 BC; *See also* 40,000 BC COM

c 120,000 BC

c 110,000 BC

c 90,000 BC

A significant amount of evidence exists in the fossil record that an archaic form of *Homo sapiens* is present in Africa

There is limited evidence of fully modern *Homo sapiens*

		The burin, thought to be the primary tool for engraving and carving such materials as bone, antler, ivory, and wood, becomes common, as do end-scrapers (*grattoirs*), thought to have been used for hide scraping or wood-working	**c 400,000 BC**

The first navigators

The first proof we have that our ancestors crossed a substantial body of water is stone tools in Japan from 100,000 years BP. Not much is known about the makers of the tools, but the mainland at that time was probably inhabited by a late population of *Homo erectus*. Japan is about 150 km (90 mi) from Korea, but there are several islands along the route. Japan is not thought to have been linked by land to Asia as the British Isles were linked to Europe during the most recent Ice Age.

The earliest known example of a wooden spear, the tip of a yew spear, is left in a site at Clacton-on-Sea in England; *See also* 120,000 BC TOO	**c 300,000 BC**

The earliest evidence of *Homo sapiens* crossing a substantial body of water is found in the desert of central Australia, where traces of human occupation have been dated at 60,000 years BP. Probably, early people paddled across the Indian Ocean in short hops from one island to another, although during the periods of low sea level 18,000 years BP, when the oceans were about 130 m (425 ft) lower than at present, they could have walked most of the way across what is now Indonesia.

Well-made side scrapers become common, suggesting that *Homo erectus* is now manufacturing clothing	**c 200,000 BC**
Although characteristic of the much later Middle Paleolithic industries, the first Levallois prepared-core flake tools begin to appear; in Levallois technology, the core is carefully shaped into a regular block of stone from which very large, thin, nearly uniform flakes that can be used without further shaping are struck	

There is scattered evidence that suggests that a few humans crossed the Pacific some 20,000 to 40,000 years ago, traveling from Asia to South America. Most authorities, however, believe that all regions except for Australia and islands were populated by land. In this view, South America received its first people from North Americans whose ancestors traveled from Asia when lowered sea levels created a land bridge (about 12,000 to 18,000 years ago).

The main islands of the Mediterranean, including Sicily, Crete, Cyprus, Rhodes, and many others, were all settled in Neolithic times, if not before. By the Bronze Age several of these islands, notably Crete and the Cyclades, contained some of the most advanced civilizations of the day, surely based mainly on trade and shipping.

The earliest known Mousterian tool industries start at such sites as Biache in northwestern France; this toolmaking tradition, associated with Neandertals, is characterized by Levallois and discoidal-core manufacturing of side-scrapers, backed knives, hand axes, and points	**c 150,000 BC**

Early evidence of trade includes obsidian imported into Greece as early as the 11th century BC from the island of Melos. Six thousand years later there is increasingly good evidence of trade between Mesopotamia and ports along the Persian Gulf and the west coast of the Indian subcontinent, along with the island of Sri Lanka. In another thousand years, by the 3rd millennium BC, Egypt was actively involved in trade in the eastern Mediterranean.

	c 130,000 BC

Early shipping stayed close to land in the eastern Mediterranean. Gradually, however, explorers from Europe moved throughout the Mediterranean and into coastal Atlantic waters along the western parts of Europe and northern Africa. Egyptian pharaohs sent expeditions down the eastern coast of Africa as well, while regular trade developed across parts of the Indian Ocean between Arabia and Africa and between Mesopotamia and India. Some of the Indian Ocean voyages may have been across the open sea, helped by reliable monsoon winds.

The earliest known example of a charred wooden spear is associated with an elephant carcass found at a site called Lehringen in Germany; *See also* 300,000 BC TOO	**c 120,000 BC**

Chinese ships and navigation tools were more advanced than any others on Earth until at least the European Renaissance, so Chinese navigators also are presumed to have ventured across open seas from early on. The most notable voyage that is well-documented, however, comes in the 15th century CE, much later than the period of early navigation.

A carved mammoth tooth and other incised bone objects from the Tata site in Hungary are among the earliest carvings and incised objects known; ground pigments are also found at the site	**c 110,000 BC**

	c 90,000 BC

c 80,000 BC

Neandertal humans bury their dead in graves with symbolic use of red ocher dyes, as seen in the Regourdou site in southwestern France

c 60,000 BC

c 40,000 BC

A perforated part of the bone of an antelope or other bovine animal dated at 45,000 BC is claimed by some to be the oldest known musical instrument, a form of whistle; indisputable whistles appear more than 15,000 years later; *See also* 30,000 BC GEN

Simple forms of stone lamps fueled with animal fat and using lichens, moss, or juniper for a wick, are in use; *See also* 1100 BC GEN

Construction of hearths in northern latitudes begins to improve noticeably; stone borders are arranged for heat retention and to direct air flow; *See also* 600,000 BC GEN

In cold climates large dwellings are built of mammoth bones, some houses covering as much as 30 sq m (300 sq ft)

c 35,000 BC

About this time there is plentiful evidence of fully modern *Homo sapiens* as the only member of the genus left on Earth, since the Neandertals disappear; from this time also technological change becomes increasingly swift

People cross a land bridge from Australia into Tasmania; changing sea levels eventually make Tasmania an isolated island, with its last land connection to Australia submerged about 10,000 BP

c 30,000 BC

The earliest known unequivocally identified musical instruments are flutes and whistles made from bird or bear bones or from reindeer antlers that date from 28,000 BC forward; the instruments have been found from France deep into central Europe and the Russian Plain; *See also* 40,000 BC GEN

			c **80,000** BC
Based on dating of bone fragments from central Australia, early humans reach that continent			c **60,000** BC

Humans or their ancestors at the Apollo-11 site in the Orange River Valley of Namibia c 44,000 BC paint slabs, the oldest "art" found in Africa; *See also* 20,000 BC COM

Humans in Eurasia wear beads, bracelets, pins, rings, pendants, diadems, and pectorals, although the earliest recognizable jewelry of any type consists of ostrich-shell beads from the Border Cave in South Africa and may date from as early as 45,000 BP

Some hearths are enclosed in clay walls and may have been used as kilns for ceramic production, although they may also have been used for baking food or other purposes

The earliest blade stone tool industry in Europe is known from the Bacho Kiro site in what is now Bulgaria

c **40,000** BC

c **35,000** BC

Paleolithic peoples in central Europe and France use tallies on bone, ivory, and stone to record numbers; for example, a wolf bone from this period shows 55 cuts arranged in groups of five

In what is now Germany, the earliest recovered animal images are made; these are detailed and sophisticated three-dimensional carvings of lions, horses, bison, and mammoths

In France people carve and engrave vulvas and, more rarely, phalluses c 28,000 BC

The earliest remains of bone, antler, and ivory objects, especially tubular beads, are found in central and southwestern France and northeastern Spain c 33,000 BC; the tool assembly is known as the Chatelperronian

In Europe a toolmaking industry termed the Aurignacian produces the characteristic toolkit of the Upper Paleolithic (later Old Stone Age) c 32,000 BC; the toolkit features blades, points, scrapers, and burins of stone as well as tools worked from bone and antler

The earliest fired ceramics, found at sites in the Pavlov Hills of Moravia (the Czech Republic) c 28,000 BC and made from the local loess, are manufactured, a development that continues for the next 6000 years; one theory is that the small statues and variously shaped blobs are intended to explode on refiring, since most are found shattered, with the shards used in divination; animal statuettes are nearly all of predators, while human statuettes occur later; *See also* 25,000 BC TOO

c **30,000** BC

c 25,000 BC

Gravettian cultures in Europe burn bone in deep pit hearths for heat and probably for cooking

Perigordian sites in southwestern France show complex arrangements of hearths, slabs, and postholes, implying complex living or storage structures

Gravettian cultures at sites such as Pavlov and Dolni Vestonice, in the Czech Republic, and Kostenki, in Russia, build huts that use mammoth bone for support when wood is not available

Sites in the Pavlov Hills of Moravia in the Czech Republic show that round, oval, and oblong surface and semisubterranean dwellings are constructed c 23,000 BC

c 20,000 BC

The first ceramics

Ceramics originally were any materials produced by heating natural earth until it changed form (without melting—glasses are formed by earth heated until it melts and then cools). Ceramics are different from merely dried earth or clay, which can be rewet and softened. The higher heat at which ceramics are produced drives off water chemically bound to the earth as well as any water that has soaked into it. The result of such heating, if a fine clay is used, can be pottery, terra-cotta, china, porcelain, brick, or tile. Today, modern ceramics imitate the type of bonding found in natural ceramics, but they are produced from purer chemicals and not always by heating.

Glasses are not only similar to ceramics, but many ceramics also include glasses as integral parts. Porcelains are in between true ceramics and true glasses, for example, as are many glazes that are important to ceramics. One role for the glass or glaze component is to make the ceramic less porous than it would naturally be.

True ceramics appear rarely in nature, but are sometimes the result of lightning strikes and forest fires. It could be argued that some metamorphic rocks are essentially ceramics. Glasses are easily recognized when they appear as rocks, chiefly as obsidian produced in volcanic eruptions.

From the control of fire by *Homo erectus* to the accidental production of ceramics is a very short step. Apparently, however, the deliberate production of ceramics had to wait until the more inventive *Homo sapiens* arrived on the scene.

At one time archaeologists believed that deliberate ceramics were a fairly recent discovery, 10,000 years old at the most. A popular theory was that basketry was invented first, but baskets do not hold liquids well. According to this theory, early people solved this problem by lining baskets with clay, which is impermeable. Sometimes baskets so lined got burned for one reason or another. The clay lining was left behind as a pot. Eventually, people found that they did not have to start with the baskets. This theory is reminiscent of Charles Lamb's famous essay on the discovery of roast pig *via* burning down the house.

Ceramics may or may not precede basketry (which is, of course, biodegradable and easily lost from the archaeological record), but they certainly date much before 10,000 BC. Furthermore, ceramics were being deliberately made well before the first known ceramic pot. About 26,000 BP, in the region now known as the Czech Republic, people built kilns and produced small ceramic figures and beads. Ovens that may have been kilns as well go back another 14,000 years.

However, practical uses for ceramics, such as pottery and brick, do seem to be a part of the Neolithic Revolution. But the first bricks were not ceramics; they were adobe—clay or mud hardened by drying but without the chemically bound water driven off by heat. When kiln-dried bricks became available, their expense caused them to be reserved for special monumental buildings, while the common people continued to build houses with sun-dried bricks.

Pottery was shaped by hand during the last part of the Stone Age. Sometimes a large pot would be built and fired in sections that were then glued together with clay and fired again. The invention of the potter's wheel about the start of the Metal Ages was one of the great steps toward civilization, leading not only to better pottery but also to the general principle of the wheel for use in transportation and machinery.

Pits dug at sites in the East European Plain c 24,000 BC constitute the first unequivocal evidence of food storage over winter; the same pits are apparently used during summer for storage of such nonfood items as fuel and raw materials for manufacture of tools and jewelry

Single-backed points and burins characterize the Gravettian tool assemblage found throughout much of Europe c 26,000 BC

c 25,000 BC

People in what is now Poland are the first to use the boomerang, about 13,000 years before the first Australian boomerang; the early Polish boomerang is made from mammoth tusk

The bow and arrow are invented, according to evidence from sites at Parpallo, Spain, and the Sahara; stone points from Parpallo appear to be tips of arrows; drawings of archers are found at the North African site; other evidence, however, suggests a later origin, perhaps as late as 8000 BC

The Venus figurines, small statues of faceless pregnant women with large breasts and buttocks, are made in Europe c 24,000 BC; they continue to be manufactured for the next 2000 years and are among the oldest ceramic objects known to be made by humans; other fired ceramic figures of animals or of unknown blobs, nearly all shattered, are also found in sites in the Pavlov Hills of Moravia (the Czech Republic), molded from wet loess; *See also* 30,000 BC TOO

The earliest animal engravings are made in the Altamira Cave in northwestern Spain, although the famous red-and-black paintings in the cave are from a later date; *See also* 10,000 BC COM

Genetic evidence suggests that the domesticated cat diverges from the wild cats of Europe c 18,000 BC

The Solutrean toolmaking industry in France and Spain c 19,000 BC is characterized by large symmetrical flint tools known as laurel-leaf points; the exact purpose of these points is unknown, as they are too thin and well-made to be used for normal tasks; some archaeologists have suggested that they were made for their aesthetic value; a Solutrean technological innovation is treating the flint with heat, giving Solutrean laurel-leaf blades a sheen like porcelain and also making the stone easier to work

c 20,000 BC

People in the Périgord region of France c 16,000 BC make pendants that have what appear to be allegorical scenes carved into them, such as a bison's head surrounded by a schematic representation of six or more people (found at Raymonden-Chancelade)

Bone and antler needles are made in France and Spain, suggesting that clothing sewn together with threads of sinew or plant material may have come into use

People begin to make smaller and smaller stone blades, called microblades, in central Africa and Eurasia c 18,000 BC; it is thought that the microblades and equally small micropoints are combined with each other by hafting several to the same piece of wood or bone to form more complex tools, such as saws

**c
15,000
BC**

Figurines from around 13,000 BC are sometimes made wearing such clothing as skirts, aprons, headdresses, or parkas

At Mezhirich in Ukraine, storage pits near mammoth-bone dwellings are distributed unequally, suggesting that some family units had more resources than others; this is the first known evidence of a social hierarchy

At Mezhirich in Ukraine, huts made from mammoth bones use specific bones for particular parts of dwellings c 13,000 BC; hundreds of bones are used to make the frame, and the weight of the bones alone is estimated at 22 tons; in 1992 workers find that a hut had been constructed with an underground entrance about 3000 to 5000 years prior to any other known underground entrances

---

Stone technology of the Middle Stone Age and Neolithic

Many anthropologists recognize that the end of the Old Stone Age was a significant transition period and label this era the Middle Stone Age. In any case, there is a dramatic change observable in stone tools, which in the modal system is labeled Mode V. This tradition involves the reduction of points and other tools to such a small size that they are called microliths; this tendency actually began with the previous Magdalenian industry.

The first industry to be based largely on microliths, dated from about 11,500 BP to about 9000 BP, is known from its French manifestation as Azilian. There are clues that suggest that microlithic technology was inspired at least in part by the growing human population and its drain upon local resources, including good stone for tools. Similar to Azilian, although somewhat different in detail, microlithic industries that followed in France are called the Sauveterrian (to 7500 BP) and Tardenosian (to 6000 BP). Other parts of Europe as well as other continents had corresponding industries with different names. By the end of the Mode V tradition, there are so many individual names given that only specialists know them.

Coexistent with Mode V as well as following it, Mode VI involves a number of ground-stone tools,

**c
10,000
BC**

especially axes and adzes (an adz is essentially an axe with its blade turned 90 degrees, a tool used mainly to shape large pieces of wood and to fill awkward spots in crossword puzzles). The modern re-creators of stone technology have shown that, although flaked tools are just as sharp as ground ones, ground edges hold up much better in hard use, in chopping down trees. Furthermore, if they lose their edge, ground tools are easily reground.

At this point our story leaves the Old Stone Age and lodges firmly in the New Stone Age, or Neolithic. The Neolithic Revolution coincides with the Agricultural Revolution. It is thought that Neolithic farmers replaced flaked stone edges with ground ones to have the kind of axes needed in clearing land.

One of the main differences between the New Stone Age and the Old is that the Neolithic used stone as only one of a great variety of materials, even including native copper.

Throughout the story there has been an interesting progression in efficiency. The length of the cutting edge obtained from one kg (2.2 lb) of stone is 60 cm (24 in.) for the early pebble tools, but progresses to 2000 cm (790 in.) by the microliths of Modes V and VI.

The Magdalenian culture, which ranges from northeastern Spain deep into central Europe, produces, in addition to the famous cave paintings of its southern portion, such as at Lascaux, France, many engravings on bone, antler, and slate, including animal figures and abstracted female human figures

The first known artifact with a map on it is made on bone c 13,000 BC at Mezhirich, Ukraine; the map appears to show the region immediately around the site at which it was found

At Mezhirich, marine shells from the Black Sea, some 580 km (360 mi) to the south, demonstrate that the people who lived there were part of an extensive trade network

Grinding stones in the Near East are probably used c 16,000 BC for processing the seeds of wild grasses, a prelude to plant domestication

At Mezhirich in Ukraine, as at other sites of the same region c 13,000 BC, pits are dug into permafrost to store food, offering up the first known form of cold storage

Bone harpoons are left at the Ishango site in Zaire, dated as being from c 16,000 BC; these are the earliest known examples of harpoons

A tool made of antler (usually from reindeer), known today as the bâton de commandement, a part of the main body and two major branches of the antler with a hole bored near their junction, becomes common; it is not clear today what the bâton was used for, although shaft straighteners for spears and softeners for leather thongs are among the uses that have been suggested; often designs or illustrations are carved into the bâton; the name bâton de commandement comes from the idea that the tool may have been a symbol of authority

The Magdalenian tool industry in Europe over the next 3500 years produces fine harpoons with both single and double rows of barbs, needles, awls, spear throwers, and points, all made from bone or antler, as well as microliths and unretouched flake points

Rope is in use, according to evidence at Lascaux, France

The spear thrower (sometimes called the atlatl, its Aztec name) is invented

c 15,000 BC

The famous polychrome red-and-black paintings of standing and lying bison are made on the ceiling of the Altamira Cave in northwestern Spain c 12,000 BC; *See also* 20,000 BC COM

Greek sailors are able to import obsidian from the island of Melos to the mainland c 11,000 BC

Bison are painted in the cave of El Castillo in Spain and at the Black Room of Niaux c 10,900 BC

People of the Azilian toolmaking industry in the Mesolithic (Middle Stone Age) c 9500 BC in France paint geometric designs on pebbles, with no cave paintings, sculptures, jewelry, or other art forms

The first forms of fired clay tokens are used c 8000 BC by Neolithic people to record the products of farming, such as jars of oil and measures of grain, at sites in present-day Syria and Iran; tokens are believed to have helped the separate ideas of number and written word to evolve over the next 5000 years; at a few sites tokens continue to be used until about 1500 BC; *See also* 30,000 BC COM

The dog is domesticated in Mesopotamia (Iraqi Kurdistan) and the Middle East (Israel) c 12,000 BC in the first example of animal domestication; earliest known remains of domestic dogs are from the Zarsian site at Palegawra, Iraq

Flax is harvested, possibly for food, by people living near lakes in Switzerland; domestic flax probably derived from the wild flax that grows around the Mediterranean several thousand years later

The first known pottery is made by the preagricultural Jomons, the second wave of immigrants to settle in Japan; *See also* 7000 BC TOO

Houses of sun-dried brick held together without mortar are built in Jericho (Israeli-occupied Jordan)

c 10,000 BC

New materials: Tooth, bone, and horn

Who can forget the australopithecine in the movie *2001: A Space Odyssey* that beat up on invaders with a bone and then tossed it triumphantly into the air, where it turned into a spaceship?

Before bone, however, ivory (that is, the material of animal teeth) was sometimes used, especially for ornaments. Ivory is both denser and less fibrous than bone, making it easier to work.

Horn, especially reindeer antler, was also popular. Reindeer antlers are a form of bone that grows directly out of animals' foreheads. Reindeer are the only form of deer in which both sexes have antlers. Like other deer, reindeer shed their antlers in the spring and grow new ones. Thus, where there are many reindeer, horn is abundant.

Paleoarchaeologists find that people using the Perigordian industry at the start of Mode IV some 34,000 years ago infrequently produced bone or ivory tools, but the frequency increased with the long-term development of the industry. Around the same time, another Mode IV industry, known as the Aurignacian, had a much greater emphasis on bone; it also utilized reindeer antler as an important raw material. Bone points (objects that appear to be arrowheads or spear points) were especially prominent and underwent steady development throughout the Aurignacian. North of the Aurignacian toolmakers, and slightly later, a tool industry called the Gravettian developed, one that relied as heavily on mammoth bone as the Aurignacian did on reindeer bone and antler.

The Solutrean tool industry that started in France and Spain about 21,000 BP was marked not so much by increased use of bone or antler, but by decreased use of stone. Needles made from bone were common, suggesting that a lot of sewing was going on.

The real jump from stone to bone (and other materials) came with the next industry, the Magdalenian, which started around 17,000 BP in France and northern Europe. This toolkit was characterized by bone harpoons, some with one and some with two rows of barbs. Other bone tools that were common included not only needles, but also awls, polishers, and spear throwers. A mysterious tool made from reindeer antler, called the bâton de commandement, may have been a spear straightener.

Following the Magdalenian there was a transition period of about 5000 years to the Neolithic. Industries were based on microlithic stone tools and a variety of bone and antler tools. When cattle were domesticated, cattle horn came into enough of a supply to be used as well.

Pre-Pottery Neolithic A people, living in Jericho, build dome-shaped houses from sunbaked brick that is formed entirely by hand (without molds)

The place now known as Tell Mureybit in Syria has stone houses, although the villagers are hunters, not farmers; anthropologists believe hunters normally move from site to site and therefore do not build permanent housing

The oldest wall known to us is built in Jericho from boulders set in place without mortar; it is more than 3.6 m (12 ft) high and is 2 m (6 ft 6 in.) thick at the base

A group living in Jericho build rectangular houses from sunbaked brick and mortar formed by heating limestone and mixing the product with sand and water; they also use the transformed lime (known as "burned lime" to archaeologists) to plaster walls and floors

The Yangshao people along the Huang He (Yellow River) in China live in underground, circular mud-and-timber huts, in communities that feature large kilns for making painted pottery

			c 9000 BC

Einkorn wheat is domesticated by the Natufians, who live at the north end of the Dead Sea in what is now Israel; *See also* 7000 BC FOO

Goats and sheep are domesticated in Persia (Iran) and Afghanistan

Pumpkins and related squash are domesticated c 8750 BC in what is now Mexico and Central America

---

c 8000 BC

The dog is the only domesticated animal in North America c 8400 BC, according to evidence from Jaguar Cave ID; this situation persists until at least 1500 BC

Floodwater agriculture is used in the Nile Valley and in southwestern Asia

Potatoes are domesticated in the Andean Highlands of Peru

Beer is brewed in Mesopotamia

The woolly mammoth becomes extinct; *See also* 2000 BC FOO

Arrow shafts are left at a site at Stellmoor, Germany, c 8500 BC the earliest direct evidence of the bow and arrow known today; *See also* 25,000 BC TOO

Stone tools known as Folsom points and probably used as arrowheads begin to be manufactured in North America by Native Americans c 8400 BC; among the earliest examples known are those from Blackwater Draw NM, although the points are named for the site of first discovery, Folsom NM

Although it is impossible to be certain, it appears that copper was occasionally used by this time; it was worked like a soft stone, not cast; *See also* 6000 BC TOO

---

c 7000 BC

The first sledges come into use

Chiles (peppers) are grown in the Andean Highlands c 7400 BC; since they are a tropical plant, they were probably domesticated previously in the lowlands

Sugarcane is grown in New Guinea

The pig is domesticated, according to evidence found at a site called Cayönü in present-day Turkey

The water buffalo is domesticated in eastern Asia and China

Flax is cultivated in southwestern Asia

Cattle are domesticated in southeastern Anatolia (Turkey) and independently in Africa

The chicken is domesticated in southern Asia

Yams, bananas, and coconuts are grown in present-day Indonesia

Durum wheat is cultivated in Anatolia; *See also* 9000 BC FOO

The Sauveterrian tool industry of inland France c 7500 BC produces some of the smallest microliths, especially microblades with two sharpened edges

The Maglemosian culture of the northern European plain c 7500 BC, in addition to stone axes and microliths, uses wooden canoes with paddles to engage in a fishing industry that involves both nets and fishhooks; many of these organic items, including wooden fishhooks and harpoons, are preserved in bogs and other wet sites; *See also* 4500 BC COM

Potters in southeast Asia (Burma to Vietnam) begin to fire pottery in kilns; *See also* 10,000 BC TOO

The oldest known woven mats are made in Beidha, Jordan; basketry probably began much earlier

People in various sites in the Near East and what is now Turkey make pottery from clay; *See also* 10,000 BC TOO

c 6000
BC

Some buildings in the Near East are rectilinear with several rooms, some with paintings on the walls; internal furniture includes benches and platforms made from cattle horns and plaster

The first machines

Simple machines are devices that do nothing but change the direction, duration, or size of a force; duration or size is changed as a result of an inverse relationship between amount and distance traveled. The single pulley is the dullest simple machine, changing only the direction of a force. Most of the other simple machines are variations on the lever or the inclined plane; for example, the wheel and axle (or crank handle) is a rotary lever, while the screw is a helical inclined plane. Perhaps the most sophisticated simple machine is the compound pulley, in which mechanical advantage is cleverly obtained with no visible levers. The compound pulley appears to have been invented in Hellenistic times, about 200 BC.

Which simple machines were used by early humans? Almost any large creature climbing a hill uses a form of inclined plane to avoid the force required in going straight up, so that hardly counts. It is also easy to believe, although difficult to prove, that early humans used levers to turn or lift heavy objects. Similarly, wedges undoubtedly go back to early times, but we have better evidence for these, since many early wedges were made from rock (early levers would have been made of biodegradable wood). The earliest stone tool is a form of wedge, as are most stone tools. The handle of an axe or hammer is a form of lever, so hafted axe heads (in use by the middle of the Old Stone Age) qualify as simple machines. Other early evidence of thoughtful use of simple machines before Neolithic times is hard to come by.

An important application of the lever from about 15,000 BP is the spear thrower, or atlatl, an extension of the human arm used to translate a small motion near the shoulder into a large motion near the end of the spear thrower. Since the time of the motion does not change while the length of the motion increases, the result is a higher velocity for the spear thrown with such a device. The higher velocity gives the spear greater momentum, useful either for distance or for penetrating power.

c 5000
BC

A well is built of wood in northwest Germany c 5300 BC at the site now known as Kuckhoven; because of local conditions, it is preserved and becomes the oldest known wooden structure

Early farmers in the delta of the Euphrates, in present-day Iraq, build houses of reeds that have been lashed together

Wooden sledges for travel over snow are in use in what is now Finland c 6500 BC People settle the island of Crete	Haricot beans (common green and dried beans) and lima beans are grown in the Andean Highlands c 6400 BC; since they are tropical plants, they were probably domesticated previously in the lowlands Bulrush millet is cultivated in what is now southern Algeria Finger millet is cultivated in Ethiopia Wine-making begins in northern Mesopotamia (now northern Iran and Iraq) or in the Levant Modern-type domesticated bread wheat and lentils are cultivated in southwestern Asia; *See also* 7000 BC FOO Irrigation is in use in Mesopotamia Peaches are grown in central China Citrus fruit is cultivated in Indonesia The grain quinoa, now becoming popular internationally, is grown in the Andean Highlands c 5600 BC	Small lead beads are produced at Catal Hüyük (in Turkey) c 6500 BC, but most likely the smelter failed to recognize that lead is a specific metal The first definitely known use of copper occurs in what is now Turkey; c 6400 BC people learn that copper can be melted and cast In Santarém on the floodplain of the Amazon in Brazil, people from a fishing culture make decorated red-brown pottery; the remains they leave are the oldest known pottery fragments in the Americas; *See also* 7000 BC TOO People in what is now western Turkey begin to make bricks in molds and then sun-dry the bricks; the same practice can be observed in Crete several hundred years later Weaving of cloth is known in Anatolia (Turkey); the first known samples of cloth are from the early city Catal Hüyük	**c 6000 BC**
A cave painting in Zalavroug, near the White Sea (a large bay of the Arctic Ocean in Russia near Finland), depicts people walking on planks attached to their feet, an early form of skis	The avocado is grown in Mexico and Central America c 5300 BC Egyptians begin to practice agriculture in the lower Nile Valley c 5200 BC, thousands of years after the Agricultural Revolution comes to other peoples in the Near East Date palms are cultivated in India The horse is domesticated for food purposes in Ukraine Rice is cultivated in the Yangtze Delta of China	In Egypt, weapons and implements made from native copper are occasionally deposited in graves; *See also* 6000 BC TOO; 4000 BC TOO Axes in Mesopotamia are made with stone heads inserted between the cleft ends of a stick and bound into place	**c 5000 BC**

c 4500
BC

Stone is used to construct buildings in Guernsey, an island in the English Channel

Machines that go around

We often picture machines with rotating parts. Rotation at less than the size of astronomical bodies and greater than the size of molecules is difficult to come by in nature. Pulleys are the primary simple machines that use rotation, but they were probably not the first use for rotation. Drilling holes by hand may be the first human use for rotary motion.

Spinning was developed early in the Neolithic, before civilization, but that does not mean that spinning wheels had been invented. The first spinning used a simple stick called a spindle as a storage device, but soon people were turning the spindle as part of the spinning process. The spindle's speed of rotation could be improved by adding a heavy disk, called a whorl. This may have been the second application of rotary motion.

One of the earliest machines using rotation is a rather complex device, the bow drill. A rod, such an an arrow shaft, is placed in a loop of a bowstring; as the bow is moved back and forth, the linear motion is transformed into rapid rotary motion. Today we are most familiar with the bow drill for starting fires. The presence of many beads with what appear to be drilled holes suggests that the device was known in Neolithic times or even in the Late Stone Age. Drilling is difficult to do without some machine help, so mass-produced beads with drilled holes suggest some sort of drilling machine.

The key development in the early history of machines is rotary motion in the form of a wheel. It is almost certain that the first wheels were simple potter's wheels rotated by hand. Potter's wheels were not invented until thousands of years after the first fired pots were made, however. Very soon after the introduction of potter's wheels, there is evidence for wheeled vehicles and simple pulleys.

The next round of more sophisticated machines involved devices for turning rotary motion into back-and-forth motion (the reverse of the way a bow drill works). Treadmills were one of the main ways to provide rotary motion, with people or animals running inside of wheels as pet hamsters and gerbils do today.

c 4000
BC

The Danish village of Köln-Lindenthal, thought to be a typical northern European community of its time, is rebuilt for the last time c 4200 BC, but for the first time with a tall palisade and earth wall around it, suggesting the increase of warfare at that time; it is settled and abandoned at least seven times over a period of 370 years

Uruk (Erech in the Old Testament and Warka to Arabs), perhaps the first great city, is founded in Mesopotamia (Iraq) on the banks of the Euphrates River

COMMUNICATIONS/
TRANSPORTATION

FOOD &
AGRICULTURE

TOOLS/
MATERIALS

Archaeological evidence suggests trade between Mesopotamia and port regions on the Indian Ocean; while this could have been overland, it seems more likely to have been conducted in boats

The simple tokens that have been used for keeping numerical records since about 8000 BC are first supplemented by complex tokens that have marks on them or that are in new and varied shapes c 4400 BC (the earlier, unmarked tokens were almost all simple geometric shapes or representations of jars or animals); *See also* 10,000 BC COM

Oak canoes are used on the Seine at what is now Paris, France c 4300 BC; workers discover three such canoes, the oldest wooden boats ever found in Europe, while digging foundations for a new building in Paris in 1992; the largest of the canoes is nearly 5 m (16 ft) long; *See also* 7000 BC TOO

The Egyptians mine copper ores and smelt them; *See also* 3800 BC MAT

A material now known as "Egyptian faience," essentially an imitation lapis lazuli, is made in Mesopotamia by covering the surface of a talc stone or soapstone with a powder made from a copper ore such as azurite or malachite and then heating the combination to a high temperature; the result is a blue-coated glass; the faience is used for the manufacture of beads and other small ornaments; later the talc substrate is replaced with a material made by fusing quartz sand and soda, perhaps the first artificial substance ever made by humans; *See also* 3000 BC MAT

c 4500 BC

People in the Near East begin to use seals to identify property by stamping the seals into wet clay

Egyptians build boats made from planks joined together; previously, boats were dugout canoes and possibly rafts of reeds bound together or skins stretched over a framework

Horses are ridden in what is now Ukraine

The guinea pig is domesticated for food in Peru c 4400 BC

Olives are cultivated in Crete

Maize, called corn in the United States, is domesticated in what is now the Tehuacán Valley or nearby Oaxaca, Mexico

Rice is cultivated in Thailand, probably having spread from a center in the Yangtze Delta

The ard, a primitive form of plow, is in use in China; plows pulled by cattle are known in northern Mesopotamia

The Egyptians and Sumerians smelt silver and gold (make the metals from their ores)

A standard kiln is developed in Mesopotamia; the fire is in a hearth below a perforated separation from the kiln chamber where the pottery is held while it is being fired

Iron beads are used in what is now Cairo, Egypt, as evidenced by oxidized remains of a string; this is the earliest direct evidence of human use of iron

c 4000 BC

OVERVIEW

The Metal Ages

It is common to speak of the Bronze Age and the Iron Age (and occasionally of a Copper Age); but the Bronze "Age" was a very short period in human history, while the end of the Iron Age is strictly a matter of definition. We still use iron and its alloys for most tools and major construction, but that has long since ceased to be the dominant thrust of technology. By about 1000 CE, even though increased use of iron continued to be a factor in people's lives, the turn to new sources of energy was more important still. Later this trend in energy use along with new ways to produce and use alloys of iron became two of the strands of the Industrial Revolution. Before that revolution was complete electricity, and later electronics, came to dominate technology. In this book, we combine the Copper, Bronze, and Iron ages into the Metal Ages to signify the period when the transition from stone, bone, and wood was more important to technology than the other factors that came later.

Metal not the only new technology

The Metal Ages are a significant time in human history that could be called equally well by various other names, many of them reflecting the emergence of new technologies along with other social factors. From the point of view of architecture, the Metal Ages include the beginning of cities, planned and otherwise, and a time of monumental construction of all kinds. This period also saw the dawn of history, for writing was developed. Farming during this period entered the age of true technology, as farmers increasingly turned to irrigation to improve crops, especially since so much of the growing human population lived in dry regions of the world. It would also be equally appropriate to call this the ceramics age, since it was also the period when pottery and other ceramics, along with glass, were dominant; for the average person during most of this period, the development of pottery was far more important than that of iron or other metals, which mostly were used by soldiers and specialized technicians, such as carpenters and masons. From another point of view, it was the first age of the wheel, although as with iron we have never given up that useful device. Finally, this was the age of another great advance in transportation—the sail, our first power source that did not depend on biological input—but the great days of sailing ships were left to come.

Civilizations of the Metal Ages

Perhaps the overall catchword for this time of amazing change in the way people lived is *civilization*. People entered this period as either simple farmers or (in some ways more complicated) hunter-gatherers. Although the vast majority of humans continued to be farmers or hunter-gatherers or both after civilization started, a significant minority became full-time warriors, traders, merchants, manufacturers, accountants, builders, or rulers.

Scholars influenced by Mesopotamia and Egypt have defined a civilization as a society that includes towns of at least 5000 people, a written language, and monumental religious works produced in service of a state religion. Yet various civilized societies that we recognize today may fail in any one of these criteria. It is not clear that the Maya had towns, that the Incas or Aztecs had a true written language, or that Chinese monuments were religious.

A different definition of civilization (owed to Robert McC. Adams in 1960) is that a civilized people is divided into classes, society is organized by the state, and labor is specialized into different trades or professions. One may wonder what Karl Marx would have thought of that definition. Actually, Marx and Friedrich Engels were pioneers in thinking about the origin of the civilized state, which they tied to division into classes; but remember that they thought that civilization would lead to eventual withering away of the state.

Astonishingly, civilization does not seem to be an idea

that arose in one place, although Sumer can lay claim to being the first as far as we know today. The Egyptian civilization that began only a few hundred years later and only a few hundred miles away seems to come from different sources and to have evolved along its own lines. Certainly, the civilizations of the Far East and the Americas are based on a different technology from either of those of the Middle East.

Despite differences in technology or society, all civilizations seem to have arisen politically from similar roots. Villages acquire rulers. Eventually one of those rulers sets out successfully to rule over ethnically similar villages in a larger region, effectively becoming a king or chieftan. By a similar process, one of the kings extends sway over a larger region, now encompassing ethnically dissimilar peoples, and acquires an empire. The empire lasts for a time and then disintegrates under pressure from outside or as a result of bad management, often both.

This general pattern has some direct effects on the type of technology possible. Kings and emperors can order large numbers of people to work together on projects. Even without advances in tools or materials, construction projects—pyramids, temples, canals, roads, and so forth—can be accomplished that would be difficult without unified control. Where we observe construction of large projects without the obvious presence of kings, we suspect that either the kings were there and failed to leave a record or, failing kings, that religious leaders were able to command people for some purposes that resulted in the monuments. It is hard to picture large numbers of people digging or moving large rocks without some central direction.

The process of forming kingdoms and empires involves warfare in nearly all cases. Thus, military technology is often at the forefront of change. War chariots preceded carts used for hauling goods by hundreds of years. Copper, bronze, and iron were employed in battle axes and arrowheads well before they became common as the tips of plows or in needles.

The influence of size

This period we call the Metal Ages is marked by great strides in technology over short periods of time in a few localities, combined with long stretches of space and time when not much seems to happen. After the initial advances of the Sumerians and early Egyptians, much of the Middle East fell into a period of technological extension—incremental advances if any, rather than large leaps forwards. Progress came swiftly in some periods in China and the Far East or in the Americas, but these periods were preceded and followed by long stretches that depended on earlier technological change. Not much seemed to happen during the later empires of Mesopotamia or Persia (Iran). The Roman Empire is notorious for its limited number of advances over such a long period of time, especially when compared with the Greek and Hellenic periods that immediately preceded it. When it comes to technology in Europe it can be argued that more happened in the half-millennium after the fall of the Roman Empire than during the half-millennium that the empire existed.

A possible reason for this pattern may be found in looking at some recent variations on the theory of evolution. Evolutionary change often occurs in smaller populations and then spreads throughout larger ones. Islands, either real ones surrounded by water or metaphorical islands of forest in the plains or valleys in the mountains, include populations that are large enough for evolutionary change but small enough for the change to spread throughout. Because of their isolation from a larger population, the change has time to spread throughout the island. Thus, when the island rejoins the mainland, the change is not swamped by unchanged genes. Alternatively, the island can produce migrants that gradually change the population on the mainland. Another idea that seems related to this one is the theory that evolution proceeds with long periods of stasis interspersed with short periods of change.

Although these theories of evolution seem to fit technological growth in the Metal Ages, they begin to fail in later periods of human history. In more recent times, worldwide communication means that there are no more islands. This does not necessarily mean that there is no further evolution.

Major advances

Architecture and construction. The appearance of what we call civilization was so coincident with the rise of cities that some archaeologists prefer to label this period the Age of Urbanization. Urbanization for this period is discussed in the essay on page 32. Urbanization generally brings with it more serious structures, not only houses for multiple families, but also various public buildings. Urban rulers tend to want monuments and even great tombs for themselves.

Along with tombs, monuments, and buildings, architects and civil engineers of the Metal Ages built many practical structures. The first stone bridges, large tunnels, dams, aqueducts, and canals were built during this period. Some of the practical structures built during this period have continued in use until today, although often repaired—rather like the axe that lasted 50 years and only needed two new heads and seven new handles during this time.

Among the Roman structures that have had a lasting effect are the roads, built largely as routes for the Roman infantry. The Roman highway engineers favored straight paths to cover the shortest distance; since the roads were largely intended for foot soldiers, they followed grades steeper than one would want for vehicles. Even today you can drive the Roman Fosse Way across southwestern England, where it is now the A429, and be startled by grades as steep as 14 percent. Some Roman roads had grades as steep as 20 percent. While Roman builders did not like to make cuts at the tops of hills or ridges (cuts tend to fill with water and are generally difficult to maintain), they did alleviate the deepest parts of valley with fills. A Roman highway was built to last, using four or five layers of sand, mortar, gravel, and concrete, topped with hard rock set in more concrete. The result was more than 1 m (about 4 ft) thick and 2 or 3m (up to 20 ft) wide. Since the road was often harder than the surrounding soil, Roman roadbeds tended to erode less than their surroundings and to become a long ridge across the land.

Although we call this time the *Metal* Ages, most structures were built with no metal except occasional metal clamps or lead mortar, but Chinese architecture provides an exception. Having learned how to make cast iron, the Chinese used it for, among other purposes, construction of suspension bridges, temples, and monuments, although not for routine building.

Communications/transportation. The height of communication that has survived from the Stone Age consists of the great cave paintings from the end of the period and language itself, probably from near the beginning. The Metal Ages offer additional peaks that are hard to match—the first written language, the alphabet, numeration systems, and the first great literary works, oral and written, that delight and instruct even today.

Other advances from this period include gradual improvement of maps. Although some engravings have been interpreted as even earlier maps, the first definitely identifiable maps start with Sumerian ones that follow hard on the heels of the invention of writing. By the time of the early Ionian scientists/philosophers, maps purporting to show the entire Earth were being made. Charts of some sort for mariners must have existed even earlier, although none have survived, probably because they were kept as state or commercial secrets. Latitude, longitude, and map grids were all invented, the first two by sailors.

In China, paper and printing were developed, although they did not spread to the rest of the world until the 15th century CE.

Just as this was the period of great advances in communication, it was also the period of the first wheeled vehicles, the first sailing ships, canals used for transportation (some with locks), and various ways to use the horse and mule in transportation. The basic methods of transportation that were introduced during this period persisted until the 18th century.

Food and agriculture. The cultivation of crops became more efficient with the invention of the plow and the use of draft animals. With draft animals come better ways to hitch them, including the rigid horse collar, first used in China and much later in Europe. Irrigation was introduced at the start of this period in Mesopotamia. It also became important to civilization in Egypt, India, Peru, and Meso-America (see the essay "Irrigation and the rise of civilization," p 27). A few new crops, such as cotton, peanuts, yams and sweet potatoes, and alfalfa were domesticated. But by far the most important new crop was maize (called corn in the US), which joins rice and wheat among the main staple grasses that are to this day the basis of human nutrition, either directly or indirectly as animal feed. The Chinese during this period introduced more advanced cultivation practices, such as the use of rows, manure, and weeding, enabling them to produce multiple crops in a single year.

There are many reasons why humans and other animals live in greater numbers near rivers. Water for drinking is the first and most important reason. Some tasty plants, such as cattails, and edible animals, such as fish and freshwater mussels, are only part of the attraction. Even before humans learned to travel by boat, river valleys provided paths; after boats were invented, the rivers themselves often became the best paths.

After the Agricultural Revolution, some rivers took on new roles. Farmer quickly learn the importance of water and sunshine to crops. Sunlight can be obtained by burning away part of the forest, which is the method used in most places with abundant water to this day. Often, however, broad reaches of land already exist with no forest cover; but this is because the region is short of water. If a farmer can get water to fields in such sunny lands, the resulting crops are prolific.

Sometimes a river runs through such a land. The river obtains its water somewhere that is rainy, usually in high mountains unsuitable for farming. It then carries the water through desert or semidesert as it flows to the sea. Rivers of this type include the Nile, Tigris, Euphrates, Hwang He (Yellow), Indus, and about 50 waterways in Peru that run from the Andes to the Pacific. Merely listing the rivers suggests that there is a connection between rivers running through dry lands and the development of civilization, for these are the rivers of Egypt, Sumer, China, Harappa, and the Chavín, virtually all of the earliest civilizations known.

The connection between rivers in deserts and civilization is thought to be irrigation. The political structure necessary to organize people for canal digging and fair use of river water for farming is also the political structure that supports other types of construction, city planning, state religion, education, and a class with the leisure to pursue mathematics, science, and philosophy. This "theory of the hydraulic society" stems from such thinkers as Friedrich Engels, Karl Marx, Max Weber, and Karl Wittfogel. While not everyone agrees that irrigation caused the state (some think that class divisions started first, enabling irrigation systems), even most critics of the hydraulic theory think that it applies to some degree to the civilizations listed in the previous paragraph. It especially seems to apply to Sumer, which was, after all, the first civilization.

In Mesopotamia, streams are at a low level during the fall planting season, requiring very deep ditches for irrigation. When the high water arrives in the spring, it not only irrigates the plants, it also lays down a lot of silt in the canals, so the deep digging needs to be repeated. Both maintaining and building the system require organization and that seems to mandate state control.

Where the hydraulic theory seems to break down is in civilizations such as the Olmec and early Maya, civilizations that developed in regions where a lot of water falls from the skies on a regular basis. In the past few years, however, aerial studies have shown that the Maya did possess large irrigation systems. Such systems were needed because most of the rain that falls over Maya sites immediately passes deep into the soil and underlying rock.

The Olmec civilization presents tougher problem. One careful study, by William Rathje, focuses on the lack of hard rock and obsidian in the lowland rain forests of eastern Mexico, where the Olmec civilization began. Rathje has proposed that the state must have arisen there to facilitate long-distance trade, a recognized part of Olmec life.

However, if you return to the main early civilizations, it is hard to find any but the Olmec not heavily involved with irrigation right from the start. Even if war, classes, or population growth came first, large-scale irrigation *requires* record keeping, organization, different classes of workers, central planning, foresight, and so forth—in short, civilization and the state.

Unfortunately for most civilizations in arid regions, irrigation from a river also contains the seeds of destruction for the civilization (see the essay "Salt and the fall of civilizations," p 71).

New farm tools enabled people to harvest, store, and prepare food from plants in better ways. But the first basic tool was the plow. Growing from a pulled digging stick at the start of this period to today's multibladed implement of agribusiness, the plow was mainly needed to soften the ground for planting. Gradually, farmers learned that it also helped fertilize the soil and remove weeds by turning the weeds into "green manure." Eventually, however, 20th-century farmers began to realize that plowing also prepared soil to dry out and be blown away by the wind.

Harvesting improvements for this period came last, with the first known harvesting machinery not used until the Roman Empire. Storage was promoted by turning grain into beer, wine, and even distilled spirits, although salting, drying, and pickling were also available and in wide use. Extraction of liquids from foods (such as oil from olives) was first handled by the bag press, but 1500 years later the Greeks invented the more efficient beam press.

Animals, ranging from insects (the silkworm and honey bee) to giant mammals (Indian elephants), also continued to be domesticated during the Metal Ages.

A few metals can be found in nature as elements, not compounds; these are called "native." Native gold is relatively common, especially in small flakes separated by weathering and transport; it is segregated somewhat from other minerals because of its density. Native silver and copper are much rarer, although modern miners have sometimes found large masses of silver (hundreds of kilograms in the rarest instances). Copper is actually the most widely distributed native metal, often precipitating from its ores as a result of chemical reactions; but amounts in the native form are always small. Native iron can rarely be found as a result of interactions between igneous rocks and coal seams, but large masses of iron (or nickel and iron) are the substance of some meteorites. The other native metals, arsenic, antimony, and bismuth, occur mostly in hydrothermal vents and are of little importance. Of the common native metals, only copper is hard enough to be useful, and it is only hard when compared with gold and silver. Nevertheless, the advantages of copper tools over stone, bone, and wood tools were apparent to early humans, and native copper was probably used from near the beginning of Neolithic times.

Although native copper is hard to come by in large amounts, copper ores are quite common and some of them produce metal easily when burned in wood or charcoal fires. Smelting of other ores, notably gold and silver, and probably even iron, seems to have preceded copper smelting, however.

Learning how to smelt copper vastly increased its supply in the early years of civilization. A Copper Age that lasted about 500 years in the Near East followed the discovery of copper smelting; it ended when bronze (copper alloyed with tin, or sometimes with other metals, ranging from silver and lead to arsenic), a better metal, was discovered. In parts of the world where copper ores were plentiful, but tin was rare, the Copper Age persisted until trade routes for tin or for finished bronze replaced copper with bronze. In Egypt, where tin cannot be found, copper hardened with arsenic was used until about 2000 BC.

While copper ores in small amounts are common worldwide, there are a few places where good copper ore in easily smelted form is abundant. For the ancient world, the island of Cyprus was the main supplier of copper—indeed our word *copper* derives from the name *Cyprus*. Cyprus not only had the ores, but as an island in the eastern Mediterranean, it was easily reached by shipping from Egypt, Greece, and somewhat more distant Rome.

Tin was more of a problem. The only known large deposits of tin in the ancient world were in western Great Britain, an island like Cyprus, but in the Atlantic instead of the Mediterranean and far to the west of Egypt and Greece. Long trade routes for tin began by crossing the

Materials/medical technology. During this period humans first began to produce materials that were quite a bit different from those found in nature. Although some use of ceramics can be traced well back into the Stone Age, the potter's wheel, production of glass, and fired brick belong to the Metal Ages, as, of course, does the smelting of metals. Animals other than humans do not strictly speaking change their environment in this way, for their creations—such as shell, chiton, silk, honey, wax, and paper (from wasps)—are mostly processed by their bodies from foods consumed. Also, materials produced by humans reflect their unique control of fire. Ceramics, glass, and smelting absolutely depend on fire.

Not all the new materials produced on a regular basis from this time were synthetics, however. Cotton and silk are certainly natural, but the production of thread or yarn from these natural fibers was new. Wool and linen were being spun earlier, near the end of the Stone Age. Many think that cotton and wool weaving reached its all-time height in this era with the work of several early Peruvian cultures, notably the one known as Paracas Necropolis. New dyes, notably the royal purple from murex shellfish, also became important. According to one clay tablet, two Mesopotamian women developed the first perfumes during this period.

Most of the basic concepts for the metals we still use today were discovered at this time, although many of the metals themselves were often in short supply. Near the end of this era, for example, ordinary farmers could not afford enough iron for iron plows or even iron spades. Instead, they used wooden plows and spades that had iron tips and cutting edges.

Two other material advances from the later part of this period have had a lasting influence—the development of paper and gunpowder. Both were invented by the Chinese, but their greatest use and impact came with their spread to Europe after the end of this period.

Recorded medical science has its beginning at this time, although there was undoubtedly Stone Age medicine that has largely been lost from the archaeological record. Ironically, the Egyptian cult of the dead made the greatest early contributions, since embalmers and morticians from as early as the third millennium needed to understand not only human anatomy (with plenty of opportunity to observe any visible causes of death), but also organic chemicals that they used to preserve organs and whole bodies. Thus, early Egyptian medicine

English Channel to Gaul (France) and continued overland to Mediterranean ports such as Massalia (Marseilles) before loading goods onto sailing freighters. Such
travel contributed to the high cost of tin, and therefore of
bronze. When the Romans near the end of the republic
occupied Britain, an occupation that lasted for about 400
years, their main goal is thought to have been control of
the tin trade.

Bronze is a relatively strong and corrosion-resistant
metal, but the Classical age also used the other main alloy
of copper, brass. Today brass is known to be made from
copper and zinc, but zinc was not discovered until the 13th
century in India and not produced in the west until its
rediscovery in 1746, although as early as the 16th century
some alchemists were aware that there was a metal at the
heart of the mineral known by the ancients as *calamine*—
recognized today as two different zinc-based compounds,
zinc carbonate and zinc sulfate. The earliest brass was
made by smelting copper ores or copper with calamine;
according to one early source, brass smelting was first
accomplished in Anatolia (Turkey) by a people called the
Mossynoikoi. It is not clear how early this occurred, but
the earliest brass item known today dates from 20 CE.

Another metal known from early times was the easily
smelted and relatively common lead. It had many purposes, including those based on a property for which lead
is often used today—it is denser (weighs more per unit of
volume) than other inexpensive materials. Thus, one
early and still common use for lead by the Egyptians was
as weights on fishing nets. From a mathematics problem
in the Rhind Papyrus (c 1650 BC) we know that lead at
that time was not really very cheap. It had half the value
of silver and a quarter that of gold.

Although iron was sporadically used when found native
or in meteorites or made from easily smelted ores—one
theory is that the earliest iron manufacture came as a by-
product of fires burned while obtaining red ocher to use
as paint—it was not important for making weapons or
utensils. The main reason is that when iron is smelted in
the same way as copper or tin, the resulting metal is not in
a form that is very strong or hard. The earliest iron samples from Egypt, however, are all of meteoric origin,
according to chemical analysis.

Iron had been known for thousands of years before people living in what is now Armenia found that reheating
and hammering produced a material vastly superior to
copper or bronze for most purposes. Those who knew the
secret of making iron weapons soon cut a great swathe
through the Near East and the eastern Mediterranean,
although they did not win all the battles. By about 1200 BC
the "secret" was no longer a secret at all and the Iron Age
had begun.

---

involved surgery as well as internal medications, while
early Mesopotamian medicine revolved around external
application of medication.

Around 400 BC a physician named Hippocrates, of
whom there is scant mention in early Greek sources,
apparently founded a school of medicine that was
extremely influential. Even if we know little about Hippocrates himself, various writings attributed to his
school survive as the *Hippocratic Collection,* written
from perhaps as early as 500 BC through about 300 CE.
These writings, all in Ionic Greek, are notable for their
attribution of disease solely to natural causes, although
the actual causes were often mistaken. Around 150 CE a
second major figure, Galen, emerged in Pergamum,
Alexandria, and Rome. Many of Galen's writings in
clear Attic Greek survive. Galen was the major influence
on European medicine until the Renaissance. His primary contribution was to have carefully dissected and
observed many mammals, including Barbary apes but
not apparently humans, and to have accurately (for the
most part) described such structures as the nervous system, the heart, and the kidneys. Galen believed that God
designed living creatures to function perfectly and that
study of those creatures would reveal God's purpose.

Although Galen was more Stoic than Christian, this atti-
tude helped maintain the popularity of his works
throughout the Christian world of the Middle Ages.

Tools and devices This period begins with perhaps the
seminal tool of antiquity, the potter's wheel—from
which various other devices that use rotary motion
flow. The wheel not only made pots better, but in Greek
hands became a tool for painting, smoothing, and
grooving. The last two operations do the same for clay
as a lathe does for wood, and it is easy to see how the
lathe evolved from the potter's wheel.

Trade and business propelled some of the other
devices of the age. It is difficult to trade in materials
unless you can measure them, and for many materials
their mass is the main measure. The invention of the
balance scale by the Egyptians enabled merchants to
measure mass. Business is difficult to arrange without
the parties operating on the same time, and this period
saw the introduction of various forms of clock and sundial. Finally, traders need to be able to navigate; the
magnetic compass and the astrolabe from this period
helped solve the navigation problem, although a complete solution would not come until later periods.

"Man's first invention, and one of the most important in history, was the wheel."
Jerome S. Meyer

The wheel symbolizes the first invention so much that the unnecessary rediscovery of any simple idea earns the cliché "reinventing the wheel." But the wheel is far from the earliest invention. In the history of humanity, the wheel is recent. It is not known for certain what the first invention was, but our ancestors were making a vast array of tools for a couple of million years before anyone got around to the wheel.

As with most inventions, the wheel not only had predecessors, but it also required a number of related inventions before it could be useful. This concept is exploited for humor in Jon Hartmann's comic strip "B.C." An early inventor appears with his wheel, although he has failed to invent the cart, to build the road, or to domesticate the animal that would make his wheel of any use. Instead, he travels on it by standing on projecting axles and simply rolling along, downhill one presumes.

Before there were wheels, people dragged things across the ground. To help them in this task they first devised various forms of yoke, so that two or more persons or animals could work together to drag the same heavy load. A later invention, the sled, could do a better job of dragging something that was too heavy to lift. One virtue of a sled is that it has nothing to catch on uneven surfaces as it is dragged. Where there was ice and snow, sleds—even in their highly individual form of skis (one sled for each foot)—were especially effective. Mesolithic rock carvings from Scandinavia show people skiing. A sled is clearly depicted in a pictograph dating from about 3500 BC in Uruk in Mesopotamia. Just as clearly, one of the earliest wheeled vehicles, looking like a sled on wheels, is in the same drawing. Similarly, in one early cuneiform script the symbol for *sledge* exists first, a virtual pictogram of a sled with turned up runners. At a later date, the same symbol was used with wheels attached to mean *cart*. Thus, the cart was an easy step from the sled.

It is widely assumed that an immediate predecessor of the wheel was logs used as rollers for moving such heavy objects as the stones used in building the pyramids. There is no evidence for this. Early wheels are from Mesopotamia, where there were few logs in any case. The Mesopotamian version featured three planks cut and joined to make a wheel. Because of the grain in wood, wheels made by slicing a round section from a log tend to fall apart rather rapidly.

The wheel was invented before it was used for transportation. The first wheels were potter's wheels. Before the potter's wheel, nature was essentially without wheels of any kind, as only a few microscopic animals, certainly unknown to early humans, possess anything like wheels for any purpose, especially not transportation.

The first use of the wheel was probably not utilitarian at all but ceremonial: Carts were used to transport effigies of deities and important people. Since important dead people were carried in the first such carts, it might be said that the hearse was invented before other forms of the cart. The use of the cart for the transport of goods seems to have appeared about a thousand years after its invention.

Wheeled vehicles were used in war from early on. In Mesopotamia, four-wheeled wagons served as platforms for javelin throwers; two-wheeled war chariots also appeared first in Mesopotamia. Chariots were easily maneuverable because of the use of light spoked wheels, first known from Egypt about 2000 BC

Wheels can be attached to the sides of carts simply with short axles on which the wheels turn independently, but early wheels were fixed to long axles that rotated as the cart moved.

The oldest known representation of a wheel is found on a clay tablet of Ur, in Mesopotamia, dating from about 3500 BC. The use of the wheel spread from Mesopotamia quickly into Northwest Europe. Wheels also came into use around 3500 BC in India and China. In Egypt, the wheel became known about 2500 BC. However, the use of the wheel remained unknown in large parts of the world, including Southeast Asia, Africa south of the Sahara, and Australia and Polynesia, until much more recent times. Carts for transport disappeared subsequently in many areas, including the Far and Middle East, around the beginning of the common era because of the introduction of the camel for transport. The camel was far better suited for travel through desert areas than oxen drawing carts; oxen are slow and require abundant water. Camels are not suitable for drawing vehicles.

Sleds and their close relatives continued in use in the Americas until after the arrival of the Europeans, although wheeled toys are known from pre-Columbian Mexico as early as 300 CE. In the mountains of Mexico and the Andes, goods were transported by carriers traveling along trails unusable by wheeled vehicles.

Historians are not certain when wheels became part of mechanical devices. However, such use is older than in transport, since the potter's wheel preceded the appearance of wheeled vehicles by a thousand years. Eventually, of course, the wheel found many uses in cogs, gears, pulleys, and all sorts of machines.

The oldest known description of a cogwheel used in connection with a chain, about 280 BC, is from the Greek writer Ctesibus. Finds dating from about 230 BC have shown that cogwheels were also known in China around that time. Archimedes described the worm gear. Gears appeared shortly afterward. Around 25 BC Vitruvius described cogwheels used for changing the axis of rotation engaging each other at angles.

Even specialists have trouble figuring out what many stone tools were used for, but any modern do-it-yourselfer would be right at home with most of the metal tools from this period. Certainly a modern carpenter or mason would recognize nearly all the tools from the Metal Ages, although a few would be different and, of course, many known to today's workman would be missing. A display of tools from a Roman workshop includes familiar hammers, hatchets, chisels, saws, awls, and pliers (pincers). Notably missing are the screwdriver and an array of wrenches, since the machine tools for cutting good screws and bolts were not invented until the start of the Industrial Revolution.

The most important devices of the time for future periods were the ones producing energy. Early waterwheels and windmills were not nearly so efficient as those of later periods, but they nonetheless contributed to life at the end of this era, mostly by providing the power to grind grain into flour. The earliest steam engine, however, progressed at most to the status of toy.

Inventing and writing numbers

Virtually all numeration systems start as simple tallies, using single strokes to represent each additional unit—/ for one, // for two, /////// for seven, and so forth. Evidence of such systems have been found as marks on bone dating from as early as 15,000 BC. Studies of modern peoples with limited words for numbers—often just one, two, and many—show that they also use simple tallies, or at least concrete objects such as sticks, to show specific numbers greater than two. Thus, the tally system for representing numbers can exist even when language has not developed words for numbers. Linguistic evidence suggests that concrete words, such as *twin* for two people and *brace* for two dead birds, precedes the concept of twoness in language. Yet each could have been indicated with two strokes or two sticks used as tallies.

In the Near East of Neolithic times the number language was a typical "one, two, many" system (as in our *monogamous, bigamous, polygamous* classification scheme), while tallies were used to represent specific numbers. When tokens were developed, however, some seem to have stood for sets. By the late 4th millennium BC, some tokens appear to have meant "ten sheep," while other tokens meant "one sheep." Thus, an envelope with two of the ten-sheep tokens and three of the single-sheep tokens would indicate a flock of twenty-three sheep. At this point in time in most of Mesopotamia, however, a different set of tokens would indicate numbers or measures of commodities different from sheep. Around this time in Uruk, traders were discovering that the same number could be used to mean ten sheep, ten bags of grain, or ten talents of copper. While this idea already existed for simple tallies, the extension to a more sophisticated numeration system represents the true creation of the abstract concept of number.

In Egypt, about a thousand years later, the tallies were also grouped at ten, but ten of those tallies were regrouped at a hundred, and ten of the hundreds at a thousand. This seems more familiar, as it is closer to the system we use. The Greeks adapted the alphabet for numerals. Many other alphabet-using peoples followed the Greek example.

Roman numerals today are also alphabetical, but they did not originate as such. Early representations show that the X for ten originated in a manner similar to the use of a slanting stroke to connect four tally marks to indicate five—after each nine tallies, the Romans drew a slant mark across the tenth one, forming something similar to an X. This led to half an X for five, for which we use a V today. The L that we use for fifty has as its ancestor a symbol that was like an upside-down capital T. Number historian Karl Menninger has proposed that early Roman counters circled the tenth X to indicate a hundred or possibly to show a thousand; the left half of the circle remains as the C that is used to indicate a hundred. Similarly, the right half of the circled X became the D for five hundred. The whole symbol somehow became a thousand. Since the Latin words for hundred and thousand start with C and M, the connection with those letters was reinforced.

All of these systems are no longer in use, except for Roman numerals in some traditional uses. A better system that arose in India near the end of this period in one form or another eventually replaced all other numeration systems. While the Indians used a derivative of the alphabet, the Greek idea of alphabetic numerals never reached them. Instead, they used horizontal tallies for one, two, three, and special symbols for four through nine. Originally, this system continued in a way similar to the Greek alphabet system, with special symbols for ten, twenty, and so forth. But around 600 CE (or some say as early as 200 BC) the Indians apparently started using place value, so that instead of writing the equivalent of 100 + 80 + 7 they wrote symbols together, as we do 187, to mean the same thing. Only the first nine digits had to be used along with a symbol for zero, which they probably appropriated from astronomers' marking of empty places. A famous inscription dated 870 CE contains the first zero that has survived. Arabs picked up the new system from the Indians and it was soon known as Arabic numerals in Europe. At the very end of the Metal Ages, the pope commanded that all Christians use what we now call Hindu-Arabic numeration.

The Mesopotamians had also invented a place-value system, but one based on sixty instead of ten. Although that did not survive to the present in its original form, it was used by astronomers and perhaps by others in technical fields. From it we inherited sixty minutes to the hour and to the degree, as well as sixty seconds to the minute and the 360° circle.

c3800 BC

c3700 BC

The first known example of a fired clay envelope used to hold counting tokens is left at a site called Farukhabad in what is now Iran

c3500 BC

Sumer, the world's first civilization, flourishes on the banks of the lower Tigris and Euphrates rivers, called Mesopotamia (present-day Iraq), for "between the rivers"; it is formed from such city-states as Uruk, Ur, and Lagash, with no overall central government linking them

Small provinces called nomes are united to form the single kingdom of Upper Egypt

Candles are in use; *See also* 250 BC GEN

A ziggurat in Ur (Iraq), 12 m (36 ft) high, shows that Sumerians are familiar with columns, domes, arches, and vaults

Complex counting tokens, including such geometric shapes as paraboloids, as well as miniature tools, furniture, fruit, and even human figures, become common in the Near East; all are made from fired clay; *See also* 4500 BC COM

The first known examples of clay envelopes marked on the outside to denote the kind or amount of counting tokens inside are left at sites in what is now Syria

Sailing ships are used by Egyptians and Sumerians

Wheeled vehicles are used in Mesopotamia, as evinced by a pictograph from Uruk; next to the wheeled vehicle is a sledge that, except for the wheels, is exactly the same design

c3300 BC

City life

Sometimes it seems that the only fashion in human society that has persisted throughout the past 6000 years has been the city. People since Sumer have left the farm or village and moved into town. If it were not for the steadily increasing human population, the countryside would have become barren long since. Instead, cities have grown from a few thousand to millions.

The primary cause for moving to the city is nearly always economic. People find that they can make money in cities; also, farms can only support so many people. A family farm may be able to carry members of three generations, but seldom can it support everyone or even every male in those three generations.

The earliest civilization, in Mesopotamia, consisted of small city-states, unified by Sargon of Akkadia after 1500 years of urban life. In Egypt the process was reversed; the first great city, Memphis, was founded after unification, about a thousand years after urbanization had struck Mesopotamia and hundreds of years before Sargon. About the time of Sargon a pair of great cities were built on the Indus River in what is now Pakistan, each laid out in the now familiar pattern of a rectangular grid. Early in the second millennium BC, cities were also built at Knossos in Crete and by the earliest Chinese emperors. In the Americas, large cities did not appear until at least the first millennium BC.

Pictographs, some of which can be interpreted today, are used in horizontal strips on baked clay tablets in Sumeria; *See also* 3000 BC COM

The Egyptians begin to write using hieroglyphic signs on papyrus, although the earliest hieroglyphs known are from an object called King Narmer's Palette, also from around this time

	Records from the court of Egyptian Pharaoh Seneferu refer to copper mines on the Sinai Peninsula; *See also* 4500 BC TOO; 3500 BC MAT		**c 3800** BC
	The oldest known use of bronze is a rod found in the Egyptian pyramid of Meduin and probably deposited about this time		**c 3700** BC
Donkeys and mules are domesticated in what is now Israel The llama and alpaca are domesticated in Peru Wine is known from around this time, as evinced by the residues found in a jar from Godin Tepe, Iran, an outpost of the Uruk (Iraq) urban center	People in Mesopotamia fire bricks in kilns, although sun-dried brick continues to be used for ordinary purposes The Egyptians mine and process iron, using it mostly for ornamental or ceremonial purposes Europeans start producing metallic copper from copper pyrites by reducing the ore in wood or charcoal fires	A bowstave is left at Gwisho Springs in what is now Zambia; it is among the earliest known pieces of direct evidence for the bow and arrow; other evidence includes Mesolithic (Middle Stone Age) bows deposited in Scandinavia and arrow shafts from about 8500 BC; *See also* 8000 BC TOO The first form of the potter's wheel, essentially a turntable that rotates only when pushed and therefore used in a start-stop fashion, is invented in Mesopotamia; *See also* 700 BC TOO	**c 3500** BC
Cotton is grown in Mexico	A man, probably a cattle herder on an expedition to replace a broken bow, becomes trapped in snow and freezes to death in the Tyrolean Alps on what is now the border between Italy and Austria; when his body is recovered from a melting glacier in 1991 it reveals much about material used in Europe at this time—leather sewn together with sinews was used for clothing and for shoes, which were stuffed with grass for insulation; grass was also used in place of sinews, apparently as a temporary measure; copper was used for a flanged ax head, although other tools were of imported Italian flint; a birch-bark box contained charcoal wrapped in maple leaves (possibly a method of carrying smoldering coals to start a fire, since no fire-starting flint was found); the bow, arrow shafts, and a rucksack frame were all of wood	A man trapped in a glacier, later known as the Ice Man, carries in his quiver two finished arrows, with feathers to cause spin, and twelve unfinished arrows, all made from viburnum and dogwood; one of the finished arrows is of a type called a composite; it has a short shaft attached to a longer one and is designed so the longer part can be broken off by a wounded animal while the short part remains; the Ice Man also carries a small flint dagger with a wooden handle, the oldest such handle known; a sharpened flint that appears to have been for cutting grass or sharpening arrowheads; another piece of flint that may have been a drill; a bone tool similar to an awl; canisters, possibly for carrying embers; and a grass-net bag; all tools are carried in a rucksack with a wooden frame	**c 3300** BC

c 3100 BC

Traders in Uruk begin to use symbols for abstract numbers to represent the same amount of objects of any type; while numbers less than ten are represented by simple tallies, ten is a separate symbol that seems to derive from the token used for a small flock of ten sheep; sixty is depicted by a symbol that appears to derive from a token representing a measure of grain

c 3000 BC

People in Mesopotamia are already tossing spilled salt over their shoulders to prevent bad luck

Semimythical ruler Menes of Upper Egypt succeeds in uniting his kingdom with Lower Egypt

An early form of abacus consisting of beads strung on wires is used in the Orient; *See also* 300 CE TOO

Farmers settled at Skara Brae in the Orkney Islands off Great Britain build squarish stone houses with rounded corners, about 6 m (20 ft) square, with corbeled roofs completed with whalebone rafters; because of a lack of wood on the islands, furniture is also of stone, including a bed with stone slabs, a stone two-shelved dresser, and built-in stone storage cupboards; the beds also feature stone posts and a canopy of leather

The first recorded architectural work of ancient Egypt, a wall around the capital of Memphis, is built under the orders of Pharaoh Menes; originally brick-coated with gypsum plaster, it is replaced in later times with stone

Clay tablets used for pictographic writing in Sumer are turned on their sides to produce a "page" that is in what modern computer users call "portrait" format instead of the horizontal "landscape" format; *See also* 3300 BC COM; 1800 BC COM

Wheels used with vehicles are strengthed by nailing a wooden rim around the outside

Ruins from this period in Mesopotamia sometimes include seashells and minerals from India and Sri Lanka, suggesting that trade by boat may have taken place between the people of Mesopotamia and those of the Indus Valley (known from ruins at Harappa and Mohenjo-Daro)

Boats built in Egypt or Mesopotamia are paddled or sailed with a simple square sail; rowing has not yet been discovered; Egyptian boats are essentially papyrus rafts at this time, although shaped with upturned ends

c 2900 BC

Egyptian scribes devise a form of writing now called hieratic script, a simplified form of hieroglyphic

c 2800 BC

Kings, called lugals ("big men"), replace councils of elders in the government of the city-states of Sumer

Astronomical evidence shows that by 2773 BC the Egyptians have instituted a 365-day calendar, although the evidence can also indicate that the calendar was introduced as early as 4228 BC

Mycenaean invaders from the plains of the Black Sea and from Asia Minor begin an 800-year-long conquest of the native peoples occupying Greece

Corbeled arches and domes are common in Mesopotamia; corbeling is building an arch or dome with layers of brick or stone set up so that each layer projects beyond the one beneath it, like an upside-down staircase; contrary to popular belief, the true arch was invented around this time

The first version of Stonehenge is constructed on the Salisbury Plain, in England, c 2750 BC; it consists of an earthen bank and a ditch, along with 56 pits known as the Aubrey Holes, after John Aubrey, who discovered them, and only three stones, including the Heel Stone

Domesticated peanuts are grown along the west coast of South America (Peru)

c 3100 BC

The height of the annual flood of the Nile begins to be measured so that crop quotas can be based on the availability of water; seeds and cattle for plowing are provided to farmers by the pharaoh, the cattle on loan

The bag press for extracting oil from olives and juice from grapes is invented in Egypt; essentially, it is a fabric bag with two sticks arranged so that, as the sticks are turned in opposite directions by four people, any substance in the bag is squeezed so that the liquid part runs out; *See also* 1500 BC FOO

Camels are domesticated in the Middle East

Elephants are domesticated in what is now India

Cotton is cultivated in India

A simple type of wooden plow is used in Egypt; plows in Mesopotamia are made with a piece of pointed timber formed into a share that actually cuts the soil and a sole that pushes soil aside, creating a deep, wide furrow

Metallurgists in the Near East (most likely in what is now Syria or eastern Turkey) discover shortly before this time that addition of tin ore to copper ore before smelting produces a harder and more useful metal than copper and that is also easier to cast; the utility of the new metal, bronze, is such that the whole subsequent era is known as the Bronze Age

"Egyptian faience" is found in the lower Nile Valley for the first time; probably it was imported from Mesopotamia, although Egypt eventually was to develop a thriving faience industry of its own; *See also* 4500 BC TOO

The rich deposits of copper ore on Cyprus are discovered and from this date forward become the main source of copper for the ancient world; *See also* 3500 BC MAT; 2500 BC MAT

Tooth filling occurs in Sumeria

Axes in Mesopotamia are made with bronze or copper heads that have a hole in them where a shaft can be inserted

Sumerians use pins made from iron or bone to hold clothes together

c 3000 BC

The sweet potato is grown along the west coast of South America in Peru

Early sailing

Anyone who has ever tried to paddle a canoe in a stiff breeze can recognize the power of wind to move boats through water, often not in the direction one wants to travel. The earliest drawing showing a sail dates from about 3200 BC, although known trade patterns suggest that sails had been in existence for at least 300 years when this drawing was made.

Any flat, stiff object can be used as a sail to propel a canoe or small boat in the direction the wind is blowing. It takes adjustable fabric sails, more than one, to move a large ship into the wind. The first flat sails were soon replaced by bags for the wind. A bag-of-wind sail is a loose sail held at its corners, in contrast to the tight, nearly flat sail most common today.

The single bag-of-wind sail persisted for about 3000 years before a second sail was added. This soon led to ships with three masts and the beginning of masts that carried more than one sail. By the time of the early Roman Empire, sailing ships existed with sails rigged more like those of a modern small pleasure boat, with large, fairly flat triangular sails jutting away from the mast instead of square or triangular bags of wind centered on the mast. For the next thousand years, the ships of the Mediterranean followed the main sail designs that had been worked out earlier. The triangular lateen sail and variations on it became especially popular.

c 2900 BC

c 2800 BC

c 2700 BC	Written contracts are being used in Sumer	The city of Uruk, ruling over 76 outlying villages, is based in an urban area that extends over 400 hectares (1000 acres), surrounded by a 10-km (6-mi) wall; its population has grown to nearly 50,000	Bowls found on Crete appear to have been made in Egypt, suggesting seagoing trade between the two; it is likely that the Minoan ships were even more venturous, trading all over the Mediterranean by this period
c 2600 BC		Pharaoh Djoser commissions c 2650 BC architect Imhotep to build a stone tomb that will be used for Djoser's body after his death; the result is the first step pyramid, 62 m (204 ft) high with a base that measures 126 m (413 ft) by 105 m (344 ft); a step pyramid consists of several flat structures on top of each other Pharaoh Snefru (reigned c 2614 BC to 2591 BC) c 2590 BC orders the eight steps of the step pyramid that is to be his tomb filled in, resulting in the first true pyramid tomb in Egypt The Great Pyramid of Giza c 2575 BC is started as a tomb for Egyptian pharaoh Khufu (known in Greek as Cheops); the base is an almost perfect square, with the greatest deviation from a right angle only 0.05 percent; the orientation of its sides is exactly north-south and east-west; completion takes about a quarter century	Wheeled vehicles are in use in China; soon the Xue clan will be especially identified with carriage building A command from the Egyptian pharaoh Snefru to bring "40 ships filled with cedar logs" to Egypt from Lebanon is the first written record of the existence of boats and shipping
c 2500 BC	Standard weights, used in trade, are developed by the Sumerians; they are based on the shekel of 8.36 g (0.29 oz) and the mina, 60 times as heavy The Longshan people of the Huang He (Yellow River) become the first civilization of China, living in large communities, using the potter's wheel, and dividing into social classes In Egypt the tools of war include wheeled ladders used for scaling walls during a siege In Egypt, the Fifth Dynasty is established c 2475 BC	A dam about 84 m (276 ft) long, 110 m (361 ft) wide, and 12 m (39 ft) high is built at Wadi Gerrawe (El Kofaro) in Egypt sometime before this date; the dam's apparent purpose is to catch the water from the occasional flooding of the Wadi that occurs during the winter rains; the first large flood destroys it People we know as Harappans begin to construct cities on the banks of the Indus River and its tributaries in what is now Pakistan; at Harappa itself they build a huge citadel nearly 425 m (1400 ft) long and 180 m (600 ft) wide along the floodplain of the Ravi, a major tributary of the Indus Civilizations in Crete, in the Cyclades (islands in the Aegean Sea), and in parts of Greece build tile-roofed houses that contain several rooms Less than a hundred years after the building of the Great Pyramid of Khufu, when the largest stones moved were granite slabs weighing about 50 tons, workers c 2490 BC are able to move stones weighing as much as 220 tons to build the mortuary temple of Pharaoh Menkaure	Clay tablets record imports of stone to southern Mesopotamia from either Magan or Makran (both ports on the Persian Gulf); Magan developed a reputation as a port, and the stone was probably transported by boat to the mouth of the Euphrates at the head of the gulf and then up the Tigris-Euphrates river system Boats in Egypt are now made of wood, instead of being papyrus rafts with upturned ends; oars have probably been invented by this time Wheeled vehicles, basically chariots, are used by Sumerian armies

FOOD & AGRICULTURE	MATERIALS/ MEDICAL TECHNOLOGY	TOOLS & DEVICES	
Culture of silkworms is started in China; according to legend, Se Ling-she, the wife of Emperor Huang-ti, is the first to unroll a cocoon and make silk			**c 2700** BC
	Egyptian embalmers take the first steps toward mummification of bodies by removing the internal organs of dead pharaohs; these organs are preserved in jars in a salt solution, while resin-soaked linens replace them in the body		**c 2600** BC
Yams are cultivated in western Africa The cat is kept in people's houses and on farms in Egypt; it is never completely domesticated; Bubastis, Lower Egypt, is dedicated to the cat goddess Ducks are domesticated in the Near East The yak is domesticated in Tibet	Tin is mined and smelted by a low-temperature, complex process at what is now Goltepe, Turkey; Goltepe is the earliest known source of the tin used to make bronze for the Bronze Age A form of soldering to join sheets of gold is used by the Chaldeans in Ur (Iraq) Metal mirrors are used in Egypt Chinese epics mention copper, although it is not clear how developed copper mining or smelting is in China at this time; *See also* 3000 BC MAT; 100 CE MAT Animal skin is used for writing upon, but the practice is sporadic and the material is not treated as much as later forms of parchment are; *See also* 250 BC MAT Egyptian carvings show a surgical operation in progress	By this time many of the tools used by masons have been developed, including plumb lines to find the vertical, levels to find the horizontal, squares to find right angles, and mallets	**c 2500** BC

c 2400 BC
The oldest preserved weight, found in the Mesopotamian city of Lagash, is 477 g (about 17 oz)

Egyptian Pharaoh Sahure c 2450 BC orders for his pyramid a representation of the fleet he had built to ferry troops to the Levant coast; this is the earliest known depiction of seagoing ships that has been preserved and the earliest recorded use of ships for military purposes (although they were undoubtedly used in war earlier)

Sargon of Akkad produces maps in Mesopotamia for land taxation purposes

c 2300 BC
Sargon the Great, one of the Akkadians (a Semitic people from the Arabian Peninsula who settled north of Sumer), is born c 2335 BC; he conquers all of Sumer, uniting it under one ruler for the first time

At its height, the Indus River civilization we call Harappan is led by two nearly identical cities laid out with straight streets on north-south grids—Harappa on an inland tributary of the Indus and Mohenjo-Daro, 560 km (350 mi) to the south, on the lower Indus; both cities have public buildings of fired brick, while personal dwellings are of sun-dried brick, uniformly made in each city; houses are constructed around interior courtyards with the rooms plastered inside with mud or gypsum; most houses feature indoor toilets and places where people can wash themselves with water carried in pitchers

A map of the city of Lagash in Mesopotamia is carved in stone in the lap of a statue of a god; it is the oldest surviving map of a city

c 2200 BC
In Egypt, the Old Kingdom period comes to an end c 2160 BC and 100 years of rapid political change follows

Eighty bluestones are set up at Stonehenge, England, in the form of two concentric circles

According to legend, Queen Semiramis builds the first tunnel below a river (the Euphrates), linking the royal palace in Babylon with the Temple of Jupiter—but present-day engineers think that her workers did not have the technology to accomplish this

c 2100 BC
The oldest preserved standard for length is the foot of the statue of Lagash, ruler of Gudea; it is divided in sixteen parts and is 26.45 c (10.41 in.) long

By this time the megaron house that becomes the model for much of the housing around the Mediterranean is developed, as revealed by excavations at the lowest level of Troy (Hissarlik); a columned porch leads into a narrow hall that passes into the large main room (a megaron); later developments in Greece include another small room behind the megaron for sleeping or storage and a circular hearth

c 2400 BC

The honeybee is domesticated in Egypt earlier than this date, for a temple drawing from this time shows the earliest known depiction of bee-keeping and honey preparation

People living on the coast of what is now Peru begin to divert river water into their fields

Manioc (cassava; also prepared as tapioca) is grown in South America along the Peruvian coast, although it probably was domesticated earlier in the tropical inland region

Records indicate that bitumen, derived from petroleum seepage, is being used for waterproofing in great quantities by the inhabitants of Mesopotamia, with recorded shipments to Ur of over a hundred tons from sites such as Hit on the Euphrates; bitumen mixed with sand or fiber (called mastic) is also used as a building material until Classical times

c 2300 BC

A clay figurine of a domestic horse from Tell Es-Sweyhat, about 320 km (200 mi) north of Damascus, Syria, is the best early horse sculpture known and confirms domestication by this time (other evidence suggests that the horse was domesticated much earlier); Thomas Holland, who directed the expedition that found the horse in 1992, believes that the principal use of the horse in the Middle East at this time in history was in the breeding of mules, favored for pulling chariots (although horses also pulled chariots at the time as well); *See also* 3500 BC FOO

People in Central America make pottery

c 2200 BC

The first nonornamental use of iron is a dagger blade found in a grave at Alaca Hüyük in Anatolia (Turkey), although it probably was more ceremonial than practical

c 2100 BC

The oldest medical text that has been preserved in its original form is a cuneiform tablet that lists a sequence of recipes for various external poultices and plasters

c 2000 BC

Aryans from the Eurasian steppes and the Iranian plateau region cross the Hindu Kush mountains and settle in what is now northern India and Pakistan

The Egyptian fortress at Buhen in Nubia (southern Egypt and northern Sudan) has a drawbridge that moves on rollers

The Sumerian system of measures includes, besides the shekel and mina, units of capacity—the log (541 mL or 33 cu in.) and the homer, equal to 720 logs—as well as units of length—the cubit and the foot, which is two-thirds of a cubit

Sandals, probably in use long before this, are buried in an Egyptian tomb; they are made from braided papyrus

A copper bar from Nippur c 1950 BC weighing 41.5 kg (91.3 lb) and 110.35 cm (43.44 in.) long is an early standard measure; it is divided into 4 "feet" of 16 "inches" each

The Elamites (from Iran) conquer Sumer c 1950 BC

The Cretan palace of Minos introduces interior bathrooms with a water supply

A picture from about this time shows 172 men moving a statue known to weigh 60 tons on a sled, with a man pouring liquid in front of the sled to make it slide more easily

Sewage systems appear in Crete

A postal system for royal and administrative messages exists in Egypt; messages are relayed from one messenger to another

Paved roads appear in Crete

Rims around wooden wheels are frequently made from copper instead of a strip of wood, as previously

People in Mesopotamia build the first wheels with spokes, using four perpendicular spokes for each wheel

According to Strabo, the Egyptians under Senusret I (Sesostris) build a canal from Lake Timsaeh (the Nile) to the Red Sea; it is not certain that this canal was completed; if so, it was later sanded in; this purported canal precedes several versions of a canal begun by Pharaoh Necho and eventually completed by either Darius the Persian or Ptolemy II

In Egypt the tools of war include a pointed battering ram that can be operated from a portable hut, for use in breaching walls of sun-dried brick

---

c 1900 BC

Amorites from western Mesopotamia conquer Elamite-controlled Sumer and move the capital to the already ancient city of Babylon

The Minoan civilization on Crete begins to be powerful

A mud-brick arch built by Canaanites in what later becomes the Philistine city of Ashkelon (in present-day Israel) is the oldest load-bearing true arch known; the ruins of the arch were discovered in 1992

---

c 1800 BC

The sixth Amorite king of Sumer, Hammurabi, is born about this time; he is known for his Code of Laws, but also vastly extends the empire based in Babylon; from this time and for many centuries thereafter, the Mesopotamian region is properly referred to as Babylonia instead of Sumeria

The Harappan civilization along the Indus River mysteriously comes to an end c 1750 BC

The Shang dynasty in China begins c 1750 BC on the floodplain of the Huang He (Yellow River), which is utilized for an irrigation system

---

The wedge-shaped point on the stylus used for inscribing pictographs on clay tables has by this time changed the writing style in Babylonia to an early form of cuneiform, which means "wedge-shaped"; the pictographs are very stylized and might more properly be called ideograms; *See also* 3000 BC COM

c 2000 BC

The first seed drills are made by adding a funnel that directs seed into a hole in a wooden plow; seeds poured into the funnel are deposited into the plowed furrow; *See also* 1500 BC FOO

In both Egypt and Mesopotamia the shaduf is introduced for watering fields; essentially, a shaduf is a long pole on a pivot with a bucket suspended from one end of the pole and a counterweight at the other, making lifting a heavy bucket of water from a river or canal a fairly easy task; the shaduf is still used today in Egypt and Asia

Domestic cats are kept by the Lake Dwellers of Switzerland; it is probable that various species of wild cats were domesticated separately; *See also* 2500 BC FOO

Alfalfa is cultivated in what is now Iran

Paddy culture of rice exists in southeastern Asia

Pygmy mammoths on Wrangell's Island survive until this time; *See also* 8000 BC FOO

Two Mesopotamian women, Tapputi-Belatekallim and (. . .) ninu (first part of name is lost) develop perfumes, (. . .)-ninu writes a text on perfumery

Metalworkers on the Khorat Plateau (northeast Thailand) become experts in casting tools and ornaments of bronze; they are aided by a plentiful supply of tin ore in close proximity to copper ore; however, this process may have begun as much as 500 years earlier

Sometime before this date a manufacturer of "Egyptian faience" overheats his core of quartz sand and sodium carbonate (naturally occurring as natron), melting the whole mass; this is the first instance of production of glass by humans; glassmakers soon find that lime—usually found in natron or in the sand they use—is needed for glass and that lead makes it brighter, but it is a long time before they realize that glass can be molded while hot; instead, early glassmakers allow it to cool and then work the glass as they would obsidian, naturally occurring volcanic glass, cutting and polishing it to make useful or attractive objects; *See also* 1500 BC MAT

Contraceptives are introduced in Egypt

In Egypt people use a drill turned by a bow to drill holes in stone; the string is wrapped around the drill and the bow is moved back and forth

Balance scales are used in Egypt and Mesopotamia, although the evidence for such balances—consisting of drawings and paintings—is not sufficiently detailed to show how the scales are pivoted or whether they can be adjusted

Early Assyrian locks have pin tumblers and are installed using large wooden bolts on doors

The bellows is invented

Looms are depicted in Assyrian and Egyptian murals

c 1900 BC

An enormous irrigation project for the Fayum desert region is begun by Pharaoh Sesotris II and completed by Ammenemes III; a canal 90 m (300 ft) wide carries water from the Nile to a 1735-sq-km (670-sq-mi) natural depression in the Fayum; flow of water is controlled by a dam with sluice gates

c 1800 BC

Jute is cultivated in India

In the Rimac Valley of Peru, near a temple and settlement now known as La Florida, villagers construct c 1750 BC a canal that is between 4 km (2.5 mi) and 10 km (4 mi) long; it quadruples the amount of arable land in the region

	GENERAL	ARCHITECTURE & CONSTRUCTION	COMMUNICATIONS/ TRANSPORTATION
c 1700 BC	The Hittites, members of the Indo-European language group from the mountains between the Black Sea and the Mediterranean, move into the plains of Anatolia (Turkey)		
c 1600 BC		The last pyramid tombs of the pharaohs of Egypt are built; some authorities think Pharaoh Ahmose I built the last one	Wheels with spokes, which originated in Mesopotamia around 2000 BC, are in use in Egypt
c 1500 BC		One of the earliest monumental building sites in the Americas, a place in coastal Peru known as El Paraíso, or as Chuquitanta, about 2 km (1.25 mi) from the Pacific, consists of thirteen or fourteen collapsed platforms made from about 90,000 metric tons (100,000 short tons) of quarried rock thought to have come from temples; one platform measures about 50 m (160 ft) on each side	One of the first alphabets is developed in Ugarit (Syria) by stripping down Mesopotamian cuneiform characters to only 30 signs; each sign stands for a different sound; elsewhere in the Middle East, scribes are also developing alphabets using symbols that are easier to write on papyrus than the wedge-shaped letters of cuneiform script
c 1450 BC	A balance with a pointer for weighing is developed in Egypt c 1450 BC	Stonehenge achieves the form in which it is known today About this time a 32-m- (105-ft-) tall obelisk, the largest from ancient times, is made for Pharaoh Thutmose III; it stands today at the Church of San Giovanni in Laterano in Rome	Barges about 60 m (200 ft) long, constructed for carrying obelisks, such as the 32 m (105 ft) one made for Thutmose III, are the largest ships to date
c 1400 BC	The Olmec civilization starts in the rain forests along the west coast of the Gulf of Mexico Pharaoh Amenhotep IV ascends the throne of Egypt in 1365 BC; with his wife Nefertiti he promotes monotheistic worship of the sun god Aten, represented as a disk, even changing his own name to Akhenaten (the spirit of Aten)		The Olmec of the Gulf Coast of what is now Mexico become the "mother civilization" of Meso-America (Central America and what is now Mexico); in particular, they bequeath a numeration system and calendar, as well as the beginnings of written language, to the Maya; later the Aztec use a simplified form of the calendar Seagoing ships in the Mediterranean are built by first joining planks together to make a hull

			c 1700 **BC**

	Bellows are used in the manufacture of glass and in metallurgy	True plows made of bronze are in use in Vietnam	**c 1600** **BC**
	The Edwin Smith Surgical Papyrus is written c 1550 BC, although it appears to be a copy of a manuscript written about 2500 BC		
	The Papyrus Ebers, also c 1550 BC, gives a description of 700 medications; it also shows that physicians prescribe diets or fasts and massage		

The beam press, which uses a lever action to press the liquid from substances such as olives and grapes, is invented in the Aegean region; *See also* 3000 BC FOO	Iron smelting—that is, producing metal from its ore—begins a period of improvement, mainly in the Mitanni kingdom in Armenia and under the people who later conquer the Mitanni, the Hittites	The earliest known fragment of a sundial dates from this time in Egypt; *See also* 800 BC TOO	**c 1500** **BC**
Bone inscriptions in China refer to the brewing of beer	Artisans in the Near East develop the art of lost-wax casting, in which a wax "positive" is encased in a clay "negative"; when hot metal is poured into the clay outer coat, the wax melts and runs out, being replaced by the metal	From the archaeological record, bone tools, such as needles, awls, and reamers, which had been common in Tasmania 3500 years earlier, have vanished after a long decline in use	
The soybean is cultivated in Manchuria			
The Sumerians invent the single-tube seed drill; *See also* 2000 BC FOO; 100 CE FOO			
The first professional millers appear about this time in Egypt; previously all grain was ground on hand mills	Glass begins to be shaped while in its hot plastic state; previously, glass had been worked after cooling, as if it were a rock; *See also* 2000 BC MAT		
The dingo, a semidomesticated dog, arrives in Australia from Indonesia			

			c 1450 **BC**

Sunflowers are cultivated in North America	Dorian invaders from the north with iron weapons conquer the Minoans, who have only bronze	An Egyptian water clock, or clepsydra, made from alabaster is in existence from c 1380 BC; Egyptians probably had been using such clocks since about 1450 BC; *See also* 200 BC TOO	**c 1400** **BC**
Multiple cropping within the same year is in use in China	The Hittites obtain the secret of making usable iron (heating and hammering it) when they conquer the Mitanni kingdom in Armenia in 1370		
	Pots of perfume sealed in the tomb of Pharoah Tutankhamen c 1352 are still fragrant when opened in 1922; they are the product of a powerful perfume industry in Egypt	The windlass is in use	

43

c 1350 **BC**	Tree ring dating shows that a large platform constructed of oak in the shallow waters of Flag Fen (near Peterborough, England) is built about this time; although the purpose of the platform is not certain today, it is thought that it guarded access to land claimed by the builders for farming	

Building with brick and stone

The principal early building material of civilization was brick, most often unfired. Even thousands of years before civilization, builders in Jericho used sun-dried mud brick, sometimes called pisé. Adobe is a somewhat better sun-dried brick made from clay and straw. Adobe brick was also commonly used in Old World cities. Fired brick was usually reserved for public buildings or for facing structures of sun-dried brick.

c 1300 **BC**	The Hittites use their superior iron weapons in war with the Egyptians in 1286 BC; this gives the Hittites an advantage but also makes the Near Eastern world aware of the potential of iron; soon the Hittite secrets of manufacture are found out	During the reign of Egyptian Pharaoh Seti I, a stone-filled dam faced with basalt blocks about 150 m (500 ft) long and 6 m (20 ft) high is built at the northern end of a low, flat valley called Orontes; it creates the Lake of Homes that is 5 m (17 ft) deep, and that exists even today

Brick, especially unfired, weathers away quickly. As a result, the best known architecture of ancient times was of stone, the material of choice for temples, palaces, and other large public buildings. Stonehenge and the limestone pyramids and sphinx of Egypt offer the principal images of truly ancient structures, followed in our minds (although much later in fact) by the marble temples of Greece, the marble amphitheaters and baths of the Roman Empire, and yet more limestone-faced pyramids in Meso-America.

By the time of Ramses II, c 1285 BC, Egyptian construction crews are able to move statues weighing as much as 1000 tons from place to place

The huge Temple of Abu Simbel on the Upper Nile in Egypt is built c 1260 BC in honor of Ramses II; cut out of the living rock, Abu Simbel is sited so that on two days of the year (one in mid-October and the other in mid-February) the sun illuminates the inner sanctuary

c 1250 **BC**		The Olmecs build one of the first large ceremonial centers in the Americas at a place on the Gulf of Mexico now known as San Lorenzo; it is constructed on a large raised platform, 54 m (150 ft) high

Simple structures of giant stones arranged in a pattern are called megaliths. The megaliths that were erected across the western European continent, on Great Britain, and on Malta include tombs (called dolmens), but also appear to include temples and observatories. Similar groups of erect giant stones were constructed as tombs in India and other places in Asia thousands of years later. From a technological point of view, quarrying, moving, and erecting the stones, which often weighed 40 tons or more, seem to many to be beyond the abilities of the Neolithic or Bronze Age farmers. Diffusionist theories have called for Minoan or Egyptian engineers, but it is now clear that the megalith

c 1200 **BC**	By this time Egyptian surveyors use a 3-4-5 right triangle to establish right angles, but the account of Greek historian Herodotus puts this practice back at least as far as 1850 BC, while the construction of unusually good right angles in the major pyramids suggests that the method may go back to before 2500 BC	

Ducts beneath the floors of the palace of the King of Arzawa, built about this time in southwestern Anatolia (Turkey), suggest that the palace had central heating; not only is this the earliest known evidence of central heating, but there is no other evidence of the practice for the next thousand years

According to tradition, Troy falls to the Greeks in 1183 BC

FOOD &
AGRICULTURE

MATERIALS/
MEDICAL TECHNOLOGY

TOOLS &
DEVICES

c 1350
BC

Building with brick and stone (Continued)

builders came earlier than the Minoans and that some of the monuments predate Egyptian engineering. The megalith builders used fire and the swelling of wet wood along with wedges to quarry their stone. Lacking wheeled vehicles, it is thought that they simply dragged the stones, perhaps with the aid of sledges, or sometimes floated them partway on rafts. An inclined plane and a predug hole helped set the stones upright. For the most part, these were the same techniques used by the Egyptians in building the pyramids, but the megalith builders probably discovered the methods independently.

The rulers of Mesopotamia did not have stones to work with. Nevertheless, they built tall structures reaching toward the heavens. Called ziggurats, these temple towers were made from brick, often cemented with tar (bitumen) or tar mixed with sand and gravel (mastic).

Sun-dried brick and adobe were usually cemented with more of the mud used to make the bricks. Tar was available in a few places. If limestone or gypsum was present, it could be heated and then ground to form lime plaster or cement that hardened when water was added. True cement hardens by incorporating the water molecule into its structure, not as a result of evaporation, but this was not known in very early times. Plaster, the more usual result of making lime, depends on reaction with carbon dioxide in the air to return to a material that is like its parent limestone. Although some plaster survives from ancient times, it is easily damaged by water and most has been lost. Lime plaster was used in ancient Egypt and by most early civilizations of both the Old and New Worlds.

In the absence of other, cheaper materials, molten lead could be used as mortar. Another dodge, practiced both by Egyptian engineers and early temple builders in South America, was to dress the stones so that they fit tightly together and to shape them so that gravity held them in place. The megalith builders who put up the last stage at Stonehenge used a version of this method, quarrying the stones so that they could dovetail mortises and tenons, forming rigid joints. Greek architects and early Romans also used iron cramps (right-angled bars) to hold stones in place.

c 1300
BC

Sometime around the 3rd century BC the Romans began to use volcanic ash found in Italy, notably huge beds near Puteoli (Pozzuoli) on the Bay of Naples. These had been produced during eruptions of Mt Vesuvius. Ground ash could be added to lime mortar to produce cement as hard as rock, cement that even hardened underwater. The silica in the ash combined with the calcium carbonate of the lime and with the water. By the early Roman Empire the new form of cement, mixed with sand and rock to form concrete, had become the principal building material, almost as cheap as brick and often better than stone. Although the Roman builders never abandoned stone and brick completely, they used more and more concrete and less and less stone and brick. Often the stone or brick ceased to have a structural purpose and was used as ornamentation for the concrete walls of buildings.

c 1250
BC

Ironworkers are reported in Jaffa, in Israel, c 1250 BC, suggesting that the Hittite secrets for iron making are known

Iron is in common use throughout the Near East by this time, so this year is commonly taken as the beginning of the Iron Age for the region; however, iron has been known for millennia

Despite their iron weapons, the Hittites are conquered by the Peoples of the Sea in 1185 BC and the Hittite secrets of ironworking spread even farther; *See also* 1300 BC GEN

The Babylonians develop c 1250 BC an instrument that can determine when a star or planet is due south; *See also* 1000 BC TOO

c 1200
BC

Mesopotamians learn to convert rotary motion into back-and-forth motion

Bells cast in bronze appear in China

	GENERAL	ARCHITECTURE & CONSTRUCTION	COMMUNICATIONS/ TRANSPORTATION
c 1100 BC	Dorian invaders begin a process, complete by about 1000 BC, of driving out the indigenous tribes of Epirus and Thessaly and then replacing the Mycenaeans throughout Greece c 1140 BC Pastoral people from the Wei River Valley conquer the Shang in China and begin the Zhou dynasty About this time people in Europe and western Asia use lamps that burn olive or nut oil with vegetable-fiber wicks; *See also* 40,000 BC GEN		The basic chariot of Mesopotamia is redesigned, using wheels with six spokes instead of four and with the axle moved from the middle of the chariot to a position at the rear; both changes result in a chariot that can travel better over rough ground
c 1000 BC		According to a clay tablet from Moab (Jordan), the ruler Mesho builds two aqueducts to supply the city of Karcho; this is the earliest known reference to the use of aqueducts to supply water	The Phoenicians develop their alphabet of 22 signs for consonants; although not the first alphabet, it is the one adapted by both Greeks and Israelites to their own needs A map on a clay tablet shows the world with Babylon as its center Chinese counting boards originate Spokes on wheeled vehicles, first used in Mesopotamia in 2000 BC, are in use in Scandinavia; *See also* 2000 BC COM The Phoenicians begin their exploration and colonization of the shores of the Mediterranean by building a port at Tyre; the region between two small offshore islands is filled with rubble, connecting the two into a single island; seawalls are built to form two harbors
c 900 BC	Natural gas from wells is used in China	The Olmec capital of San Lorenzo is destroyed by war or revolt	
c 850 BC	The first great civilization of Peru, known today as the Chavín, which arises in the north, reaches central Peru and establishes itself in the region, possibly by conquest but possibly as a result of a religious revival (one theory is that such a revival may have begun in response to a great natural disaster, such as a tsunami or extensive earthquake, around 500 BC)	The 14-m- (45-ft-) high Chavín de Huantar Temple, now called the Castillo, is begun near the continental divide in what is now Peru; over a period of 500 years it is completed with several levels of interior galleries; it is built with stones without mortar and ceilings partially supported by cantilevered stones projecting from the walls; the earliest known stone edifice in Peru, it is about 75 by 72 m (245 by 235 ft) square at its base; vents bring air into the interior rooms; a system of canals and conduits carries water from a nearby river through the temple The first known arched bridge is built in Smyrna (Izmir, Turkey)	The ram is added to warships by the Greeks (or possibly by the Phoenicians); previously the only strategy for fighting at sea had been to board an enemy ship and engage opponents hand to hand; with the ram, it is possible to sink enemy vessels

The rotary quern, or rotary hand mill, is invented for grinding grain; it replaces the saddle quern, a form of mortar and pestle; the rotary quern demonstrates one of the first applications of rotary motion; a handle is used to rotate one stone that is placed on top of a stationary stone; this device is the ancestor of the large mills operated by animal, water, or wind power that began to be built in Europe in Hellenistic and Roman times

The Assyrians capture iron-making centers when they invade what is now Armenia; still, it is not until 900 BC that the formidable Assyrian army is re-equipped with iron weapons; *See also* 1200 BC MAT

c 1100 BC

Oats are cultivated in central Europe

Reindeer are domesticated in the Pazyryk Valley of Siberia

Several food preservation techniques exist in China: salting, using spices, drying and smoking, and fermentation in wine (vinegar)

Mesopotamian craftspeople begin using lead glazes on brick and tile

Lead is recognized as a metal, although people at this time have no particular use for it in metallic form

Dyes made from a shellfish, the purple murex, are introduced by the Phoenicians

Etruscan gold workers make dental bridgework, mostly for cosmetic purposes

Silk thread found in an Egyptian mummy's hair demonstrates that silk is being made by this time, although probably only in China. from which small amounts of it reach Egypt by way of Persia (Iran); *See also* 50 BC MAT

The Duke of Chou in China builds c 1050 BC either an early "south-pointing carriage" or a magnetic compass; a south-pointing carriage uses a differential gear to keep a part of the carriage pointing in the same direction even though the carriage turns; *See also* 1200 BC TOO; 300 BC GEN

c 1000 BC

Iron objects appear in Italy among the people called Villanovan; they are also left in graves in Greece and Crete; *See also* 1100 BC MAT; 700 BC MAT

c 900 BC

c 850 BC

	GENERAL	ARCHITECTURE & CONSTRUCTION	COMMUNICATIONS/ TRANSPORTATION
c 800 BC	According to tradition, Carthage is founded in North Africa by the Phoenicians in 814 BC; various evidence confirms that a date around 800 to 850 BC is likely According to tradition, the Olympic Games are started in 776 BC with a single event, a foot race of one stade (about 200 m or 650 ft)	The Olmec build pyramids in La Venta (Tabasco, Mexico)	Phoenicians exploring the Atlantic Ocean's European shore establish Gadir (Cadiz, Spain) as the farthest western end of their trade route
c 750 BC	According to tradition, Rome is founded on Apr 21, 753 BC on the banks of the Tiber by Romulus		The Phoenician alphabet reaches Greek Attica and is quickly adapted to the Greek language
c 700 BC	About this time the Etruscans of the Italian peninsula join the Phoenicians in colonizing the shores of the Mediterranean Sea The first standard coinage is in use in Lydia (western Turkey), reputedly issued by King Croesus In central Europe, various Indo-European and other peoples gradually evolve into the collection of tribes we call the Celts (and the Romans called the Gauls)	About this time Hezekiah's Tunnel is built as part of the waterworks for Jerusalem, primarily to provide water during an anticipated siege; two teams worked from opposite ends of the tunnel, but modern research suggests that they widened and straightened a natural passageway through the limestone that underlies the city rather than cut through unbroken rock King Sennacherib of Assyria c 690 BC builds several aqueducts to supply Nineveh, mostly in the form of open canals; the most remarkable carries water about 48 km (30 mi) to an artificial reservoir formed by damming the Tebitu River from another river, the Gomel; the canal is 19.8 m (65 ft) wide and at one point raises on corbeled arches 9.8 m (30 ft) for 274.3 m (900 ft) so it can cross a stream; *See also* 550 BC ARC	The Phoenicians, or possibly the Greeks, introduce war galleys that have two banks of oars (biremes); these are more compact, sturdier, possibly more seaworthy, and less of a target to opposing rams, since the same number of rowers are used as in previous warships with single banks of rowers; some suggest that since the length of the ship was not increased by the Phoenicians, the bireme was top-heavy and easily capsized
c 650 BC			An unknown Greek naval architect designs a ship with a lateral projection above the gunwale that provides room for a third bank of oars; the result, called a trireme, becomes the common military vessel throughout the early Classic period; some sources claim the trireme is invented in 704 BC by Ameinokles of Corinth, but this seems to be too early The Greeks c 640 BC build a ship canal across the peninsula of Leukas (Levkás), which today is an island in the Ionian Sea

| | | Querns with what appear to have the first known form of crank operation appear in Syria; a peg is inserted into a hole drilled near the outer edge of the upper stone of a rotary quern, enabling the user to turn the stone in a continuous motion by moving the peg in a circle; previous querns used radial spokes for mechanical advantage, but the peg handle is about a dozen times as efficient; *See also* 100 BC TOO | **c 800** BC |

Egyptian sundials use six time divisions; *See also* 1500 BC TOO; 680 CE TOO

| | | | **c 750** BC |

In Assyria (Iraq) the practice of digging underground tunnels, called qanats, to subterranean aquifers for irrigation purposes begins; the Romans and Arabs later spread this practice to the Mediterranean region

Incubators are used in Egypt to hatch chicken eggs

Glaucus of Chios invents soldering with an alloy that melts easily

A large-scale iron-manufacturing center develops at what is now Hallstatt, Austria; another such center is in Meroë, Nubia (Sudan); *See also* 900 BC MAT; 500 BC MAT

The true potter's wheel that can be rotated continuously is introduced in Mesopotamia

The Assyrians use a sort of pulley

c 700 BC

Inventing writing and the alphabet

In recent times the origin of the first known writing system—from Mesopotamia, Anatolia, and the Levant (the Near East)—has been worked out in detail. Writing in the Near East has its origins in the Stone Age, shortly after the Neolithic Revolution. Fired clay was used to form small objects, today called *tokens,* that represented different kinds of trade goods. The tokens gradually evolved in complexity and became a universal trade language for the region, complete with standard fired clay envelopes to hold them. The tokens inside a clay envelope could be indicated by pressing each token into the wet clay before the envelope was sealed and baked. Eventually, it became apparent that the tokens themselves were not needed, only their images in the clay. The envelopes evolved into the clay tablets that survive in hundreds of thousands, while at the same time the images of the tokens evolved into true writing.

The original versions of full-scale written languages seem to contain a different symbol for each word, although inflected languages may from the beginning combine a word symbol with an inflection symbol. Puns and near puns often are introduced to create a simpler written language, somewhat like rebus writing. These can be standardized into syllables.

A separate sign for each consonant sound used in the language, with the reader left to guess the vowels, is the basis of the earliest alphabet, which arose about 1600 BC in the region now called Israel and Jordan. Over several hundred years, that alphabet split into several varieties. Trade and the movement of peoples shifted the varieties in various ways. One version developed in Arabia and in India, leading to the curvy writing still used in Muslim regions.

Because the Phoenicians were great traders, they spread their version of such an alphabet around the Mediterranean. One descendant, the Greek alphabet, like the Phoenician but with some letters interpreted as vowel sounds, became the system used by the Etruscans. This Etruscan alphabet was taken up in modified form by the Romans, who eventually made it the dominant alphabet of western Europe. After printing was invented, a form of the Roman alphabet from Italy became the standard printed alphabet which you see before you now.

Unlike most other inventions of merit, the alphabet seems to have arisen only once and been spread by diffusion. Languages that are not written in a descendant of the first Semitic alphabet are not based on some other alphabet.

c 650 BC

c 600 BC

The Assyrian Empire collapses in 612 BC

Massalia (Marseilles, France) is founded by the Phoenicians

The Babylonian Captivity of the Israelites begins when Nebuchadnezzar II of Babylon conquers Jerusalem in 587 BC and removes the inhabitants to Babylon

Pharaoh Necho II (ruled 610 BC to 594 BC) abandons a plan to dig a canal from the Nile to the Red Sea in favor of sending a group of Phoenician ships around Africa from the Red Sea to the Mediterranean; according to Herodotus, the only source for this tale, the ships take 3 years, planting and harvesting crops along the way, but succeed in making the circumnavigation; many today are skeptical of the account

The Chaldean king Nabopolassar builds the oldest known stone bridge over a major river; spanning the Euphrates, the drawbridge is 116 m (380 ft) long and rests on seven pillars of stone, brick, and wood, with a wooden frame that has a section drawn up at night to keep people from crossing it

King Sennacherib of Assyria (Mesopotamia) builds an 80-km (50-mi) canal connecting Nineveh with Bavia

Periander, ruler of Corinth, has a road of 7.4 km (4.6 mi) built over the isthmus of Corinth to transport ships across land on wheeled platforms that travel in limestone tracks; Periander develops this rail transport in lieu of digging a canal, which would have been too expensive; the ships, probably pulled by oxen, are mainly lightweight warships and perhaps small merchant ships; *See also* 60 CE COM

Pharaoh Necho II begins to dredge the Wadi Tumilat to connect the Nile with the Red Sea, but does not complete the canal; it is completed about a hundred years later by Darius the Persian king or around 285 by Ptolemy II; *See also* 2000 BC COM

c 550 BC

The Buddha is born as Siddhartha Gautama c 560 BC in the Ganges River Valley; he is the son of the ruler of the small kingdom of Kosala in what is now India

Karush, whom we know as Cyrus the Great [b c 600 BC, d c 530 BC], becomes ruler of Persia (Iran) in 559 BC

The Chinese sage Kongfuzi, known as Confucius in the Western world, is born in 552 BC

Persia gains control of much of the Near East c 540 BC; Allies of Cyrus the Great of Persia capture Babylon on Oct 13 539 BC

Nebuchadnezzar of Babylon builds a high dam that is roughly 25 km (16 mi) long, joining the Tigris to the Euphrates and creating a giant lake behind it; the dam is earth-filled and lined with fired brick using mastic (bitumen and sand) mortar; the primary purpose of the lake is to act as a giant moat on one side of Babylon for defensive purposes

Greek architect and engineer Eupalinus of Megara [b c 600] builds c 530 BC an aqueduct and water-supply system for Megara (Greece); however he is much more famous for the water supply system he later builds on Samos; *See also* 500 BC ARC

Anaximander of Miletus draws up the first known map of the inhabited Earth—actually only the part known to the Ionians around this time; he puts the map on a cylinder to reflect the curvature of Earth

c 500 BC

According to tradition, Rome succeeds in winning its independence from the Etruscans in 509 BC and becomes a republic

Modern estimates are that there is one slave for every two free adults in Athens.

Greeks win the Battle of Marathon on Sep 12 490 BC largely because Persian arrows fail to penetrate Greek armor

In the Battle of Salamis in Sep 480 BC 380 Greek ships commanded by Themistocles defeat a fleet of about 1000 ships assembled under the command of the Persian ruler Xerxes

Architect and engineer Eupalinus of Megara constructs c 522 BC a 1.7-m-high by 1.7-m-wide (5.5-ft by 5.5-ft) tunnel 1006 m (3300 ft) long under 275-m- (900-ft-) high Mt Castro on the island of Samos (Greece) to supply water; the tunnel is dug from both ends and although the two parts connect, they are off by about 6 m (20 ft) horizontally and 1 m (3 ft) vertically (known from archaeological measurements made in the 1870s and 1880s); *See also* 550 BC ARC

Darius I of Persia has the Samian Mandrokles build a floating pontoon bridge across the Bosporus in 512 BC to let his army invade Macedonia

Greek traveler and historian Hecataeus of Miletus (Turkey) [b c 550] develops a map of the world showing Europe and Asia surrounded by ocean

Darius the Persian king [b Persia, c 550 BC, d 486 BC] has the canal of Pharaoh Necho from the Nile to the Red Sea completed, effectively linking the Mediterranean with the Indian Ocean; the canal is 145 km (90 mi) long and 45 m (150 ft) wide; eventually, like most ancient canals, it gradually fills and falls into disuse although it is reopened about 285 BC by Ptolemy II; *See also* 600 BC COM; 300 BC COM

The Chinese develop the art of fumigating houses to rid them of pests

Greek sculptors use marble from the islands of Paros and Naxos in carving statues

Greek potters make several improvements; the potter's wheel is enlarged and made heavier, so that it acts as a flywheel; pottery is shaved on the wheel after it has been dried to give a finer surface; kilns are improved; an elaborate sequence of firing at specific temperatures and with various masks produces the characteristic black-and-red ware of the time

According to tradition, Anaximander [b Miletus, Greece, 610 BC, d c 546 BC] invents the first sundial; in fact, simple sundials existed earlier in Mesopotamia and Egypt; *See also* 800 BC TOO

Anarcharis the Scythian is believed to invent the anchor

In 598, the second year of the reign of King Chao of Yen, whale-oil lamps with asbestos wicks are used; however, this may have been in 308 BC, as there were two King Chao's of Yen

c 600 BC

Paddles and oars

Transportation over water by means of rafts appears to go back to earliest times. Early rafts would have been propelled by the first forms of paddle.

Unwieldy rafts were eventually replaced by canoes, almost certainly the first form of boat. Although still paddled, the canoe is shaped to make for easier steering and passage through water than a raft. The earliest known canoes are from nearly 10,000 BP, although similar craft probably were built much before. The small canoe is still a useful boat. But, although Polynesians and South American Indians eventually built giant war and cargo canoes, larger paddled vessels have been replaced by ships with more powerful means of propulsion.

An oar is just a paddle in which the fulcrum that for the paddle is one hand is replaced with a fulcrum (the oarlock) attached to the hull of the boat. As a result, only one arm is needed to use an oar, so one person can operate oars on both sides of a boat at once. Thus, rowing takes a lot less skill than paddling, which needs two hands and is used on one side at a time. Further-

more, as the size of a boat is increased, separate rowers can be assigned to several sets of oars on each side; also, the mechanical advantage of the lever can be increased by lengthening the oar.

The ability to use several sets of oars was most important during war. Although sails had been invented earlier even than oars, designers of warships preferred rowing to sailing in battle. Sails are wonderful for transporting large amounts of goods with small crews. For fast maneuvering with no dependence on the vagaries of the winds, ancient military planners turned to ships powered by oars.

From about 1000 BC until about 200 BC shipbuilders gradually increased the number of oars and the number of rowers per oar in an effort to produce ships that were faster and more powerful than those of their enemies. This trend culminated in a ship that used 4000 rowers, but that was too unwieldy to make it into battle. Rowing with variations in methods of attack continued to be an important part of naval practice until the invention of the cannon.

c 550 BC

Chinese farmers use practices, such as planting crops in rows, hoeing weeds, and applying manure, that will not be used in the West until the eighteenth century

Steel is made in India

Ironworking is common throughout the area that today is Germany and Scandinavia; *See also* 700 BC MAT; 450 BC MAT

Stonemasons' tools, mostly punches and chisels, are being made from iron, as are the saws and chisels of woodworkers

Well before this time the Persians develop composite bows made from animal tendons and horn instead of wood; such bows do not deteriorate in hot weather and can be left strung for long periods of time without losing elasticity

The early Iron Age, from 1100 to 500 BC, sees the invention of lathes, saws, pegs, shears, scythes, iron axes, picks, shovels, and the rotary quern used to grind flour or pulverize ore; particularly good evidence for the presence of the lathe is a bas-relief of Darius I at Persepolis showing him on a throne that has legs and rungs clearly turned on a lathe

c 500 BC

c 500 BC cont

Greek theaters begin to be constructed with the wave nature of sound in mind; among other uses of acoustic principles, open vases are placed around the theaters as resonators to enhance sound levels

Zapotecs near what is now Oaxaca, Mexico, begin to erect stone pyramids, notably at Monte Albán, 500 m (1300 ft) above the countryside

The astronomer Harpales plans and has built for the army of Xerxes of Persia in 480 BC a floating bridge across the Bosporus that uses 674 ships as its pontoons, all laced together with flaxen cables; the army successfully marches across the strait, but is defeated by the Greeks

A picture on the wall of an Etruscan tomb from about this time is the earliest known depiction of a two-masted sailing vessel; the foremast carries a smaller sail than the mainmast and slants forward over the bows

c 450 BC

Pericles [b Athens, Greece, c 495 BC, d Athens, 429 BC] orders construction of the Parthenon in 447 BC to commemorate the triumph of the Greek army over the Persians at the Battle of Plataea; the sculptor Phidias, working with architects Actinus and Callicrates, completes the Parthenon in Athens in 438 to the extent that it can be used for the dedication of Phidias' gold-and-ivory statue of Athena, although further construction continues; construction of the Parthenon is completed in 432 shortly before the death of Pericles; *See also* 1680–89 ARC

An optical telegraph using torches to signal from hilltop to hilltop operates in Greece c 430 BC; it uses combinations of five torches to indicate each letter of the Greek alphabet; *See also* 1793 COM

c 400 BC

Domes, beams, and columns

Most early buildings were domes—the roof and walls were a single entity. Probably this was in imitation of huts built by bending saplings. In any case, it was not clear where the walls stopped and the roof began. Structurally, the main problem for any building is how to keep the roof up. Larger buildings also have problems with forces on walls that tend to knock them over. Forces on domed buildings are not directed as much outward as on buildings with real roofs and vertical walls.

The first rectangular buildings were sized by the length of available wooden beams that could be used as the basis of a flat roof. Several rooms of this size could also be built adjacent to each other to make a larger building.

Later, builders introduced columns in the interiors of rooms to hold up the ends of some beams, allowing construction of larger rooms. Horizontal beams above windows and doors that support walls are called lintels, so construction that relies on beams held up by columns is often called post-and-lintel construction. Post-and-lintel construction was the basis of most Egyptian and Greek architecture, even the famous Greek temples; it is also commonly used today, along with other construction techniques.

Carthaginians begin to build quadriremes c 410 BC, ships with four banks of rowers, for their navy with the intention of invading Syracuse; Diodorus Siculus, however, says that workers hired by Dionysius the Elder of Syracuse [b Syracuse, c 430 BC, d 367 BC] were the "first to think of construction of such ships," in 399 BC

Dionysius the Elder of Syracuse in 399 BC introduces the quinquereme (a ship with five banks of rowers) to his navy, which already includes quadriremes; as a result, Syracuse becomes a major naval power, sweeping the Tyrrhenian and Adriatic seas free of pirates; Athens and other Hellenic cities soon begin adding the two new types of ship to their own navies; by 325 BC Athens has 43 quadriremes and 7 quinqueremes along with its main fleet of 360 triremes

Semilegendary Chinese artisan Lu Pan produces the first known kite

Romans develop the first safety pins, but the idea is lost with the fall of the Roman Empire and not revived until the pin is reinvented in 19th-century US; See also 1842 TOO

c 500 BC cont

King Shu Jse in China builds a moving and singing magpie of bamboo, and a wooden horse that moves with the help of springs

Arches

Early dome houses built from bricks had each layer of more-or-less rectangular brick run in a circle of a somewhat smaller radius than the one below it, with each brick resting partly on the lower course and partly cantilevered off into space. A section across such a building is called a corbeled arch. A true arch uses wedge-shaped bricks or blocks; the result in its simplest form is a semicircle. Although both the corbeled arch and the true arch were known to the builders of the earliest civilizations, a devotion to mass-produced rectangular solids for most construction meant that the corbeled version was much more common. The Romans, on the other hand, preferred the greater structural strength of the true arch, and also tended, even for rectilinear structures, to use bricks and stones that were wedge shaped.

The arch's common use is one of the hallmarks of Roman architecture. Even when concrete replaced brick and stone as the basic building material, Roman builders favored arches, despite the fact that, unlike an arch built from stone wedges, a concrete arch offers no structural advantage. Because arches direct some force laterally, as well as vertically, they need some support from the sides. While this can be supplied in many ways by one sort of buttress or another, Roman builders often used additional arches, resulting in several arches in a row (called a *course*). The columns on the outside of such a course still needed some sort of buttress to withstand the lateral forces.

Although early Greek temples are constructed of sunbaked brick and wood, marble eventually replaces it; the finding of an excellent source of marble on Mt Pentelicus, a short distance north of Athens, enables Greek builders to construct the famous white marble buildings of the Acropolis; because such marble can be cut in beams up to 4.5 m (15 ft) long, Greek architects do not need to utilize arches, corbels, or domes on roofs, employing mostly post-and-lintel construction

An ironworking industry develops in Britain

c 450 BC

Hippocrates of Cos [b Cos, Greece, 460 BC, d Thessaly, c 370 BC] observes that a man involved in lead mining has developed abdominal cramps

Locks operated with keys are invented in Sparta, Greece

c 400 BC

Catapults operating on the principle of torsion (double-armed catapults) are introduced by engineers working for Dionysius the Elder of Syracuse in 397 BC; these devices, which use hair ropes to hold potential energy, are used to propel darts at an enemy; See also 350 BC TOO

Archytas of Tarentum (Tarento, Italy) [b c 420 BC, d c 350 BC] builds c 390 BC a series of toys, among them a mechanical pigeon propelled by a steam jet; he is also credited by some with the invention of the pulley

Plato [b Athens, c 427 BC, d Athens, c 347 BC] is said to invent c 380 BC a water clock with an alarm, probably by using two jars and a siphon; in such a scheme, water slowly empties through the night until it reaches the siphon, whereupon it swiftly is transported via the siphon to the other jar; water rising in the other jar forces air through a whistle

c 350 BC

GENERAL	ARCHITECTURE & CONSTRUCTION	COMMUNICATIONS/ TRANSPORTATION
Alexander the Great [b Pella, Macedon, 356 BC, d Babylon, 323 BC] and his army cross the Hindu Kush mountains into the Indus River Valley in 327 BC, extending his empire the farthest east	Celtic chiefs begin building Maiden Castle in south Dorset, one of the great fortified castles of Britain At the direction of Alexander the Great, the Macedonian architect Deinokrates lays out the new city of Alexandria on the Nile delta in 331 BC; its central north-south and east-west streets are each 14 m (46 ft) wide, dividing the city into four quarters, each with its own character; the plan is completed later by Ptolemy I, who joins several rocky islands and builds a causeway to form two harbors	The geographer Syclax the Younger publishes the first *Periplus* ("Coast pilot"), a guide giving distances between points, locations for fresh water, and other information useful to sailors About 340 BC or shortly afterward an unknown shipbuilder, probably a Phoenician, builds ships with one or two banks of oars with each oar propelled by multiple rowers; while earlier ships have three or more *banks* of oars, most subsequent ships have three or more *files* of rowers

c 300 BC

GENERAL	ARCHITECTURE & CONSTRUCTION	COMMUNICATIONS/ TRANSPORTATION
Athens's navy, outnumbered by Alexander the Great's navy, which continues to fight after his death, loses the Battle of Amorgos in 322 BC; it never becomes a major sea power again The Chinese *Book of the devil valley master* contains the first known reference to a lodestone's alignment with Earth's magnetic field; the lodestone is called a "south-pointer" although it is evidently used for divination, not for finding south; *See also* 1000 BC TOO; 80 CE TOO About this time, Aineias the Tactician describes the portcullis, a gate overlaid with iron (later entirely of iron) that can be lowered at the entrance to a fortress to prevent attackers from gaining entrance	The Via Appia, from Rome to Alba Longa (later extended to Capua and Brindisi), and the first Roman aqueduct, the Aqua Appia, bringing water underground to Rome from springs 16 km (10 mi) away, are built by Appius Claudius Caecus in 312 BC People now known as the Adena build the Great Serpent Mound in the Ohio Valley, a mound in the shape of a serpent holding an egg in its mouth; the mound, measured along the body of the serpent, is 328 m (1254 ft) long Romans and Hellenic Greeks begin to add flower gardens to the inner courts of their Mediterranean-type houses; the concept of a garden was introduced by the Persians Parts of the Great Wall of China are built by various warlords The Pharos lighthouse is built at Alexandria by Sostratos of Knidos in 283 BC under the direction of the first two Ptolemies; it stands between 116 and 134 m (380 and 440 ft) tall and is built from hard stone with lead used as mortar; the lower section contains about 50 rooms; surmounted on this is a tower with an octagonal cross-section; a fire burns in the top tower night and day Chares of Lindos completes the Colossus of Rhodes c 280 BC, one of the Seven Wonders of the World; it is about the height of the modern Statue of Liberty in New York Harbor—between 27 and 37 m (90 and 120 ft)—and is built from brass plates each 2 cm (about an in.) thick attached to an iron framework supported by stone columns; the work is built in stages from a growing mound of earth so that the workers are always at the level of their work; *See also* 200 BC ARC	The successor of Alexander in Macedon, Antigonus the One-Eyed [b 382 BC, d 301 BC], and his son Demetrius, in charge of his navy, arrange c 315 BC to have Phoenician shipyards build warships with six and even seven files of rowers Demetrios Phalereus persuades Ptolemy I (Ptolemaios Soter) in 307 BC to collect copies of all known books at Alexandria in an institution known as the Library; the project is furthered by nearly all of the remaining Ptolemaic rulers of Egypt and, at its height, probably contains almost 750,000 books on papyrus scrolls; however, some may have been duplicates since it is unlikely that many works were written in Classical times; *See also* 50 BC COM By 301 BC the navy of Demetrius of Macedon includes ships that have eight to thirteen files of rowers Dicaerchus of Messina (Sicily) [b c 355], a student of Aristotle's, develops a map of Earth that is on a sphere; it has lines of latitude based correctly on the position of the noonday sun; *See also* 550 BC COM The Chinese invent a form of harness that is pushed by a horse's chest instead of a horse's throat, not adopted in the West until the eighth century CE; *See also* 100 BC COM Under Ptolemy II, the canal linking the Nile to the Red Sea is completed or reopened c 280 BC; it was begun by Pharaoh Necho about 300 years earlier and was either completed or brought near completion by Darius the Persian, who, according to Diodorus Siculus, stopped short of finishing it because he feared that the higher Red Sea would flood the Mediterranean; Ptolemy installed a dam that prevented any such flooding; *See also* 500 BC COM; 100 CE COM

| | | Catapults that throw stones instead of darts are mentioned for the first time in 332 BC in connection with the siege of Tyre by Alexander the Great; *See also* 400 BC TOO; 100 CE TOO | **c 350 BC** |

| Mohist writings in China describe the first known use of poison gas in warfare; the Mohists were followers of Mo-tzu, a 5th-century BC Chinese philosopher | The Chinese c 310 BC invent a form of bellows, known as the double-acting piston bellows, that produces a continuous stream of air; such a bellows was not known in the West until the 4th century; *See also* 2000 BC TOO | **c 300 BC** |

The Chinese invent cast iron

About this time the screw is invented for fastening and other uses, but it is little used since all screws must be made by hand; *See also* 60 CE TOO

Convex lenses are introduced in Carthage (Tunisia)

A type of sundial known today as the hemicycle of Berosus is invented; it is a bowl with lines on its inner surface to mark the daylight hours when the shadow of a horizontal projection (gnomen) falls on them; for correct solar time, the hemicycle had to be level and face due south

---

Cast iron (part I)

Cast iron is not really iron and not all iron that is melted and cast is cast iron. It is any one of a number of alloys of iron with carbon (and less importantly silicon, manganese, and other elements). The origin of the name *cast iron* comes from the fact that when the metal ancients knew as iron became completely melted, it could be cast.

Pure iron melts at 1535°C (2795°F). Early smelters produced a product close to pure iron, but the fires of the first foundries never reached high temperatures. Although many iron ores smelt into iron without melting, impurities from the ores stay in place as small clumps in the relatively pure metal and have to be worked out (*wrought* is an old past participle of *work*) of the iron lump, oddly called a *bloom*, by pounding. In the process, the iron can be shaped into useful forms, such as blades or shields, but since wrought iron is never melted, the shapes must be simple. An iron pot is possible but difficult.

Ironworkers in China discovered that burning iron ore mixed with charcoal produced a thick liquid instead of a bloom. Carbon from the charcoal mixed with the iron to produce an alloy with a melting point as low as 1130°C (2066°F), a temperature that could be exceeded by the hot-burning charcoal. The Chinese quickly appreciated the advantages of cast iron over wrought iron. They also soon learned to vary the recipe to produce cast iron with different qualities, such as forms that were less brittle when cooled. The availability of cast iron also meant that all of the techniques previously developed for working with bronze or gold could be adapted.

c 250 BC

Asoka becomes the third Mauryan emperor in 269 BC; he completes the unification of India that his grandfather started after the death of Alexander the Great, leaving only three small provinces outside the empire; these are preserved when Asoka converts to Buddhist nonviolence in 259 BC

The first Punic War between Rome and Carthage begins in 264 BC; *See also* 150 BC GEN

Although the origin of candles is obscure, they are mentioned by Philon of Byzantium about this time; some think candles were invented by the Etruscans; tapers (wicks dipped in wax once) and wax-impregnated reeds precede candles, but the earliest form of artificial lighting device is the oil lamp; even at this time, and for the next 2000 years, lamps are preferred to candles as light sources; *See also* 3500 BC GEN

The 64-km- (39.5-mi-) long Aqua Anio (Anio Vetus, or "old Anio," so called after the Anio Novus is constructed) aqueduct is built largely below ground from stone in 272 BC; *See also* 300 BC ARC; 150 BC ARC

Archimedes

Archimedes was born in southeastern Sicily in Syracuse, the principal Greek city-state on the island. However, Alexandria, in Egypt, was the center of learning for the Hellenic world, and Archimedes is said to have studied there.

While in Egypt, he invented the device that is mostly closely associated with him by name, the Archimedean screw. This is a helix in a tube that, when turned in the proper direction, raises a fluid in which the bottom of the tube is immersed to the top of the tube. Unlike a vacuum pump, the Archimedean screw is not dependent on air pressure and can raise water higher than the 10-m (30-ft) limit of vacuum pumps. The device is still used for irrigation and other purposes today.

Archimedes is often said to have discovered the lever. He did work out the mathematics of simple machines and may have discovered the compound pulley. According to the story, after he remarked, "Give me a place to stand on and with a lever I will move the whole world," Hiero challenged Archimedes to a demonstration. Archimedes obliged by using compound pulleys to launch single-handedly one of the largest ships made to that time, complete with crew aboard.

c 200 BC

The Zhou dynasty disintegrates in China in 221 BC and the Qin dynasty forms; it is the first to unify China into a single empire but it does not last long after the death of its founder, Shi Huangdi; *See also* 1100 BC GEN

Liu Bang establishes the Han dynasty in China in 202 BC

Ctesibius of Alexandria builds a water organ, called in Latin a *hydraulis*; the *hydraulis* is a pipe organ in which air is supplied at constant pressure by water in a bell that obtains additional air from a bellows; models (uncovered in Carthage in 1885) and portions of actual instruments (found near Budapest) from Classical times still exist; the *hydraulis* is the instrument that Emperor Nero played on several infamous occasions; *See also* 120 CE GEN

The Colossus of Rhodes, one of the Seven Wonders of the World, is toppled during an earthquake in 224 BC; its remains lie in place until 656 CE when Arabs cart it away for scrap metal; *See also* 300 BC ARC

General Meng Tian, working under the orders of the Chinese Qin emperor Shi huangdi, completes building the 2400-km- (1500-mi-) long Great Wall in the north in 214 BC; it extends from the Yellow Sea to the deserts of what is now Turkestan; Meng Tian's workers take just 7 years (this date is estimated; Shi huangdi is now thought to have taken power in 221 BC)

An aqueduct is built in 180 BC to supply the citadel of Pergamon, built on a knoll with an elevation of 332 m (1089 ft), from a spring in the mountains at an elevation of 1173 m (3850 ft); along the way, the aqueduct passes through two valleys with low points of 172 m (564 ft) and 195 m (639 ft), as well as a ridge of 233 m (764 ft) between them; this is accomplished by using a bronze or lead pipe to cross the low points of the aqueduct and using water pressure (as high as 20 atmospheres in the lowest point) and a siphon effect

Emperor Shi huangdi in 219 BC constructs the Magic Canal linking two rivers, one flowing south and the other north; although only 32 km (20 mi) long, the canal enables a ship to sail from Canton (or anywhere else on the China Sea) to the latitude of present-day Beijing

Emperor Shi huangdi constructs for his tomb a relief map of his empire with the rivers formed from flowing mercury

The largest naval vessel of the Classical age is built by Ptolemy IV (Philopater) of Egypt; it has 4000 rowers in 40 banks, and carries as many as 3250 others as crew and fighting marines; this supergalley is a catamaran over 120 m (400 ft) long; it is never used in battle, however

As a result of the building of the largest naval vessel by Ptolemy IV (Philopater) of Egypt, the dry dock is invented to launch it, since the usual means of launching by having crews of workers drag a ship across land would be too difficult; instead, the ship is built in a channel that is connected to the sea; when the ship is completed, the channel fills with water and launches it

FOOD &
AGRICULTURE

MATERIALS/
MEDICAL TECHNOLOGY

TOOLS &
DEVICES

Archimedes (Continued)

During Archimedes' time, Syracuse was ruled by his cousin, King Hiero II. The detection of a fraudulent amount of gold in a crown Hiero had ordered was accomplished by measuring the crown's density using the water displaced when the crown was submerged to find the volume. Despite the detailed story related by Vitruvius of this event, it seems that, like the lever, the concept of density was generally known previously. Instead, Archimedes probably discovered that a body immersed partly in a fluid displaces a weight of the fluid equal to the weight of the body. This principle extends beyond the concept needed to detect fraud in the manufacture of a crown.

The details of Archimedes' defense of Syracuse are less generally known. Apparently he devised a number of improved catapults and crossbows that pushed back ordinary waves of attackers. When the Roman general Marcellus brought out his own "secret weapon," a kind of seagoing siege vehicle, Archimedes used levers to drop huge boulders on the attackers, sinking the ships. Another story is that with mirrors he focused the Sun's rays on the ships to set them afire, but this is unlikely.

Marcellus' army finally managed to find a weak spot in the defense of the city and overran it. Although Marcellus had instructed his soldiers to spare Archimedes, one of them encountered him contemplating a geometric figure drawn in the sand (the common way to do geometry at the time). When the soldier damaged the figure and Archimedes protested, the soldier killed him.

Philon of Byzantium (Turkey) advocates the use of bronze springs in catapults and experiments with the expansion of air by heat

According to tradition, parchment is invented in Pergamum (Bergama, Turkey), from which its name derives; parchment is animal skin treated so that it can be written on on both sides; vellum, which may have originated later, is often taken to be parchment made from very young or unborn animals

Archimedes [b Syracuse, Sicily, c 287 BC, d Syracuse, 212 BC) works out the principle of the lever and other simple machines; he demonstrates his work by launching a large ship by himself

Philon of Byzantium [b c 300 BC] builds a cardan joint, a form of universal joint using two interlocked forks, although traditionally this device has been attributed Girolamo Cardano; Philon also describes a primitive form of the chain-and-sprocket drive (now used in bicycles) and an early application of a human-powered treadmill to power a bucket-and-chain water hoist; Philon also designs a chain drive for use in repeated loading of a catapult; in his book *On the lifting of heavy weights* he describes a system of gears

Pottery jars lined with copper discovered since 1936 in and around Baghdad and at the Baghdad Museum may be the oldest known electric batteries, apparently used for electroplating c 230 BC; these jars close with an asphalt plug and contain remains of iron rods; they could explain the production of extremely thin layers of silver and gold on certain objects from antiquity; *See also* 1837 MAT

c 250 BC

The Chinese develop a malleable form of cast iron

In Peru and Ecuador, metalworkers use cylindrical clay furnaces and blowpipes (instead of bellows) to cast gold and to alloy it with silver

Concrete is used in the Roman town of Palestrina in Italy

The development of gears leads to the ox-powered waterwheel, used to raise water to the level of fields for irrigation purposes

The astrolabe, used for determining the level of elevation of the Sun or another star, is invented and introduced to Hellenic sailors; some have attributed its invention to Appolonius of Perga (shortly before 200 BC) and others to Hipparchus (shortly after 200 BC); Hipparchus [b Nicaea, c 190 BC, d c 120 BC] certainly used some tool of this type in his astronomical research

Alexandrian engineer Ctesibius [b c 250 BC] improves the water clock, making it the most accurate time-keeping device available for the next 2000 years; he also invents a water pump consisting of a cylinder and piston and two valves, one at the intake and one at the outlet; when the piston is lifted, the valve at the inlet is open, admitting water, while the valve at the outlet is closed; when lowering the piston, the valve at the inlet is closed and the one at the outlet is opened; he also invents a form of air gun, called a "wind gun"; *See also* 1400 BC TOO; 720 CE TOO

c 200 BC

c 200 BC cont

Many buildings in Rome are three-story apartment houses called *insulae,* or "islands"

Shipbuilders introduce three-masted vessels (a foremast called an artemon, the main, and a mizzenmast at the rear); a large ship, the *Lady of Syracuse,* whose building is supervised by Archimedes (although designed by Archias of Corinth) may have been the first of these

c 150 BC

The Punic Wars between Rome and Carthage end in 146 BC with Rome the victor and Carthage destroyed; *See also* 250 BC GEN

Reports from an expedition to the West in 126 BC by Zhang Qian are among the many influences that lead to the establishment of the Silk Road, the trade corridor from China to the West

The 92-km- (56.7-mi-) long Aqua Marcia aqueduct, completed from 144 to 140 BC, is the first Roman aqueduct with substantial elevated portions, about 11.1 km (6.9 mi) being built entirely from stone on arches; it carries about 194,500,000 L (51.4 million gal) of water a day

The 20-km- (12-mi-) long Aqua Tepula aqueduct in 125 BC is the first to use the arches of the Aqua Marcia as part of its route so that it can serve more elevated portions of the city of Rome; it carries only 18,000,000 L (4.8 million gal) of water a day; *See also* 250 BC ARC; 50 BC ARC

The first paved roads are built in Rome c 170 BC

The Greek historian Polybius describes communication by means of fire signals; *See also* 450 BC COM

The capital of China at this time, Chang'an (later Xi'an), is linked in 133 BC by a 145-km (90-mi) canal to the Huang He (Yellow River); *See also* 200 BC COM

c 100 BC

Antipater of Sidon, probably using earlier Hellenic guidebooks, compiles the earliest known list of the Seven Wonders of the World (a slightly different list is given somewhat later by Philon of Byzantium); the list consists of the Pyramids of Egypt, the Hanging Gardens of Babylon, the giant statue of Zeus at Olympia, the Temple of Diana at Ephesus, the Mausoleum at Halicarnassus, the Colossus of Rhodes, and the lighthouse of Pharos; of these, only the Pyramids of Egypt stand today

Apartment houses in Rome, called *insulae,* or "islands," commonly reach five stories

Roman engineers learn shortly after this date how to rotate the arch to produce a dome, which they use thereafter extensively in construction of public buildings; *See also* 100 CE ARC

Gaius Sergius Orata c 85 BC begins to install a system for producing heated water in otherwise neglected villas; he then resells the villas for vastly more than they cost him; similar systems are soon installed at public baths throughout Italy

The Chinese invent the collar harness for horses c 110 BC, the most efficient form of harness to this day; it is not used in the West until the Middle Ages; *See also* 300 BC COM

The notebook, or *codex,* consisting of leaves of parchment sewn together on one side, is in use in Rome; *See also* 870 CE COM

The Celts invent a swiveling pair of front wheels (sometimes called a bogie) to make steering easier on carts

Ssuma Ch'ien's *Historical records* written c 90 BC includes the first known reference to parachutes

The first known written record of a professional baker identifies the origin of baking for the public as occurring in 170 BC in Rome while it is warring with Macedonia

In 126 Chang Ch'ien brings wine grapes—*vinifera*—to China from the West

Silver is so rare and costly in Rome that a visiting delegation from Carthage entertained by all the richest families, encounters the same set of silver dinnerware each night

The Chinese make paper c 140 BC; it is used as a packing material, for clothing, and for personal hygiene, but not for writing; *See also* 100 CE MAT

A Chinese text c 140 BC describes an incense burner suspended in a system of concentric rings so that it remains horizontal; this is the first Chinese description of such an universal joint, said to have been discovered by Fang Feng

Gaius Sergius Orata develops a system for heating seawater by circulating hot air beneath tanks that he uses to promote faster growth in oysters that c 80 BC he then sells to wealthy Romans

Roman consul Marcus Aemilius Scaurus has the Po Valley drained in 109 BC to make new land for veteran soldiers, using the canals also to connect the Po to Parma for boat passage

The Han dynasty in China nationalizes cast iron and salt production in 119 BC

Glassblowing is developed, probably somewhere in the Levant (Syria, Lebanon, Jordan, or Israel), most likely by Phoenicians

The people of the culture known as Paracas Necropolis from their burial ground on the Paracas Peninsula of Peru produce what many archaeologists believe to be the finest textiles ever made by humans, embroidering on cotton and on the wool of the llama and its relative the vicuña

Small, round glass skylights, the ancestors of windowpanes, are installed in houses of the Roman Republic; *See also* 100 CE MAT

The Romans discover that volcanic ash from Puteoli, near Naples, called pozzolana, makes excellent concrete that will set and keep its integrity even under fresh or salt water

In Illyria (Yugoslavia and Albania), and probably in western Anatolia (Turkey), water-powered mills are used for grinding grain; these are horizontal turbines that only work on swiftly running streams; with no gears, the upper horizontal stone turns at the same speed as the turbine, while the lower one is stationary; *See also* 0 TOO

The Chinese invent the crank handle for turning wheels; *See also* 50 CE TOO

The Antikythera device, a complex bronze orrery (astronomical computing machine) of great accuracy, based on an intricate system of 24 gears and 13 axles, is built; it contains the first known differential gear and has sometimes been called "the first computer"; its existence was revealed in 1900 when it was found by Greek sponge divers in a wreck off the island of Antikythera, although it was not explained until 75 years later by Derek de Solla Price; *See also* 130 CE TOO

Roller bearings from wood and bronze are used on the wheels of carts in what is now Denmark; *See also* 50 CE TOO

c 50 BC

Acting on the advice of Greek astronomer Sosigenes [b c 90 BC], the Julian calendar of three 365-day years followed by one of 366 days is introduced in Rome by Julius Caesar in 46 BC; as a result of changes to make the seasons correct, the year 46 has 445 days, making it the longest year on record

Octavian takes the title Augustus, or "revered one," in 27 BC to symbolize his office as the first Roman emperor

Pliny the Elder [b Como, Italy, 23, d near Mt Vesuvius, Aug 25 79 CE] mentions that mineral oil found in the Adriatic region is used in lamps there instead of olive oil; *See also* 100 CE GEN

The troops of Julius Caesar [b Rome, Jul 12 100 BC, d Rome, Mar 15 44 BC] build a wooden trestle bridge in 55 BC that spans the Rhine, taking 10 days for the feat, which suggests the state of the art

Lucius Cocceius Auctus designs and builds two tunnels in 40 BC to ease traffic bottlenecks around Naples, each about 3 m (10 ft) wide and varying in height from 2.5 m (9 ft) to 21 m (70 ft); according to a report by Seneca, they are dark and dusty

The 23-km- (14.2-mi-) long Aqua Julia aqueduct of Marcus Agrippa [b Italy, 63 BC, d 12 BC], is built in 33 BC; like the earlier Aqua Tepula, it uses the arches of the Aqua Marcia as part of its route so that it can serve more elevated portions of the city of Rome; it carries 50,000,000 L (13.2 million gal) of water daily; *See also* 150 BC ARC; 40 CE ARC

Marcus Tullius Tiron develops a shorthand system c 70 BC, used by Cicero, Julius Caesar, and others; it will remain in use for about ten centuries; *See also* 1580–89 COM

Lucretius describes c 60 BC how the illusion of motion can be created by a sequential display of frames

Julius Caesar is said to have invented the newspaper c 60 BC; news items, such as nominations, battles, etc., are written up and posted on main buildings in towns

Julius Caesar removes hundreds of thousands of volumes in 48 BC from the Library at Alexandria; during the fighting over control of Alexandria in the wake of the assassination of Julius Caesar, either part of the Library at Alexandria is burned or the books that Cleopatra had given to Caesar are lost

c 0

Roman general Marcus Agrippa builds in 19 BC the aqueduct at Nemausus (Nîmes) which features the Pont du Gard bridge over the Gardon River; the bridge has three tiers of arches 49 m (160 ft) high and is 274 m (900 ft) long; it is built of stone without cement or mortar and still stands today

On Jun 9, 19 BC Marcus Agrippa completes the 21-km- (13-mi-) long Aqua Virgo aqueduct, which runs almost completely underground from springs on the estate of Lucullus to the Campus Martius; it brings 104,000,000 L (27.5 million gal) of water a day to Rome

Emperor Augustus sets a limit of about 21 m (70 ft) on the heights of buildings in Rome

Herod the Great has the first large harbor constructed in the open sea c 10 BC to support his new town of Caesarea Palestinae (near present-day Haifa, Israel); the harbor is constructed of giant blocks of concrete poured into wooden forms

Roman Emperor Augustus has the 33-km- (20.3-mi-) long Aqua Alsietina aqueduct built underground to Trastevere in 2 BC; its water is not drinkable, however, for it is built to supply a 360-m (1181-ft) by 540-m (1800-ft) artificial lake designed for mock sea battles to amuse the Romans; *See also* 50 BC ARC; 340 CE ARC

Roman general Nero Claudius Drusus [b 38 BC, d Germany, 9 BC] in 12 BC joins the largest lake in what is now the Netherlands, the Flevo Lacus, to the Rhine with a canal (the Fossa Drusiana) that also uses the Yssl River for part of the passage; *See also* 40 CE COM

Sometime before the start of the Roman Empire, some ships are fitted with a small triangular topsail above the mainsail

The earliest known depiction of a ship's rudder is made in China

The Chinese build suspension bridges of cast iron that are strong enough for vehicles to cross

c 50 BC

Roman consul Manius Curius Dentatus, conqueror of the Sabines, orders the marshes at Sabine Rieti drained in 28 BC by means of an 800-m (2624-ft) canal partly cut in rock and routed so that it forms a dramatic waterfall into the river Nera

Roman soldiers fighting the Parthians of western Asia see silk in 53 BC for the first time, used for Parthian military banners; their reports eventually lead to the creation of the Silk Road system for the trade of goods between China and the Roman Empire; *See also* 500 CE MAT

The *Ayurveda* is compiled; it becomes the basic Hindu medical treatise c 40 BC

c 0

Vitruvius describes gold amalgam c 25 BC, that is, gold dissolved in mercury

A book on architecture and engineering c 25 BC by Roman architect Vitruvius (Marcus Vitruvius Pollio) [b c 70 BC] also discusses astronomy, acoustics, sundials, and waterwheels; it remains the main source of knowledge about Roman construction until the Renaissance; the Vitruvian waterwheel is the first known vertical, undershot waterwheel (the water runs beneath the wheel); gearing is used to make the speed of the millstone one-fifth the speed of the wheel, a great improvement for grain grinding; *See also* 1410–19 GEN

Yang Hsiung's *Dictionary of local expressions* indicates that the Chinese have invented the belt drive before 15 BC, which will not be known in Europe until 1430

The Chinese invent methods for drilling deep wells to obtain salt water and natural gas; the wells reach 1460 m (4800 ft)

The semilegendary Ko Yu invents the wheelbarrow, although this Chinese invention is also attributed to Jugo Lyang (Chu-Ko Liang) around 200 CE

Hero of Alexandria

The best and nearly the only evidence that exists as to when Hero of Alexandria (also known in English as Heron) lived is a description of an eclipse of the Moon that could only have taken place in 62 CE. In one of his books, he called himself Hero Ctesibius, but since Ctesibius lived several hundred years before that eclipse, one has to assume that any connection is intellectual or coincidental. Aside from much of his writing, we know nothing further about Hero.

Hero wrote on all the important technological and mathematical topics of the day. The surviving works in Greek, Arabic, or Latin translation are *Automaton-making, Geometrica, Mechanics, Metrica, Mirrors, Pneumatics, Siegecraft,* and *The Surveyor's Transit,* as well as a fragment on water clocks and parts of a dictionary of technology. Perhaps more books are yet to be found: *Metrica* was not known until an 1896 discovery of a version from about 1100.

Various devices are associated with Hero, including a form of jet-propelled steam engine known as the aeolipile, the first known windmills (actually more like pinwheels), and water clocks and water organs. Other gadgets relate to temple "miracles," such as doors that open when a sacrificial fire is lit, mirrors that produce ghosts, a coin-operated dispenser of holy water, and a device that appears to turn water into wine. Hero's most innovative work, however, included some clever but incorrect laws of mechanics and advanced ideas for surveying that were probably too far ahead to have been actually employed.

What probably mattered most in practice were not the gadgets, but Hero's descriptions of such useful devices as wine presses, cranes, and compound pulleys, although these tools were not likely to have been invented by Hero.

c 10 CE

c 30 CE

c 40 CE

Caligula [b Antium, Aug 31 12, d Rome, Jan 24 41] initiates building of the 69-km- (20.3-mi-) long Aqua Claudia and the 87.5-km- (54.0-mi-) long Anio Novus aqueducts, both built on arches across the Roman Campanga, and both completed in 52 CE; *See also* 0 ARC; 110 CE ARC

Caligula orders a lighthouse some 38-m (124-ft) tall built at Boulogne, France

Roman general Gnaeus Domitius Corbulo [d c 67 CE] digs a ship's canal in 45 joining the Rhine with the Meuse River in what is now Germany; *See also* 0 COM

c 50 CE

Under Emperor Claudius, a 60-km (32-mi) aqueduct built to supply Lyon with water includes three inverted siphons, structures in which the aqueduct descends into a valley and rises on the far side due to the pressure of the descending water; one siphon consists of eighteen lead pipes that cross a ravine 66 m (217 ft) deep on 16-m (52-ft) arches

c 60 CE

The first large-scale persecution of Christians in Rome takes place under the emperor Nero in 64 CE

Nero starts to build a canal across the isthmus of Corinth in 65, a project contemplated at least as early as 600 BC, but his increasing political difficulties cause him to abandon it; *See also* 600 BC COM; 1893 TRA

	Thaddeus of Florence describes the medical use of alcohol in *De virtutibus aquae vitae* (On the virtues of alcohol)	**c 10** CE
	Tu Shih invents the water-powered bellows for use in working cast iron	**c 30** CE
Roman Emperor Claudius in 41 orders work to begin on a 5.6-km (3.5-mi) tunnel through rock to drain Lacus Fucinus (Lago Fucino) to make new farmland; the project is directed by the aristocrat Narcissus	*De materia medica* by Greek physician Pedanius Dioscorides of Anazarbus [b c 20 CE] deals with the medical properties of about 600 plants and nearly a thousand drugs	**c 40** CE
Farmers in the Middle East and China begin to store grain in cylindrical, airtight storage bins, similar to modern grain silos; *See also* 1873 FOO The project to drain Lacus Fucinus (Lago Fucino) is completed after 11 years in 52 and Emperor Claudius attends opening ceremonies on the bank of the Liri River, where the tunnel from the lake is supposed to empty; when the gates are opened, however, no water emerges as a result of an incorrect grade; another ceremony is held in 53; this time the tunnel is too successful, as the flood of water released when the gates are opened sweeps away the banquet tables; Narcissus is imprisoned and dies soon after; *See also* 40 CE ARC		Two floating Roman temples found in Lake Nemi in Italy, built by Caligula or Claudius sometime between 37 and 59, contain what appears to be a bilge pump with the first known evidence of a crank handle for operating it; hand mills of the time evidently also used handles that were essentially cranks; *See also* 100 BC TOO; 300 CE TOO The two floating temples from Lake Nemi contain rotating wooden platforms; one platform is mounted on eight bronze balls with small projections used as axles, the oldest prototypes of ball bearings known (although these are not strictly ball bearings, since true ones rotate freely); the other is mounted on a similar device using cylinders, and stands the same relation to roller bearings as the other does to ball bearings; *See also* 100 BC TOO	**c 50** CE
Pliny the Elder describes a harvesting machine that, pulled by oxen, cuts off stalks of grain		Hero of Alexandria (also known as Heron) [b about 20] builds a toy, later called an aeolipile, that is powered by steam; a closed kettle containing water rotates when jets of steam issue from the kettle as the water boils; there is apparently no idea of putting it to practical use Hero of Alexandria describes (and perhaps invents) a device for cutting screws with female threads; male screw threads still have to be cut by hand; *See also* 300 BC TOO	**c 60** CE

c 70 CE

Roman emperor Vespasian sends his son Titus to quell a revolt in Judea (Israel); Titus not only succeeds, but completely destroys Jerusalem; the last rebels commit mass suicide in the fortress Masada in 73 rather than be captured

Roman Emperor Vespasian orders the building of the Colosseum in Rome; at about 180 m (600 ft) long and 50 m (175 ft) high, it will remain the largest amphitheater in the world until the construction of the Yale Bowl in 1914

The Grand Canal of China, eventually 965 km (600 mi) long, is started

c 80 CE

c 90 CE

A two-volume work on Roman aqueducts by Sextus Julius Frontinus [b c 30 CE, d 104] summarizes major advances in construction since ancient times

c 100 CE

The Moche people of Peru experience a period of culture and expansion starting about this time, featuring giant pyramids, the rediscovery of lost-wax casting, and some of the most realistic pottery representations of people and animals of the ancient world

Hero of Alexandria describes his invention of a lamp in which a column of salt water supplies pressure to raise the level of oil into the wick

Roman engineers learn how to use an ordinary round dome (not a cloister vault) to roof a square building; they insert slanting triangles, called pendentives, at the corners of the square to provide a connection between the circular dome and the square walls; this method becomes an important feature of Byzantine architecture; they also learn that two cylindrical vaults (essentially extended arches) can cross each other at right angles, creating a roof that can cover a large square; such cross vaults or groined vaults become common in both Roman and late medieval architecture; *See also* 100 BC ARC

Emperor Trajan reduces the limit on height for buildings in Rome to about 18 m (60 ft) for greater safety

Apollodoros of Damascus builds in 105 a famous bridge across the Danube for Roman Emperor Trajan to use in conquering the Dacians (who live in what is now Romania); the depiction of it on the column Trajan erects to commemorate this victory shows that the bridge uses a truss, or diagonal, load-bearing member; in time, the truss becomes the foundation of construction, especially in the 19th century, but this is the first recorded use; *See also* 130 CE ARC

The Roman emperor Trajan, like Ptolemy II almost 400 years earlier, restores the canal begun by Pharaoh Necho II about 600 BC and completed a hundred years later by Darius the Persian; unless constantly dredged, the canal soon fills up with blown sand; *See also* 300 BC COM; 640 CE COM

The Chinese discover that horses can be hitched with one horse in front of another, allowing a heavier load without need for a wider road; the tandem hitch spreads through Europe in the Middle Ages; *See also* 2300 BC FOO

Zhang Heng develops the method of using a grid to locate points on a map c 110 CE

| | | According to Pliny the Elder, who makes the first known reference to it, the wine press was invented by Greeks during the previous century | **c 70** CE |

| | Wire cable is being made by this time, as evidenced by a 5-m (15-ft) length of 2-cm (1-in.) bronze cable found in the ruins of Pompeii, destroyed by Vesuvius in 79 | Lou-en Heng describes a compass consisting of a spoon carved from a magnetic iron mineral placed on a polished surface of bronze; the handle of the spoon points north-south; *See also* 300 BC GEN; 270 CE TOO

Wang Ch'ung's *Discourses weighed in the balance* makes the first known Chinese reference to the chain pump, a device that raises water from rivers or lakes | **c 80** CE |

| The Chinese invent a device that winnows grain using a rotating fan to separate the grain from the chaff | | | **c 90** CE |

| The Chinese invent the multitube seed drill; *See also* 1500 BC FOO; 1700–09 FOO

The Chinese discover that dried, powdered chrysanthemum flowers can be used to kill insects; this is believed to be the first insecticide; the active ingredient, pyrethrum, is widely used today, especially on vegetables, since it is biodegradable and virtually harmless to mammals | Copper is in use in the Americas, with copper metallurgy well developed by the Moche (Mochica) in Peru (200 to 600 CE); *See also* 2500 BC MAT

According to 2nd-century Hellenic author Pausanias, "Theodorus of Samos [is] the first to discover how to pour or melt iron and make statues of it"

Glass windowpanes of the modern type appear in houses of the early Roman Empire (Principiate); writer Columella even suggests the first known cold frames, glass (or transparent gypsum or mica) boxes in which to raise plants sensitive to cold; he suggests putting them on wheels to roll outside on sunny days; *See also* 100 BC MAT

Chinese tradition has it that paper is invented by the eunuch Tsai Lun [b Kueiyand, Kweichow, c 50 CE, d c 118] in 105; however, archaeological evidence suggests that paper is invented at least 250 years earlier, although it is used for packing and other purposes, not for writing upon; *See also* 150 BC MAT; 110 CE MAT | The double-armed catapult—rather like a large crossbow—is replaced by the single-armed catapult, which is more like a torsion-propelled spear thrower than like a bow; the new catapult is nicknamed the onager after the Asiatic wild ass, an animal that, legend has it, defends itself by kicking stones at pursuers; *See also* 350 BC TOO; 1100–09 CE TOO | **c 100 CE** |

	GENERAL	ARCHITECTURE & CONSTRUCTION	COMMUNICATIONS/ TRANSPORTATION
c 110 CE		The 57.3-km- (35.4-mi-) long Aqua Trajana, built in 109, carries water underground to Rome; it is the tenth major aqueduct in Rome's water supply; *See also* 40 CE ARC; 230 CE ARC	
c 120 CE	A reference to "brazen pipes blown from below...by bellows" in Pollux's *Onomasticon* is considered to be the first mention of a pipe organ, although the water organ, its ancestor, goes back to the time of Ctesibius; *See also* 200 BC GEN	Roman Emperor Hadrian [b Spain, Jan 24 76, d near Naples, Jul 138] builds a 24-km- (15-mi-) long aqueduct c 115 to supply Athens; it is still in use today, although supplemented by a new system built between 1920 and 1931; *See also* 230 CE ARC Roman engineers and soldiers build Hadrian's Wall in 122 to protect the occupied area of Britain from northern tribes	
c 130 CE		Roman Emperor Hadrian builds the Pantheon, a magnificent domed building honoring seven gods; it still stands; in addition to its remarkable dome, it also uses a bronze truss, but that is removed and melted down for cannon in 1625; *See also* 110 CE ARC; 1570–79 ARC	
c 150 CE			Lucian of Samosata describes in *True history* how a Greek ship could travel to the Moon by winds and waterspouts
c 170 CE	The first actual contact between Roman and Chinese emissaries occurs in 166, although the empires have been trading indirectly for a couple of hundred years along the Silk Road; the emissaries that reach the Han court are representing the Roman emperor Marcus Aurelius Antoninius		
c 180 CE			A statue of Roman Emperor Marcus Aurelius from about this time shows him using a hand stirrup; *See also* 480 CE COM
c 190 CE			According to carbon-14 dates, the Nazca lines and figures, giant drawings on the pavement of the world's driest desert (near Nazca, Peru) that include various animals, spirals, and geometric figures, all clearly visible only from the air, are started at this time; the latest dates obtained are about 600 CE

	The oldest known piece of paper used for writing is in existence; *See also* 100 CE MAT; 590 CE MAT	**c 110 CE**
		c 120 CE
	Zhang Heng invents in 132 the first seismograph, a device that indicates the direction of an earthquake by dropping a ball from the mouth of a bronze dragon into the mouth of a bronze frog Zhang Heng combines a water clock with an armillary to produce a device that keeps track of where stars are expected to be in the sky; *See also* 1070–79 CE TOO	**c 130 CE**
		c 150 CE
Roman physician and anatomist Galen of Pergamum [b Pergamum (Bergama, Turkey) c 130, d Sicily?, c 200] becomes the first to use the pulse as a diagnostic aid		**c 170 CE**
		c 180 CE
	Ding Huan, who is believed to be the inventor of the magic lantern, describes in 189 a cardan suspension for incense burners, saying that the secret of its first Chinese inventor, Fang Feng, was lost	**c 190 CE**

c 200 CE

At Moche, the apparent capital of the Moche (Mochica) civilization of coastal Peru, two huge adobe structures are built; the Huaca del Sol is a terraced platform 2400 by 160 m (1120 by 530 ft) at the base and 41 m (135 ft) high, built from about 130 million mold-made adobe bricks; the Huaca de la Luna is on a platform about half that size

The Maya center at Tikal, built about this time, has shrines at the peaks of pyramids that are roofed with corbel arches, the only arches of any kind known in the Americas

The Zapotec of Monte Albán (near Oaxaca, Mexico) develop the crude pictographic writing of their Olmec predecessors into a full-fledged hieroglyphic script that is used to record dates and events

c 230 CE

The 22.2-km- (13.7-mi-) long Aqua Alexandriana aqueduct, built in 226, is the eleventh and last of the major aqueducts serving ancient Rome, a water-supply system that began in 312 BC with the underground Aqua Appia; the Aqua Alexandriana is built almost entirely on arches; the complete aqueduct system is able to supply about 750 million L (200 million gal) of water to the city during each 24-hour period; *See also* 110 CE ARC

c 250 CE

c 260 CE

c 270 CE

A series of wars and revolts are thought to result in partial destruction of the Library at Alexandria; the first of these occurs when Septima Zenobia, queen of Palmyra, captures Egypt in 269; Roman Emperor Aurelianus [b c 215, d 275] puts down a revolt in Alexandria and probably destroys part of the Library in 272

c 300 CE

On Feb 12, 303, the Roman emperor Diocletian closes Christian churches throughout the empire and commands the Christian clergy to make sacrifices to Roman gods or die

The palace of Roman Emperor Diocletian at Spalatum uses arches supported by freestanding columns, a device that later becomes common in medieval architecture

Roman Emperor Diocletian puts down yet another revolt in Alexandria in 295 and probably destroys part of the Library; *See also* 270 CE COM; 390 CE COM

The Maya develop the day-count calendar, which combines the 365-day Olmec calendar that has a 52-year cycle with the 260-day cycle known as the tzolkin, which has 13 cycles of 20 days each; this calendar dates events back to 3000 BC

FOOD & AGRICULTURE	MATERIALS/ MEDICAL TECHNOLOGY	TOOLS & DEVICES	
	The Chinese develop porcelain	The Chinese invent the whippletree (whiffletree), a device that allows two oxen to pull a single cart together	**c 200** CE
			c 230 CE
	Parchment supersedes papyrus as a writing material for the first time in the lands around the Mediterranean; *See also* 2500 BC COM		**c 250** CE
		Ma Chün builds a "south-pointing carriage" using differential gears	**c 260** CE
		In China, the first form of a compass is probably in use for finding south; earlier applications of magnetic lodestones were more magical than practical; *See also* 80 CE TOO; 1040–49 TRA	**c 270** CE
An agricultural text by Xi Han in 304 mentions the use of carnivorous ants to control insects attacking vegetable gardens	The Chinese learn to use coal instead of wood as fuel in making cast iron Bellows are in use in Europe to produce higher quality iron; bellows had been producing cast iron in China for about 600 years and had been used in the Near East as early as 1600 BC; *See also* 300 BC TOO The earliest known knitted fabrics exist, although net making, the ancestor of knitting, seems to go back to Neolithic times	The physician Oreibasios mentions cranks as a device helpful in setting broken bones; this is the first known mention of a crank in Western writings; *See also* 50 CE TOO; 830 CE TOO The Chinese may have begun to develop the abacus	**c 300** CE

	GENERAL	ARCHITECTURE & CONSTRUCTION	COMMUNICATIONS/ TRANSPORTATION
c 330 CE	On May 11, 330 a date deemed propitious by astrologers—the new city that will come to be known as Constantinople (Istanbul) is dedicated, after having been laid out by Roman emperor Constantine as the new capital of the Roman Empire	At Antioch (in Turkey), Emperor Constantine [b c 280, d May 22 337] orders an octagonal church with the altar and pulpit in the middle and aisles radiating from the altar like spokes; the church, the Domus Aurea, built in 327, sets the style for Romanesque churches	
c 340 CE		Emperor Constantine of Rome repairs the Aqua Virgo aqueduct in 338; *See also* 540 CE ARC; 1450–59 ARC	
c 350 CE	The city of Antioch (in Turkey) installs the world's first system of public lighting; *See also* 1811 GEN		
c 370 CE			A Roman writer known as Anonymous suggests to Roman emperor Valens that warships be built with three pairs of paddle wheels, each wheel powered by oxen walking capstan spokes about the deck; within the next 200 years such boats are actually in use; *See also* 520 CE COM Roman Emperor Valens in 373 orders the burning of all non-Christian books
c 380 CE			
c 390 CE			The final destruction of the Library at Alexandria is attributed by some to a Christian mob led by Bishop Theophilus in 391; other sources say the mob simply burned the temple of Serapis, where some of the books were kept; *See also* 300 CE COM; 650 CE COM
c 400 CE		The Buddhist monastery Bamiyan is carved into the sandstone cliffs of the foothills of the Hindu Kush; it houses a thousand monks and a 37-m- (120-ft-) high statue of Buddha The early Anasazi people enter what is now called the "four corners" region (where CO, UT, AZ, and NM come together at right angles) as hunter-gatherers; they live in caves along rocky cliffs and canyon walls	The arrival in 405 of the Chinese scribe Wani to teach the Japanese crown princes to read and write signals the official adoption of Chinese ideograms for Japanese writing

FOOD &
AGRICULTURE

MATERIALS/
MEDICAL TECHNOLOGY

TOOLS &
DEVICES

c 330
CE

Salt and the fall of civilizations

A common theory about the rise of civilization is that the need for state organization of irrigation became the prime mover that led to all other changes. Irrigation may also be implicated in the *fall* of civilizations.

Mesopotamia provides the main example. When large-scale irrigation began in the Tigris and Euphrates river valleys, it was accompanied by urban growth, writing, empires, monumental construction, and such technological innovations as the wheel. The former desert prospered and was central to the growth of human population and influence for about 4000 years. The former desert land became known as the Fertile Crescent. Today the Fertile Crescent is mostly desert again.

Although there are several factors involved in the decline back into desert, salt in the soil appears to be the major one. The process of irrigation reverses the normal flow of water through the soil. Outside the desert, soil is watered mainly by rain that carries salt in the soil deeper into the ground. Evaporation in an irrigated desert causes water in the soil to move from lower levels toward the top of the soil as the Sun dries out upper layers. Such movement carries salt toward the surface.

Most plants cannot grow in salty soil. Romans, appreciating this, sowed salt on the ruins of Carthage after its defeat to prevent it from returning to power.

Flooding can erase the effects of irrigation by washing the salt back into lower regions of the soil. Egypt and China, subject to floods, have had fewer problems with salty soil than has Mesopotamia. In Egypt, however, damming of the Nile in 1964 prevented flooding, but also started a buildup of salt in the soil.

The Maya practiced irrigation in a rain forest, where salt buildup would seem less likely than in a desert. Nevertheless, it appears that they too managed with irrigation to reverse the normal flow of minerals through the porous limestone soil of Central America, contributing to the collapse of their civilization as well.

Today farmers in other irrigated regions, such as Pakistan or the San Joaquin Valley in the US, face similar buildup of salt in the soil.

c 340
CE

c 350
CE

c 370
CE

Verses by Decimus Magnus Ausonius [b Gaul, c 310, d Gaul, c 395] written in Gaul (northern Europe) c 375 CE refer to a water-powered sawmill

c 380
CE

---

c 390
CE

The Chinese make steel by forging together cast and wrought iron

The Chinese invent the umbrella

The overshot waterwheel, in which water that has been retained in a mill pond is directed by means of a chute over the top of a waterwheel, is developed in the Roman Empire, although it probably originated about 200 years earlier

Hypatia develops the hydroscope (also called the hydrometer), an instrument for determining the specific gravity of a liquid; it consists of a sealed tube weighted at one end; the depth the tube sinks into a liquid is the measure of its specific gravity; she also improves the astrolabe

c 400
CE

c 420
CE

c 480
CE

Archaeological evidence suggests that the stirrup for mounting horses and for stability while riding was invented in China or Korea c 475; in the later Roman Empire only hand stirrups were used; *See also* 180 CE COM

c 490
CE

Giant statues of Buddha, some 50 m (160 ft) tall, are chiseled from rock cliffs at Yungang, China, in 489; after the capital of the Wei Empire in China is returned in 494 to Luoyang, where it had been until it was sacked by the Huns in 311, Buddhist monks move the carving of statues of Buddha in river cliffs from Yungang to Longmen, just south of Luoyang; there, over a period of about 300 years, they carve from 100,000 to 150,000 statues into the rock

c 500
CE

According to tradition, although the date applies only to the Western Empire, the Roman Empire falls; the emperor is replaced by one of the empire's Germanic generals, who, in turn, is slain by Theodoric the Goth; *See also* 50 BC GEN

The Mogollon people of NM make pit houses, with living areas sunk a meter or two (3 to 5 ft) below grade and a single room covered over with a sloping roof of mud-coated branches and thatch supported by timbers

After the Anasazi occupy the "four corners" region, they begin to make their first true houses, domes of logs laid horizontally around a saucerlike excavation and plastered with mud

Because of an increase in rainfall in Europe, low-lying regions, such as what is now the Netherlands, begin to flood; where before people lived and farmed on "islands" in swamps, water rising as much as 10 cm (4 in.) per century causes inhabitants to build dikes around their islands to maintain their land; gradually this process leads to diking and draining swamps to make arable land

The abacus is used in Europe, although counting boards based on the same principle were used by the ancient Greeks and Romans as much as a thousand years earlier

c 520
CE

The first European paddle-wheel boats are designed, to be powered by oxen walking in circles, as in a mill; it is unlikely that these were built; *See also* 370 CE COM

FOOD &
AGRICULTURE

MATERIALS/
MEDICAL TECHNOLOGY

TOOLS &
DEVICES

A 7-m (24-ft) pillar of iron is erected at Delhi c 415, India, as the base of a statue; although the statue is long gone, the pillar remains, largely as a result of the extreme purity of the iron; iron chains are regularly used in suspension bridges at this time in India as well

c 420
CE

c 480
CE

c 490
CE

Silk is cultivated in Byzantium after Christian monks smuggle silkworms from China; however, most silk continues to be imported from China, even though local cultivation continues; *See also* 50 BC MAT; 1220–29 MAT

The Manteño culture in what is now Ecuador manufactures thin copper ax heads thought to be for use as money up and down the west coast of the Americas (some are found as far north as Mexico); the axes are too thin to be used as tools or weapons

The most elaborate water clock ever constructed in the West is built at Gaza near the frontier between Egypt and Palestine (Israel); as a statue of Helios moves along below a series of doors to mark the time, the hours are indicated by other statues beating gongs and popping from open doors; at day's end, a mechanical trumpet blows

c 500
CE

c 520
CE

c 530
CE

Eastern Roman Emperor Justinian orders the building of Santa Sophia (also called the Hagia Sophia), designed by Isidore of Miletus (Turkey), the first building with a dome large enough to cover a town square; the dome is actually a pendentive, formed by cutting off the sides of a hemisphere to produce a figure that covers a square and then slicing off the top, producing four upside-down "triangles" with curved sides joined at their upper bases; the pendentive dome itself is 37 m (120 ft) across and 14 m (46 ft) high, but rests on the top of a building, giving it a total height of 61 m (200 ft)

c 540
CE

The Aqua Virgo aqueduct to Rome is destroyed by the Goths in 537; *See also* 1450–59 ARC

c 550
CE

Chryses of Alexandria builds a flood-control dam on a small river at Daras on the Persian border of the Eastern Roman Empire; it is the first horizontal arch dam, and the last until modern times, as the dam curves upward into the current

The Liang emperor Luan's *Book of the Golden Hall master* describes wind-driven land vehicles used in China; a wind-driven carriage can carry 30 people and travel hundreds of kilometers in a single day; sails are also applied to wheelbarrows, and wheelbarrows with sails become a common symbol of Chinese culture in eighteenth-century Europe

c 560
CE

The earliest known pointed arch, called an ogive in cathedral architecture, to which we can assign a date is built at Qasr-ibn-Wardan in Syria in 561, although an older pointed arch on a Roman colonnade at Diarbekr (in Turkey) also exists

c 570
CE

The Avars, an Eastern people who eventually settle in the Caucasus, invade Hungary in 568, bringing with them the trace harness and the stirrup for use with horses; both were originally developed in China; *See also* 480 CE COM

c 580
CE

Women in Northern Ch'i under siege from neighboring kingdoms in 577 in China invent matches so they can start fires for cooking and heating

The Sui dynasty in China begins with the reign of Yang Jian in 581

The water-powered flour sifting and shaking machine invented in China is mentioned for the first time; this device works like a steam engine in reverse, changing rotary motion into the back-and-forth motion needed for sifting; it is the first machine capable of doing this

c 530 CE

Roman General Belisarium installs waterwheels on boats on the Tiber River during a siege of Rome by the Ostrogoths in 537; this type of water-wheel becomes very common in European towns during the Middle Ages

c 540 CE

c 550 CE

c 560 CE

c 570 CE

c 580 CE

What caused the Agricultural Revolution? (part 1)

The Agricultural Revolution took place at different times in different places. Yet anthropologists continue to look for a single factor that caused peoples in various regions to turn to farming as a primary source of food.

The most obvious idea was used at first: Farming is better than hunting and gathering; so when people found out about farming, they stopped hunting and gathering and settled down in villages. However, this elegant idea may have been completely wrong.

Many scientists today believe that farming is not better than hunting and gathering. Careful studies have shown that hunter-gatherers have more leisure and a better diet than farmers. People knew how to farm for a long time before they made it a main way of living and many who knew how to farm never stopped hunting or gathering. People also settled down into permanent villages before they started farming. In addition, it is odd that the Agricultural Revolution occurred at quite close to the same time in the Near East, in Southeast Asia, in what is now Mexico, in South America, and in what is now China. By the early 1950s, various archaeologists began to offer different explanations for the Agricultural Revolution.

The Climate Changed. Gordon Childe proposed in 1951 that a drier climate produced the Agricultural Revolution, reducing game and wild food plants. People had to move to oases, where domestication was essential for survival. This idea ran afoul of subsequent studies showing that the climate did not become drier at the right time.

A different view is that when the ice caps retreated, people were forced to abandon their reliance on reindeer and mammoths. With that major resource gone, some other source of food needed to be found. Similarly, the rise in sea level that accompanied the melting of the glaciers caused people in Southeast Asia to live on less land, resulting in the invention of agriculture there.

The problem with the end-of-the-ice-age theory is that the ice age ended 5000 years too soon; and it was not the people living near the edge of the ice or tundra who first started farming.

There Were Too Many People. Mark Cohen of the State University of New York is among those who put the blame on population pressure causing resources to be exhausted. There is some evidence that people had known for at least 20,000 years that a single seed planted in the ground could grow into a plant with many seeds on it. Diminishing resources caused by population pressure finally forced people to do the hard work of farming. To replace lost game, they also began to breed their own animals for meat and fiber (with milk as a side benefit).

The problem with this theory is that some people, in what is now Mexico, for instance, started farming when there were not very many people.

Other ideas were then tried, as discussed on page 77.

c 590
CE

c 600
CE

The earliest known windmills are built in Persia (Iran); they use a vertical shaft and horizontal sails, partially protected from the wind, to grind grain; *See also* 1100–09 ENE

Over 4250 m (14,000 ft) high in the Bolivian Andes, 20 km (12 mi) south of Lake Titicaca, people now known as the Tiahuanacans build the large shrine of Tiahuanaco, featuring a temple platform that is 15-m (50-ft) high and 200-m (656-ft) wide at its base, as well as the large, decorated Gateway of the Sun; the site is known for the excellence of its stone construction without mortar, famous for being so closely fitted that a knife blade cannot be inserted between the stones, and the size of the stone blocks, some weighing more than 90 metric tons (100 short tons); these blocks were quarried 5 km (3 mi) from the site; although most of the construction is held together by its own weight, some stones are joined by bronze or copper clamps, apparently to earthquake-proof the structure

The Anasazi begin to make pit houses like those of the Mogollons, but Anasazi pit houses have paved floors and use stone to line the pits; some houses are partitioned into several rooms

A Maya ceramic vessel shows what appear to be pages from a complex book, suggesting that Maya books had been developing for several centuries; this is the oldest direct evidence of Maya books, of which only three are known today, all from around the 14th century

The Chinese print whole pages with woodblocks, although the earliest known surviving pages are from the 8th century; *See also* 870 CE COM

About this time Saint Isadore of Seville makes perhaps the first allusion to a quill pen, although the quill had probably been used for this purpose earlier; *See also* 1809 COM

Chinese Emperor Yangdi completes a major section of the Grand Canal c 610 ce, linking the Yangtze and the Huang He (Yellow River); this round of construction starts in 605, although construction of the first segments dates back at least until 70 CE and some parts may have been based on canals dug as early as 500 BC ; *See also* 70 CE COM

c 610
CE

Japanese texts mention "burning water" used in lamps, which is thought to refer to petroleum; *See also* 100 CE GEN

Li Ch'un builds the Great Stone Bridge over the Chiao Shui River in China; it is the first example of a segmental arch bridge and it survives intact

c 620
CE

The Arab prophet Muhammad flees Mecca (the hegira) in 622, establishing the founding date of the Islamic religion

China starts the publication of a court newspaper in 618, with several dozens of copies printed; this form of news reporting at the court lasts until 1911; *See also* 50 BC COM

c 630
CE

The accession of Emperor Taizong in 626 institutes the beginning of the Tang dynasty in China

c 640
CE

For the last time, the canal begun by Pharaoh Necho II about 1200 years earlier is restored to use, this time by the Arab general 'Amr ibn-al-'As; it is covered in sand for good sometime in the 8th century; *See also* 100 CE COM; 1856 TRA

	The oldest known reference to paper for use as toilet paper is made in China in 589; *See also* 110 CE MAT; 750 CE MAT	**c 590** CE

In the Nazca region of what is now Peru, filtration galleries called *puquios* are built for irrigation purposes; these are tunnels of the same general purpose and design as the qanats built to tap underground aquifers by the Assyrians and many subsequent inhabitants of dry regions, including the Spanish; when Spain colonizes the Nazca region, the conquistadores and their followers reintroduce the qanat to the region

c 600 CE

It is believed that forks are introduced for dining by the royal courts in the Middle East; such early forks had two tines and were a replacement for a second knife, previously used to skewer food while the first knife was used for cutting and lifting food; *See also* 1100–09 FOO

What caused the Agricultural Revolution? (part 2)

Scientists have continued to seek the cause of the Agricultural Revolution.

Society Became Complex. The longer people were around, the more they developed complex societies that included specialists of all kinds, including people in charge. Such a society, with division of labor and wealth that can be accumulated (by the people in charge), is forced into farming.

Some present-day or recent hunter-gatherer societies, such as the Native Americans of the Northwest coast, developed complex societies without farming—but they had salmon or some similar resource, so they did not need to farm.

People Had Time to Do It. In Carl Sauer's theory of 1952, people, having mastered hunting and gathering, developed a lot of free time. So they took up farming as a hobby. This theory, however, calls for agriculture to get its start in regions with abundant resources, which does not accord with the archaeological record.

It's What to Do in the Off-Season. Hunting and gathering, like farming, are seasonal. The seasons of good hunting and gathering often do not cover the whole year. During the time when not much else is going on, people can improve their food supply by planting crops. Crops that would be harvested after the good gathering season and before the good hunting season would presumably be favored. This fails to explain why people did not farm earlier or why so many started farming in such a short time period.

People Moved to Town. One good reason for calling the change the Neolithic Revolution is that more than farming was involved. About a thousand years before agriculture started, people, especially those dependent on trade or on the storage and processing of wild grass seeds (such as wheat), began to live in permanent communities. Even if a region as a whole still had good food resources, the immediate vicinity of such communities would soon run short of both wild grasses and game. Domestication of plants and animals saved village life. The problem here is showing why people settled down.

Plants Grow in Garbage. According to Edgar Anderson in 1956, plant remains tossed out in the garbage sprouted. This produced new crops from the discarded materials, crops that could be easily harvested. People noticed this and developed the systematic way of throwing out parts of plants that we today term agriculture.

It Did Not Happen. The last refuge of the historian or archaeologist faced with a major shift in society is to say that it started much earlier, went on much longer, and lacks any moment in time that can be singled out. In this view, people gradually over tens of thousands of years replaced hunting and gathering with farming. This is essentially the idea Robert Braidwood put forward in several books and stated fairly explicitly in 1951.

Additional theories are presented on page 79.

c 610 CE

c 620 CE

c 630 CE

c 640 CE

c 650 CE

The Taika Reform is instituted in Japan in 646, putting rice lands in the hands of the emperor and abolishing workers' guilds

The Buddhist king Harsha Vardhana of northern India is assassinated in 647 by the army after one conspiracy of Hindu Brahmans against him fails; his empire disintegrates

The final destruction of the Library at Alexandria probably takes place in 646; a famous story, perhaps true, says that when the muslim general `Amr ibn-al-`As captured the city he asked the caliph what to do with the Library; he was informed that if the books in it conformed to the Koran they would be superfluous; if they disagreed, they would be blasphemous; therefore, all books should be destroyed; the general followed his orders

c 660 CE

Wu Shao, a concubine in the harem of Emperor Taizong, after years of intrigue, is given the title empress by the third Tang emperor, Gaizong in 655; as Empress Wu she rules China through Gaizong and her sons and on her own until 705

c 670 CE

c 680 CE

c 690 CE

Caliph Abd al-Malik orders the Dome of the Rock built in 687 in Jerusalem over the rock where Muslims believe Muhammad ascended to heaven and Abraham offered his son as a sacrifice to God; the large wood-and-brass dome covered with gold over an octagonal base is completed in 691; one of the largest structures of this period, with a basal circumference of 161 m (528 ft) and a height of 34 m (112 ft), it survives in essentially unchanged form into the 1990s, although the gold is stolen from the dome much earlier; in 1992 plans are made to replace the dome with a new one of brass covered once again by a thin coating of gold

In 688 Empress Wu of China builds a pagoda 90 m (294 ft) tall of cast iron

In 695 the empress Wu has a cast-iron column made from about 1325 tons of iron to commemorate the Zhou dynasty in China

	Greek physician Paul of Aegina lists the symptoms of lead poisoning	**c 650** CE
		c 660 CE
	Callincus [b Heliopolis, Egypt, c 620] invents in 673 a substance that will burn in water and thus act as a weapon against wooden ships; it is known as sea fire, wet fire, or Greek fire; one school of thought holds that it may have contained quicklime, which produces heat in contact with water Churches in England are equipped with glass windows starting in 674	**c 670** CE
	The first sundial in England is built in Newcastle c 675	**c 680** CE

c 690
CE

What caused the Agricultural Revolution? (part 3)

A final set of theories on the cause.

It Just Happened. People did not set out to domesticate wheat or goats: Domestication was the inevitable result of the farming and hunting practices of the early Neolithic. Harvesting wheat over a period of years changed the nature of the wheat, since the seeds left behind by stone sickles grew on the plants that hold tightly to their seed. Evolution then resulted in fields of wheat where most of the wheat clung tightly to the seeds. Such seeds would not produce next year's crop. As a result, people were forced to help the wheat by planting some of the seeds.

Similarly, hunters killing larger goats produced an evolutionary shift to smaller goats. It became necessary to take steps to breed these smaller goats to produce enough meat.

This theory is fairly reasonable for wheat, but not much good for peas and lentils, which seem to have been domesticated about the same time. There are logical gaps that are hard to fill in for the explanation of animal domestication.

Farming Was a Gift from Great Beings. One day beings who knew how to farm decided that humans should also have this gift. These beings taught humans to plant and to reap and how to use the produce of the farm. Then the beings went away and left the humans to manage for themselves.

Recently this theory has been espoused by various popular writers, although it has won little acceptance by scientists. Nearer the time of the Agricultural Revolution, however, this was the most popular theory and often even had official backing by the leaders of society.

As noted in part one of this essay, however, the main flaw in the theory that farming is a gift from benevolent beings is that there is reason to think people were better off before farming.

	GENERAL	ARCHITECTURE & CONSTRUCTION	COMMUNICATIONS/ TRANSPORTATION
c 700 CE		In Cholula, just east of the Valley of Mexico, the largest pyramid ever is built; it is larger than both the Pyramid of the Sun in nearby Teotihuacán and the Great Pyramid	Between 704 and 751 the earliest printed text known, a Buddhist charm scroll, is created using woodblocks to produce images; *See also* 870 CE COM
c 720 CE			
c 730 CE	Charles Martel [b 689, d France, Oct 22 741] stops the Arab expansion at the Battle of Poitiers in 732, 160 km (100 mi) from Paris, France		
c 750 CE	The city of Teotihuacán (in Mexico) falls to unknown invaders about this time, after 150 years of decline	The Anasazi leave their pit houses at least in summer and move into masonry buildings that consist of long lines of rooms that face the old pit houses; rooms in the masonry houses are used for different purposes, such as sleeping, cooking, or storage; the pit houses are probably retained for ceremonial use or simply because they are warmer than the row houses in winter	The first printed newspaper appears in Beijing, China, in 748 About this time the Hohokam people living in what is now NM discover etching; by covering part of a shell with pitch and exposing the rest to acidic cactus juice, they etch decorations into shells; etching will be rediscovered in Europe in the 15th century
c 770 CE	Charles, later Charlemagne [b Aachen, Germany, c 742, d Aachen, Jan 28 814], is crowned king of the Franks in 768 along with his brother Carloman, but Carloman dies shortly afterward; *See also* 800 CE GEN		
c 790 CE	Charlemagne introduces the royal foot as the unit of length and the "Karlspfund" (about 365 g) as a unit of weight Viking raiders strike Great Britain for the first time on Jun 8, 793	The Japanese Emperor Kammu [b 737, d 806] orders the capital at Nara to be abandoned in 793 because it is dominated by Buddhist monasteries; the new capital, now known as Kyoto, built 24 km (15 mi) away, is a masterpiece of urban planning, designed for a population of 100,000; it is modeled somewhat after the Chinese city of Chagan	
c 800 CE	Charlemagne is crowned Emperor of Rome by Pope Leo III in 800; *See also* 770 CE GEN	The only building of any importance erected in western Europe between the fall of Rome and 1000 CE is the minster in Aix-la-Chapelle, constructed as a tomb for Charlemagne The Anasazi begin to construct the first of the multistory apartment dwellings for which they are famous; built with masonry walls more than a meter (3 ft) thick, the pueblos, as the Spanish later name them, contain dozens of families	The phonetic script called kana is developed for writing Japanese; the simplified script is much easier to learn than Chinese ideograms, and women as well as men become literate; the women of the next couple of centuries write some of the great works in the Japanese language, notably *Tale of Genji* by Lady Murasaki

FOOD & AGRICULTURE	MATERIALS/ MEDICAL TECHNOLOGY	TOOLS & DEVICES	
		The first printed reference to "ball arithmetic," which may have been an early form of the abacus, appears in Chinese texts; it implies that the abacus goes back to the 2nd century CE	**c 700** CE
		Buddhist monk I-Xing and Chinese engineer Lyang Lingdzan build in 724 a water clock that has an escapement, the device that causes a clock to tick; the clockwork is used to power various astronomical devices rather than to indicate the hour; *See also* 500 CE TOO; 1080-89 TOO	**c 720** CE
			c 730 CE
	In Egypt, the practice of using a thin film of a metal oxide, called a luster glaze, is developed for use on glass The Arabs learn paper manufacture from the Chinese at Samarkand in 751 in central Asia (present-day Uzbekistan)		**c 750** CE
			c 770 CE
		Wang Chu in China builds a wooden otter that can catch fish	**c 790** CE
Using water from the Gila and Salt rivers, the Hohokam of Arizona construct by this time one of the most extensive irrigation networks of dams and canals in the Americas The rigid horse collar in which the main load is on the chest and skeletal system of the horse is introduced in Europe, roughly a thousand years after its first use in China; horse collars are depicted on the Bayeux tapestry, probably woven before 1092	The art of paper manufacture spreads from Samarkand to Baghdad in 793; *See also* 750 CE MAT; 900 CE MAT Blast furnaces for making cast iron are built in Scandinavia During the Tang dynasty in China, stoneware pottery and perhaps porcelain are made		**c 800** CE

c 810
CE
In 810, the Chinese government takes over the issuing of paper bank drafts, the ancestor of paper money; *See also* 900 CE GEN

c 820
CE

c 830
CE
Emperor Theophilus of Byzantium, who reigns 829 to 842, possesses a throne that is adorned with mechanical lions that move their heads and roar and mechanical birds that peep

c 850
CE
The modern saddle harness is in use in Europe

c 860
CE
Vikings discover Iceland about this time; *See also* 980 CE GEN

c 870
CE
Basil the Macedonian is declared coemperor of Byzantium in 866, only to murder the man who made the proclamation, his coemperor Michael III, somewhat more than a year later

The *Diamond Sutra* is printed on May 11, 868, according to a notice from Wang Jye, its printer; this is the earliest complete printed book (actually a scroll) extant; found in 1900 by Aurel Stein in a cave in Kansu, the *Diamond Sutra* consists of seven large sheets, one with a woodcut picture, pasted together to make the scroll; *See also* 700 CE COM

c 890
CE

c 900
CE
Real paper money—that is, printed paper used as a medium of exchange—is in use in Szechuan Province in China; *See also* 810 CE GEN

The Mississippian city now called Cahokia, near the confluence of the Mississippi and Ohio rivers, features a shrine atop a 6-hectare (15-acre), 30-m- (100-ft-) high temple mound and a local population of about 10,000; the temple mound is the third largest structure of its time in the Americas

The symbol for zero is used in an inscription dated 876 CE in India, the first known reference to this symbol, although the concept may have originated earlier

The horseshoe is introduced; *See also* 1835 FOO

82

c 810 CE

The first great explorers

Pytheas of Marseilles made the greatest recorded voyage of exploration of ancient times, probably about 310 BC, circumnavigating Great Britain and reaching Norway (or possibly Iceland). Not only did Pytheas travel about as far as Columbus did during his first voyage, but he returned with careful reports on all he had seen and with excellent navigational data. The primary purpose of the trip was probably to locate the source of tin, which he found in Cornwall, but Pytheas clearly had scientific goals in mind as well.

Explorers also traveled along the coast of western Asia in those days, but their names have been lost. These routes were pioneered by Indian and Arab navigators perhaps as early as 3000 BC. The traders that followed the explorers eventually were displaced by Egyptians under the Ptolemies, starting with two voyages to India around 120 BC by the explorer Eudoxus of Cyzicus.

According to a legend dating from at least the 6th century CE, a Chinese junk captained by Xi and He around 2640 BC sailed east from Japan for thousands of miles and found a continent. If true, it could only have been one of the Americas. The best evidence that something of this nature could occur are well-authenticated accounts of Japanese boats blown off course and landing in the Americas in the late 18th and early 19th centuries.

c 820 CE

Muhammad ibn Musa al-Khwarizmi [b Khwarizm (Uzbekistan), c 780, d c 850] describes the astrolabe, which he probably learned of from Hellenic sources; it is used to find rising and setting times for a celestial body, the altitude of the body above the horizon, and the body's azimuth

c 830 CE

The *Utrecht psalter* provides the oldest pictorial evidence of a crank handle for turning wheels in the West, nearly a thousand years after the Chinese use such handles; the illustration shows a man sharpening a sword on a grindstone; there is some archaeological and written evidence that the Romans knew of the crank, but crank handles were not in regular use in Roman times; *See also* 300 CE TOO

c 850 CE

In China, *Classified essentials of the mysterious Tao of the true origin of things*, attributed to Zheng Yin, describes a primitive form of gunpowder; it warns against mixing it because of the danger of burning; *See also* 1040–49 MAT

c 860 CE

c 870 CE

c 890 CE

Han Chih-Ho builds a wooden cat that catches rats and dancing tiger-flies

c 900 CE

Arab chemists and physicians prepare alcohol by distilling wine

The art of paper manufacture spreads from Baghdad to Cairo; *See also* 800 CE MAT; 1100–09 MAT

c 950 CE

c 960 CE

General Zhao Kuangyin heads a revolt that initiates the Song dynasty in China; he becomes Emperor Taizu

c 980 CE

The Viking Erik the Red, banished from Iceland for feuding, discovers Greenland in 982; *See also* 860 CE GEN; 990 CE COM

Qiao Wei-yao, concerned about the amount of stealing taking place as boats are hauled over spillways, invents the lock for canals in 983; previously, boats frequently broke up as they were being hauled; people waiting for just such a moment would dash into the wreckage and steal as much cargo as possible

c 990 CE

Bjarni Herjolfsson, sailing from Iceland to Greenland in 986, gets lost and sights the Americas, but does not land; eventually he finds Greenland; *See also* 980 CE GEN

c 1000 CE

The church of Hosios Loukas is built at Phocis, near the Gulf of Corinth; made of stone, brick, and tile in the shape of a cross within a square, it typifies the fortresslike nature of the churches of Byzantium

Alhazen [b al-Hasan Ibn al-Haytham, Basra in Iraq, c 965, d Egypt, c 1039] describes the camera obscura about this time, although he is not the inventor; a real image is projected into a darkened box; it can be viewed from inside the box or can be projected onto ground glass to be seen from outside; Alhazen's optical works are influential in Europe after translation of his work into Latin as *Opticae thesaurus* in 1572; *See also* 1550–59 COM

Pope Sylvester II (Gerbert) imposes the Arabic numerical system, including the use of the zero, on Christians

A monk from Malmesbury Abbey in England reportedly builds a glider that flies some 200 m (600 ft) before crashing, perhaps because it is not stabilized by a tail

---

How the Egyptians did NOT build the pyramids

The Great Pyramid of Khufu (Cheops in Greek history) was one of the Seven Wonders of the World to the ancients and we still marvel at it today. Although it is not the largest pyramid of the past—honors go to an otherwise undistinguished pyramid from Cholula, Mexico, that was built about half as far back—Khufu's pyramid amazes because of its antiquity, the size of its stones, the perfection of its orientation, and its many hidden chambers within. Outwardly about as simple as an edifice can be, the Great Pyramid is actually a complex and interesting object. Furthermore, it is flanked by two nearby pyramids, each almost as large, that would be wonders by themselves if the Great Pyramid had never been built.

The Greek historian Herodotus, writing about two millennia after construction of the pyramids, claimed that it took teams of 100,000 men at a time working for 20 years to build them. Herodotus also claimed that the ancient Egyptians had marvelous machines that they used to lift the stones into place. Neither of these claims seems to be true.

A current misconception is that the Egyptians moved the large stones on rollers made from tree trunks. According to this theory, by having teams constantly replenish the front rollers with those from the rear, it should have been possible to move a stone almost as if the wheel had been invented. This is idle speculation with no evidence in its favor and much circumstantial evidence against it.

FOOD & AGRICULTURE	MATERIALS/ MEDICAL TECHNOLOGY	TOOLS & DEVICES	
	In 954 the emperor Shih Tsung has the largest single piece of cast iron in ancient China made to celebrate his campaign against the Tartars; known as the Great Lion of Tsang-chou, it weighs about 40 tons	Gerbert of Aurillac, who becomes pope in 999, invents a type of escapement consisting of an oscillating horizontal arm that controls the speed of a gear	c 950 CE
			c 960 CE
		Chang Ssu-Hsün invents in 976 the chain drive for use in a mechanical clock	c 980 CE
			c 990 CE
The Arabs introduce the lemon plant to Sicily and Spain	People along the coast of Ecuador learn to smelt platinum before this time, although how they obtain the high temperatures required remains a mystery; *See also* 1550–59 MAT	Gears used with waterwheels and water clocks become common in the Arab world	c 1000 CE

How the Egyptians DID build the pyramids

Using wall paintings and related texts, along with practices employed in more recent times and a few artifacts, we can put together the following scenario:

Quarrying Limestone was cut in blocks by making notches with hard rock, inserting wooden wedges, and wetting the wedges. When the wedges swelled, the stone fractured. Sometimes copper wedges in increasing sizes were used instead. Granite, used for plugs and caps, may have been cut by pounding it with a harder stone.

Transporting Even later than the pyramids, very heavy objects, some ten times the weight of pyramid stones, are shown in wall paintings as being moved on sledges, not on rollers or wheels. The sledges were aided by pouring a liquid in their path to lubricate the way.

Leveling Ancient Egyptians probably used shallow ponds. A brick or mud wall can be used to enclose the site. Introduce water to make a pond; its surface will be level. Then holes can be drilled to a fixed depth below the surface. Remove the water, and the bases of the holes mark the level surface; just remove everything down to this base.

Slanting Keeping the slope of the sides the same over an area of 31,000 sq m (7.7 acres) was another difficult task. It seems likely that the masons used the diagonals of the square to locate its center, and erected a vertical pole (using their ability to find right angles) to the planned height of the finished pyramid. They could then sight to find the correct slope.

Lifting This is the easy part. Instead of lifting the stones, some as heavy as 150 tons (although the average limestone block was only about 2.5 tons), the Egyptians built long ramps to the working level so that stones could be moved up on sledges. Then levers and wedges were used to put the stones in place.

Lowering The outer blocks were cut in final form and needed to be lowered very carefully or the process would damage the facing. It is thought that this was accomplished by lowering from one set of wooden braces to another, with each set of braces only about a centimeter (half an inch) lower than the one above it.

OVERVIEW

The Age of Water and Wind

This was the age of water and wind in several senses of the phrase. Sailing ships linked the world together, with only Antarctica and a few isolated islands unvisited by 1734. Also, in Europe especially, it was the age of water and wind power. Both windmills and waterwheels were invented during the Metal Ages, but their extensive use in Europe was not until the fall of the Roman Empire and the rise of a new social and political structure in Western Europe.

Although Europe was the region where the extensive sources of water and wind energy changed life the most, it did not contain the most advanced society of its time, especially in the beginning of the period. Europeans who encountered Chinese civilization, notably Marco Polo in the 13th century, recognized that they were in the presence of a technology that could accomplish much that went beyond the medieval technology of the region dominated by the Roman church. Arab scholars and scientists also were more advanced than Europeans in the beginning of this era. The period of exploration by ships from Portugal, Spain, Holland (the Netherlands), England, and China revealed the civilizations of Africa and the Americas as well.

By the beginning of the 18th century, most of the governments of Asia had calcified into top-heavy empires and kingdoms that were unable to face change. In Africa, the slave trade by Arabs and Europeans had completed the destruction of large states, leaving poorly organized tribes in their wake. The native population of the Americas was in even more disarray, largely as a result of depopulation from disease, slavery, and war. In both the Americas and parts of Africa, the growing population of European colonists and their descendants had formed new states modeled on those of Western Europe, their former home. These states were at first vassals of Western Europe, contributing raw materials and buying finished goods. Meanwhile, the eastern portions of Europe and Western Asia did not so much fall as become gradually moribund.

As a result of all these trends, the people of Western Europe had become by the end of this period the richest and most technologically advanced in the world. In 700 years they had progressed from one of the least advanced societies to the most advanced, certainly in their own view of themselves. For most of the next 300 years they would remain technologically superior.

The Fall and Rise of Western Europe

The fall of the Roman Empire was followed by the extinction of the administrative and military infrastructure that had been established throughout Western Europe by the Romans. Roads, bridges, and water-supply systems built by the Romans fell into disrepair, postal services ceased, the use of currency became replaced by barter, and agricultural production diminished considerably. Trade almost stopped entirely, and people relied on the local production of food and goods.

During the first centuries after the fall of the Roman Empire the standard of living in Europe returned to levels comparable to those just after the Agricultural Revolution. Poverty was endemic and people suffered from wars, piracy, famine, and epidemics. During the last centuries of the first millennium, a reversal of this decline gradually took place. Trade and food production began to increase, starting in the northern part of Europe. Water was more abundant in the north than in the Mediterranean countries, and the soil was more fertile.

According to many historians, the first important technological revolution took place during these years. It was a revolution mainly because several technologies from antiquity that had been forgotten became used again on a large scale, and a series of new inventions brought medieval technology to a more advanced stage than possessed by the Romans. Many of the tools and machines developed during these times remained practically unchanged until the Industrial Revolution.

The revival in Europe took several forms. Landowners reintroduced the rotation of crops, as practiced by the Romans, increasing food supplies substantially. Roads were paved again, and transportation and trade profited from important inventions, such as the steerable carriage and the padded shoulder collar for horses. Previously, horses carried a girth band to pull carriages, and because of its choking effect, the horses could pull only small weights. Horseshoes, previously used by the Romans, came into general use during the 10th century, allowing horses to travel over longer distances.

A key technology taken over from antiquity was that of the waterwheel. In Northern Europe, in contrast to Mediterranean countries, water was abundant and the water supply in rivers was dependable. As a result, the waterwheel became the most important source of mechanical power throughout Northern Europe for many centuries.

A few centuries after the disappearance of the administrative infrastructure of the Romans, a new form of organization of society was gradually introduced in Europe. Monasteries played an important role in this restructuring. They were the first to start making large parts of forested land cultivable. Eventually, monasteries became entirely self-supporting, relying on the work of the brethren; therefore they were also the first to reintroduce technologies, such as waterwheels to drive mills and other machines.

Many of the ideas from antiquity that impeded the full development of technology, such as low esteem for manual labor, still existed. Christianity in the Roman Empire was originally the religion of the slaves and the poorer working classes. Even when its adherents became the rulers and the rich, Christian philosophy advocated the idea that manual labor is not degrading, and this opened the way to the reintroduction of devices and machines to help laborers. The Christians also rejected slave labor, and as cheap labor became scarce in Europe, landowners were forced to invest in labor-saving machines, following the example set by the monasteries.

Much of the artistic work of the period was in the service of the church. Artisans built organs, cast bells, and created stained-glass windows. Religion was also the inspiration when later in the Middle Ages great numbers of people and artisans banded together to build the large cathedrals.

During the 12th century, Western Europe lived in relative calm after invasions by Vikings ceased. Medieval society entered a second period of important changes. Trade developed and markets appeared, increasing the standard of living. Cities grew in which people made their living from trade and manufacture instead of from farming.

Artisans associated themselves in guilds that, after the example of the trade corporations that had existed in Rome, protected guild members' rights, established trade rules, and took on the training of apprentices. The guilds contributed to the spread of techniques of manufacture because of the requirement that apprentice artisans travel abroad to work as journeymen to improve their skills. On the other hand, because of the many fixed rules they imposed on their trades, guilds sometimes discouraged technical innovation.

The end of the Middle Ages in Western Europe was marked by a number of crises. The famine of 1315 to 1317, the Hundred Years' War between 1337 and 1453, the bubonic plague that killed one fourth of the people during that same period, and the resulting financial collapse, slowed down development. During these years the population of England dropped from 3.7 to 2.2 million.

Financial stability and an economic revival returned to Western Europe around 1470, first south of the Alps and in Spain, and later in the North. The crisis of the 14th and 15th centuries had also destroyed the feudal system of the Middle Ages, allowing the artisans and traders in the cities to gain power. A new type of owner took over. Land, mills, and manufacturing facilities were not anymore owned by the churches or vassals, but by entrepreneurs, the early capitalists.

The 15th century was marked by a return to concerns of the world, away from the mysticism of the Middle Ages. This new atmosphere encouraged the development of the arts, science, and technology. People searched to improve the utilitarian aspects of technology. States recognized the importance of technology for defense and trade and encouraged technological development. Kings engaged engineers to improve their fortifications and weapons. Technology was now fully accepted, and many artists (the best example is Leonardo da Vinci) became architects and technologists. The teaching of mathematics, now viewed as the basis of the arts (perspective) and technology, was introduced into universities. This became the period we call the Renaissance.

Exploration

Before the age of water and wind began, Viking explorers had already settled in Greenland and begun a destructive dialog with the natives they encountered. Soon after 1000 CE the Viking rovings reached the Americas. Later Basque whalers probably also visited the Americas. These contacts, however, did not result in any significant exchanges or expansions of technology.

Africa during the beginning part of this period included a number of independent kingdoms, notably Ghana and Mali in the western Savanna region and the Yoruba Empire of Oyo, famed for its excellent metalworking. Along the east coast the trading kingdoms and city-states of Mogadishu (Somalia), Brava, Malindi (Kenya), and Zanzibar flourished.

Independence in Africa was for a time snuffed out. First the Arabs came from the east and the north, trading in goods and in slaves and also converting one ruler after another to Islam. Then the Portuguese and other Europeans came from the west and took over vast regions of the continent.

The European thrust eventually (1497) passed around the tip of Africa to reach India, linking Asia and Europe by water. By that time, Columbus had reached the Americas (1492), completing the linkage of most of the globe.

The technology that made this linkage possible began with the Viking ships, the first Western ships with rudders; it continued with improvements in sail rigging and ship construction. By about 1450 all the elements were in place to produce the caravels of the Portuguese, quickly imitated in one element or another by most of the Western seafaring nations (and distinctly improved upon by the Dutch).

The technological interchange that resulted from the new worldwide linkage was surprisingly small. The biggest influence during this period was the passage of Chinese technology into the West—but that had come over land before and during the great age of exploration. By the time European ships reached China, they were already using improved versions of Chinese techniques. Although Africa had strong metalworking and mining traditions, the African technology did not influence that of Europe; rather, it was the other way around.

The biggest influence in both directions came with the opening of the Americas to European conquest and colonization. Although the Europeans possessed better transportation, harder metals, and firearms, the key to Europe's complete dominance came from the more destructive European bacteria and viruses (especially the latter). Native populations were devastated by smallpox, measles, and other unfamiliar viral diseases. The main technology from the Americas to be adopted in Europe was horticultural (see "Old and New World plants meet," p 128).

Technology and Science

The first center of technology during this period was the Far East, but the vast improvements made in everything from communication (printing) to transportation (the rudder) in China during this period seem not to have had a basis in or relationship to science. Gunpowder, for instance, did not arise as a result of understanding the chemistry of combustion and it did not lead to studies in how chemicals combine or what about chemical reactions produces heat or gases.

During this period, the center of technology gradually moved toward the west (some would argue that it has continued to do so since then, moving from Asia to Europe, from Europe to North America, and—in our day—from North America back to Asia again). As the center shifted from China to Persia (Iran) under the Arabs and the Arab world in general, pure science began to arise and become something of an influence on the growth of technology. The Arabs, after all, preserved the knowledge of antiquity that had been lost after the fall of the Roman Empire in the West. Exposed to such scientists as Aristotle, the best Arabian and Persian thinkers became scientists and mathematicians themselves. It is perhaps not surprising that a people whose rulers lived in tents made little contribution to the arts of architecture and construction. Instead, the Arabs followed the traditions of Hippocrates and Galen, contributing to the development of medicine and chemistry. Influenced by Alexandrian science, some Arab scholars used water-driven clocks to drive toys and temple apparatuses. The Arabs and Persians were very careful astronomers and made many technical improvements in the main astronomical instruments of the day, astrolabes and sundials. Otherwise, the Arabs largely transmitted the technology of the past to the West, not changing it very much.

Although some European philosophers, such as Hugues de Saint Victor, Vincent de Beauvais, and Raymond Lulle associated technology with science, there was very little contact between science and technology during the European Middle Ages. It was only during the Scientific Revolution of the 16th and 17th centuries that science started playing a role in technical development.

The 17th-century French philosopher René Descartes, for example, considered the laws of nature to be the foundation of machines. In a letter to Christiaan Huygens, Descartes wrote that "you have to explain the laws of Nature and how She acts under normal circumstances before you can teach how She can be applied to effects to which She is not accustomed." Huygens soon did exactly that: Knowing that the period of a pendulum depends mainly on its length for small swings, he used the pendulum, modified to keep better time, to control the speed of clocks.

During the 16th century the first treatises on technology that describe experimentation, explain the use of mathematics in science and technology, and call for a more scientific approach to technology were published. A new type of technical professional arose, the "engineer" who went beyond the crafts of the artisans by using mathematics and science in building and inventing. Scientists became more interested in technological matters, and many technological problems were discussed by the members of the Royal Society in England and the Academie Royale des Sciences in France.

Hints of Things to Come

In technology, hints of a new world rising were evident in the 15th century. A few preindustrial factories operated. The earliest known assembly lines and industries based on interchangeable parts appeared (although some smelting and casting operations, even manufacture of uniform bricks, go back to late Neolithic times). Patent laws to protect inventions and the stock market, eventually to become a major means for financing technology, began. Printed books became the means for spreading technological developments with unprecedented swiftness. Less obviously, the development of perspective was a step toward better drawings of machine parts.

In the 16th century, similar hints of the future included the first machines for turning fiber into cloth (knitting in this case) and the development of better machine tools.

By the 17th century, a number of inventors were experimenting with steam, and the potential of steam as a source of power was recognized.

At the end of the age of water and wind, the Industrial Revolution was already quietly under way.

Major Advances

Architecture and Construction. Architecture in Europe, which had been on a decline since Roman times, flowered again in cities, especially with the building of the large cathedrals in Northern Europe. New breeds of professional workers—architects, stone masons, and technicians—traveled from building site to building site, bringing their expertise with them. Because of this wandering existence, masons did not belong to guilds or corporations as did the other trades. Since the cathedrals were the first truly large buildings of the Middle Ages, their builders developed a number of construction technologies, such as pulleys, cranes, and other lifting devices, and methods for transporting heavy loads. The building of cathedrals was accompanied by the development of stone quarries. Transport of stone from the quarries even influenced the development of canals and locks.

Much of the monumental architecture of this time is still available for visiting or even in use today. Europe is dotted not only with cathedrals, but also with a great many castles, nearly all constructed during this period. In Asia such giant temples as Angkor Wat in Cambodia were built. In North America the pueblos of the US Southwest eclipsed the apartment houses of the Roman Empire.

Rigid bridges, as opposed to suspension bridges, made several significant advances in the period as engineers struggled with the problem of lengthening a span without damming too much of a river. The increased use of trusses helped improve bridge stability and also aided the construction of buildings in general.

Communications. The next major development in communication after the invention of writing, printing with movable type, came during this period. In many ways this invention immensely sped up technological development around the world. Just as the mass production of the written word made a big difference, another set of inventions made writing more personal—quill pens and the first pencils. What was written had to stay, for the rubber eraser was not invented until after this period.

An offshoot of printing in an entirely different direction was the newspaper. Getting started largely around the 17th century, newspapers took advantage of printed words so cheap that they could be read once and discarded.

The world itself was successfully mapped for the first time during the age of wind and water (although there had been some precursors in Classical times). New maps and globes helped people understand the discoveries of the great explorers of the age.

This was also a period in which several important precursors of photography were invented or discovered,

Building technology during the early Middle Ages in Europe lagged behind that of the Romans, the Muslim peoples, and the Chinese. But a technical revolution in construction took place during the years 1100 to 1270 in Northern Europe, especially in England and France. During that period, a number of cathedrals of exceptional size and height appeared in several cities.

To achieve these stunning dimensions, particularly noticeable in the spacious interiors, the cathedral builders employed three technical innovations: the ribbed vault, the pointed arch, and the flying buttress. These three elements were first used together at the cathedral at Durham, England.

A vault is, so to speak, the three-dimensional equivalent of the two-dimensional arch. It is ribbed if it is provided with independent arches for stronger supports. The main purpose of a stone vault in a cathedral is to protect the building from fire by isolating the timber roof from the rest of the building. The vault has also an esthetic function, giving the interior a soaring appearance, in contrast to that of a flat, wooden ceiling.

Structurally, the vault at Durham, and in later churches, was made self-supporting by a system of diagonal and transversal ribs, which, because of their pointed shape, deflected vertical pressure caused by the weight of the vault sideways. The ribs were constructed first, and the space in between ribs was then filled in. The ribs had, besides their structural function, an esthetic function.

Because the ribs supporting the vault translated a vertical force into a transversal force, pushing the piers and supporting walls outward, the builders had to find a way to neutralize this force. They achieved this by building flying buttresses that in subsequent years would become the most remarkable feature in cathedrals throughout Europe. These were additional walls on the outside that were perpendicular to the walls of the vault.

The basic structure of the Durham Cathedral was replicated shortly afterward in a number of cathedrals in Europe. Religious fervor, but also competition between cities, pushed toward ever higher structures. In France, the "Cathedral crusade" was sparked by Suger, the abbot of the great monastery at Saint Denis, near Paris. He was the main designer for the rebuilding of the church at St. Denis, which was completed in 1144. His friend, Bishop Henry of Sens, built the Cathedral at Sens, which was completed in 1160. Both church buildings incorporated the three new elements of Durham Cathedral, but the structures were far lighter and less massive than those of Durham Cathedral. William of Sens went in 1174 to England, where he rebuilt the Cathedral of Canterbury.

Subsequently, the role of the flying buttresses became increasingly important in that they took over most of the support of the vault, allowing side walls with huge windows. Notre Dame Cathedral in Paris, which was started in the 1160s, is one of the examples of this new trend.

The vault of Notre Dame, at 33.5 m (110 ft), was higher than anything built before. Soon an element of competition between cities came to the fore, and cathedrals with even higher vaults were built. Chartres Cathedral has a vault at the same height as that of Notre Dame, but the vaults at Rouen and Reims are 38 m (125 ft), the vault of Amiens Cathedral is 42.5 m (140 ft), and that of Beauvais is nearly 49 m (157 ft).

The cathedrals at Beauvais and at Cologne are the best examples of Gothic architecture in its most perfected form. All the walls are replaced by huge expanses of stained glass, the entire structures being supported by a complex system of buttresses at the outside.

In France alone, about 80 cathedrals were built between 1150 and 1280. Later the cathedral fever abated, mainly because of the huge costs, and many of the cathedrals were left unfinished. Construction of some—Cologne is an example—was not completed until the 19th century.

---

notably the camera obscura, the magic lantern, and the fact that silver salts turn black when exposed to light.

Energy. To a large extent this period is defined by its use of energy. Although water mills had been invented earlier and were used by the Romans, their use greatly expanded during this time. Similarly, the use of wind to power ships was hardly a new idea, but new riggings and ways to steer ships into the wind made wind power more useful than ever. Furthermore, the idea of the waterwheel was combined with the idea of the sail to produce the first European-style windmills.

Coal increasingly became the fuel of choice, first in China and then in Europe, replacing wood as forests dwindled in the face of increasing population. Coal had been known in antiquity and then forgotten, so it seemed remarkable to Marco Polo that the Chinese burned black rocks. Coal found lying around on the ground or in streams was burned for a long time before anyone thought of mining it. In Europe, people in what is now the Netherlands were the first to think of looking for coal underground. News of their success soon reached England, where coal mining became a major industry in the 17th century. Early in that century, also, the Englishman Hugh Platt became the first to heat coal and produce coke, a more satisfactory fuel for many purposes. Coke burns with a higher temperature than coal, for one thing.

Coke and coal may have been the trigger for the extensive experiments with steam that took place during the 17th century, although cast-iron containers may also have had something to do with the new respect for steam. In any case, by the end of the 17th century the first primitive steam engines had been built, usually for the operation of pumps at coal mines, providing a useful synergy right there. With some other useful synergies, the Industrial Revolution was getting started.

Food and Agriculture. Several factors contributed to the revival of agriculture in Europe around the 10th century. Population density, which had become very low partly as a result of a major epidemic during 742 to 743, started to increase again. A warmer climate during the first part of the age of water and wind also contributed to more successful harvests and to the population increase, although this was followed by the rigors of the Little Ice Age, a period of colder climate that extended from about 1500 through the middle of the 19th century.

Farmers introduced several new techniques that made agriculture much more effective. The plow replaced the much simpler ard and took on the form it has today. The most important improvement was that the new plows turned the earth over instead of merely cutting a groove in the soil. Turning over soil increases the fertility of the land by plowing in organic matter.

The second improvement in farming was the use of horses instead of oxen for working the fields. New breeds of horse, stronger and more resistant to disease, were introduced by the Berbers and spread throughout Europe. The padded shoulder collar and horseshoe made the horse much more efficient than the oxen.

The most important innovation in medieval agriculture was the introduction of crop rotation based on a 3-year cycle. Traditionally, a 2-year rotation cycle had been in use: One year the land was cultivated and the next year it was left alone. In the 3-year cycle, the first year the land carried a winter crop, the second year a spring crop, and the third year it was left uncultivated.

Among the precursors of the Industrial Revolution was the roller mill, an improved form of water-powered mill for producing flour from grain.

Materials. The reduction of iron ore to metal during this period used the same methods as metallurgists had developed in antiquity, but the size of furnaces increased. Some nations produced better iron than others; this became a factor in several major battles of the period. Productivity of iron making and working was increased with the introduction, during the 12th century, of mills driven by animal power or waterwheels. The waterwheels or squirrel cages put in motion by running dogs were used to drive bellows, allowing an increase in size and temperature of furnaces.

The 14th century saw major changes. Cast iron, long used in China, became common throughout Europe. Blast furnaces appeared around the 14th and 15th centuries in the Liège area (in Belgium). The demand for iron and other metals increased substantially with the introduction of the cannon during the 14th century. However, basic methods for preparing steel, cast iron, and wrought iron remained unchanged until coke was substituted for charcoal at the beginning of the 18th century.

With the new emphasis on both iron and coal, the first works on mining were published during this period.

Although iron in various forms was important, wood was the basic material of the age. Besides its use in the production of iron and in heating, wood was also the main construction material for houses, means of transport, ships, and even machines. Waterwheels, windmills, gears, and camshafts were usually made from wood as well, although iron was used for reinforcement and for the reduction of friction in bearings. Tools such as shovels and plows continued to use iron edges on wooden bodies. Barrels made of wood were used for storage instead of the earthenware storage jars that had been used in antiquity.

Papermaking, which had originated in China, spread throughout Europe during the first part of this era. Early paper was made from used fibers such as linen rags, not from wood, the source of most of today's paper. The manufacture of paper preceded the development of printing and made the invention of printing useful.

Gunpowder, another material that had originated in China, had a great influence on Europe and the rest of the world during this period.

Medical Technology. Perhaps the single greatest advance in medical technology of the period started during the 13th century when people found that glass lenses could be used to correct vision defects. Spectacles for correction of the almost inevitable farsightedness of old age were introduced first, followed by glasses that corrected the common ailment of myopia (nearsightedness). Although these instruments seem commonplace today, spectacles were like a miracle for the time.

Another significant advance of the time is more problematical. Opium became the first painkiller since alcohol to enter general use. Like alcohol, and like most painkillers of any power since, opium was to prove a mixed blessing.

Toward the end of the period people began using inoculation, although not yet vaccination, to protect against smallpox—that is, they would deliberately expose themselves to smallpox so that later they would not catch it.

Surgery is perhaps the oldest form of medicine. The practice of making holes in the skull, or trephining, is prehistoric, while one of most important early Egyptian papyruses, the Edwin Smith Papyrus of about 1700 BC, is a list of surgical techniques.

When young Galen learned his surgery in Pergamum around 150 CE, he acquired many of the skills that had been taught since the time of Hippocrates and that are still employed by surgeons today. He learned to set fractures and replace dislocations, to stitch or strap torn muscles or skin (tying off blood vessels along the way), and to cut or tie off hernias. The common invasive techniques used included removal of growths of all kinds and of stones from the easily accessible bladder, trephining to relieve pressure (still used), and drainage of fluid that had collected in various body cavities.

After Galen moved to Rome in 162, he dissected many animals and developed a clever (but mistaken in important details) theory of how the body functions—advancing knowledge much more than practice. In the 150 years after Galen's death, his reputation quietly grew; by the 4th century, he was the recognized authority for all medicine in the West and soon in the Arab world as well. Although Galen's view of medicine prevailed, it did not stop the slow, anonymous growth of surgical practice. By the 7th century, for example, surgeons had learned to perform tracheotomies when needed and to handle aneurysms if they were located in accessible places.

Arab doctors advanced medical standards of treatment, including roles for immunization, anesthesia (hashish), diet, and cleanliness, but they failed to go beyond a limited understanding of Galen in surgery, largely because Islamic practice was against deliberate bloodshed (except in war or as punishment). The Arabs separated surgery from medical practice, a division that passed over to the West. Surgeons were the lower class of practitioners, while physicians, whose closest approach to surgery was to let blood, were the elite. For a time the barbers who cut hair and beards were also the surgeons who cut skin and organs. These barber-surgeons of the West were hampered by a lack of anesthetic beyond strong drink (although some knew of opium-based concoctions that were better) and a belief that pus was good for healing. Speed was essential because of the patient's pain, and the most successful procedures were the most external, such as lancing boils or removing cataracts. Amputation of limbs was frequently needed because of the lack of cleanliness in treating minor wounds, and surgeons developed considerable skill at the process. Cauterization—sealing off blood vessels by searing flesh—was found to be easier and more commonly practiced than use of ligatures to tie off arteries and veins.

In the East, the surgical traditions were not much different from those in the West. In India, surgeons performed much the same kind of operations as Galen learned and at about the same time. In China, despite advanced ideas on circulation of the blood and use of biological substances (from animals as well as the better recognized herbal medicines), acupuncture was more emphasized than invasive surgery.

About 1275 surgery in the West began a slow improvement. The scalpel, which had been replaced by ordinary knives, was revived. In 1543 the great work of Vesalius corrected many misunderstandings about human anatomy, with excellent and accurate illustrations as well. Still, it was not until the 18th century that the barber-surgeons in France and England were completely replaced by medical surgeons. Shortly before that time Ambrose Paré introduced boiling oil to control infection in wounds, which surprisingly was a great improvement over previous practices. The widespread use of firearms probably contributed to increased knowledge of how to remove segments of gangrenous intestine.

(Smallpox vaccination involves exposure to a different, related virus, one that does not itself cause smallpox).

Smallpox produced a pandemic, but not *the* pandemic, of the period, which came earlier—the Black Death, a pandemic of plague, so reduced the population of Europe that it is thought to have changed history.

Tools and Devices. Virtually unchanged in methods from antiquity to the early Middle Ages, the textile industry underwent a series of innovations during the 13th century. The most important was the mechanization of fulling and spinning. Fulling is a process in cloth manufacture in which the weight and bulk of cloth is increased by beating on it. It is tedious work by hand. Fulling mills, in which the hammers were tripped by cams mounted on the shaft of a waterwheel, appeared in many places, replacing workers that used to beat the cloth with hammers. Because fulling mills depended on waterpower, they often were built outside cities and thus contributed to the spread of the textile industry outside towns. Spinning, performed during the early Middle Ages with the so-called "rock" or distaff, improved considerably with the introduction of the spinning wheel in the 13th century. The spinning wheel was gradually improved and made more "mechanical" during the next several centuries. Looms similar to the ones used in antiquity remained throughout the Middle Ages, but also underwent improvements during the late Middle Ages and Renaissance.

Firearms were invented near the start of the period

and developed throughout, from the first cannons to the familiar flintlocks and muskets of the 18th century. The mechanical clock also appeared during this period and reached a high degree of accuracy by the 18th century. Although the suction pump might seem a minor achievement today, it was viewed with amazement when first used to remove air from a vessel. The vacuum led to great advances in science, although pumps were first more prosaically engaged in the vital task of pumping fluids from wells and tunnels. The first mechanical calculators were based on Napier's bones, a form of mechanical logarithm table, but Pascal and Leibniz developed mechanical calculators that were based on gear ratios and that were similar in operation to those to come in the 19th and early 20th centuries. These early calculators did not have the benefit of parts made to high tolerances by modern machine tools.

Several of the devices that would make mechanical calculators and machines with close tolerances of all kinds possible were invented during this period. Measurement was greatly improved, not only of lengths with the Vernier scale and the micrometer, but also of other qualities such as heat (various thermometers, including the pyrometer), humidity (the hygrometer), and air pressure (the barometer). Important auxiliary methods of transmitting motion or force included crank handles, drive belts, and foot treadles, all of which were to be of great importance in factories until the advent of small electric motors late in the 19th century. Better machine tools, especially vastly improved lathes, also began to appear during the 16th century.

During the Renaissance the design of machines developed into an art. Several treatises appeared during that period, the so-called "machine books" or *theatrum machinorum*. Although still deeply rooted in the technical tradition of antiquity and the Middle Ages, these books explored a wide variety of technical solutions, some of them realizable, some just dreams. Many of the proposed machines were what we would call gadgets, although some were of very large size. Automatons were very popular; Leonardo da Vinci, for example, designed and built a number of them.

Among the devices that were not strictly machines or tools, the most notable were various adaptations of the lenses that had become well known since their 14th century application to improving failing vision. As good lenses were developed, inventors put them together in new ways to make microscopes and telescopes.

Transportation. The roads built by the Romans remained in use throughout Europe during the Middle Ages, although they degraded continuously because very little was done to maintain them. Transport of goods was made difficult by the many tolls along the roadways, but the income from tolls was generally not used to repair the roads. The road situation improved during the second half of the 15th century when, because of an improving economy, traffic and trade between different countries increased. In cities some roads were paved as early as the 12th century, but throughout this period highways between cities were never as good as the old paved Roman roads had been.

During the early Middle Ages most goods were transported on pack animals and on small two-wheeled carts. Carts with four wheels and two axles were used for heavy loads, but they remained impractical until the invention of the mobile front axle sometime in the 14th century.

Transport on rivers and canals, especially of heavy loads, developed early, mainly because of the bad shape of roads. The construction of canals started all over Europe during the 12th century, and the first European locks to make rivers and canals navigable over longer distances appeared by the 14th century. From the 13th century on, the sea became an important road for traffic; the Hanseatic ports start developing around that time. Soon intense traffic developed between the Hanseatic ports from the north and the ports of Italy, passing through the Strait of Gibraltar.

Shipbuilding improved throughout these times. There is direct evidence from the beginning of the period that more modern methods had been introduced. Even with better methods, the ships produced at the beginning of the era were of the type used by the Vikings, with one mast and a square sail. The rudder, however, was introduced in the West in Viking ships; it was mounted on a vertical axis at the aft of the ship, as in Chinese ships. Later, near the beginning of the 15th century, lateen (triangular) sails were reintroduced, allowing ships to travel against the direction of the wind more easily. The basic way that boards were put together to make vessels was changed. Ships were equipped with superstructures and three masts.

Navigation became much more reliable with the introduction of the compass, at first a magnetized needle placed on a straw floating on water. Astronomical navigation was improved around 1480 with increased use of the astrolabe by sailors to measure the positions of stars.

More reliable maps included the Mercator projection, a projection from a spherical surface onto a flat surface in such a way that angles and shapes remain preserved. However, throughout the Middle Ages and the Renaissance, the impossibility of keeping accurate time limited navigation. This problem was not solved until the introduction during the 18th century of clocks that could run on time on moving ships.

	GENERAL	ARCHITECTURE & CONSTRUCTION	COMMUNICATION / ENERGY
1000 to 1009	Following reports from Bjarni Herjolfsson, the Viking Leif Erikson sails from Greenland to the Americas, where he spends the winter of 1001, probably at L'Anse aux Meadows in Newfoundland	The Château de Josselin is built in 1008 on a cliff above the Oust River in France	**Waterpower** Waterpower has a surprisingly long history. Developed in antiquity, the waterwheel became the main source of power during the Middle Ages and remained so until the 19th century, when it was replaced by the steam engine. Waterpower still plays an important role today in many countries where it is used for generating electricity, although the waterwheel has now been replaced by the much more efficient water turbine.
1010 to 1019	Danish King Canute takes over the thrones of both England and Denmark in 1017; *See also* 1060–69 GEN	The al-Hakim Mosque is completed in Cairo in 1013; like all traditional mosques, it is modeled on Muhammad's home in Mecca, although greatly enlarged; the al-Hakim Mosque is designed to have room for half the male population of Cairo in its large interior courtyard, which contains nothing but a fountain used for washing before prayer	The first application of waterpower was in the grist mill. Water mills appeared throughout the Roman Empire during the 3rd and 4th centuries, but the fall of the Roman Empire ended prospects for widespread use. Much of the technology of waterwheels from the Roman and Classic civilizations was taken over and perfected in the Islamic world, in which several types of irrigation systems were powered by waterwheels. In Europe, waterwheels came into widespread use only during the 10th century. The *Domesday book* (11th century) lists 5624 water mills in England, which corresponds to about one water mill per 400 inhabitants.
1020 to 1029	Chinese authorities nationalize banks in 1023 and replace them by a central bank that has the authority to issue money; *See also* 1100–09 GEN	The castle of Fougères in northern Brittany (in France) is built in 1024 of red masonry	
1030 to 1039		A Romanesque Cathedral of Notre Dame is built in Chartres, France, by Bishop Fulbert [d 1028] shortly before 1030; the nave is not vaulted but covered with a wooden roof that burns during 1030; two transcepts are added after the fire; *See also* 1130–39 ARC	
1040 to 1049		The Abbey at Jumièges in Normandy, started by Lefranc at this time, becomes the inspiration for the Abbey of Westminster in England; *See also* 1060–69 ARC	Sometime between 1041 and 1048 Bi Sheng, an obscure commoner in China, invents movable type; he bakes ideographs out of clay and stores the pottery characters in wooden cases according to a rhyme scheme since China has no alphabet; for printing, the ideographs are arranged in a frame, embedded in resin and wedged in with slivers of bamboo; *See also* 870 CE COM; 1440–49 COM

1000 to 1009

Waterpower (Continued)

The early waterwheels were mainly of the undershot type; water passed under the wheel, driving the lower paddles. Undershot wheels were easy to install and they appeared in many locations on rivers and streams. Many of these water mills were mounted on barges, making their operation independent of the water level of the river. This type of floating mill was first invented by the Roman general Belisarius in 537. He installed floating mills on the Tiber when the Goths, besieging Rome, had cut the city's water supplies. In cities it was preferable to install floating mills under bridges because of the faster flow of water between the arches. The number of water mills built under bridges in Paris was so great at one time that they seriously impeded boat traffic on the Seine.

During the late Middle Ages, waterwheels became the power source for a wider range of applications, including pumps, hammers, grinding wheels, saws, and lathes.

The overshot wheel also appeared in the late Middle Ages. Overshot wheels are more efficient because they use more energy in the flow of water. Their efficiency was further increased by replacing the paddles with buckets, whereby the weight of water collected in these buckets added to the driving force acting on the wheel. John Smeaton calculated in the 18th century that an undershot wheel uses 22 percent of the energy in the flow of water, while an overshot wheel uses 63 percent.

1010 to 1019

In China, wells are being drilled by a method using ropes so that a person does not have to travel to the bottom of a well while digging it; the drill bit is a heavy pointed weight attached by ropes to a rocking bar at the top of the well; the rocker is used to raise and drop the bit repeatedly; *See also* 1420–29 TOO

1020 to 1029

A vessel wrecked off Serçe Limani, Turkey (opposite Rhodes), about this time offers the earliest known evidence that seagoing wooden ships are being built in the modern way, starting with a keel and framework to which planking is added; *See also* 1300–09 TRA

1030 to 1039

A painting from 1035 shows a spinning wheel in use in China; *See also* 1280–89 TOO

1040 to 1049

In China Ceng Gong-liang publishes the first recipes for three varieties of gunpowder in 1044; *See also* 850 CE MAT; 1260–69 MAT

Tsêng Kung-liang's *Wu ching tsung yao* (Compendium of important military techniques) of 1044 describes magnetized iron "fish" that float in water and can be used for finding south; about this time the Chinese begin to use the compass for navigation, most likely using this device; *See also* 270 CE TOO; 1080–89 TRA

	GENERAL	ARCHITECTURE & CONSTRUCTION	COMMUNICATION	ENERGY
1050 to 1059		Arundel Castle is constructed in Sussex, England	Some Chinese books are printed with movable type; *See also* 1040–49 COM	
1060 to 1069	On Sep 28 1066, William of Normandy and about 7000 soldiers and associated workers in about 700 ships land at Pevensey, England, to begin the successful Norman conquest of England; *See also* 1010–19 GEN; 1080–89 GEN	English King Edward the Confessor has the Abbey of Westminster built; it is the first monumental work in stone in England and the first Norman-style Romanesque building there (Edward had been raised in Normandy); consecrated on Dec 28 1065 while Edward was ill and unable to attend, nothing of the church remains today; *See also* 1040–49 ARC; 1240–49 ARC The Abbey Church of Jumièges is dedicated in 1067; construction had taken nearly 30 years; *See also* 1040–49 ARC		
1070 to 1079	The success of the Seljuk Turks over Byzantine forces at the 1071 Battle of Malazgirt, Armenia (in Turkey), ensures Seljuk rule of all of Anatolia (Asian Turkey) in addition to their empire in Asia; *See also* 1390–99 GEN	Construction begins under William the Conqueror on Windsor Castle, today the largest inhabited castle in Europe; *See also* 1180–89 ARC Construction begins under William the Conqueror of the Tower of London's "white tower" c 1078; various other parts and walls are added at later dates		
1080 to 1089	The "Domesday" survey of England, ordered by William the Conqueror, is completed in 1087; it shows an English population of about 1.5 million subservient to about 10,000 Normans and their allies; *See also* 1060–69 GEN; 1210–19 GEN	The Church of St. Semin in Toulouse, France, is built The Benedictine Abbey Church at Cluny is started in 1088; it will be the largest church on Earth until St. Peter's in Rome is consecrated; *See also* 1620–29 ARC		The *Domesday Book* of 1086 lists 5624 water-wheel-driven mills in England south of the Trent River, or about one mill for each 400 persons

FOOD & AGRICULTURE	MATERIALS	MEDICAL TECHNOLOGY	TOOLS & DEVICES	TRANSPORTATION	
			Prince Bhoja [b 1018, d 1060] writes *Samarangana-sutradhara* in 1050; the work deals with the construction of automata		1050 to 1059
					1060 to 1069
Europeans reinvent the whippletree (whiffletree), a device that allows the use of two animals—horses in Europe; the whippletree had originally been developed for oxen by the Chinese in the 3rd century CE		Trotula of Salerno advocates cleanliness, a balanced diet, exercise, and avoidance of stress for maintaining health in *Practica brevis* and *De compositione medicamentorum*; her *De passionibus mulierum curandarum* (The diseases of women) deals with the medical needs of women	Abu Ishaq al-Zarqali constructs a series of complex water clocks at Toledo (in Spain), some of which show the phases of the Moon; *See also* 130 CE TOO; 1080–89 TOO		1070 to 1079
		In China, Su Sung builds a giant water clock, orrery, and mechanical armillary sphere between 1088 and 1092; 10 m (35 ft) tall, it is considered the finest mechanical achievement of its time; a constant flow of water turns escapement wheels that are geared to power the clock and orrery (a mechanical representation of the heavens, somewhat like a planetarium); the armillary sphere is available for use in checking the accuracy of both the clock and orrery by observing the Sun and planets; *See also* 720 CE TOO; 1360–69 TOO	Chinese scientist Shen Kua's *Dream pool essays* of 1086 contains the first known reference to the use of a magnetic compass for navigation; *See also* 1040–49 TRA		1080 to 1089

	GENERAL	ARCHITECTURE & CONSTRUCTION	COMMUNICATION	ENERGY
1090 to 1099	On Nov 27 Pope Urban addresses a council of bishops and abbots of France at Clermont in 1095, starting the First Crusade; Christian forces gather and attempt to drive the Muslims from Jerusalem and the Holy Land; *See also* 1140–49 GEN	The last of the *aulae regis*, or great halls of medieval England, is Westminster Hall, the largest in Europe; it is 73 m (240 ft) long and 20.5 m (67.5 ft) wide; the hall still stands today, despite its shoddy construction by masons just learning their craft		
1100 to 1109	China introduces bank notes printed in six colors in 1107; *See also* 1020–29 GEN; 1120–29 GEN	The Anasazi of Chaco Canyon construct the building we know today as Pueblo Bonito; it has 800 rooms that could be home to as many as 1000 people; as an apartment house, it is not to be surpassed in size until a large apartment complex is built in New York City in the 19th century; *See also* 1882 ARC The earliest pointed arch, or ogive, on a cathedral is used in the Romanesque cathedral of Durham, built in 1104; *See also* 1120–29 ARC	Woodcuts are used in Europe for block printing capital letters; *See also* 1290–99 COM The Chinese invent multicolor printing in 1107, mainly to make paper money harder to counterfeit; *See also* 1450–59 COM	The papal bull of 1105 mentions windmills in Europe for the first time; it grants a concession to the abbot of Savigny to build windmills in the French dioceses of Bayeux, Coutances, and Evreux; *See also* 600 CE GEN; 1180–89 ENE
1110 to 1119		King Suryavarman II of Khmer (Cambodia) comes to the throne; his regime sees the beginning of construction of the great temple complex of Angkor Wat, which eventually requires an estimated 350 million m³ (455 million cu yd) of building material transported to the forest site by rafts; the completed temple covers over 120 hectares (300 acres) and is part of a much larger temple complex		
1120 to 1129	Seventy million bank notes are issued in China in 1126 without being guaranteed by gold; the consequence is disastrous inflation; *See also* 1100–09 GEN The first artesian well in Europe is drilled in 1126; in China, artesian wells are known before the Christian era	The first cross-ribbed vault still in existence is built in 1122 over the ambulatory at Morienval Cathedral; *See also* 1100–09 ARC Abbé Suger [b Flanders or St. Denis or Toury, France, c 1081, d St. Denis, Jan 31 1151] starts construction in 1129 on the first portion of the Abbey Church of St. Denis, which will become the first Gothic church with flying buttresses		

Pope Urban II finds a vine at Marmoutiers, founded by St. Martin, supposedly planted some six centuries before by the saint himself

1090 to 1099

Forks for dining reach Italy from the Middle East, making the first inroads for the implement in Europe; *See also* 600 CE FOO; 1370–79 FOO

Italians learn to distill wine to make brandy; *See also* 1270–79 FOO

The art of paper manufacture spreads from Cairo to Morocco; *See also* 900 CE MAT; 1150–59 MAT

During the Song dynasty in China, true porcelain pottery is made, along with improved glazed stoneware called celadon; *See also* 1370–79 MAT

The gravity-powered catapult is invented; the principle of the lever is used to give great speed to a missile attached to the long arm of the catapult when a heavy weight brings down the short arm; the new catapult is more reliable than the torsion-powered catapult, in use for about 1400 years, since it is not affected by humidity

1100 to 1109

Al-Jazari

One can compare the use of water in the technology of the Middle Ages to the use of electricity in present-day technology. Just as electricity is now used for transporting energy and for controlling machines, so water served in the Middle Ages for powering machines and for controlling them. One of the engineers from that time who perfected the use of water for running and controlling machines was al-Jazari.

Ibn al-Razzas al-Jazari was born around the middle of the 12th century in the north of Mesopotamia, between the Tigris and the Euphrates rivers. He grew up in the scientific tradition of Islam, which encouraged the study of science. His *The science of ingenious mechanisms*, completed in 1206, describes several types of machines for lifting and pumping water and several types of water clocks. Several designs were invented by Al-Jazari himself. One of his best-known designs is the *saqiya*. It is a mechanism for lifting water that consists of a chain with pots attached to it, driven by a waterwheel.

Al-Jazari is best known for his devices powered by water draining through a small hole. The flow of water changes with the amount remaining in the vessel. In one design al-Jazari kept this flow constant by a conical valve mounted on a float in a smaller vessel. If this smaller vessel filled up too fast, the valve closed, stopping the flow. The speed of the machinery could be adjusted by rotating a drum with a small hole in it through which water escaped from the small vessel. Rotating the drum such that the hole would be in a lower position increased the flow of water and the speed of operation.

The first European stanch, a predecessor of the lock, is thought to have been installed in 1116 on the river Scarpe in Flanders; a stanch is a gate placed in a weir (a dam used for raising water level) that can allow boats to pass; *See also* 1190–99 TRA

Chu Yu's *P'ingchow table talk* of 1117 contains the first mention of a compass used for navigation at sea; *See also* 1080–89 TRA; 1190–99 TRA

1110 to 1119

1120 to 1129

1130 to 1139

Trading on La Bourse in Paris is taking place as early as 1138, although it does not yet involve stock in corporations; *See also* 1460–69 GEN

Cluniac style, featuring wavy curves and ornamentations and exemplified by the cathedrals of Reims and Winchester, is replaced in Europe by Cisterian style, favoring straight lines and right angles

The new choir of the Canterbury Cathedral in England is dedicated in 1130

Archbishop Henry, the Boar, [d 1142] begins in 1130 the Cathedral of Sens, the first Gothic cathedral; it is believed that the choir is complete by 1140; *See also* 1120–29 ARC; 1140–49 ARC

A fire in the Romanesque cathedral built by Bishop Fulbert at Chartres is the immediate stimulus for building new towers at Chartres Cathedral (officially the Cathedral of Notre Dame at Chartres, France); the northern tower is started in 1134 and the Royal Portal and southern tower are begun a few years later; *See also* 1190–99 ARC

The harbor lighthouse at Genoa, Italy, is built in 1139 or in 1161; Christopher Columbus's Uncle Antonio Colombo becomes the lighthouse keeper in 1449, but the light does not survive into the present

1140 to 1149

The Second Crusade begins in 1147 when Muslim forces recapture Edessa in Syria (Urfa, Turkey); *See also* 1090–99 GEN; 1190–99 GEN

The cathedral of San Ruffino is built (in Italy)

Abbé Suger embarks on the building of the first Gothic church, the choir portion of the Abbey Church of St. Denis, completed and consecrated in 1144; this new Gothic style may have been influenced by Neoplatonic thought, as at Chartres, where architecture is viewed as applied geometry; *See also* 1130–39 ARC

The best surviving example of early Cisterian architecture, the Abbey of Fontenay, begun in 1130, is completed in 1147; it emphasizes the "perfect ratio" of one to two espoused earlier by St. Augustine

Scholars in Toledo begin translating the writings of Arabic scientist Avicenna [b Bukhara, 980, d Hamadan, Iran, Jun 1037] into Latin; his works cover almost every aspect of health and medicine; *See also* 1470–79 MED

1130 to 1139

Why is there no Classic steam engine?

The Classic civilization forms a continuous society with different state organizations stretching from the time of Homer and Hesiod in Greece (c 800 BC) through the fall of the Western Roman Empire, often dated as 476 CE, a period considerably longer than a millennium. Although Classical science was of the highest order found in ancient times—the Ionians of the 7th century BC virtually invented science—there are few technological advances attributed to it. The beam wine/olive press of the Greeks is perhaps the most characteristic mechanical invention, although there were a few others. The Romans mainly exploited known engineering principles more completely than in the past, as in Roman roads and aqueducts. By far the most advanced technology invented under Roman rule was the waterwheel, but it was not employed systematically until after the fall of the Western Empire. During this period virtually all significant advances in technology were made in China and apparently not communicated to the Romans, despite steady trade with the Chinese in various goods.

The later Classical inventors had sufficient means to produce machinery as advanced in concept as that of the early Industrial Revolution. Hero of Alexandria, who lived about the 1st century CE, described devices that included all the elements needed to build a functional and useful steam engine, either turbine or piston driven. But he used these devices as either toys or installed in temples to make it seem that small miracles were taking place. He did not employ them as practical pumps (the first use for piston steam engines in the 18th century) or to power mills or move vehicles. In the 1670s, before truly practical steam engines, Ferdinand Verbiest made a version of Hero's steam turbine; he promptly installed it in a cart and drove the cart a few inches.

Thus, we can boil down the problem of no Classical machines to the specific lack of steam power. Common reasons put forward include the following:

Slavery Greek and Roman upper classes had an ample supply of barbarian slaves, so they not only did not need steam power, but they equated anything related to work as definitely lower class and to be avoided.

Poor natural resources The Mediterranean region does not contain large deposits of either coal or iron ore. Indeed, when Marco Polo reported that people in China burned black rocks as fuel, Italians thought he was making things up.

Lack of scientific background Science and technology evolve, just as organisms do. You could no more expect a steam engine before the invention of the thermometer, the microscope, and so forth than you could expect flowering plants before algae, mosses, and ferns.

Lack of technical background Gear-cutting and screw-cutting machines were not devised until the 15th and 16th centuries, when they were used by clock makers. Without metal lathes and similar devices, even the best designed machine would have been impossible to build well enough to function.

Greeks and Romans had other things on their mind There was a great religious movement for hundreds of years in Hellenic and Roman times, the most likely times for progress on steam engines. For the most part, people did not care about the material world, only the spiritual one. Evidence for this is that the cleverest devices of the period were devoted to causing temple doors to open mysteriously or to produce other apparently miraculous effects.

1140 to 1149

1150 to 1159

The light tower at Meloria, Italy, is often considered the first lighthouse in Europe since Roman times, although the lighthouse at Genoa may have preceded it; lighthouses are being erected all along the Italian coast in the 12th century, including towers at Venice, Tino, and near the Straits of Messina; *See also* 1130–39 ARC; 1300–09 ARC

1160 to 1169

The earliest surviving engineering plan for a building's water supply system is a drawing of the plans for Christ Church in Canterbury, built about 1160; the plan shows water lines from outside in red and a separate system for handling rainwater in yellow, with tanks used to settle the water and hold it for eventual use by varied parts of the church, including the latrines

The cornerstone of the Cathedral of Notre-Dame in Paris is laid in 1163

1170 to 1179

The original form of the stone London Bridge, known today as Old London Bridge, is started by Peter Colechurch in 1176; after it is completed in 1209, houses are built on it, ruining what little structural integrity the bridge has (there are 19 arches of different sizes and a drawbridge); the houses, many three or four stories high, cause the bridge to continually lose pieces, as in the nursery song; *See also* 1832 ARC

According to tradition Saint Bénezét starts construction of the Pont de Avignon (Avignon bridge) in 1178, probably copied from the nearby Roman Pont du Gard aqueduct bridge

The art of paper manufacture spreads in 1150 to Spain from Morocco; *See also* 1100–09 MAT; 1270–79 MAT

Potters in Persia (Iran), with the aid of potters from Egypt, develop a new low-fired pottery resembling soft-paste porcelain, but based on powdered silica sand, frit, and white clay

In Damascus Sharaf al-Din al-Tusi invents the linear astrolabe, a simple device based on a marked rod, a plumb line, and a cord; although not equal in precision to a regular astrolabe, it is simple to construct and use; *See also* 200 BC TOO; 1390–99 TOO

The Chinese develop the first rockets; *See also* 1380–89 TRA

1150 to 1159

1160 to 1169

Wind power

Sailboats are probably the oldest form of wind power, and it is likely that the sail inspired the first builders of windmills. Wind power that provides rotary motion is a more recent invention than waterpower. Some historians believe that the first windmills appeared in the Greek islands at the beginning of the Christian era, but firm proof of this has yet to be found. During the 7th century windmills probably appeared in areas with no flowing water, such as the western part of Afghanistan. Early windmills consisted of a rotor turning on a vertical axis to which were attached curved sails made of fabric. The wind was admitted through funnel-shaped openings facing the direction of the wind. Because no gears were required for driving the millstones, losses from friction were minimal. By the 10th century such mills were used throughout the Arab world.

Windmills were also in use during that time in Iran, and in China, where they were employed for irrigation and land drainage.

In Europe windmills appeared about 1150. These windmills had horizontal axes or windshafts that had to be placed in the direction of the wind. The first type used was the postmill: The entire mill, including the millstones, was mounted on a fixed foundation, or "post," on which it could rotate. A second type of mill, the tower mill, appeared at the end of the 14th century. Only the cap, housing the windshaft, of a tower mill rotates. In some later types of tower mills, an auxiliary windshaft with blades placed parallel to the main windshaft automatically rotated the top cap into the required position.

Windmills in Europe were first used for milling grain, but soon their use widened to included many other activities, such as powering sawmills and mills for crushing ores. In Holland, one very important application was pumping water for the reclaiming of land. The first windmill in Holland started pumping in 1414. Windmills drove both Archimedian screws and waterwheels for pumping water. Those equipped with waterwheels could pump water up only a few feet, and therefore several windmills in succession were required to pump water to greater heights. At one time, 7500 windmills were in use in Holland for water pumping and industrial applications. Their success was so great that even in the 18th century, John Smeaton considered increasing the use of windmills in England. In 1759 Smeaton investigated the performance of windmills by building models, but eventually concluded that they could not compete with the more powerful steam engines then in use.

1170 to 1179

1180 to 1189

ARCHITECTURE & CONSTRUCTION

The ancient castle at Dover, which may date to Roman times with various Saxon expansions, is rebuilt with massive stone walls and turrets; *See also* 1070–79 ARC; 1280–89 ARC

The Marco Polo Bridge across the Yongding River in southern China is built in 1189; 213 m (700 ft) long, the bridge is still heavily used by bus and truck traffic

ENERGY

Windmills with horizontal axes begin to be built in the North Sea region; the windmills referred to in the papal bull of 1105 were probably of this type, since there is no evidence that windmills of the Persian type, with vertical axes, ever reach Europe; *See also* 1100–09 ENE; 1340–49 ENE

1190 to 1199

GENERAL

Recognized as the real power in Japan, Yoritomo [b Japan, 1147, d Japan, 1199] is accorded the title of shogun (short for "barbarian-subduing generalissimo") in 1192, beginning the period of Japanese history called the Shogunate; although the Emperor still rules nominally, the shoguns are actually in control

The disastrous Fourth Crusade begins in 1198; Christian forces are diverted from Jerusalem and instead sack the Christian city of Constantinople, completing the division of Christendom; *See also* 1140–49 GEN

ARCHITECTURE & CONSTRUCTION

On the night of Jun 10-11 1194, a great fire destroys most of the town of Chartres and all of the cathedral except for the west facade; the people of the village immediately decide to rebuild the cathedral and work is started on a new design that incorporates many mathematical relationships based on the Golden Ratio and other mystical ideas; *See also* 1130–39 ARC; 1220–29 ARC

Richard the Lion-Hearted [b Oxford, England, Sep 8 1157, d Chalus, France, Apr 6 1199] builds the Château Gaillard castle in 1197

1200 to 1209

GENERAL

Ibn al-Razzas al-Jazari completes in 1206 his *Treatise on the theory and practice of the mechanical arts*, giving descriptions of automatons and clepsydras, several of which he has developed himself; *See also* 1290–99 GEN

The Mongol chieftain Temujin [b c 1167] is renamed Genghis Khan, meaning universal ruler of all Mongol peoples in 1206; he soon leads his people on a wave of conquest that eventually includes most of Asia and a large part of Europe; *See also* 1260–69 GEN

ARCHITECTURE & CONSTRUCTION

At Mesa Verde in what is now southern CO, the Anasazi build pueblos in the face of large cliffs—probably for defensive purposes; these survive today, protected by the overhang of the cliffs, but they are not as well built as earlier pueblos, such as those at Chaco Canyon NM, which date from 100 to 200 years earlier; *See also* 1100–09 ARC

British architects develop the hammer-beam truss that supports and strengthens a timber roof without blocking the windows

ENERGY

Coal mines are operating in pits in the Firth of Forth, near Edinburgh, and in Northumberland in Great Britain; *See also* 1300–09 ENE

Houses with chimneys become common, although chimneys had been in use earlier for bakers' ovens and for smelting

	A paper mill is established in Hérault, France, in 1189; *See also* 1150–59 MAT; 1270–79 MAT		Ibn al-Razzas al-Jazari publishes in 1180 a treatise on automatons, *The science of ingenious mechanisms*; it contains a description of a clock that uses the loss of weight in a burning candle to keep track of time	The oldest Western evidence for the use of a rudder is in carvings made about this time; *See also* 1190–99 TRA King Philippe Auguste of France orders the roads of Paris paved for the first time in 1184	**1180 to 1189**

The oldest known depiction of a fishing reel is made in China; *See also* 1410–19 FOO			Jordanus Nemorarius describes in 1190 in *Liber de ratione ponderis* several simple machines, such as inclined planes and levers, and gives the first mathematical proof for the law of the inclined plane	The sternpost rudder, possibly borrowed from the Chinese, replaces the steering oars that have been used in Europe and the near East since antiquity; *See also* 1180–89 TRA *De naturis rerum* (On natural things) by Alexander Neckam [b St. Albans, England, Sep 8 1157, d Worcestershire, 1217] contains the first known Western reference to the magnetic compass; *See also* 1080–89 TRA The first documented European stanch, considered a predecessor of the lock, is installed over a 10-year period on the river Micio in Italy; *See also* 1110–19 TRA; 1370–79 TRA	**1190 to 1199**

Buckwheat is grown in Europe	The Catalan forge for smelting iron emerges in Catalonia (Spain); using a column of falling water to push air into a mixture of iron ore and burning charcoal, the forge is now considered a predecessor to the blast furnace; the resulting iron is used as the basis for wrought iron; *See also* 1340–49 MAT			The first major canal to be built in the West since Roman times is the Ticinello of 1209, which carries water 26 km (16 mi) from the Ticino River to Milan, where the water is used for irrigation and to fill a moat; later, as the Milanese realize its value in navigation, the canal is enlarged and renamed the Naviglio Grande (Grand Canal); its waters are also used for fish weirs, to the detriment of navigability; *See also* 1380–89 TRA	**1200 to 1209**

	GENERAL	ARCHITECTURE & CONSTRUCTION	COMMUNICATION	ENERGY
1210 to 1219	English barons in Jun 1215 force King John to sign the oath known as the Magna Carta, the basis of English law to this day; in addition to asserting the supremacy of law over the monarchy, the great charter also attempts to standardize weights and measures			
1220 to 1229		The vault of the present cathedral at Chartres is completed; *See also* 1190–99 ARC; 1240–49 ARC Work starts on Beauvais Cathedral c 1225, planned to be one of the largest Gothic cathedrals, with a vault 45 m (148 ft) tall, the highest ever built; *See also* 1280–89 ARC		
1230 to 1239			During a Mongol siege in 1232, the Chinese use kites to send messages behind enemy lines	
1240 to 1249		The cathedral at Chartres is completed in essentially the same form as we know it today, although there have been minor changes and one fire (in 1836) since this date; *See also* 1220–29 ARC King Henry III [b 1207, d 1272] starts building Westminster Abbey in London in 1245, officially the Abbey of St. Peter, as an extension and replacement for an earlier structure that may go back as far as 785; *See also* 1060–69 ARC		
1250 to 1259	Buttons with matching buttonholes come into use, replacing pins and laces for closing up shirt fronts and other clothing; buttons themselves had been known since Roman times, but they were closed with loops, not buttonholes Hulagu Khan conquers Mesopotamia in 1258 and orders the irrigation system destroyed to aid in subjugating the population	The Teutonic Knights build the *Schloss* castle of Königsberg (now Kaliningrad, Lithuania) in 1257	Communication between the Mamluk Empire's twin capitals of Cairo and Damascus is carried in part by carrier pigeons flying in relays and taking about one day; ordinary mail by a mounted postal service, also using relays, covers the 640-km (400-mi) distance between the capitals in only four days	

FOOD & AGRICULTURE	MATERIALS	MEDICAL TECHNOLOGY	TOOLS & DEVICES	TRANSPORTATION	
					1210 to 1209
	The *Dictionary* of Jean de Garlande mentions techniques for producing silk thread; *See also* CE 500 MAT		In 1221 the Chinese use bombs that produce shrapnel and cause considerable damage; previous uses of gunpowder relied mostly on the loud explosions *Le livre des métiers* by Jean de Garlande of 1221 lists several textile machines	A window at Chartres Cathedral shows the earliest known Western wheelbarrow	**1220 to 1229**
			A 33-page manuscript by Villard de Honnecourt in 1235 contains descriptions of a variety of machines, including a hydraulic sawing machine with automatic advance of the work piece		**1230 to 1239**
		A 1240 decree of the Holy Roman Empire permits the dissection of human cadavers; *See also* 1310–19 MED	Some type of firearm may have been used at the siege of Seville in 1247, but the evidence for this is scanty		**1240 to 1249**
Shortly before this time, Holy Roman Emperor Frederick II writes *De arte venandi cum avibus* (The art of hunting with birds), summarizing his lifetime of hawking with peregrines, gyrfalcons, and saker falcons; in it he claims to have introduced into Europe the practice of hooding hawks, which he learned from the Arabs				According to some, a canal with a simple lock is built in 1253 in Sparendam, the Netherlands; if so, it would be the first lock in Europe; *See also* 980 COM; 1370–79 TRA	**1250 to 1259**

1260 to 1269

The Mongol advance to the west, halted in Europe in 1241 by internal forces (after Mongols captured what are now Russia, Ukraine, Poland, and Hungary), is finally stopped in 1260 by the armies of the Mamluks in Egypt; the Mamluks not only defeat the Mongols in a decisive battle on Sep 3 at Ain Jalut (in Israel), but also advance to establish rule over Syria; *See also* 1200–09 GEN;1270–79 GEN

The Cathedral of Pisa is built to commemorate a naval victory in 1263 over the Saracens at Palmero on Sicily

1270 to 1279

Mongol ruler Kublai Khan [b Mongolia, 1215, d 1294] has himself declared emperor of China in 1271, the start of the Yuan dynasty; *See also* 1260–69 GEN

Marco Polo [b Venice (in Italy), c 1254, d Venice, Jan 9 1324] is brought in 1275 to the capital of Kublai Khan by his father and uncle, Niccolo and Matteo, who had previously visited in 1266

A great storm in 1277 breaks through the sand dunes of what is now the Netherlands; the sea pours in, drowning thousands of people and enlarging the Flevo Lacus into the Zuider Zee

1280 to 1289

The roof of Beauvais Cathedral collapses in 1284, demonstrating that the trial-and-error method of Gothic construction sometimes fails; the roof of the cathedral collapses twice in all and the tower, built in the 14th century, collapses once

Edward I [b 1239, d 1307] builds Conway Castle in Wales in 1284 as part of his campaign to subdue the Welsh; Caernarvon Castle in Wales is constructed in 1285; *See also* 1180–89 ARC; 1290–99 ARC

				1260 to 1269
	Etienne Boileau's *Book of trades* describes several techniques used in textile manufacture In *Opus majus*, written in 1267 and 1268 by Roger Bacon [b Ilchester, England, c 1220, d Oxford, Jun 11 1292], but not published until 1733, gunpowder is mentioned for the first time by a European; *See also* 1040–49 MAT; 1280–89 MAT	Roger Bacon discusses spectacles for the farsighted in his 1268 work, *Opus majus*; *See also* 1280–89 MED		

				1270 to 1279
Soon after his return from the Crusades in 1274, Edward I establishes a garden at the Palace of Westminster; the influence of Islamic gardens is felt not only from Edward's experiences in the Holy Land, but also his first queen, Eleanor of Castile (in Spain), has been influenced by Moorish traditions; *See also* 1570–79 FOO The first official whiskey distillery in Ireland begins operation in 1276; *See also* 1100–09 FOO; 1520–29 FOO	A paper mill operates in Montefano (in Italy) in 1277; *See also* 1180–89 MAT; 1390–99 MAT		Borghesano of Bologna in 1272 invents a machine for throwing silk (forming thread from silk); it remains a trade secret until 1538, when its construction becomes known in Florence	

				1280 to 1289
	The Syrian al-Hasan ar-Rammah writes *The book of fighting on horseback and with war engines* in 1280, in which he mentions the importance of saltpeter as an ingredient in compounds that produce fire and rockets, suggesting knowledge of gunpowder; *See also* 1260–69 MAT; 1290–99 MAT	Mamluk Sultan Qalawin builds Mansuri Maristan in Cairo, Egypt in 1284; it is the most sophisticated medical center of its time, with separate wards for fevers, eye diseases, surgery, dysentery, and mental illness Allesandro della Spina is said to make use in 1286 of the invention of a friend, Salvino degli Armati: eyeglasses made with convex lenses to correct nearsightedness; *See also* 1260–69 MED; 1300–09 MED	The earliest known Western reference to a spinning wheel is made in the 1280 statutes of a guild in Speyer, Germany; *See also* 1030–39 TOO The first mechanical clocks appear in Europe, apparently in response to stories about the existence of mechanical clocks in China; the European clocks are driven by a weight whose descent is controlled by an escapement; *See also* 720 CE TOO	

1280 to 1289 cont

1290 to 1299

Raymond Lulle [b Majorca c 1235, d 1315] in 1296 discusses different technologies in his book *Arbor scientiae*; he covers metallurgy, building, clothing, agriculture, trade, navigation, and the military arts, and insists on the theoretical knowledge required by the practioners; *See also* 1200–09 GEN; 1420–29 GEN

The Castle of Beaumaris, built on the shores of Menai Straits in 1295, is the last of the Welsh castles to be started; it is completed in 1320; *See also* 1280–89 ARC

Block printing is used in Europe for printing complete pages with woodcuts; *See also* 1460–69 COM

1300 to 1309

A lighthouse 49-m (161-ft) high is built in 1304 in the harbor at Leghorn (Livorno, Italy); *See also* 1150–59 ARC; 1610–19 ARC

The practice of burning coal increases in England to the point where the lime burners of London are forbidden in 1307 to use coal because of the smoke, which is much more noxious than wood smoke; *See also* 1200–09 ENE; 1600–09 ENE

1310 to 1319

Mansu Musa ascends the throne of the Mali Empire in 1312, the most powerful state in western Africa, including most of Nigeria and the coastal nations from Benin through Senegal

Scotland's largest cathedral, St. Andrew's, is consecrated in 1318; construction began in 1160

Wang Chen develops new printing techniques and has over 60,000 Chinese characters made from hardwood at his disposal; using these, he prints his *Treatise of agriculture* in 1313; *See also* 1040–49 COM; 1440–49 COM

FOOD & AGRICULTURE	MATERIALS	MEDICAL TECHNOLOGY	TOOLS & DEVICES	TRANSPORTATION	
			According to some sources, one of the first mechanical clocks in Europe is installed at Westminster Abbey in 1288 to mark the hours; *See also* 1290–99 TOO The first known gun is made in China in 1288; a small cannon, it probably has predecessors that go back 10 years or more; *See also* 1240–49 TOO; 1310–19 TOO		**1280 to 1289 cont**
	A Japanese woodcut from 1292 shows a bomb exploding, indicating that gunpowder has reached Japan; *See also* 1280–89 MAT; 1320–29 MAT		A clock is reportedly installed at Canterbury Cathedral in 1292; *See also* 1280–89 TOO; 1330–39 TOO		**1290 to 1299**
Methods of milling rice are developed in Lombardy (in Italy); *See also* 1580–89 FOO Sometime during this century the tulip is domesticated (in Turkey); *See also* 1550–59 FOO		Eyeglasses become common, being produced in Venice, the glassmaking center of Italy; but lenses often are made from inferior glass that refracts light unevenly; *See also* 1280–89 MED; 1450–59 MED	The hourglass, which measures short intervals of time by the passage of sand from one glass vessel to another, is invented Wire drawing is accomplished by a worker seated on a water-powered swing; at the peak of each forward swing the worker grabs the wire with tongs; during the backward swing the wire is drawn through a small hole in a plate, thinning it	Carvel construction, in which boards are placed edge to edge, begins to be used in shipbuilding alongside the older clinker-planking technique, in which boards are overlapped; carvel construction can be used to make lighter and larger ships, although clinker-planking produces sturdier ships; *See also* 1020–29 TRA; 1400–09 TRA	**1300 to 1309**
		The first recorded case of body snatching (grave robbing) for medical dissection is prosecuted in 1319; *See also* 1240–49 MED	The first description of a firearm in Western writing is a 1313 account of an "iron pot" or "vase," about which little else is known; the manufacturer may have been Bernard Schwarz, a German monk; it is thought that these early protocannons were mostly made from wooden or iron staves bound together with hoops, as in a barrel; *See also* 1280–89 TOO; 1340–49 TOO		**1310 to 1319**

1320 to 1329

Guy de Vigevano writes a treatise on war machines for King Philip V of Valois that describes a number of devices consisting of prefabricated parts that can be quickly assembled, a new concept at the time; one of the devices described is a ship powered by propellers turned by cranks; *See also* 1775 TRA

1330 to 1339

On Oct 19 1337 Edward III of England [b Windsor, Nov 13 1312, d Richmond, Jun 21 1377] claims the French throne, held by Philip of Valois, touching off what comes to be known as the Hundred Years' War; although the main fighting lasts only 59 years, English sovereigns continue to claim the French throne until 1802; *See also* 1390–99 GEN

Francis Bacon and the scientific method

Francis Bacon is one of the important thinkers of the scientific revolution of the 17th century. Although Bacon was neither a mathematician nor an experimental scientist, he exerted great influence on his contemporaries by introducing a method in which science is based on observation and experimentation.

He argued that the sciences should follow the example of the "mechanical arts" by being "founded on nature." But he also asserted that the mechanical arts should have science as their master. Bacon is one of the first thinkers to bridge the gap that existed between technology and the sciences. Since antiquity, technical invention was unrelated to science; it was the domain of the artisan. Artisans built machines based on empirical experience only. Science was thought to deal only with ideas. Bacon strongly opposed this concept and asserted that the knowledge of practical things also belongs to science. Because Bacon was not a scientist himself (although he died from a cold caught while stuffing a chicken with snow in an experiment with refrigeration), he extended his ideas on science to every aspect of human life. In Bacon's view, science not only serves to fulfill our intellectual curiosity, but also is useful in all phases of humanity.

Bacon was one of the first to fully understand that knowledge is power. He believed science would serve to improve the human condition and create a better world, and stated that the final goal of science is "the relief of man's estate" and the "effecting of all things possible."

Bacon described a model of such a better world in his book *New Atlantis*. New Atlantis is an island with a utopian society whose well-being is entirely based on science and technology. In this book Bacon gave his contemporaries an extraordinary look into the future of technology. He mentions the skyscraper ("High Towers, the Highest about half a Mile in height"), the refrigeration of food, air-conditioning ("Chambers of Health, wher wee qualifie the Aire"), telephones ("meanes to convey Sounds in Trunks and Pipes"), airplanes, and submarines.

1340 to 1349

Taddeo Gaddi [b Florence (in Italy), c 1300, d Florence 1366] designs and builds in 1345 the bridge we now call the Ponte Vecchio (old bridge) in Florence; the Ponte Vecchio is a segmental arch bridge, an improvement over the pure semicircular arch bridges previously built; *See also* 1350–59 ARC

Windmills are in use in Holland (the Netherlands) to carry water from inland areas out to sea to reclaim land; *See also* 1180–89 ENE; 1450–59 ENE

FOOD & AGRICULTURE	MATERIALS	MEDICAL TECHNOLOGY	TOOLS & DEVICES	TRANSPORTATION	
	A treatise with this date mentions gunpowder in the West, although Roger Bacon had written of it earlier; there is also some written evidence of firearms before this; *See also* 1290–99 MAT; 1620–29 MAT	Henri de Mondeville [b Normandy, France, 1260, d 1320] in his 1320 work *Chirurgia* (Surgery) advocates sutures and cleansing of wounds; *See also* 1360–69 MED	A 1323 manuscript from Briey in France describes bellows driven by a waterwheel A drawing of a small cannon that fires arrows, which appears in Walter de Milemete's *De officiis regum* of 1326, offers the first pictured evidence of a gun in Europe; *See also* 1310–19 TOO; 1340–49 TOO	China's Grand Canal, 1770 km (1100 mi) long, is completed in 1327; built over many centuries, starting in 70 CE, it connects Beijing to many parts of northern China and to the Yangtze; *See also* 610 CE COM	1320 to 1329
Zen priest Muso-Soseki introduces in 1339 the use of smooth, flat-topped stones in his garden in Kyoto, Japan			Richard of Wallingford [b Wallingford, England, c 1292, d St. Albans, May 23 1336] makes an elaborate clock in 1330 or in 1326 for St. Albans' Abbey in England; it shows many astronomical configurations and is among the early weight-driven mechanical clocks for which evidence still exists; *See also* 1290–99 TOO The compound crank handle is invented in 1335; *See also* 830 CE TOO; 1430–39 TOO A mechanical clock built by Gugliemo Zelandino is mounted in a tower of the San Gottardo Chapel in Milan (in Italy) in 1335; it is the first public clock that strikes the hour; *See also* 1350–59 TOO		1330 to 1339
The first blast furnace is developed in or around Liège, Belgium; *See also* 1200–09 MAT	Plague spread by fleas on rats, endemic in parts of central Asia, breaks out as trade between Asia and Europe increases; it spreads to Astrakhan (in Russia) on the Volga and Caffa (Feodosiya, Ukraine) in the Crimea; it is the beginning of the epidemic known as the Black Death	The English use ten cannons at the siege of Calais, which takes place in 1346-47; although there is evidence of earlier use of cannons, it is not as clear-cut; *See also* 1320–29 TOO; 1350–59 TOO Giovanni di Dondi starts building his famous astronomical clock in 1348, completing it in 1364	A German miniature representing the Flight from Egypt shows a cart with a suspension for absorbing shocks; such suspensions are believed to have been introduced about this time; *See also* 1450–59 TRA	1340 to 1349

1350 to 1359

In Segovia, Castile (Spain), the Alcazar is built in 1352 (*Alcazar* is Moorish for "the castle")

The segmental arch bridge Ponte Castelvecchio is built in 1356 over the river Adige at Verona, Italy; it has a major span of 49 m (160 ft); *See also* 1340–49 ARC; 1560–69 ARC

1360 to 1369

Nikolaus Faber builds for the Cathedral of St. Stephen in Halberstadt (in Germany), the first pipe organ to have a complete scale including semitones; it uses three keyboards as well as foot pedals; *See also* 120 CE GEN

Peasant Zhu Yuanzhang leads a revolt in Jan 1368 that results in his being named emperor of China, starting the Ming dynasty; *See also* 1270–79 GEN

Gloucester Cathedral initiates the period of the Perpendicular style of Gothic architecture in England, which becomes very popular and persists until the English Renaissance of the 16th and 17th centuries

1370 to 1379

Gabriel da Lavinda writes a manual on cryptography in 1379; *See also* 1620–29 COM

FOOD & AGRICULTURE	MATERIALS	MEDICAL TECHNOLOGY	TOOLS & DEVICES	TRANSPORTATION	
	The University of Paris, interested in an alternative to parchment in 1355, licenses the construction of paper mills in Troyes and Essonnes; *See also* 1270–79 MAT; 1390–99 MAT	The Black Death reaches northern Europe in 1352, including Scandinavia and northern Russia; *See also* 1340–49 MED	The first bronze cannons are introduced, although they are still not cast as a single piece, but are fashioned by joining several parts; *See also* 1370–79 TOO The mechanical clock in Strasbourg Cathedral, France, is built in 1354; on top of the tower is a mechanical cock that flaps its wings and crows at twelve o'clock (the cock remained active until 1789 and still exists in the Strasbourg Museum); around this time clocks are also installed in Padua, Genoa, Bologna, and Ferrara (in Italy); *See also* 1330–39 TOO; 1380–89 TOO		**1350 to 1359**
		Chirurgia magna by Guy de Chauliac [b c 1300, d c 1368] describes in 1360 how to treat fractures and hernias; *See also* 1320–29 MED; 1460–69 MED	Giovanni de Dondi publishes in 1364 the first description of a modern clock, weight-powered, escapement-regulated, and with a balance wheel; this form of clock is invented by Dondi or by members of his family, although key elements, such as the escapement, are no doubt borrowed from the Chinese or Western clockmakers (who use weights instead of water to drive the clock); *See also* 1340–49 TOO; 1400–09 TOO		**1360 to 1369**
Inventories of the possessions of Charles V of France reveal that he has gold and silver forks, although the forks were "only used for eating mulberries and foods likely to stain the fingers" according to the inventory; *See also* 1100–09 FOO; 1530–39 FOO	German potters achieve production of true stoneware and a salt glaze, creating the first such pottery in Europe; *See also* 1100–09 MAT		An arbalest, a steel crossbow powered with a crank or with a windlass and pulley, is introduced as a weapon of war The first cast-bronze cannon made in a single piece in the West is cast at Augsburg, Bavaria (Germany) in 1378; *See also* 1350–59 TOO; 1540–49 TOO	The first record of a canal lock in the West describes a lock at Vreeswijk in Holland (the Netherlands) built this year, although others think the first Western locks were constructed a few years later in Italy or over 100 years earlier at Sparendam, Holland; *See also* 1250–59 TRA; 1390–99 TRA	**1370 to 1379**

1380 to 1389

Construction of the cathedral in Milan begins in 1386; its design with four towers at the corners of the crossing tower is intended to represent Christ surrounded by the four evangelists; after a few years of construction, however, it becomes apparent that there are structural flaws and French and German advisers are called in; they recommend more geometry and less "art"

1390 to 1399

On Oct 1 1396 the Hundred Years' War is effectively over, although there is no peace treaty; *See also* 1330–39 GEN

The Battle of Nicopolis (Nikopol, Bulgaria) on Sep 25 1396 confirms Ottoman rule of the Balkans; *See also* 1070–79 GEN; 1450–59 GEN

Metal type is used for printing in Korea; *See also* 1310–19 COM; 1440–49 COM

1380 to 1389

FOOD & AGRICULTURE

MATERIALS

Cast iron becomes generally available in Europe; *See also* 1340–49 MAT

MEDICAL TECHNOLOGY

TOOLS & DEVICES

The first small arms made in the West, known as fire sticks, are mentioned in writing in 1381; *See also* 1420–29 TOO

The earliest built mechanical clock that survives today is a device in Salisbury Cathedral in England built in 1386; the clock does not tell time with hands, but sounds the hour and indicates astronomical events; *See also* 1350–59 TOO

TRANSPORTATION

Rockets are used for the first time in Europe in the Battle of Chioggia, between the Genoese and the Venetians; *See also* 1150–59 TRA; 1650–59 TRA

Work begins on the Milan cathedral in 1386; the bishop of Milan has branch canals built from the moat supplied by the Naviglio Grande; he uses the branch canals to ferry stone to the site of the cathedral being constructed; *See also* 1200–09 TRA; 1430–39 TRA

1390 to 1399

A paper mill is established in Nuremberg (in Germany) in 1391; *See also* 1350–59 MAT; 1490–99 MAT

In 1391 *A treatise on the astrolabe* by Geoffrey Chaucer [b 1343?, d London, Oct 25 1400] shows how to construct an astrolabe and how to use such a device to compute the elevation of a star; *See also* 1150–59 TOO; 1420–29 TOO

A 24-km (15-mi) canal is constructed from Lake Mölln to the river Delvenau (in Germany) in 1391; traffic passes from the Baltic up the river Stecknitz, through Lake Mölln, along the canal, down the river Delvenau to the river Elbe and on to the North Sea; a boat would go over a dozen stanches on the rivers and pass through two locks on the canal; *See also* 1380–89 TRA

Locks are installed on the Po river in 1395, making it possible for barges to transport marble for the construction of the cathedral in Milan

The first representation of a four-wheeled cart with a turning front axle is found on a seal of François of Carrara made in 1396; *See also* 1340–49 TRA; 1450–59 TRA

1400 to 1409

The dulcimer is mentioned for the first time

Hughes Aubriot, provost of Paris and builder of the Bastille, roofs over a stretch of the open ditches that serve as sewers in Paris (as in most medieval cities); eventually all the drainage ditches of Paris are covered, giving some semblance of a sewer system; *See also* 1500–09 GEN

Konrad Kyeser's *Bellifortis* of 1405 discusses military technology, including fortifications and war machinery; *See also* 1450–59 GEN

The *Geography* of Ptolemy (Claudius Ptolemaeus), written around 100 CE, is translated into Latin in 1406; it greatly influences the European view of the world; *See also* 1540–49 COM

Leonardo da Vinci

The place of Leonardo in history as an artist and even as a scientist is well known. Many of his inventions, such as the helicopter, were only on paper, however. His actual achievements and contributions to the technology of his time are less well known and somewhat problematical. On the one hand, it is said that none of his ideas were practical, while on the other, it is said that his improvements in canal locks have been imitated ever since.

After a striking success as a painter in his youth, Leonardo at about the age of 30 began to take commissions in fields that we would today identify as architecture or engineering. From then on, he made at least part of his living designing improvements on buildings and canals, some of which he then built and some of which languished. Increasingly he designed in secret devices that anticipated flight, steam power, and much more.

Much of the reason for the apparent mystery of his ideas about technology is that he wrote and drew them in coded notebooks that, although he worked on them for much of his adult life, were never sufficiently organized to decode and publish. Indeed, most pages of the notebooks were not published until late in the 19th century. All his life he found it difficult to complete a task, even abandoning paintings for which he had been paid. Single pages of the famous notebooks, although nearly all written so that each deals with a separate chain of thought, often slip from one topic to another.

Among the public achievements that contributed to the technology of his time are the maps of Italy he made at the beginning of the 16th century. Although none of his architectural plans was executed, many were known at the time and became a major influence on such buildings as St. Peter's in Rome. In addition to canal lock gates, replacements in 1497 for worn-out gates on the oldest canal in Italy, it is thought that he actually built a screw-cutting machine and a turret windmill, both advanced for the time and not followed by others until much later. Some give Leonardo partial credit (along with Palladio) for the truss bridge, but trusses were occasionally used by the Romans and did not become common in bridges until centuries after Leonardo's death.

It is thought that some may have seen his notebooks and been influenced, notably Girolamo Cardano, but that is not certain. If they had been, we might have earlier had the advantage of Leonardo's experiments with tensile strength, his analysis of forces produced by arches and hoisting devices, and his work on the strength of beams and columns. A half century after Leonardo's death, another great Italian, Galileo, repeated such experiments and analyses—and published them in a clear, coherent form.

Coffee, which grows wild in Ethiopia, is made into a beverage there

Oil is used as a base for paints

The suction pump is invented during this century; *See also* 1640–49 TOO

An illuminated manuscript from France from 1406 shows one of the earliest known mechanical, weight-driven clocks to use a balance wheel instead of a weighted bar (foliot) as part of the escapement; *See also* 1360–69 TOO

By this date carvel-constructed ships weighing up to 1000 tons are in existence in Europe; *See also* 1300–09 TRA; 1450–59 TRA

A lateen sail rigged fore and aft and carried on a mizzenmast begins to be used in the Basque region around the Bay of Biscay; its use spreads rapidly; *See also* 1570–79 TRA

The Chinese junk has four permanent masts and two temporary ones, a central rudder, and a hold divided into watertight compartments; fleets of junks patrol the Indian Ocean as far away as Sri Lanka

At the command of Emperor Yong'le a giant exploring expedition sets out from China in 1414 under the leadership of the grand eunuch Admiral Zheng He; 62 giant junks, the largest more than 120 m (400 ft) long and 45 m (150 ft) wide, travel to what are now India, Sri Lanka, the middle East, and Africa; the fleet returns safely with trade goods and gifts from the Africans in 1415; similar exploration continues for about 20 years; *See also* 1400–09 TRA; 1430–39 TRA

Beginning a long period of naval exploration, Portuguese sailors discover and claim Madeira in 1418, which had been discovered about 70 years earlier by sailors from Genoa, but forgotten; *See also* 1420–29 TRA

Johann Gutenberg

Little is known about Gutenberg the person. Johann Gensfleisch zum Gutenberg was a goldsmith from Mainz, Hesse (Germany), born just at the end of the 14th century. He was on the losing side of a political conflict in Mainz when he was in his early thirties and, as a result, moved to Strasbourg, then a free city. There is a tiny bit of evidence that by 1435 he had turned from goldsmithing to work at least part time on the development of printing. Other goldsmiths and people connected with books had similar ideas, for paper was being manufactured in France and Germany as early as the 14th century. One reason goldsmiths may have become involved is that they were accustomed to making stamps to put on their products, and such stamps are very close to a single piece of type.

Gutenberg's actual inventions were two. Although he did not invent the casting of metal type—the Chinese had tried it and found it too difficult to do properly—he developed the first system for casting type so that the letters could be arranged to form a flat surface. And he invented printers' ink that would function with metal type. The arrangement he developed to print with a modified wine press was good enough not to change in any substantial way for about 300 years. From the very first, Gutenberg produced what we still recognize as printing of the highest order.

The date of 1440 that is often given for Gutenberg's invention is inferred in part because by 1450 he was borrowing money to build equipment and to start printing broadsides, pamphlets, and eventually at least one, and probably three, books. Unfortunately for Gutenberg, his lender had a son-in-law who liked the idea of being a printer, so the lender foreclosed and the son-in-law took over Gutenberg's equipment. (The son-in-law, Peter Schoeffer, turned out to be an even better printer than Gutenberg and an innovator in his own right.) Gutenberg's greatest and most definite achievement, the 1282-page, 42-line Bible, was set in type around 1452 and published in an edition of 300 copies before Aug 1456.

When Gutenberg printed a popular encyclopedia in 1460, the *Catholicon,* he described the achievement as "printed and accomplished without the help of reed, stylus or pen but by the wondrous agreement, proportion and harmony of punches and types."

	GENERAL	ARCHITECTURE & CONSTRUCTION	COMMUNICATION	ENERGY
1410 to 1419	Papal secretary Poggio rediscovers the lost manuscript of *De architectura* by Vitruvius, believed to have been written about 25 BC; *See also* 0 TOO; 1480–89 ARC	Filippo Brunelleschi [b 1379, d 1446] starts construction in 1419 of his design for a dome over the roofless cathedral of Santa Maria del Fiore in Florence (in Italy); the dome, completed in 1436 and bridging a gap of 42 m (138 ft), still stands and is known simply as il Duomo		
1420 to 1429	Conrad Mendel publishes between 1423 and 1429 a book describing late medieval crafts and depicting 355 craftsmen; the work is known as the *Mendel book*; *See also* 1290–99 GEN; 1490–99 GEN		The painters Hubert [d 1426?] and Jan Van Eyck [b Flanders (Netherlands), c 1390, d Bruges, Belgium, Jul 9 1441] are the first to use oil paints systematically; before them, painters usually used paint based on albumin (egg whites), although oil paints had been used on a few previous occasions	
1430 to 1439	Shipyards in Venice in 1436 introduce interchangeable parts for ships and build galleys on an assembly line; a visitor from Spain observes the assembly line in operation and notes that ten galleys are produced in six hours		Leon Battista Alberti [b Genoa (in Italy), Feb 18 1404, d Rome, Apr 25 1472] writes *On painting* in 1436, which explains how to calculate the correct proportions of figures viewed at different distances and how to make the planes of a painting converge on a point; *See also* 1470–79 COM	

1410 to 1419

Dutch fishers become the first to use drift nets; *See also* 1190–99 FOO

On Oct 25 1415 English longbow forces under Henry V defeat heavily armored French armies because English arrows, propelled from stiff bows with flaxen strings, can penetrate French iron armor that contains a lot of carbon

Benedetto Rinio's *Liber de simplicibus* of 1410 describes and illustrates 440 plants that have medicinal uses; *See also* 1490–99 MED

1420 to 1429

The *Book of fireworks*, a handbook for gunners, is published in Germany in 1420; it remains in print until the 16th century; *See also* 1530–39 GEN

The bit and brace system for drilling holes is invented; *See also* 1010–19 TOO

Muhammad Taragay, known as Ulugh Beg (Great Ruler) [b Soltaniyeh (in Iran) Mar 22, 1394, d Samarkand (in Uzbekistan), Oct 27 1449] in 1424 builds an observatory at Samarkand that contains the world's largest astrolabe; the astrolabe has a radius of 40 m (132 ft), giving it a precision of from 2 to 4 arc seconds; *See also* 1390–99 TOO; 1490–99 TOO

The first genuine flintlock small arms are introduced; *See also* 1380–89 TOO

Portugal's Prince Henry the Navigator [b Oporto, Portugal, 1394, d 1460] in 1420 sets up an informal clearinghouse of naval knowledge and a center for exploration at Sagres, on the southwestern tip of Portugal; from this port city Portuguese ships set forth to explore the Atlantic and the African coastlines; *See also* 1410–19 TRA; 1430–39 TRA

Filippo Brunelleschi in 1421 receives the first known patent from the Republic of Florence (Italy); it is for a canal boat equipped with cranes; *See also* 1470–79 GEN

1430 to 1439

The earliest representation of a drive belt in Europe shows it being used to turn a grindstone; *See also* 1330–39 TOO

Noting the Azores marked on an Italian map of 1351, Prince Henry the Navigator of Portugal instructs Gonzalo Cabral to find the now lost islands, which he does in 1432; *See also* 1420–29 TRA; 1450–59 TRA

1430 to 1439 cont

1440 to 1449

The European slave trade in Africa begins when one of the Portuguese explorers sent out by Prince Henry the Navigator returns in 1440 with a dozen Africans; within the decade hundreds of Africans are sold at public auction in Portugal

Johann Gutenberg [b Mainz (in Germany) c 1396, d Mainz, Feb 3 1468] and Laurens Janszoon Koster invent printing with movable type, probably independently of the Chinese; however, the inventions of paper and printing with blocks, essential to printing with movable type, were probably learned as a result of diffusion to Europe from China; Gutenberg is likely responsible for two main innovations, the method of casting metal type and the development of an ink that adheres to cast metal type; *Speculum Nostrae Salutis* (Mirror of our salvation), printed by Koster, is probably the earliest printed book; *See also* 1040–49 COM; 1450–59 COM

1450 to 1459

On May 29 1453 the Ottoman Turks capture Constantinople; the flight of scholars from the city is often considered the defining event of the start of the Renaissance period in Europe; *See also* 1390–99 GEN

Roberto Valturio completes in 1455 his *De re militari*, a treatise on military machines, most of which were conceived by his employer, Sigismond Malatesta; printed at Verona in 1482, it is the first printed book with technical illustrations; *See also* 1400–09 GEN

Leon Battista Alberti writes his *De re aedificatoria* (On buildings) about 1452 or perhaps as early as 1440; it discusses problems of urbanization, stressing harmonious proportions and simplicity of form in design; it also gives a detailed description of a canal lock; his book becomes known as the Bible of architects; *See also* 1480–89 ARC

The Aqua Virgo aqueduct to Rome is put back in service by Pope Nicholas V in 1453; it continues to need major repairs from time to time; *See also* 540 CE ARC

Johann Gutenberg prints the 42-line Bible at Mainz (in Germany) in 1454, inaugurating the era of movable type; type began to be set in 1452 and is published before Aug 1456; *See also* 1440–49 COM

Peter Schoeffer [b 1420, d c 1502] prints his Psalter in 1457, using letters that incorporate both blue and red inks; his method remains a mystery until 1830, when William Congreve shows that Schoeffer uses type made of two parts that fit together, but are inked separately; Schoeffer is also a pioneer in printing with Greek characters; *See also* 1440–49 COM; 1480–89 COM

The Dutch invent the *wip-molen*, a form of windmill in which the top portion, bearing the sails, can be turned to face the wind; *See also* 1340–49 ENE; 1500–09 ENE

1430 to 1439 cont

In 1438 Filippo da Modena and Fioravante da Bologna install a lock (and later a second one) on the Naviglio Grande, replacing the stanch where branch canals to the cathedral connect to the moat around Milan; this is the oldest recorded lock in Italy, *See also* 1380–89 TRA; 1490–99 TRA

1440 to 1449

Willem Beukelsz introduces a method for curing herring in brine aboard ships; the herring is thus preserved and can be sold directly upon arrival ashore

1450 to 1459

Johanson Funcken develops a method in 1451 for separating silver from lead and copper; *See also* 1520–29 MAT

Henry VI in 1452 accepts immigration of Hungarian and Bohemian miners for the improvement of mining practices in England

Gold is used for filling teeth; *See also* 1828 MED

Nicholas Krebs, known as Nicholas of Cusa [b Kues, Germany, 1401], constructs spectacles for the nearsighted; *See also* 1280–89 MED; 1784 MED

The Ottoman sultan Muhammad II in 1453 uses the biggest cannon of its time; it fires balls weighing 270 kg (600 lb) to take Constantinople; *See also* 1370–79 TOO; 1540–49 TOO

Portuguese shipbuilders develop the caravel, the type of ship that will be used by the great explorers of the 15th and 16th centuries; the basic design is of carvel construction, with two or three masts, lateen rigging, and a sternpost rudder; *See also* 1400–09 TRA; 1530–39 TRA

The first successful passenger coach, a four-wheeled wagon with a strap suspension, is built in Kocs, Hungary in 1457; its design spread throughout Europe in the 15th and 16th centuries; the name Kocs for the wagon became "coach" when it reached England; *See also* 1340–49 TRA

	GENERAL	ARCHITECTURE & CONSTRUCTION	COMMUNICATION
1460 to 1469	A form of the stock exchange is founded in Antwerp; *See also* 1130–39 GEN; 1600–09 GEN		Albert Pfister becomes the first printer to combine woodcuts with movable type on one page, with both the text and images printed in a single pass through the press; *See also* 1290–99 COM
			Sweynheym and Pannartz in 1465 print the first book in Italy, at Subiaco; *See also* 1450–59 COM; 1470–79 COM
1470 to 1479	The Republic of Venice (Italy) adopts a formal patent law in 1474, providing the first legal protection for inventors; *See also* 1420–29 TRA; 1620–29 GEN	Francesco di Giorgio Martini writes in 1470 his *Trattati di architettura, ingegneria e arte militare,* mainly known for its architectural content	The Italian painter Piero della Francesca [b Borgo san Sepolcro, Italy, c 1410, d Borgo, Oct 12 1492] starts his book on perspective, *De perspectiva pigendi* in 1470; *See also* 1430–39 COM
			Gasparini's *Pergamensis epistolarum libri* in 1470 is the first book printed in France; *See also* 1460–69 COM
			William Caxton [b Kent, England, 1422?, d London, 1491] prints the first book in English in 1474
1480 to 1489		*De re aedificatoria* by Leon Battista Alberti, on construction methods and architecture, written some 30 or 40 years earlier, is published in 1485; *See also* 1450–59 ARC	Erhard Ratdolt produces the first printed edition of Euclid's *Elements* in 1482; it is the first printed book illustrated with geometric figures
		The lost manuscript of *De architectura* by Roman architect Vitruvius is published in 1486 and becomes the key influence on architecture in the 16th century; *See also* 1410–19 GEN	The first decrees for the prohibition of books perceived as dangerous appear in Mainz (in Germany) in 1485; *See also* 1550–59 COM
			Erhard Ratdolt prints Johannes de Sacrobosco's *De spheara* in 1485, the first printed book to use more than two differently colored inks on the same page; *See also* 1450–59 COM; 1710–19 COM

FOOD & AGRICULTURE	MATERIALS	MEDICAL TECHNOLOGY	TOOLS & DEVICES	TRANSPORTATION	
		Heinrich von Pfolspeundt in 1460 writes *Bündt-Ertzney*, the first book on surgery to be published in Germany; *See also* 1360–69 MED; 1490–99 MED	The first small breech-loading cannons are built in 1461, although their design proves inadequate for larger cannons; breech-loaded cannons do not become common until the latter part of the 19th century; *See also* 1845 TOO		**1460 to 1469**
		The first complete edition of the Avicenna's *Canon of medicine* is printed in 1473 in Milan; *See also* 1130–39 MED	The first attempt is made in 1478 to develop a form of cannon-fired projectile that will explode on impact by using hollow balls filled with gunpowder; the resulting accidents cause this line of development to be abandoned until 1634, when increased strength of cannon walls makes the projectile feasible; *See also* 1500–09 TOO	On the orders of Louis XI of France in 1478, the construction of a tunnel through Mont Viso at a height of 2000 meters is started; it connects the Dauphiné and the domain of the Marquis de Saluces, and is the first such enterprise undertaken in modern times	**1470 to 1479**
			The flyer is added to the spinning wheel; *See also* 1280–89 TOO; 1520–29 TOO	Leonardo da Vinci [b Italy, Apr 15 1452, d Amboise, France, May 2 1519] describes a workable parachute in 1480; *See also* 1783 TRA Portuguese sailors use a table of solar heights above the Iberian Peninsula compiled by Spanish astronomer Abraham Zacuto to determine latitude by measuring the height of the noonday Sun Portuguese explorer Bartholomeu Dias [b c 1450, d Atlantic Ocean 1500] rounds the Cape of Good Hope at the southern tip of Africa in Jan 1488; he becomes the first European known to have sailed around the cape; *See also* 1420–29 TRA; 1490–99 TRA	**1480 to 1489**

1490 to 1499

Leonardo da Vinci develops an oil lamp with a glass chimney that also acts as a lens, concentrating the light for use in reading at night

Polydore Vergil [b Urbino, Italy, c 1475, d Urbino, 1555] writes *De inventoribus rerum* (On the inventors of things) in 1499, one of the earliest compilations discussing inventors and their inventions; the book covers gunpowder, glass, printing, and ships; *See also* 1420–29 GEN; 1580–89 GEN

Graphite, known commonly as lead, is used for writing in England; *See also* 1560–69 COM

Martin Behaim makes the first globe map of Earth in 1492, omitting the about to be discovered Americas and Pacific Ocean; *See also* 1510–19 COM

Michel de Toulouze's *L'art et instruction de bien dancer* (The art and teaching of good dancing) of 1496 includes the first publication of music with movable type, printing red staffs on one pass through the press and notes in black type with a second pass; in 1498, Venetian printer Ottaviano die Petrucci [b 1466] also invents a way of printing music using movable type

1500 to 1509

Gravity-powered water systems for towns are in use; *See also* 1400–09 GEN; 1600–1609 GEN

The three-bay barn, later a fixture in New England farms, is introduced during this century in Britain

Pope Julius II in 1503 chooses Bramante [b Donato d'Agnolo, 1444, d 1514] to design and rebuild St. Peter's Church on the ruins of an early Christian basilica; Bramante designs hastily and with many errors, compounded by various reworkings by a succession of popes between 1514 and 1546; *See also* 1540–49 ARC

German cartographer Martin Waldseemüller [b Baden, c 1470, d Alsace, c 1518] in 1507 publishes a thousand copies of a map on which the name America is first applied to the new continent discovered by Columbus and later explored by Americus Vespucci (between 1497 and 1504); unlike Columbus, Vespucci recognizes that he is exploring a new continent and not part of Asia; *See also* 1520–29 COM

Leonardo da Vinci designs in 1502 a windmill with a revolving turret, and may have built one at Cesena (in Italy); about 50 years earlier windmills with turrets were in use by the Dutch; *See also* 1450–59 ENE; 1580–89 ENE

1490 to 1499

On his second voyage in 1493, Christopher Columbus discovers that Native Americans produce balls that bounce; the balls are made from coagulated and dried sap of a tree; today we know this product as rubber; *See also* 1490–99 TRA

The first paper mill appears in England in 1494; *See also* 1390–99 MAT; 1660–69 MAT

The first cast-iron cannonballs are used in 1495 in the conquest of the kingdom of Naples by France, according to Vannoccio Biringuccio [b Siena (in Italy), Oct 20 1480, d 1538 or 39], writing about 40 years later; *See also* 1380–89 MAT; 1540–49 MAT

The Chinese invent the modern form of toothbrush, with pig bristles at a right angle to the handle

An "anatomical theater" is opened in Padua in 1490 by A. Benedetti Da Legnano for demonstrating the dissection of corpses

Christopher Columbus finds in 1493 on his second voyage that American natives use tobacco as medicine; *See also* 1410–19 MED

Hieronymus Brunschwygk [b c 1452, d 1512] publishes in 1497 the first known book on the surgical treatment of gunshot wounds; *See also* 1460–69 MED; 1540–49 MED

Martin Behain [b Nuremberg (Germany) 1436 or 1459, d Lisbon, Portugal, Aug 8 1507] adapts the astrolabe for use in determining latitude; *See also* 1420–29 TOO

In 1494 Leonardo da Vinci makes a drawing of a clock with a pendulum; *See also* 1580–89 TOO

In 1496 Leonardo da Vinci designs roller bearings and a rolling mill; *See also* 1550–59 TOO

On Oct 12 1492 Christopher Columbus [b Cristoforo Colombo, Genoa (in Italy), 1451, d Valladolid, Spain, May 20 1506] lands on an island in the Caribbean Sea, becoming the principal European discoverer of America

Leonardo da Vinci draws his notion of a flying machine in 1492; *See also* 1500–09 TRA

Leonardo da Vinci in 1497 designs new gates for locks on the Naviglio Grande; the gates are of the miter type and can be folded back into recesses in the walls when the lock is open, creating more room for ships; with a different placement for the water discharge, the new arrangement occupies about a sixth of the space of previous gates, and is used on locks even today; *See also* 1430–39 TRA

Portuguese navigator Vasco da Gama becomes the first European to sail to India by rounding Africa, arriving in May 1498; *See also* 1480–89 TRA

1500 to 1509

Spinach is introduced in Europe

The oil palm is cultivated in Guinea

Raw sugar is refined in 1503

Hieronymus Brunschwygk's *Liber de arte distillandi desimplicibus* of 1500, known as the "Small Book," describes the construction of furnaces and stills, herbs usable for distillation, and medical applications of distillates; he publishes his "Big Book," dealing with the same subjects, in 1512

Jakob Nufer of Switzerland performs the first recorded cesarean operation on a living woman

Rifling in gun barrels is introduced; grooves are cut in spirals into the interior of the barrels, imparting a stabilizing spin to the projectile

In 1500 Leonardo da Vinci draws a wheel lock musket, the first known with this type of ignition, which comes into use about 15 years later; earlier small arms were ignited by various forms of matches; *See also* 1510–19 TOO

Leonardo da Vinci designs the first helicopter; it is not built and probably would not have flown; *See also* 1490–99 TRA

Chinese scientist Wan Hu ties 47 gunpowder rockets to the back of a chair in an effort to build a flying machine; the device explodes during testing, killing Wan, who acted as pilot; *See also* 1150–59 TRA

1500
to
1509
cont

The Pont Notre Dame in Paris, designed by the Italian Fra Giovanni Giocondo, connects La Cité island with the north bank of the Seine in 1507; *See also* 1600–09 ARC

By the end of the 15th century, about 35,000 different books have been published and printed, the total number of copies produced is estimated at 20 million; 77 percent of the books are in Latin while 45 percent deal with religion

---

Old and New World plants meet

The ecosystem around the North Pole was integrated in many ways long before any people populated the New World. Forests of pine or oak filled with brown bears dining on bramble berries and salmon, and tundras tramped by reindeer (caribou) hunted by wolves, were and are typical of both northern North America and northern Eurasia. When Old World people arrived in the northern New World—proto-Indians from Asia first and later post-Columbus explorers from Europe—there was little to surprise them.

South of this region, however, Old and New World species diverge. The most dramatic example is the marsupial order of mammals, found in southern North America and all of South America, but totally lacking in Eurasia. Old and New World insects, plants, and other life forms often bear only a distant relationship, making the division into two worlds sensible. A few odd relationships do exist, however. Many plants, for example, are native both to North America and China in closely related forms.

Technology in the Old and New worlds was directly affected by this divergence of plants and ani-

mals. One reason often advanced for the lack of development of wheeled vehicles in the Americas is that there was no suitable native fauna that could be converted into draft animals. Horses were introduced from Europe; so were oxen. Conversely, Native Americans found a vast range of suitable flora for domestication. Although Old World rice and wheat dominate the carbohydrate basis of the human diet in most of the world, it is difficult to imagine any complex cuisine not based on such New World plants as chile peppers (hot or sweet), tomatoes, potatoes, green and dried beans, pecans, cashews, peanuts, and squash. Furthermore, large populations of humans depend on New World corn and manioc as their main carbohydrate; and corn is the basis of the food pyramid for farm animals around the world. Where civilization started, in the Near East, the native plants that became important domesticates include wheat, rye, barley, and oats, cucumbers, figs, peas and lentils, olives, walnuts, dates, apricots and pears, and grapes. From Europe we have garlic and onions, beets, the cabbage family, lettuce, cherries and apples, carrots, parsley, and the minty herbs.

1510
to
1519

The beginning of the Protestant Reformation is often traced to 95 theses directed against the Roman Catholic Church's concept of indulgences; Martin Luther [b Eisleben (in Germany), Nov 10 1483, d Eisleben, Feb 18 1546] writes and, as legend has it, nails the theses to the door of the church in Wittenberg (in Germany)

On Mar 13 1519 Hernán Cortés lands in Mexico with 11 ships, 508 soldiers, 16 horses, and various Indian and African aides; by Aug 1521 Cortés is in control of virtually all of Mexico; *See also* 1530–39 GEN

Martin Waldseemüller prepares an atlas containing 200 maps in 1513; *See also* 1500–09 COM

At Fano, (in Italy), De Gregoriis prints the first book with Arab characters in 1514; *See also* 1520–29 COM

Geographer and mathematician Johannes Schöner [b Karlstadt (in Germany), Jan 16 1477, d Nuremberg, Jan 16 1547] in 1515 is the first to construct a globe showing the Americas

Henry VIII of England appoints the first Master of the Post in 1516; *See also* 1550–59 COM

Bergbüchlein (Booklet on mining), which is reprinted five times before 1540, is published in 1505; *See also* 1550–59 MAT

Peter Henlein [b Nuremberg, Germany, 1480, d 1542] builds in 1502 a spring-driven watch, one of the first clocks intended to be carried about (these were called watches because they were first used by watchmen); the Henlein clock, nicknamed "the Nuremberg egg," is about the size of a softball and has only one hand; sometimes called a pocket watch, it is on a chain like a pocket watch, but is worn with the chain around the neck; *See also* 1360–69 TOO; 1650–59 TOO

The *Codex Leicester* of 1504 describes a device for measuring the expansion of steam; the device consists of a vessel covered with cowhide and a metal top that is counterbalanced by a weight; boiling water in the vessel lifts the hide, which pushes the top up; however, no reference is made to the possibility of steam power

Old and New World plants meet (Continued)

Oddly, there is one plant that occurred and was domesticated in both the Old and New worlds—cotton. Perhaps its mobile seeds were a factor, although some archaeologists suspect that somehow cotton was transferred by humans in some very early pre-Columbian contact. Cotton does get around. One wild variety is found in Hawaii.

Sometimes New World plants provided a technology that had been completely lacking in the Old World. No one in Europe had even considered a substance like rubber, which was first encountered by Europeans on Columbus's second voyage of 1493. Native Americans knew that rubber could be used to make balls that bounced, but they had little other use for the substance. After 175 years of European knowledge of rubber, a second use was found—it could erase marks made by India ink or by a graphite pencil. Fifty years or so later a Scotsman found a way to make raincoats from it. Twenty years after that the first rubber tires were installed on wheels. By the 20th century rubber had become such a necessity that when wartime cut off supplies, vast sums were spent seeking synthetic versions.

Pope Julius de Medici hires in 1514 Leonardo da Vinci to drain the 775-sq-km (300-sq-mi) Pontine Marshes southwest of Rome; although Leonardo prepares a bold and practical plan, the pope dies before it can be carried out; *See also* 1590–99 FOO

The *Mariegola dell'arte di tintory*, a treatise on the dyeing of materials, is published anonymously in 1510; *See also* 1540–49 MAT

The Royal College of Physicians is established in London in 1518

An instrument called a polimetrum is used in 1513 to make a field survey, from which a map is drawn, of a region in the Upper Rhineland (in Germany); *See also* 1530–39 GEN

The wheel lock musket begins to be manufactured in 1515; it quickly displaces the matchlock, which is difficult to operate in damp weather, although the matchlock continues to be used in many weapons because of its simplicity; *See also* 1500–09 TOO

1520 to 1529

In Oct 1529, Ottoman troops give up their siege of Vienna, ending the expansion of their empire into Europe

Exiles from Portugal print a handbook in Hebrew at Fez, Morocco; it is the first book printed in Africa; *See also* 1510–19 COM

Peter Apian (Petrus Apianus) [b Leisnig, Germany, Apr 16 1495, d Ingolstadt, Apr 21 1552] in 1520 publishes a map that shows America and in 1524 a book on cartographical methods; *See also* 1500–09 COM

1530 to 1539

Francisco Pizarro [d Lima, Peru, Jun 26 1541] takes charge of the Inca Empire in 1533; he is later assassinated by rival Spaniards; *See also* 1510–19 GEN

Mathematician and geographer Reiner *Gemma* Frisius [b Dokkum, Friesland (in Netherlands), Dec 8 1508, d Louvain, May 25 1555] describes the method of separating a region into a series of triangular elements for surveying; *See also* 1510–19 TOO; 1610–19 GEN

Niccolò Tartaglia [b Brescia, (in Italy), 1499, d Venice, Dec 13 1577] writes *Della nova scientai* (Of the new science) in 1537, initiating the science of ballistics; *See also* 1420–29 TOO

1540 to 1549

The French Royal Ordnance introduces in 1544 six calibers for the artillery, thus rationalizing the production of ammunition

The Grand Duke of Florence decrees in 1545 that brocade workers that have left the town must return; this is an economic measure intended to protect the brocade trade; *See also* 1570–79 GEN

The Pont Neuf at Toulouse, France, built in 1540, is an early example of a bridge constructed with semielliptical arches instead of semicircular ones; *See also* 1340–49 ARC; 1560–69 ARC

Pope Paul III asks Michelangelo [b Michelangelo di Buonarroti Simoni, Caprese, Tuscany (in Italy), 1475, d Rome, Feb 18 1564] to correct the errors in Bramante's 1503 design for St. Peter's Church in Rome and to complete construction of the edifice; *See also* 1590–99 ARC

Sebastian Münster [b Ingelheim (in Germany) 1489, d Basel (in Switzerland), May 23 1552] publishes *Cosmographia universalis* in 1544 in Basel; it is the first major compendium on world geography; *See also* 1400–09 COM

FOOD & AGRICULTURE	MATERIALS	MEDICAL TECHNOLOGY	TOOLS & DEVICES	TRANSPORTATION	
Turkeys are imported to Europe from America; the Portuguese import the orange tree from South China; maize (corn) is imported into Spain from the West Indies Paracelsus's *Archidoxis* (written in 1527, but not published until 1570) reports that frozen wine will have a higher proof than liquid left unfrozen; *See also* 1100–09 FOO	*Probierbüchlein* (Little book of assays), which becomes an important guide to the assaying of metals, is published in 1520; *See also* 1540–49 MAT; 1570–79 MAT Gunpowder may be in use in mines at Chemnitz (in Germany); however, this practice may have started 100 years later; *See also* 1320–29 MAT; 1620–29 MAT	Physician and alchemist Philippus Aureolus Paracelsus (Theophrastus Bombast von Hohenheim [b Schwyz, Switzerland, May 1 1493, d Salzburg, Austria, Sep 24 1541] introduces tincture of opium, which he names laudanum; *See also* 1530–39 MED	Spinning wheels powered by foot treadles come into use; *See also* 1480–89 TOO; 1530–39 TOO		**1520 to 1529**
Catherine de Medici marries the future King Henry II of France in 1533 and brings forks, meals with separate courses, and serious cuisine from Italy into France; French cuisine as we know it today evolves from this Italian beginning; *See also* 1370–79 FOO; 1610–19 FOO Alonso Herrera publishes his *Libro di agricultura* in 1539		The first book on dentistry is published in Leipzig (in Germany) in 1530 Paracelsus's *Paragranum* of 1530 is the first book to suggest the use of chemical substances, such as compounds of mercury and antimony, as remedies *Dispensatorium* by Valerius Cordus [b Germany, 1515, d 1544], one of the first books to describe most known drugs, chemicals, and medical preparations, is published in 1535	The spinning wheel is in general use in Europe; Johann Jürgen improves on it by adding twisting rotation to the thread; *See also* 1520–29 TOO	Gemma Frisius is the first to point out, in *De principis astronomiae et cosmographie* of 1533 that knowing the correct time according to a mechanical clock and comparing it with Sun time can be used to find longitude; *See also* 1590–99 TRA Italians use a glass diving bell in 1535 to explore sunken ships in Lake Nemi; *See also* 50 CE TOO; 1620–29 TRA The earliest known book on shipbuilding, *De re navali* by Bayfius, is published in 1536	**1530 to 1539**
Leonhard Fuchs [b Bavaria (Germany), Jan 17 1501, d Tübingen, (in Germany), May 10 1566] describes in 1542 about 400 German and 100 foreign plants, including peppers, pumpkins, and maize from the New World Luigi Alamanni publishes *La coltivatione* in 1546	Vannoccio Biringuccio publishes *Pirotecnica* in 1540, a book dealing with metallurgy; it also describes a boring machine for cannons that is powered by pedals; *See also* 1774 TOO Christoph Schurer uses cobalt in the production of blue glass In 1548 Gioaventura Rosetti publishes in Venice a treatise on dyeing	A 1545 book on surgery by Ambroise Paré [b Mayenne, France, 1510, d Paris, Dec 20 1590] advocates abandoning the practice of treating wounds with boiling oil and using soothing ointments instead; *See also* 1490–99 MED The first glass eyes are probably made	Spanish arms makers develop the *Moschetto* (from which the name "musket" derives, although the name is used for earlier small arms today); the *Moschetto* requires two persons to hold and fire it, igniting the charge with a matchlock; *See also* 1500–09 TOO According to tradition, Ralph Hog of Buxted, England, invents cast-iron cannons in 1542; *See also* 1370–79 TOO; 1739 TOO Peter Bullmann builds automatons powered by spring-driven clockworks		**1540 to 1549**

1550 to 1559

Girolamo Cardano [b Pavia (in Italy), Sep 24 1501, d Rome, Sep 21 1576] makes a camera lens in 1550

The Prince of Thurn und Taxis receives the authorization of Charles V in 1550 to forward private messages, the first postal service open to the general public; *See also* 1510–19 COM

According to tradition, Giambattista della Porta [b near Naples, Oct 1535, d Naples, Feb 4 1615] invents the camera obscura in 1553—although there are clear references to the device preceding della Porta; *See also* 1000 CE COM

The Portuguese print *Concluôes publicas et doutrina Christiano* in India in 1556; it is the first book printed in that country; *See also* 1520–29 COM

The Papal authorities publish the *Index libvrorum prohibitorum*, commonly known as the Index in 1559; it is a list, regularly updated, of books forbidden to be read by Catholics, except with special authorization; *See also* 1480–89 COM; 1660–69 COM

1560 to 1569

The first research institute is founded in Naples (in Italy)

The first industrial exposition is held in 1568 at Nuremberg (in Germany)

Saint Basil's Cathedral is completed outside the Kremlin in Moscow in 1560; it has eight separate chapels clustered around a central altar, each capped with striped onion domes of different heights; according to legend, Ivan the Terrible blinds the builders when the cathedral is completed so that they cannot build another as beautiful

Bartolomeo Ammanati [b 1511, d 1592] builds the "basket-handle" or elliptical arch bridge known as the Ponte Santa Trinita, or Ponte della Trinita over the Arno just below the Ponte Vecchico in Florence (in Italy); *See also* 1340–49 ARC; 1590–99 ARC

It is known that silver salts turn black, but the association of this change with light is not yet recognized; *See also* 1720–29 COM

Conrad von Gesner [b Zürich, Mar 26 1516, d Zürich, Dec 13 1565] describes the pencil in 1565, although it may have been invented near the beginning of the 16th century; *See also* 1490–99 COM

Flemish geographer Gerhardus Mercator [b Gerhard Kremer, Rupelmonde, Belgium, Mar 5 1512, d Duisburg, Germany, Dec 2 1594] in 1568 introduces the map projection that bears his name; *See also* 1590–99 COM

The *Badianus manuscript* is compiled in 1552 by two Aztec scholars with a text in Nahuatl by Martinus de la Cruz and a Latin translation by Juannes Badianus; the catalog of medicinal plants from Central America with color illustrations is available only in the Vatican library until 1940, when the first facsimile edition is published; *See also* 1670–79 FOO

Pedro de Cieza de Leon in 1553 describes the potato in *Chronicle of Peru*; *See also* 1560–69 FOO

Count Van Egmond begins drainage of the Egmondermeer in 1556, starting a much larger drainage operations in the Low Countries than any before

The tulip is imported from Constantinople to Vienna; *See also* 1300–09 FOO; 1630–39 FOO

Spanish conquistadores and colonists recognize large deposits of platinum; although the metal is unknown in Europe, the Spanish view it largely as a potential source of counterfeit gold because of its high density; *See also* 1000 CE MAT

Venetian glassmakers develop techniques for making mirrors from glass backed with reflecting metals, such as tin or mercury amalgam; *See also* 2500 BC MAT; 1840 MAT

De re metallica by Georgius Agricola [b Glauchau, Saxony (Germany), Mar 24 1494, d Chemnitz, Saxony (Germany), Nov 21 1555], published posthumously in 1556, offers a detailed discussion of the location of minerals; it remains an important reference work until the end of the 18th century; *See also* 1500–09 MAT; 1570–79 MAT

Rolling mills are in use in Nuremberg, Germany; *See also* 1700–09 TOO

Kräuterbuch (Book on herbs) by Hieronymus Bock [b Heidesbach, Germany, 1498, d Hornbach, Feb 21 1554] is published posthumously in 1560

Military engineer Craponne builds in 1560 the 65-km (40-mi) Canal de Craponne for irrigation in the south of France; it is the earliest French canal; *See also* 1640–49 TRA

The first potatoes arrive in Spain from America in 1565, sunflowers in 1569; *See also* 1550–59 FOO

Jacques Besson publishes in 1569 his *Theatre of instruments and machines*, based largely on the work of Francesco di Giorgio; it discusses a large number of mechanisms and mechanical devices; of most importance, the work describes and illustrates a lathe Besson has designed for woodworking and what may be the first workable screw-cutting machine; *See also* 1570–79 TOO

1570 to 1579

The Grand Duke of Florence in 1575 authorizes as a further economic measure to protect the brocade trade the killing of any brocade worker who leaves town; *See also* 1540–49 GEN

Examples of prosperity in England during the Elizabethan Age, as described by William Harrison in 1577 in *Description of England*, include chimneys instead of smoke holes, glazed windows instead of wooden lattices, regular beds instead of straw pallets, pewter plates instead of wooden platters, and tin or silver spoons instead of wooden ones

The Mogul Empire, which has been expanding throughout the century, includes all of northern India by 1579

Architect Andrea Palladio [b Padua, Italy, Nov 30 1508, d Maser, (in Italy), Aug 19 1580], perhaps the first architect fully to understand how the truss works (although it was used extensively in the Middle Ages), publishes *I quattro librii dell'architettura* (The four books of architecture) in Venice in 1570; *See also* 130 CE ARC; 1792 ARC

A loose collection of maps called *Theatrum orbis terrum* is produced by Abraham Ortelius [b Antwerp, Belgium, Apr 14 1527, d Antwerp, Jul 5 1598] in 1570; *See also* 1510–19 COM; 1590–99 COM

1580 to 1589

Pockets are added to trousers; *See also* 1700–09 GEN

On the advice of astronomer Christoph Clavius [b Bamberg (in Germany), 1537, d Rome, Feb 6 1612], Pope Gregory XIII reforms the calendar in 1582, dropping the 10 days between Oct 4 and Oct 15; in the new Gregorian calendar, century years that are not divisible by 400 will no longer be leap years, as they were in the Julian calendar

Chu Tsai-yü invents equal temperament in music in 1584; *See also* 1630–39 GEN

Pope Paul V builds the Acqua Marcia-Pia aqueduct in 1585, the first new aqueduct in Italy since the Aqua Alexandrina of 226; this is the first new Italian aqueduct in 1644 years; *See also* 230 CE ARC

English poet Sir John Harington [b 1561, d Kelston, England, Nov 20 1612] designs a flush toilet at his home at Kelston, near Bath; it is described in his 1589 *The metamorphosis of Ajax;* the toilet contains water in the bowl that can be flushed with water from a cistern into a nearby pit; a similar toilet is installed in Richmond Palace by Harington's aunt, Queen Elizabeth I; *See also* 1778 ARC

Timothy Bright [b Sheffield, England, 1551, d Shrewsbury, Nov 1615] in 1588 adapts the shorthand developed by Marcus Tullius Tiron to the English language; *See also* 50 BC COM

Dutchman Peter Morice installs a water turbine in London in 1582 that powers a pump that supplies the city with water from the Thames River; *See also* 1620–29 ENE

Agostino Ramelli in 1588 is the first to draw a fully developed windmill and all its working parts, in *Livre des diverses et artificieuses machines; See also* 1500–09 ENE; 1754 MAT

Simon Stevinius in 1589 files for a patent on a windmill that provides for an alternate drive using horses when there is no wind; about this time 101 similar patents are filed in Holland (the Netherlands) for other improvements on the windmill; *See also* 1580–89 GEN

1570 to 1579

Barnaby Googe [b Alvingham, England, 1540, d 1595], in his *Four books of husbandry* in 1577, proposes the use of fallow land and stresses the importance of weeding; *See also* 1700–09 FOO

English gardens, according to William Harrison's *Description of England* in 1577, now include flowers and "rare and medicinable" herbs; *See also* 1270–79 FOO

Lazarus Ercker [b Anaberg, Saxony (Germany), c 1530, d Prague, 1594] publishes *Beschreibung Allerfürnemisten mineralischen Ertzt und Bergwerksarten* (Description of mineral ores and mining techniques); *See also* 1550–59 MAT

Bernard Palissy [b Saintes, France, c 1510, d Paris, 1590] rediscovers the enameling of pottery; *See also* 1580–89 MAT

Jacques Besson publishes in 1578 his *Théâtre des instruments mathématiques et mécaniques*; in it he describes improvements on the lathe, including adding a lead screw and nut and controlling the workpiece by templates and cams, making it possible to turn more diverse and intricate shapes; *See also* 1560–69 TOO

The ribbon loom, a loom for narrow fabrics that can weave several pieces at a time, is invented in 1579; *See also* 1590–99 TOO

A Dutch sailor introduces the first separate topmast, an extension of the mainmast that can be added or removed as weather conditions change; *See also* 1400–09 TRA

Humphry Cole reportedly invents the ship's log for keeping track of the speed of a ship with respect to the water; the first printed reference, from 1573, is to the log-and-line, in which a float attached to a line is paid out for a specified time and the length of the line is used to measure the speed

The use of the Dutchman's log for measuring the speed of a ship is known by 1577; unlike the log-and-line, it uses marks on the side of a ship and the interval between the first and last mark passing a floating object indicates the measure of the ship's speed; *See also* 1710–19 TRA

1580 to 1589

In *Livre des diverses et artificieuses machines* of 1588, Agostino Ramelli makes detailed drawings of his concept of a flour mill that would grind grain using a corrugated roller with spiral grooves, describing the first known roller mill; the feed hopper has its throughput regulated by the speed of the millstone, an early example of a feedback mechanism; Ramelli also designs a system of vibrating screens, now known as a middlings purifier, to remove bran from flour; *See also* 1300–09 FOO; 1660–69 FOO

Mathematician Thomas Hariot and metallurgist Joachim Gans set up the first European scientific laboratory in the New World, a small smelting operation designed to test ores for gold and silver; it is part of Sir Walter Raleigh's first colony on Roanoke Island (VA), a settlement that is evacuated in 1586; a second colony in the same region is started in 1587, but it disappears without a trace; remains of the Hariot-Gans laboratory were discovered in 1991; *See also* 1520–29 MAT; 1610–19 MAT

According to a widely reported legend, 17-year-old Galileo [b Galileo Galilei, Pisa (in Italy), Feb 15 1564, d Arcetri (in Italy) Jan 8 1642] observes in 1581 that the lamps in the cathedral of Pisa swing in a light breeze in the same time regardless of the size of the swing (the amplitude); although this conclusion is not accurate, it is close to true, and Galileo continues to make the claim about pendulums all his life; this observation leads to pendulum-driven clocks; *See also* 1490–99 TOO; 1600–09 TOO

The Dutchman Humphrey Bradley is brought to England in 1584 to help design and construct a harbor for Dover; *See also* 1580–89 FOO

Giambattista della Porta [b near Naples, Italy, Oct 1535, d Naples, Feb 4 1615] writes *Natural magic* in 1589, the first Western book to mention kites and kite flying

1580 to 1589 cont

Simon Stevinius [b Bruges, Flanders (the Netherlands), 1548, d The Hague or Leyden, Holland, 1620] discovers in 1586 the theorem of the triangle of forces (addition of vectors); *See also* 1580–89 ENE

Agostino Ramelli's *Livre des diverses et artificieuses machines* (The various and ingenious machines), published in 1588, is one of the most popular illustrated machine books; it is reprinted and recopied for the next four centuries; *See also* 1490–99 GEN; 1600–09 GEN

Joseph Justus Scaligier [b Lot-et-Garonne, France, Aug 5 1540] devises the Julian day count in 1583, which sets Jan 1 4713 BC as day one; he numbers all days subsequently, so Jan 1, 1994, is Julian day 2,448,987; it is called *Julian* in honor of Scaligier's father, Julius Caesar Scaligier

A 327-ton Egyptian obelisk that the Romans brought from Egypt in ancient times is raised to a vertical position in 1586 by a team led by Domenico Fontana [b Melide, Switzerland, 1543, d Naples, 1607]

1590 to 1599

Giacomo della Porta and Domenico Fontana complete the dome on St. Peter's in Rome, making it the largest church in the world; the church was designed in 1503 by Bramante, but mostly completed by Michelangelo, who also designed the dome; *See also* 1540–49 ARC; 1620–29 ARC

The Rialto bridge of 1591 in Venice (in Italy) combines the segmental and the elliptical arch style of construction; *See also* 1560–69 ARC

The Globe Theater, home to nearly half of William Shakespeare's plays in their first productions, is built in 1599 by the Lord Chamberlain's Men on the south bank of the Thames in London; most of the audience of 3000 stands in an uncovered pit, although there are also benches in roofed galleries

Francisco de Marchi [b 1504, d 1577] describes techniques of siting and construction for forts in *Della architecttura militare* (On military architecture), published posthumously in 1599

The *Mercurius gallo-belgicus*, a newspaper, starts publication in London; it reports on developments on the Continent and is published until 1610; *See also* 1600–09 COM

Mercator's *Atlas sive cosmographicae*, published posthumously in 1595, contains a collection of detailed maps of Europe

1580 to 1589 cont

Humphrey Bradley is assigned by the Privy Council of England the task of completing the drainage of the Fenn region that is traversed by the rivers Ouse, Nene, Welland, and Witham, about 28,300 hectares (70,000 acres) of waterlogged land; he proposes in 1589 a plan to drain the region using gravity, a project not actually completed until the 17th century; *See also* 1580–89 TRA

Bernard Palissy's *Discours admirables de l'art de terre, de son utilité, des esmaux et du feu* (Admirable discourse on pottery and its uses, on enamels, and on fire) covers a wide range of geological and chemical ideas; *See also* 1570–79 MAT

In England iron makers begin in 1589 to patent methods for using different forms of coal at different stages in the manufacturing process, although none of these early methods seems to have replaced charcoal in the actual smelting; *See also* 1610–19 MAT

The Reverend William Lee, an English clergyman, invents the first knitting machine in 1589, a predecessor of the stocking frame; he is probably also responsible for the spring-beard needle, one of the two main types of knitting needle; *See also* 1758 TOO

1590 to 1599

Italian engineer Ascanio Fenici, sponsored by Pope Sixtus V, who has worked for over 3 years on draining the Pontine Marshes, sees his project abandoned in 1590 because of the death of the pope; although several other plans will be developed for draining the marshes, this is the last actual work on the project until the 1930s, when Mussolini's government finally accomplishes the task; *See also* 1510–19 FOO

Galileo's *Della scienza mechanica* (On mechanics), which he writes in 1594 after he had been consulted regarding shipbuilding problems, is one of the first books to deal scientifically with the strength of materials; it is widely circulated in manuscript; *See also* 1630–39 MAT

Zacharias Janssen [b LeHaye, Netherlands, 1580, d Amsterdam, c 1609] probably invents the compound microscope about this time; *See also* 1600–09 TOO

Galileo develops a thermoscope, a primitive form of thermometer that uses air instead of a liquid; it is grossly inaccurate, but forms the basis for measuring temperature for the next 10 years at least; *See also* 1610–19 TOO; 1700–09 TOO

The Republic of Venice (Italy) grants a patent to Galileo in 1594 for a device for lifting water to a higher level; *See also* 1470–79 GEN

The inventor of the ribbon loom is strangled in Danzig (Gdansk, Poland) in 1596 when workers fear that his invention will put them out of work; *See also* 1570–79 TOO

Robert Norman's *The safegarde of saylors*, a translation from the Dutch, is a navigational manual from 1590 that contains maps of the appearance of the coast as seen from sea

The Dutch invent the *fluytschip* (flute), the most sensible cargo ship of its time, cheap to build and operate, yet able to carry a large cargo

Admiral Visunsin of Korea develops the first ironclad warship in 1596

Philip III of Spain in 1598 offers a prize and a lifetime pension to anyone who can discover a way to find longitude at sea; *See also* 1530–39 TRA; 1610–19 TRA

1600 to 1609

King Henri IV orders the first modern Parisian water supply system, with pumps raising water from the Seine from under one of the arches of Pont Neuf; the pumps, designed and installed by Flemish engineer Jean Lintlaer, are powered by an undershot waterwheel; after the royal palaces are supplied, surplus water is offered to the public; *See also* 1500–09 GEN

In 1602, shares of the Dutch East India Company are traded, claimed in Amsterdam to be the first true stock exchange; *See also* 1460–69 GEN

Vittorio Zonca [b 1568, d 1602] publishes in 1607 (posthumously) in Padua (in Italy) his *Nuovo teatro di machine et edificii* (New theatre of machines and edifices); a second edition appears in 1621 and a third in 1656; the book describes among other machines, a machine for throwing silk (twisting fibers together) using waterpower, although at the time the design for such a machine is a state secret

The Pont Neuf in Paris is completed in 1607; it is the oldest bridge still standing in Paris; *See also* 1500–09 ARC

Alexander Top proposes in *The olive leaf* that God created the Hebrew alphabet by using the first letters of 22 of the things He created during the first week; *See also* 1680–89 COM

Abraham Verhoeven starts the publication in Antwerp, Belgium, of the first newspaper on the European continent, the *Nieuwe Tijdingen*; *See also* 1590–99 COM

Hugh Platt in 1603 discovers coke, a charcoal-like substance produced by heating coal; *See also* 1300–09 ENE; 1700–09 MAT

The first attempt is made in 1609 to harness the tides in the Bay of Fundy as a source of power; small mills are successfully powered by this means

Giambattista della Porta publishes *Spiritali* in 1604, in which he describes an apparatus to empty a water container using steam pressure

1610 to 1619

Eratosthenes batavus by Willebrord Snell [b Leiden, Holland (Netherlands), 1580, d Leiden, Oct 30 1626] develops in 1617 the method of determining distances by trigonometric triangulation; *See also* 1530–39 GEN

After religious wars cool down, two bridges started much earlier are completed in 1611 in France; the Pont Henri IV, started in 1576, is completed at Châtellerault; the Pont Neuf of Toulouse, started in 1542, is completed over the Garonne; the Toulouse bridge uses elliptical arches with a maximum span of 32 m (104 ft); *See also* 1600–09 ARC; 1680–89 ARC

The 57-m (186-ft) Tour de Cordouan lighthouse at the mouth of the Garonne River in southern France, begun in 1584, completed in 1611, and designed by Louis de Foixe, is the first lighthouse to have a revolving beacon and the first to be built on rock in the open ocean; it was preceded by a simple tower with a lighted fire; *See also* 1300–09 ARC; 1690–99 ARC

Salomon de Caus [b Dieppe, France, 1576, d 1626] describes in 1615 a ball for ejecting water under steam pressure in *Les raisons des forces mouvantes* (About violent forces: A description of some useful and amusing devices); he also gives the first known description of a rolling mill and describes a number of automatons; de Caus had built some of these moving figures for the garden of Heidelberg Palace and, later, on a greater scale, for the Duke of Burgundy's palace near Paris; *See also* 1620–29 ENE

1600 to 1609

The doughnut originates in Holland (the Netherlands), although the hole is not removed from the center until the middle of the 19th century

Italian scientist Zimara recommends a mixture of snow and saltpeter as a refrigerating substance for preserving food

The Jerusalem artichoke, a sunflower tuber unrelated to the artichoke and native to North America, is mentioned in 1603 by Samuel de Champlain, founder of Quebec, as being cultivated by Native Americans

Jan Adriaenszonn [later called Leeghwater, b 1575] drains the Beemster in the Low Countries in 1608 using improved multistaged waterlifts that employ two to four scoops in tandem; See also 1550–59 FOO; 1620–29 FOO

Sanctorius Sanctorius [b Justinopolis (Kopen, Slovenia), Mar 29 1561, d Venice, Mar 6 1636] describes in 1603 his device that uses a pendulum for counting pulse beats; See also 1700–09 MED

On Nov 29 1602 Galileo writes in a letter that pendulums of the same length swing with the same time no matter what the amplitude, the first written evidence that he has developed this concept; See also 1580–89 TOO; 1640–49 TOO

Willem Diericks Van Sonnevelt develops in 1604 a machine that can produce up to 24 ribbons; it is operated by just one person

Dutch scientist Hans Lippershey [b Wesel, (in Germany), c 1570, d Middelburg, Netherlands, c 1619] invents the telescope in 1608; See also 1660–69 TOO

Hans Lippershey and, separately, Zacharias Janssen invent the compound microscope in 1609; Janssen's invention may have been as early as 1590

The first Swedish canal is built in 1601; it is one of the early canals of Europe; See also 1640–49 TRA

1610 to 1619

The first tea from Asia arrives in Europe

Thomas Coryate [b Somersetshire, England, 1577, d Surat, India, 1617] writes in 1611 *Crudities hastily gobbled up in five months*, about his travels in France, Italy, Switzerland, and Germany during 1608; in the book he advocates the use of the fork, a custom that he introduces to English society; See also 1530–39 FOO; 1630–39 FOO

Simon Sturtevant obtains in 1611 the first British patent for a process in which coal is used instead of charcoal for smelting iron; this process does not become commercially successful for a hundred years; See also 1580–89 MAT; 1620–29 MAT

Cementation steel (blister steel) is introduced in 1614; in cementation, carbon is placed into contact with iron at a high temperature, causing the carbon to penetrate into the iron and make it into steel

Dutch physicist Cornelius Drebbel [b Alkmaar, Netherlands, 1572, d London, 1634] in 1610 designs temperature regulators to control the temperature of an oven for chemical experiments and to control an incubator

The basic modern arrangement for a compound microscope, in which both images are convex and the resulting image is inverted, is developed in 1610 by Johannes Kepler [b Württemberg (in Germany), Dec 27 1571, d Regensburg, Bavaria (Germany), Nov 15 1630]; See also 1600–09 TOO

Galileo, hoping to win a prize and pension set up by Philip III of Spain for anyone who can find longitude at sea, in 1616 suggests using the satellites of Jupiter; although ignored by Philip, Galileo's method will later be used to find accurate longitude on land; See also 1590–99 TRA; 1710–19 TRA

**1610
to
1619
cont**

**1620
to
1629**

The Statute of Monopolies of 1623, which lays down the laws for granting patents for inventions, is passed in England; previously, the king had granted monopolies; subsequent efforts through the reign of Charles I to reassert the king's powers in this matter are finally and definitely overcome by the Long Parliament in 1640; *See also* 1470–79 GEN; 1790 GEN

Francis Bacon's *New Atlantis* in 1627 is a utopian tale of a society based on technology and technological progress

St. Peter's Church in Rome, originally designed by Bramante and redesigned by Michelangelo, is finally consecrated in 1626 over a hundred years after Michelangelo's death; it stands as the largest church in the world for more than 250 years; *See also* 1590–99 ARC; 1990 FOO

Benedetto Castelli [b Perugia, Italy, 1577, d Rome, 1644], one-time student of Galileo and later superintendent of waters in the region around Bologna and Ferrara, publishes in 1628 *Della misura dell'acque correnti* (On the measurement of running waters), laying the foundation of hydraulic technology

Francis Bacon [b London, Jan 22 1561, d London, Apr 9 1626] invents in 1623 a code that consists of two letters that with combinations of five characters allows the representation of the letters of the alphabet; *See also* 1370–79 COM

Giovanni Branca describes in his 1629 machine book *Le machine* a steam turbine in which steam is directed at vanes on a wheel; *See also* 1580–89 ENE; 1660–69 ENE

The first chemical works built by the European colonists is a saltpeter works in Jamestown (VA) that produces a key ingredient for gunpowder; *See also* 1580–89 MAT

Porcelain pottery is made in Japan after a deposit of kaolin is found in 1616 on Kyushu by imported Korean potters

A book in English published in 1611 contains the first mention of "Nut for a Scrue," thought to be the first indication of a threaded nut-and-bolt combination

Sanctorius Sanctorius's *Commentaria in artem medicinalem Galeni* of 1612 contains the first printed mention of the thermoscope, a primitive thermometer invented by Galileo; *See also* 1590–99 TOO; 1640–49 TOO

John Napier describes a device for multiplying that comes to be known as Napier's rods or Napier's bones; it is similar to a multiplication table, with movable rods that show a pair of factors; the product can be read from the rods; *See also* 1620–29 TOO

Pilgrim settlers in Massachusetts observe in 1621 that Native Americans bury a dead fish in each hill of maize (corn) planted to ensure a good yield; *See also* 1570–79 FOO; 1690–99 FOO

Jan Adriaenszonn (by now known as Leeghwater) is hired in 1628 by the Duc d'Epernon to drain the marshes of Gironde (in France); *See also* 1600–09 FOO

Jan Adriaenszonn (Leeghwater) in 1629 proposes draining the Haarlemmermeer (in the Netherlands), using 160 of his multistaged waterlifts, but the cost of 3.6 million guilders is too great; the Haarlemmermeer is not drained until 1852; *See also* 1600–09 FOO

The first beehive ovens for making coke are introduced in England; although inefficient, they are cheap; *See also* 1610–19 MAT; 1841 MAT

Lead mining begins in the US in 1621

Lord Dudley in 1621 obtains a patent for a method of making iron using coal instead of charcoal; the method is invented by his natural son Dud Dudley [b Worcestershire, England, 1599, d Worcester, Oct 25 1684]; *See also* 1610–19 MAT; 1660–69 MAT

Gunpowder is in use in a mine in Chemnitz (in Germany) by 1625; *See also* 1320–29 MAT

Edmund Gunter [b Hertfordshire, England, 1581, d London, Dec 10 1626] in 1620 creates a form of slide rule by locking Napier's bones on a surface; *See also* 1610–19 TOO

About this time William Oughtred [b Eton, England, Mar 5 1574, d Abury, Surrey, Jun 6 1660] invents the slide rule (according to Oughtred's statement in 1632)

In Tübingen (in Germany) Wilhelm Schickard (Schickardt) builds a mechanical calculator in 1623 based on the idea of Napier's bones; it can add, subtract, multiply, and divide, and is intended to aid in astronomical calculations; *See also* 1610–19 TOO; 1957 ELE

Cornelius Drebbel in 1620 builds a navigable submarine, powered by rowers, that can carry 24 persons; it cruises 5 m (15 ft) below the surface of the Thames in London on several occasions; the success of the craft is probably due to the production of oxygen from saltpeter by a process Drebbel kept secret; *See also* 1530–39 TRA; 1710–19 TRA

1620 to 1629 cont

1630 to 1639

Harmonie universelle (Universal harmony), published in 1636 by Marin Mersenne [b Oizé, France, Sep 8 1588, d Paris, Sep 1 1648], describes equal temperament in music for the first time in the West, although it was previously known in China; *See also* 1580–89 GEN

Stephen Day installs the first printing press on the American continent at Cambridge MA in 1638

1640 to 1649

On Dec 2 1642, Dutch skipper Abel Tasman becomes the first European to land on Tasmania, but he does not encounter any Tasmanians; *See also* 1772 GEN

Athanasius Kircher [b Geisa (in Germany), May 2 1601, d Rome, Nov 28 1680] invents the magic lantern and describes it in 1646 in his book on light and optics, *Ars magna lucis et umbrae* (The great art of light and shadows)

FOOD & AGRICULTURE	MATERIALS	MEDICAL TECHNOLOGY	TOOLS & DEVICES	TRANSPORTATION	
	Marthurin Jousse publishes *Théâtre de l'art du charpentier* in 1627 in which he discusses certain properties of materials; this serves as an early precursor of material science; *See also* 1590–99 MAT		The snaphaunce lock, a predecessor of the flintlock, is introduced in 1625 as a way of firing guns; like the flintlock, the snaphaunce creates a spark by mechanically striking flint and steel when a trigger is pulled; the flintlock is based on the same idea, but parts are arranged for more efficient and reliable operation; *See also* 1700–09 TOO Pappenheim (in Germany) in 1626 invents a gear pump that is still used as a fuel pump in automobiles		**1620 to 1629 cont**
Johann Glauber [b Karlstadt (in Germany), 1604, d Amsterdam (in the Netherlands), Mar 10 1670] proposes in 1630 the use of saltpeter as fertilizer; *See also* 1620–29 FOO; 1690–99 FOO The first dining fork in the Massachusetts Bay Colony, and possibly the only fork in North America, is imported in 1630 by Governor John Winthrop; *See also* 1530–39 FOO A craze for tulips provokes speculation in bulbs in Holland (the Netherlands); this "tulipmania" results in prices higher than $5000 for single bulbs of the most desired varieties of tulip; *See also* 1550–59 FOO	After the Church forces him to withdraw from study of the heavens in 1638, Galileo returns to an early interest in the strength of materials, which he treats along with other matters in *Discoursi e dimonstrazioni matematiche, intorno à due nuove scienze* (Dialogues concerning two new sciences); the work includes a study of the breaking strengths of beams that is flawed in parts, but extremely influential; *See also* 1590–99 MAT A glassworks is established in Plymouth (MA) in 1639; it is one of the first manufacturing plants in the American British colonies		Pierre Vernier [b Ornans, France, Aug 19 1580, d Ornans, Sep 14 1638] describes in 1631 his invention for precision measurement, known today as the Vernier scale William Gascoigne [b Middleton, England, c 1612, d Marston Moor, Yorkshire, Jul 2 1644] invents the micrometer; he places it in the focus of a telescope to measure the angular distance between stars	Wooden rails for carts to travel on are installed in the coal mines of Newcastle, England	**1630 to 1639**
Sir Richard Weston, having observed crop rotation practiced successfully in Flanders, describes the concept in 1645 to the English in the first written account of the method in Great Britain; *See also* 1570–79 FOO		The Grand Duke of Tuscany, Ferdinand II [b Italy, Jul 14 1610, d May 24 1670] in 1641 invents a thermometer that uses liquid in a glass tube that has one end sealed; this is a slight improvement on Galileo's thermoscope; *See also* 1610–19 TOO; 1650–59 TOO		The 55-km- (34-mi-) long Briare Canal, started in 1605 by contractor Hughes Cosnier, linking the Loire and the Seine in France, is opened in 1642; a trip from one end to the other uses 40 locks; *See also* 1660–69 TRA	**1640 to 1649**

Pendulum myths

The basic mechanism of clocks led to the development of better ways to cut gears and make other metal parts. No other artifact of this period required such careful workmanship. As a result, clocks became an important precursor of the Industrial Revolution.

Despite the improvement in manufacture, clocks were not nearly accurate enough to bother with minutes, much less seconds. When Galileo needed to time short intervals around the turn of the 17th century, he used his pulse because clocks were not sufficiently accurate.

Statements to the effect that Galileo invented the pendulum in 1581, thereby greatly improving the accuracy of clocks, are simply not true. Instead, sometime in the 1580s or later, Galileo, having observed lamps swinging and other motion of this type, claimed that the time of a pendulum's swing (its period) depends only on the length, not on the size (amplitude), of its swing. Some historians of science think that Galileo did not actually believe this (it is not true!), but that he claimed it because it seemed more dramatic than the truth, which is that for small amplitudes the period is very close to dependent only on the length of the pendulum. Galileo suggested that this concept might be used to make a more accurate clock, but he never got around to trying to do so (although his son experimented with pendulum clocks). Interestingly, Leonardo da Vinci seems to have had the same idea a century earlier than Galileo, but he never got around to trying to build a clock using a pendulum either.

Since the period of the pendulum varies somewhat with the amplitude, a clock using a pendulum for regulation needs some kind of compensating mechanism. Unwittingly, Galileo suggested the idea that led to the first good compensating method as well. He called the attention of mathematicians to a curve called the cycloid, which is the curve generated by a point on the rim of a rolling wheel. Mathematicians in the 17th century were very interested in the properties of the cycloid. In a famous incident in 1658, the mathematician Blaise Pascal, who had otherwise retired from mathematics for religious reasons, thought about solving the properties of the cycloid to keep his mind off a toothache. He published his conclusions, which led to the realization by Christiaan Huygens that the cycloid is a tautochrone—a curve through which an object falls in the same amount of time no matter how high it is.

Again, the myth and the reality are somewhat different. The myth is that Huygens was unable to make a good pendulum clock until he found the tautochrone. In reality, Huygens made the first good pendulum clocks with no thought of the cycloid. When he showed that the cycloid was a tautochrone—2 years later—he made a few clocks that used this idea, but they were clumsy and did not work as well as those that just used a pendulum with a small amplitude. The other pendulum clocks that began to be made in Holland (the Netherlands) about this time did not employ the cycloid either, and the first preceded proof that the cycloid is a tautochrone.

A few years later a series of little-known English craftspeople found various ways to handle the problems of the pendulum. William Clement developed an escapement that kept the amplitude of the swing small. Nearly 50 years later George Graham developed an escapement that nudged the pendulum at one point and braked it at another to make the whole operation more accurate.

In addition to the ordinary pendulum, a different kind of pendulum, called a balance wheel, was used in small clocks from early in the 15th century. Subject to the same problems as the ordinary pendulum, the balance wheel also needed compensation to make it accurate. George Graham found a way to do this, and 50 years later another obscure Englishman, Thomas Mudge, found a better way, one used in balance-wheel watches and small clocks to the present day.

Even if Galileo had been correct in asserting that the pendulum swing depended only on its length, the pendulum could not be a perfect timekeeper because its length varies with temperature. The same English craftspeople who found ways to correct the variation caused by changing amplitude also found clever ways to correct the variation caused by changes in length. By the middle of the 18th century, the best pendulum clocks kept nearly perfect time, losing seconds per month. But the accuracy of these mechanical pendulum clocks was very much superseded in the 20th century—first by accurate electric clocks; then by clocks in which the expansion and contraction of a quartz crystal acted as the "pendulum" for an electric clock; and finally by clocks that used the natural vibrations of molecules as pendulums, the so-called atomic clocks.

1640 to 1649 cont

During the 25 preceding years, the Netherlands acquires 25,514 hectares (63,046 acres) of new land for farming by diking and draining circular inland areas; *See also* 1690–99 FOO

The son of Galileo, Vincenzio Galilei, builds a clock with a pendulum in 1641, a device based on a concept of his father; *See also* 1600–09 TOO; 1650–59 TOO

Blaise Pascal [b Auvergne, France, Jun 19 1623, d Paris, Aug 19 1662] develops in 1642 a mechanical calculator that can add and subtract; he produces about 50 different versions of the calculator over the next 10 years, some for sale, starting as early as 1645; *See also* 1620–29 TOO

Evangelista Torricelli [b Faenza, Italy, Oct 15 1608, d Florence, Oct 25 1647], on a suggestion from Galileo, develops the first barometer in 1643, using mercury as a fluid in a glass column sealed at the top; when the tube is upended in a dish, the mercury sinks to about 76 cm (30 in.), leaving a partial vacuum at the top; this produces the first vacuum known to science; *See also* 1660–69 TOO

Otto von Guericke [b Magdeburg, Germany, Nov 20 1602, d Hamburg, May 11 1686] perfects the air pump; he uses it to produce vacuums with experiments beginning as early as 1645 and in public demonstrations in the 1650s and 1660s; the most famous demonstrations use teams of horses that are unable to break apart spheres held together by a vacuum; *See also* 1400–09 TOO

Athanasius Kircher in 1646 invents a distance-recording device, or milometer, for carriages

1650 to 1659

Otto von Guericke demonstrates in 1650 that electricity can be used to produce light, obtaining a luminous glow from a rotating globe of sulfur by applying pressure from his hand to the globe; *See also* 1700–1709 GEN

James Ussher [b 1581, d 1656], archbishop of Armagh, publishes in 1654 his *Annales veteris et novi testamenti*, a work on biblical chronology that dates the Creation to 4004 BC; this date is refuted in the 18th and 19th centuries, when other evidence becomes available

C'hen Yuan-lung's *Ko-chin-ching-yuan*, published in 1655, describes new inventions

The Taj Mahal, a tomb built for Mumtaz Mahal [d 1631], is completed in 1650; built of white marble in front of a reflecting pool, it is considered among the most beautiful structures ever built

Early clocks

In antiquity people told time by the Sun and stars. Other methods of keeping time could be used as backup, however. Water clocks (clepsydras) had been known since ancient Egyptian times. These depended on the relatively constant lowering of the level of water in a vessel with a deliberately made leak. Clepsydras were not very accurate, not easy to read in poor light, and in need of frequent refilling. Burning tapers and lamps might also show the passage of time, but the burning time of tapers or lamps varied greatly.

In the Middle Ages members of religious orders were expected to pray at definite times. Failure to maintain discipline because of cloudiness or variable flames was not acceptable. Classical sources (and probably rumors of Chinese inventions) referred to devices that could imitate the Sun and stars. Such devices were powered by water clocks, but the Chinese rumors may have told of a clever device, which we call an escapement, that could convert the smooth motion of flowing water into a series of short rotary motions. These motions could give a more accurate representation of the movement of astronomical bodies.

In duplicating this concept, an unknown European inventor recognized that with an escapement to slow the fall, a weight could be substituted for water. If the weight were lifted by hand at regular intervals, there would be no need to deal with the continual addition of water to operate the mechanism. The falling-weight idea, however, required something to make it more regular, since a weight accelerates, or falls faster, as it descends. If the weight could be made to fall the same short distance over and over, the motion would be more regular. This was accomplished by adding a sort of dumbbell, called a foliot, to the escapement. The weights at the end of the foliot fall a short distance with each tick of the clock, one balanced by the other one. The combined fall of the weight that powers the clock and the short stroke of the foliot is off only an hour or so each day. The earliest mechanisms of this type, around the end of the 13th and beginning of the 14th century, were used to power representations of the heavens. Thus, if the clock was off when the weight was pulled up, it could be reset on any clear day or night by comparing the representation with the actual positions of heavenly bodies.

Monks and nuns were summoned to prayer by a bell. The elaborate astronomical model was not needed; a system of striking the hour with a series of rings of the bell was sufficient. Sometime later people added a dial to show the hours with a pointer (hand). A similar pointer for minutes was not used until clocks greatly improved in accuracy. Although the first clocks were installed for use in religion, within a few years people began to keep time by the hour, since the ringing of the bell often could be heard or the dial seen all over a village.

The manufacture of clocks became a thriving industry in the 14th and 15th centuries. Every town soon had to have its own.

Dutch engineer Cornelius Vermuyden, commissioned by England's Charles I, succeeds in 1653 in draining and reclaiming 124,000 hectares (307,000 acres) in the Fenns region of England

Johann Shultes's *Armementarium chirugicum* (The hardware of the surgeon) of 1655 describes a procedure for removing a female breast

The Grand Duke of Tuscany, Ferdinand II in 1654, invents the sealed (at both ends) thermometer using liquid as the indication of temperature; improvements on this basic design by Fahrenheit about 60 years later result in the modern thermometer; *See also* 1640–49 TOO; 1700–09 TOO

Christiaan Huygens [b The Hague (in the Netherlands), Apr 14 1629, d The Hague, Jun 8 1695] develops the first accurate pendulum clock in 1656; *See also* 1640–49 TOO; 1650–59 TOO

Salomon da Coster, in The Hague (the Netherlands) begins to construct in 1657 a series of spring-driven clocks that use a pendulum instead of a foliot balance (weighted bar) or balance wheel to regulate the time; these clocks are thought to be an outgrowth of the work of Christiaan Huygens the previous year; *See also* 1670–79 TOO

Robert Hooke [b Freshwater, Isle of Wight, England, Jul 18 1635, d London, Mar 3 1703] invents the spiral spring for watches in 1658; *See also* 1500–09 TOO; 1670–79 TOO

Christiaan Huygens develops in 1658 a form of pendulum clock based on the cycloid, a curve that, when used to enclose a pendulum on a string, adjusts the beat precisely as the amplitude changes so that the beat remains the same; this turns out to be a clumsy adaptation, not so successful as his previous pendulum clocks; *See also* 1640–49 TOO; 1660–69 TOO

Writer Cyrano de Bergerac in an early work of science fiction suggests seven ways of traveling from Earth to the Moon; although six of the ways would not have worked, the seventh is by means of rockets; *See also* 1380–89 TRA; 1680–89 TRA

Christiaan Huygens in 1659 constructs a chronometer for use at sea; however, it is influenced by the motion of the ship and does not keep correct time

Gunpowder and guns in East and West

When gunpowder was invented by the Chinese, they were afraid of it, which was natural. But the Chinese did not immediately think of guns, which seems odd to people in the West to this day. It took nearly 500 years for word of gunpowder to reach Europe, but almost no time after that for Europeans to invent guns. The Chinese, and others in the East who learned of gunpowder from the Chinese, did use the explosive mixture in war, but at first they counted on its loud noise scaring the enemy. Later they developed bombs or grenades as well as military rockets. But none of these devices was to have the effect on world order that guns very quickly came to have.

An interesting example of the different reaction to guns and gunpowder comes from Japan. Like other Eastern peoples, the Japanese learned of gunpowder bombs and rockets from the Chinese, Mongols, and Koreans in the course of various wars. But on Aug 25 1543 the Japanese found out about guns for the first time when a Chinese junk with three Portuguese aboard drifted ashore. The Portuguese had guns—match-fired muskets known as arquebuses. Within a year the guns had been copied and were soon manufactured all over Japan, then in a period of intense internal wars. According to one account, by 1556 there were 300,000 arquebuses in Japan and wars were being won by arquebus-armed soldiers against cavalry armed with swords. Cannons were also introduced late in the 16th century and became important in the many local wars.

But early in the 17th century, the local wars were over and the period of stability known as the shogunate had started. Although guns were not banned (as sometimes has been written), they were no longer viewed as necessary. Swordsmanship was what was important to the samurai warriors. This period, when swords replaced guns for more than 200 years, has been the inspiration for many science-fiction adventure novels, in which the culture spurns its many powerful tools and fights with swords or primitive guns.

Western guns were among the many factors that helped ensure Europe's dominance over Africa, the Americas, Asia, and Oceania, although other elements were also important in each case. Indeed, Spanish hand weapons in Mexico were not powerful enough to penetrate cotton armor on native armies (but cannons and disease germs were effective enough). In Asia, the Japanese abandonment of gun power left the island unable to resist heavily armed Western battleships early in the 19th century.

1660 to 1669

The Sheldonian Theatre in Oxford, England, designed by Christopher Wren [b Wiltshire, England, Oct 20 1632, d London, Feb 25 1723], is completed in 1662; although superficially different from the Gothic style used for Oxford University, the techniques used for construction of the roof are based on Gothic methods

The "Printing Act," exerting strict controls on publishing by Charles II, is ratified by the English Parliament in 1662; it is "An Act for preventing the frequent abuses in printing seditious, treasonable and unlicensed books and pamphlets and for regulating of printing and of printing presses"; *See also* 1550–59 GEN

Disserto de arte combinatoria by Gottfried Wilhelm Leibniz [b Leipzig (in Germany), Jul 1 1646, d Hannover (in Germany), Nov 14 1716], published in 1666, contains his suggestion based on the work of Raymond Lully, that a mathematical language of reasoning can be developed; it will not be until the 19th century when George Boole and others develop this idea further; some of these ideas find application in computer science and artificial intelligence research; *See also* 1854 COM

Otto von Guericke in 1660 develops a way to charge a ball of sulfur with static electricity, producing the greatest amount of electricity gathered in one place to this time; *See also* 1650–59 GEN; 1700–09 ENE

The Marquis of Worcester claims in 1663 to have discovered the power of steam to raise water from wells and to burst cannons; *See also* 1620–29 ENE; 1690–99 ENE

1670 to 1679

Christiaan Huygens builds a motor driven by the explosions of gunpowder; *See also* 1620–29 MAT

Georg Andreas Bockler in 1662 changes Agostino Ramelli's design for a roller mill by adding a second corrugated, grooved roller; *See also* 1580–89 FOO; 1834 FOO

Cheddar cheese is invented in the English village of Cheddar in 1666

Cast iron is used for the pipes supplying water to the gardens at Versailles in 1664

John Clayton in 1664 discovers a pool of natural gas near Wigan, England; *See also* 1739 MAT

In his *Mettalum martis* of 1665 Dud Dudley claims to have learned the secret of making iron with coal instead of charcoal; previously, his father, Lord Dudley, had been granted a patent for this process; despite the patent and the claim, modern writers have generally assumed that the process would not have worked for iron of any usable quality since it called for coal instead of coke; *See also* 1620–29 MAT; 1700–09 MAT

Robert Hooke proposes in 1665 that artificial silk might be manufactured by extruding a solution of gum

Because almost all European paper is made from recycled cloth rags, which are becoming increasingly scarce as books and other materials are printed, the English Parliament in 1666 bans burial in cotton or linen cloth so as to preserve the cloth for paper manufacture; *See also* 1490–99 MAT; 1798 MAT

Johann Baptista van Helmont [b Brussels, Belgium, Jan 12 1580, d Brussels, Dec 30 1635 or 1644] publishes *Oriatrike* (Physic refined) in 1662, which becomes very popular

Richard Lower [b Cornwall, England, c 1631, d London, Jan 17 1691] demonstrates the direct transfusion of blood between two dogs in 1666

Physician Thomas Sydenham, "the English Hippocrates," [b Wynford Eagle, England, Sep 10 1624, d London, Dec 29 1689] publishes *Methodus curandi febres* in 1666; in this and other books, he advocates the use of opium to relieve pain, chinchona bark (quinine) to relieve malaria, and iron to relieve anemia; *See also* 1520–29 MED

In an experiment demonstrated before the Royal Society in 1667, Robert Boyle [b Ireland, Jan 25 1627, d London, Dec 30 1691] shows that an animal can be kept alive by artificial respiration

Otto von Guericke is the first to use a barometer to forecast weather in 1660; *See also* 1640–49 TOO

Christiaan Huygens in 1661 invents a manometer for measuring the elasticity of gases

Optica promota by James Gregory [b Drumoak, Scotland, Oct 1638, d Edinburgh, Oct 1675] gives the first description of a reflecting telescope in 1663

Gaspar Schott describes a universal coupling (Cardan joint) in 1664; *See also* 1670–79 TOO

In 1666 Robert Hooke designs a new type of escapement for clocks in which the moving part of the escapement is placed in the same plane as the balance wheel on which it acts, thus increasing the accuracy of the clock; *See also* 1650–59 TOO; 1720–29 TOO

Jean de Thévenot [b Paris, c 1620, d Paris, Oct 29 1692] describes in 1666 his concept of a carpenter's level, a bubble floating in a thin glass tube filled with liquid

In 1667 Robert Hooke invents the anemometer, an instrument for measuring the force or speed of the wind

Blaise Pascal proposes in 1662 the introduction of a public transport system in Paris; coaches would travel along predetermined routes and take passengers for a small fee; the first coach goes into service the following year

Pierre-Paul Riquet [b France, 1604, d 1680] starts construction in 1666 on the 290-km (180-mi) Canal du Midi (also known as the Languedoc Canal and the Canal of the Two Seas); it connects the Mediterranean Sea and the Bay of Biscay in the Atlantic Ocean, using the river Garonne; *See also* 1680–89 TRA

1660 to 1669

Dom Pérignon [b Sainte-Menehould, France, 1638, d Hautvillers Abbey, France, 1715] invents champagne in 1678; he uses sugar to start a second fermentation in the cask and stores the bottles so that sediment can be removed

Johannes Kunckel [b Rendsburg (in Germany), 1630, d Pernau, Mar 20 1703] invents the artificial ruby in 1679, a form of colored glass; he publishes the results of his experiments in *Ars vitraria experimentalis*

William Clement of England in 1670 invents the recoil, or anchor, escapement, which controls the amplitude of the pendulum (a smaller arc makes for a more accurate clock); he also invents the minute hand for clocks; *See also* 1650–59 TOO; 1710–19 TOO

Jesuit monk Francesco de Lana [b Brescia, Italy, Dec 10 1631, d Brescia, Feb 22 1687] in 1670 designs an airship—never built—that would be lifted by four copper spheres containing a near vacuum

1670 to 1679

1670 to 1679 cont

Perpetual motion (part 1): An old dream

Since antiquity, people have been fascinated by the motion of the stars on the celestial sphere. People hoped to imitate this motion and build a machine that would run forever without using an external force, such as wind or flowing water. Many believed that rotation is an intrinsic property of the wheel. Even the Polish astronomer Copernicus believed that anything round— according to him the perfect shape—would rotate by itself.

One of the early attempts to build a perpetually turning wheel can be found in a Sanskrit manuscript of the 5th century. It describes a wheel with sealed cavities in which mercury would flow in such a fashion that half of the wheel would always be heavier than the other half, thus keeping the wheel running. Around 1235 the French architect Villard de Honnecourt devised a wheel based on a similar principle. An odd number of hammers pivot around their attachment points on the wheel. Because half of the wheel always has a larger number of hammers than the other half, the wheel would keep turning.

It was during the Renaissance that engineers became seriously interested in building a machine that would produce power continuously. They were inspired by the large number of windmills and waterwheels that were then in use. The concepts were simple: A waterwheel drives a pump that continuously pumps up the water that runs the wheel into an elevated reservoir, or a windmill actuates giant bellows that drive the windmill.

Many other designs made their appearance as well. Some were systems in which water would keep flowing endlessly. Others were complicated mechanisms. In one type, steel balls roll down an inclined plane. The inclined plane pivots when the ball reaches its end. This motion then actuates a mechanism that brings the ball back to its beginning position on the inclined plane.

1680 to 1689

In Sep 1682, Isaac Newton's *Philosophiae naturalis principia mathematica* (The mathematical principles of natural philosophy), known as the *Principia*, establishes Newton's three laws of motion and the law of universal gravitation; the book also becomes the basis for the development of theoretical mechanics over the next 200 years

The French army introduces in 1688 bayonets that are attached to muskets, replacing bayonets used by the French for the past 30 years or so that fit into the barrels of guns

About this time German flute maker Johann Denner [b 1655, d 1707] invents the first form of the clarinet; it has no keys and is played like a recorder, with fingers covering the nine holes as needed; as the instrument develops, various extensions are added to make it possible to open and close more than 20 different holes

After two designers fail to solve problems that arise in building a bridge across the Seine between the Tuileries Palace of Louis IV and the south bank of the Seine, François Roman [b Belgium, 1646, d 1735] develops a new technique in bridge building in 1685; first he sinks a large barge loaded with stone on a dredged site, forming a caisson (from the French *caisse,* or box) in which the pier is built; the final bridge, known as the Pont Royal, earns Roman the label "first engineer of the century"; *See also* 1610–19 ARC; 1772 ARC

Venetian gunfire sets off ammunition stored by the Turks in the Parthenon in Athens in 1687, causing an explosion that destroys the roof and wrecks much of the rest of the building; *See also* 450 BC ARC

Robert Hooke in 1684 describes a system of visual telegraphy

John Wilkins, the bishop of Chester, writes in 1688 that Adam invented the Hebrew alphabet and writing sometime after God had created him; *See also* 1600–09 COM

Isaac Newton [b Woolsthorpe, Lincolnshire, England, Dec 25 1642, knighted 1705, d London, Mar 20 1727] proposes in 1680 that a jet of steam could be used (like a rocket) to power a carriage, an idea now considered one of the precursors of the development of the jet engine; *See also* 1650–59 TRA

1670 to 1679 cont

A fire in 1571 destroys a large part of the library at the Spanish palace and monastery El Escorial, taking with it the original drawings of native American plants complied in the 1570s by Francisco Hernández and Aztec artists and draftsmen; copies of the drawings, but not of the notes in the Aztec language, exist; they were first published in 1635; *See also* 1550–59 FOO

Robert Hooke's *De potentia restitution, or of a spring* of 1679 describes the rule now known as Hooke's law of the spring: Force is proportional to extension; *See also* 1941 MAT

Gottfried Wilhelm von Leibniz invents a computer in 1673 that uses Pascal's adding machine as its basis but that can also multiply and divide; he builds several copies of the device over the next 4 years; *See also* 1640–49 TOO; 1820 TOO

In 1675 Christiaan Huygens is the first to develop a practical spring-driven clock; it uses a spiral spring similar to that used in mechanical wound watches and clocks of today; *See also* 1650–59 TOO

Robert Hooke invents a form of universal joint in 1676; *See also* 1660–69 TOO

A Jesuit priest living in China reportedly develops a vehicle that is propelled by a form of steam turbine

Anthony Deane publishes his *Doctrine of naval architecture* in 1670

1680 to 1689

Denis Papin describes in 1682 the use of his "steam digester," first demonstrated in 1679; it is a pressure cooker that he uses to produce gelatin from animal bones; he also fits the device with a safety valve invented by himself; *See also* 1690–99 ENE

A lime kiln is established in Pennsylvania in 1681, setting in motion one of the first chemical industries in North America

Marcello Malpighi [b Crevalcore (in Italy), Mar 10 1628, d Rome, Nov 30 1694] undertakes the first scientific study of fingerprints; *See also* 1823 GEN

The centrifugal pump is invented; *See also* 1640–49 TOO

Guillaume Amontons [b Paris, Aug 31 1633, d Paris, Oct 11 1705] invents a hygrometer, an instrument for measuring humidity, in 1687

Giovanni Alfonso Borelli [b Giovanni Alonzo, Naples (in Italy) Jan 28 1608, d Rome, Dec 31 1679] shows in his book *De motu animalium* (Concerning animal motion) (published posthumously in 1680 and 1681) that human muscles are not strong enough in proportion to human weight for flight similar to that of birds

Pierre-Paul Riquet's Canal du Midi connecting the Mediterranean Sea and the Bay of Biscay in the Atlantic Ocean is opened in May 1681; *See also* 1660–69 TRA; 1690–99 TRA

In his 1687 *Systemate mundi* (System of the world), the third volume of the *Principia*, Isaac Newton describes how to launch an artificial satellite with a cannon and provides a diagram to illustrate

1690 to 1699

The *Collection for improvement of husbandry and trade* starts publication in England in 1691; it is a journal for artisans

Thomas Corneille [b Rouen, France, Jul 20 1625, d Les Andelys, Dec 8 1709] publishes in 1694 *Le dictionnaire des arts et des sciences* (Dictionary of arts and sciences) as a supplement to the dictionary of the French Academy; the Corneille work is one of the predecessors of modern encyclopedias; *See also* 1700–09 GEN

Christoph Polhem [b 1661, d 1751] establishes a mechanical laboratory in 1697 for the study of simple machines, or machine elements, which he terms the mechanical alphabet; the different machine elements are studied with the help of mechanical models

Henry Winstanley, after 2 years of labor, succeeds in 1698 in building the first lighthouse on the Eddystone Rocks off Plymouth, England; at 18 m (60 ft) or 24 m (80 ft), depending on whose account is followed, it is not high enough; spray continually puts out the 60 tallow candles that are the source of light; the next year he makes it 12 m (40 ft) taller to escape the ocean spray; *See also* 1610–19 ARC; 1700–09 ARC

Denis Papin describes in 1690 an apparatus in which the condensation of steam in a cylinder creates a vacuum in *Dion. Papini Nova methodus ad vires motrices validissimas levi pretio comparandas* (Papin's new method of obtaining very great moving powers at small cost); in his steam engine of 1698 the piston is moved by the pressure of steam rather than atmospheric pressure; *See also* 1660–69 ENE; 1700–09 ENE

The Miner's Friend, invented in 1698 by Thomas Savery [b Shilstone, England, c 1650, d London, May 1715], designed to pump water from coal mines, is patented; it becomes the first practical machine powered by steam; *See also* 1700–09 ENE

1700 to 1709

In England men begin wearing suspenders to hold up their trousers; *See also* 1580–89 GEN

Vincenzo Maria Coronelli [b Venice (in Italy) c 1650] begins in 1701 to publish his *Biblioteca universale sacro-profana* in Italian, the first of the great alphabetically arranged encyclopedias; he completes only seven volumes (A to Caque) by 1706, when the project is abandoned; *See also* 1690–99 GEN

Francis Hauksbee [d c 1713] produces a faint electric light by agitating mercury in a vacuum; he demonstrates his discovery to the Royal Society of England in 1703; over the next few years he continues to demonstrate various electrical experiments; he describes this in his *Physico-mechanical experiments* of 1709 and in public lectures in 1712; *See also* 1650–59 GEN

The Eddystone Light, undergoing repairs, is swept away in a storm in 1703, along with the repair ship and Henry Winstanley, its builder; *See also* 1690–99 ARC; 1755 ARC

Thomas Savery publishes *Treatise on fortifications* in 1705, translated from the Dutch book by Baron Coehoorn

The second Eddystone Light, built by John Rudyerd starting in 1706 and completed in 1709, is a 28-m (92-ft) circular, wooden tower; *See also* 1755 ARC

The English *Daily courant*, started in 1702, is thought to be the world's first daily newspaper; *See also* 1600–09 COM

England's Copyright Act of 1709 protects authors for 21 years, guaranteeing an income proportional to the sales of a book

London clergyman John Harris [b c 1667, d 1719] produces in 1704 the first alphabetical encyclopedia in English, the *Lexicon technicum, or an universal English dictionary of the arts and sciences*; it is in one volume; a second volume appears in 1710; *See also* 1720–29 GEN

Thomas Savery's *The miner's friend* of 1702 gives a description of his steam engine; *See also* 1690–99 ENE; 1700–09 ENE

Denis Papin in 1707 develops a modified version of Thomas Savery's high-pressure steam pump; *See also* 1710–19 ENE

Isaac Newton in 1709 builds an electric generator consisting of a rotating glass sphere; F. Hawksbee improves the design by using a metal chain to capture the electricity generated by friction; *See also* 1660–69 ENE

During the 150 preceding years, the Netherlands acquires 167,264 hectares (413,318 acres) of new land for farming by diking and draining land covered by the sea; *See also* 1640–49 FOO

John Woodward [b Derbyshire, England, May 1 1665, d Apr 25 1728] demonstrates in 1699 that plants grow best in water containing other substances; *See also* 1620–29 FOO

Clopton Havers [b Stambourne, England, c 1655, d Willingale, England, Apr 1702] publishes in 1691 the first complete textbook on the bones of the human body

German mathematician Gottfried Leibniz completes in 1694 a calculating machine called the "Stepped Reckoner;" it uses binary representation of numbers for its operation and can store a multiplicand in a register, thus eliminating the need of successive additions to effect a multiplication; the calculator can also divide and extract square roots; *See also* 1670–79 TOO

The 74-km- (46-mi-) long French Canal of Orléans is completed in 1692 and connected to the Loire River; *See also* 1680–89 TRA

Pierre-Paul Riquet's Canal du Midi (Languedoc Canal) is finally completed in 1692, 12 years after Riquet's death; it is 240-km (150-mi) long; 100 locks carry it to a height of 63 m (206 ft) above the Atlantic and 189 m (602 ft) above the Mediterranean, which it connects; *See also* 1680–89 TRA; 1761 TRA

Paul Hoste publishes *Théorie de la construction des vaisseaux* (Theory of the construction of vessels) in 1697, in which he describes designs of ships for which the center of gravity lies higher than the hydrostatic center of force

1690 to 1699

Jethro Tull [b Basildon, England, Mar 30 1674, d Hungerford, England, Feb 21 1741] invents in 1701 the machine drill for planting seeds in rows; this enables cultivation between rows; *See also* 1570–79 FOO; 1730–32 FOO

Nicolas Fatio de Duiller uses gems for bearings in clocks in 1704

Johann F. Böttger [b Schliez (in Germany), Feb 4 1682, d 1719] and Count Ehrenfried von Tschirnhaus discover in 1708 how to make true hard-paste porcelain from kaolin, a process that had been discovered and kept secret in the Orient

The eldest Abraham Darby [b England, 1677, d 1717] sets up an ironworks at Coalbrookdale, England, where Darby improves the manufacture of coke by heating coal in air-tight containers, which reduces the impurities in the finished product; *See also* 1660–69 MAT; 1750 MAT

Giacomo Pylarini [b 1659, d 1715], considered by some the first immunologist, in 1701 inoculates three children with smallpox in Constantinople in the hope of preventing development of more serious cases when they are older; *See also* 1710–19 MED

In 1707, *The physician's pulse watch* by John Floyer [b Hintess, England, 1649, d Litchfield, England, Feb 1 1734] introduces pulse-rate counting into medical practice and puts forth a special watch for it; *See also* 1600–09 MED

The flintlock for igniting a charge in small arms is developed from the snaphaunce lock

Christoph Polhem of Sweden in 1700 improves the rolling mill, devising mills that can produce metal plates with profiles; *See also* 1550–59 TOO

Charles Plumier [b 1646, d 1704] publishes *L'art de tourner* (The art of turning) in 1701, which gives the first detailed description of using a lathe for turning iron

Gabriel *Daniel* Fahrenheit [b Danzig, (Gdansk, Poland), May 24 1686, d The Hague (in the Netherlands), Sep 16 1736] constructs an alcohol thermometer in 1709; *See also* 1650–59 TOO; 1730–32 TOO

1700 to 1709

1710 to 1719

GENERAL

Mary Butterworth invents in 1716 a method for counterfeiting bank notes by placing a sheet of muslin cloth on a genuine bill and poking out the letters on the muslin; the image is transferred onto paper with a hot iron, and enhanced with quill pen and ink

ARCHITECTURE & CONSTRUCTION

The first North American lighthouse north of Mexico is the Boston Light on Little Brewster Island in Boston Harbor; at sunset on Sep 14 1716 it is lit for the first time; the light lasts until the Revolution, when it is destroyed first by colonists and then, more completely, by the British as they evacuate Boston in 1776; *See also* 1700–09 ARC

COMMUNICATION

Jacob Christoph Le Blon invents three-color printing in 1710; *See also* 1480–89 COM; 1710–19 COM

Henry Mill, a London engineer, in 1714 patents the first known typewriter; little is known today about how the device worked, if it did; the patent claimed his device could impress letters "so neat and exact as not to be distinguished from print"; *See also* 1867 COM

Jacob Christoph Le Blon in 1719 patents a four-color color-printing system based on the reconstruction of colors from primary colors by superimposing images obtained with plates inked in blue, yellow, red, and black; *See also* 1710–19 COM

ENERGY

Thomas Newcomen [b Dartmouth, England, Feb 24 1663, d London, Aug 5 1729] in collaboration with Thomas Savery erects in 1712 near Dudley Castle the first practical steam engine to use a piston and cylinder, bringing the engine out of the laboratory and into the workplace for the first time; it drives a pump in a mine and produces about 5.5 hp; *See also* 1700–09 ENE; 1720–29 ENE

1720 to 1729

GENERAL

The first known attempt to brighten a lighthouse light is made with a tin reflector installed in 1727 at the Cordouan Light off the coast of France; *See also* 1610–19 ARC; 1763 GEN

La science des ingénieurs by Bernard Forest de Bélidor [b Catalonia, Spain, c 1693, d Paris, 1761], published in 1729, is a popular manual of construction rules and tables that is reprinted until 1830; *See also* 1580–89 GEN

Ephraim Chambers [b Kendal, England, d May 15 1740] publishes in 1728 one of the most influential and popular works of the time, the *Cyclopedia, or an universal dictionary of art and sciences* in two volumes; in translation it is the first complete Italian encyclopedia as well as the progenitor of the great French encyclopedia of Diderot; *See also* 1700–09 GEN; 1743 GEN

ARCHITECTURE & CONSTRUCTION

Nicholas Bion [b France, 1653, d Paris, 1733] describes in 1723 the surveying instruments of the day, most of which are still in use

Bernard Forest de Bélidor publishes in 1729 *La science des ingénieurs* (Knowledge—or skill—of the engineers); it is a handbook in mechanics for civil engineers that deals with forces in beams, pressure of soil, and construction of retaining walls; because of its general usefulness, it is reissued over the next hundred years; *See also* 1737 ARC

COMMUNICATION

In 1725 William Ged [b Edinburgh, 1690, d Edinburgh, Oct 19 1749] improves stereotype, a printing technique developed during the 15th century; pages are printed from plates that are molded on type; the plates can then be used again for reprints; *See also* 1793 COM

Johann H. Schulze discovers in 1727 that silver salts turn black as a result of exposure to light; *See also* 1560–69 COM; 1810 COM

ENERGY

The first steam engine on the European continent is erected in 1720 at Königsberg (Kalingrad); *See also* 1710–19 ENE

Jacob Leupold [b Planitz, Germany, Jul 25 1674, d Leipzig, Germany, Jan 12 1727] publishes *Theatrum machinarum generale* (nine volumes, 1723-1739), offering the first systematic treatment of mechanical engineering; the work includes the design of a noncondensing, high-pressure steam engine comparable to those built at the beginning of the 19th century; *See also* 1710–19 ENE

A steam engine is operating in 1729 at a coal mine at Jemeppe-sur-Meuse, near Liège; it is one of the earliest on the Continent

John May and John Meeres build in 1726 a steam engine to provide Paris with water from the Seine at Passy near Paris; *See also* 1730–32 ENE

					1710 to 1719

FOOD & AGRICULTURE — Thomas Masters files a patent in 1715 for the invention by his wife of a machine for preparing corn using a combination of wooden cogwheels, mortars, and trying trays, powered either by a horse or a waterwheel

MATERIALS — René Antoine Ferchault de Réaumur [b La Rochelle, France, Feb 28 1683, d France, Oct 18 1757] presents to the Academy of Sciences in Paris in 1710 a material entirely woven from glass fiber that probably was produced by Carlo Riva of Venice

MEDICAL TECHNOLOGY — Emanuel Timoni describes to the British Royal Society the Turkish practice of inoculating young children with smallpox to prevent more serious cases of the disease when they get older; *See also* 1700–09 MED

Dominique Anel [b Toulouse, France, 1679, d 1730] in 1714 invents the fine-point syringe, still known by his name, for use in treating *fistula lacrymalis*

Lady Mary Wortley Montagu [b London, 1689, d London, Aug 21 1762] brings back to England the Turkish practice of inoculation; she has her own two children vaccinated in 1717 against smallpox; *See also* 1700–09 MED; 1720–29 MED

TOOLS & DEVICES — The eight-day clock, which needs to be wound only once a week, is developed in 1715

George Graham [b Cumberland, England, c 1674, d 1751] invents in 1715 the deadbeat escapement for clocks, a device that gives a tiny shove to the pendulum near the center of its swing and drags slightly on it as it goes to the extremities, improving the accuracy of the clock to a few seconds per day; *See also* 1670–79 TOO; 1720–29 TOO

J.N. de la Hire in 1716 invents the double-acting water pump, which produces a continuous stream of water

TRANSPORTATION — Thomas Savery invents a form of ship's log in 1710, but it does not catch on; *See also* 1570–79 TRA; 1802 TRA

The British Parliament in 1714 passes a bill setting up a prize of 20,000 pounds for the first person to develop a sufficiently accurate way to find longitude at sea; *See also* 1610–19 TRA

The French in 1716 establish the first national highway department

In 1716, Edmond Halley [b Haggerston, England, Nov 8 1656, d Greenwich, Jan 14 1742] develops a diving bell with a system for refreshing the air and demonstrates its use; *See also* 1620–29 TRA

					1720 to 1729

FOOD & AGRICULTURE — Joseph Foljambe takes out the first English patent on a moldboard plow sheathed in iron in 1720, although iron-sheathed plows had been used in England even before Roman times; *See also* 1819 FOO

The possibility of cross-fertilization in corn is discovered in 1724

MATERIALS — René de Réaumur discovers the role of carbon in the hardness of steel; his *L'art de convertir le fer forgé en acier* (The art of converting iron into steel) of 1722 is the first technical treatise on iron; *See also* 1660–69 MAT

Hammered, wrought-iron plate becomes commercially available

MEDICAL TECHNOLOGY — Zabdiel Boylston [b Brookline MA, Mar 9 1679, d Brookline, Mar 1 1766] introduces inoculation against smallpox into America during the Boston epidemic of 1721; *See also* 1710–19 MED

Jean Palfyn introduces in 1721 the use of forceps for facilitating birth

Drinkers from North Carolina complain that rum from Massachusetts causes stomach problems and partial paralysis; after Boston physicians attribute the problem to lead parts used in the stills that produce the rum, the Massachusetts legislature outlaws that use of lead in 1723; *See also* 1768 MED

TOOLS & DEVICES — George Graham develops in 1720 the cylinder escapement, a version of his deadbeat escapement that can be used with clocks based on balance wheels instead of pendulums; *See also* 1710–19 TOO

In 1721 George Graham develops the mercury compensating pendulum for clocks; a jar of mercury is used as a pendulum bob so that as the mercury expands with temperature it will just compensate for the expansion of the brass rod of the pendulum; *See also* 1650–59 TOO

In 1725 George Graham develops the horizontal escapement, which incorporates a rotating cylinder that replaces the function of the anchor

155

1720
to
1729
cont

Recognizing the power of steam

That steam can exert power was known by the ancient Greeks. But as was the case with many of their discoveries, they applied this knowledge only to toys. Hero of Alexandria built such a toy, the aeolipile. It consisted of a spherical vessel fitted with two jets pointing in opposite directions; when steam was admitted to the sphere, it escaped through the two jets, causing the sphere to rotate.

During the Renaissance there was renewed interest in the power of steam. Although it is not true that James Watt saw the lid rising on a pan of boiling water and got the idea for the steam engine, someone surely did. Noticing that steam in an enclosed place has the power to lift objects is the first step toward building a steam engine. Alternately lifting with steam and letting the steam cool back into water is the simplest form of a steam engine. The natural philosopher Giambattista della Porta recognized the two advantageous aspects of steam: by creating pressure, steam could, for example, force water out of a vessel, or by condensing, it could create a vacuum. He demonstrated the latter property by filling a flask with steam and plunging it underwater. As the steam cooled and condensed, a vacuum formed in the flask, drawing in water.

It was Denis Papin, who had studied medicine and physics, who first thought of using steam to move a piston in a cylinder to deliver work. He started working for the French Academy of Science and became the assistant of Christiaan Huygens, the Dutch astronomer who was one of the founding members of the academy. Huygens was then experimenting with air pumps and with a gunpowder engine. One of Huygens's aims was to create a vacuum by exploding gunpowder in a cylinder with a piston. The piston would shoot to the top of the cylinder and thus create a vacuum.

Papin, who also experimented with such devices, found that steam was not only a much better agent to create a vacuum, but could be used to deliver work. His simple steam engine of 1690, also called Papin's cylinder, consisted of a cylinder closed at the bottom, containing water, and sealed off by a movable piston. When the water boiled, the piston rose, and when the steam condensed, creating a vacuum, it descended.

Papin, being primarily a scientist—he even taught mathematics at a German university—never did build a practical engine that could deliver mechanical work. People less removed from industry, such as the British engineer Thomas Savery and later Thomas Newcomen, devised the first machines that put steam to work.

The first device that used steam for delivering work did not have a piston. It was devised by the

---

1730
to
1732

Benjamin Franklin [b Boston MA, Jan 17 1706, d Philadelphia PA, Apr 17 1790] publishes the first issue of *Poor Richard's almanack* in 1732; *See also* 1744 ENE

Steam engines are used for rotary motion in 1732 by pumping water into a tank that feeds a waterwheel; *See also* 1720–29 ENE; 1763 ENE

Recognizing the power of steam (Continued)

British engineer Thomas Savery and was patented in 1698. His design, a water pump used for pumping water out of mines, and therefore called the Miner's Friend, made use of both the ability of steam to exert pressure and to create a vacuum. First steam from a boiler was admitted to a vessel; after the steam supply valve was closed, it was condensed by a spray of water. The created vacuum in the vessel raised water through a one-way valve, and when the vessel was filled, steam was admitted into it again, forcing the water upward through a second one-way valve. Savery produced a working model of his steam pump, but there are no records of such engines being used in mines. It is known that one engine worked at the York Buildings in London, pumping water from the Thames. Its use was abandoned because of a number of technical difficulties—the main one being that the precision workmanship and strength of materials at the time were insufficient for this type of machine.

In 1726 Stephen Hales [b Bekesbourne, England, Sep 17 1677, d Teddington, England, Jan 4 1761] takes the first measurement of blood pressure, of a horse

The post crown, an artificial top portion of a tooth mounted on a post inserted into the root canal, is introduced in 1728; *See also* 1890 MED

De motu cordis et aneurysmatibus by Giovanni Lancisi [b Rome, Oct 26 1654, d Jan 20 1720] is posthumously published in 1728, discussing heart dilatation

Basile Bouchon develops in 1725 a semiautomatic weaving loom in which the index fingers (which lift and lower the warp threads) are controlled by an endless paper tape; feeding the paper tape is still done manually; the loom becomes fully automatic in later designs by Vaucanson and Jacquard; *See also* 1775 TOO

In 1726 John Harrison [b Yorkshire, England, Mar 24 1693, d London, Mar 24 1776] constructs a gridiron compensating pendulum clock to counteract the expansion in the length of a pendulum caused by temperature

Jacques de Falcon builds in 1728 a semiautomatic weaving loom in which the index fingers (which lift and lower the warp threads) are controlled by perforated cards; the cards must be fed into the machine manually

---

Jethro Tull publishes *Horsehoeing husbandry* in 1731; *See also* 1700–09 FOO

Charles Marie de la Condamine rediscovers rubber on the Amazon River (it was known from the time of Columbus's second voyage); Jacques de Vaucanson will use the material a few years later to make flexible hoses; *See also* 1490–99 MAT

George Martine [b Scotland, 1702, d 1741] performs the first tracheotomy for treatment of diphtheria in 1730

René de Réaumur in 1730 constructs an alcohol thermometer with a graduated scale from 0 to indicate freezing to 80 to indicate boiling; *See also* 1700–09 TOO

Independently, Englishman John Hadley and American Thomas Godfrey in 1731 invent the sextant, an instrument for finding the altitude of the Sun that is based on the equality of the angle of incidence and the angle of reflection; *See also* 1490–99 TOO

OVERVIEW

The Industrial Revolution

The history of science and technology is often described as exponential. From a period of millions of years it becomes reasonable to focus on one of a few thousand years and then on one of a few hundred years. The period we have labeled the Industrial Revolution is only about 150 years long, from the invention of the flying-shuttle loom in 1733 up to the first practical electric lamps in 1879.

But is is clear that it was not the loom by itself that separated the period before 1733 from that afterward. Instead, it was the coming of factories and the many related social changes during the early 18th century that were both caused by and were the cause of changes in technology. Similarly, the light bulb represented the beginning of a new set of changes that included large-scale electric networks, the giant dams to power them, the changes that electric appliances brought to lifestyles, from air-conditioning to television and beyond, factories with dispersed sources of power that helped promote the assembly line, and so forth. All these changes, like other technological trends before and after them, were massively interrelated. The Metal Ages are not simply characterized by the use of metals; the Industrial Revolution is not just about industry.

The Development of Industry

Different definitions yield entirely different ideas of what the Industrial Revolution was. Define industry in terms of energy use and perhaps the Industrial Revolution happened when people stopped using human and animal power and began using inanimate power sources; such a definition could place the Industrial Revolution as early as the time of the first extensive use of wind and water power—medieval times in the West and somewhat earlier in China. Define industry in terms of production, and the revolution began when mills began to centralize the production of textiles. This might be interpreted as the 13th century, the start of the time of fulling mills. If the date is tied to changes in the materials people use, such as the first large-scale production of iron in blast furnaces, the use of which called for the employment of workers in something like a factory system, the revolution would be postponed to the 15th and 16th centuries. Or maybe it is the factory system that is equivalent to industry. The first factory was probably a silk-thread mill built in 1719; that was six stories high and employed 300 workers, mostly women and children.

A few years after that first factory, John Kay made one of the key inventions that started this Industrial Revolution, the flying shuttle and, about that time, Abraham Darby discovered how to make cast iron using coke instead of charcoal.

But both of these inventions are too early to indicate the Industrial Revolution if you concentrate on the revolution part. Few lives were influenced by the early 18th-century inventions in weaving and steel making. It was years before other smelters succeeded in following Darby's example; and weavers afraid of losing their jobs destroyed Kay's loom and sent him packing to France.

About 1750, however, cotton workers, with less of a tradition behind them than the wool workers who attacked Kay's loom, started using the flying shuttle. This set into motion a chain of events that revolutionized the textile industry by any definition. With the flying shuttle, cotton workers were able to weave so much faster that they ran out of yarn. Seeing an opportunity, James Hargreaves invented the spinning jenny, a machine that multiplied the amount of yarn produced. This time the spinners were upset, and they destroyed some of Hargreaves's machines. The spinning jenny could make only one of the two types of yarn needed for weaving. Richard Arkwright also saw opportunity, so he invented the water frame, a machine that produced the other type of yarn. Unlike the spinning jenny, the water frame was too large and too expensive to put in a cottage. Arkwright had to build a factory to house his machine; he is considered the founder of the modern factory system.

Arkwright's factory was operating in 1769, by which time the Industrial Revolution had definitely begun; many date the start back to the 1750s, when the cotton weavers first adopted the flying shuttle.

Yet, there is another key invention not yet in place in 1769. Since at least 1629, when Giovanni Branca suggested using steam to propel a turbine, people had been experimenting with steam power. Branca was followed by the Marquis of Worcester in England, Denis Papin in France, and, in England again, Thomas Savery. Savery, at the very end of the 17th century, was the first to make a practical steam device, known as the Miner's Friend. Thomas Newcomen developed an actual steam *engine* that was more along the lines of the modern machine. Newcomen then got together with Savery, and the new engine was successfully manufactured and sold to drain mines. Over 100 Newcomen engines were installed during the 18th century.

James Watt, an engineer at Glasgow University, developed in 1765 a new device, the steam condenser, that greatly improved the efficiency of the Newcomen engine. Ten years later, Watt teamed with a manufacturer of iron products, Matthew Boulton of Birmingham, to manufacture his new engine. Boulton had access to the technology needed to make the finely machined parts that gave the Watt engine greater efficiency and durability. Boulton also convinced Watt in 1781 to convert the engine from a simple pump to a device producing rotary power, thus creating the first steam engine that could power other machinery. Four years later the first steam-powered cotton mill opened in Papplewick, Nottinghamshire. Certainly by this date, the revolution part of the Industrial Revolution had been completed. After that came consolidation of power by the revolutionaries.

Lighting the Revolution

During the Industrial Revolution, people learned to light up the dark. The use of coal and coke for making steel or powering steam engines became commonplace, and thus technical-minded people became familiar with the gases produced during the manufacture of coke. These gases burned with a pale light. Rather than waste the gases, various enterprising inventors in Germany, France, and England experimented in the early years of the 19th century with coal-gas lighting for houses, factories, and city streets. Cities were being lit this way as early as 1811, and by 1820 the major cities of Europe were all partially illuminated. One name for the mid-19th century is the gas-light era, referring to the period of somewhat more than 50 years during which gas lighting was used on a large scale and before electric lights were introduced.

Although gas light is not so bright as electric light, it is bright enough to extend the day beyond the natural dawn-to-dusk requirements of the human as a diurnal animal. Gas light also made it possible for large indoor structures to be used as factories. Before gas lighting, much work that we would now do indoors was conducted in the open, with small interior workrooms for rainy-day labor adjacent to the work site. Eventually the gas-lit, and later the electric-lit, factories would lead to second and third shifts of workers, making better use of capital if not of human beings.

Although most of the 19th century was illuminated by coal gas (natural gas was not exploited for any purposes to any extent until the 20th century), experiments with electric lighting were ongoing throughout the century as well. It was not until 1858, however, when the first arc lamp was installed in a lighthouse, that practical electric lighting existed. Much better arc lights were not invented and used until just before the first long-burning incandescent lamps were constructed by Thomas Edison and Joseph Swan.

Rise of the Engineer

The major technological breakthroughs and building enterprises before the Industrial Revolution were the work of amateurs, craftspeople, and military leaders. The profession of engineer was one of the great 18th century inventions. When we think of the great technological achievements of the 18th and 19th centuries, the names that come to mind are sometimes those of inventors, such as Cyrus McCormick, Josiah Wedgwood, or Joseph-Marie Jacquard, and sometimes those of scientists, such as Michael Faraday, Nicolas LeBlanc, William Perkin, or Joseph Henry, but primarily they are the names of engineers, especially British engineers, that come to mind: James Brindley, Marc Isambard Brunel, George Cayley, Ferdinand de Lesseps, John Ericsson, Robert Fulton, Henry Maudslay, John McAdam, William Murdock, Joseph Paxton, John Rennie, John Augustus Roebling, John Smeaton, George Stephenson, Thomas Telford, James Watt, and John Wilkinson. Today we would class some of these people as mechanical engineers and others as civil engineers, but before the Industrial Revolution few such professions existed at all. The builders of the great French canals of the 17th century and of the Italian canals about 200 years earlier were among the first civil engineers, but they did not think of themselves this way. John Smeaton was the first person to call himself a civil engineer.

Civil engineering actually started in France 3 years before Smeaton set up shop on his own when, in 1747, the French government started a school of bridges and highways. By 1771 there were enough engineers in England for them to begin to gather together on a regular basis in what was called the Smeatonian Club. France continued to provide the lead in engineering education, however, reorganizing the school of bridges and highways and starting the Ecole Polytechnique in the 1790s. In the early 19th century the Smeatonian Club led to the more formally organized Society of Civil Engineers in England, the first organization of its kind.

Science and technology were united in the same person or the same school for these new developments in engineering. Around the same time, science and technology were also united in a new form of publication. Encyclopedias, such as Ephraim Chambers's of 1728, Denis Diderot's of 1743, Jacques Buot's of 1761, the Brittanica of 1771, Friedrich Brockhaus's of 1808, and the Americana of 1847, were all largely intended to bring the new technology to the masses and to concentrate the sciences and humanities.

Major Advances

Architecture and Construction.
It is perhaps not surprising that the principal progress in architectural and construction during the Industrial Revolution centered around industry—canals, bridges, tunnels, and highways were built in profusion. Domestic, ecclesiastical, and governmental buildings were built, but with little advance in technique until James Paxton's Crystal Palace at Queen Victoria's Great Exhibition in London. The great halls and fake castles of the time used earlier construction techniques imitated those styles. The dominant building of the period, the factory, was seldom architecturally distinguished.

The view of the Industrial Revolution in the US, Canada, and England is heavily skewed toward events in England. Thus, we think of the Canal Age as starting in the second half of the 18th century, although the major canal building in France, even more extensive than in England or the US, was a century earlier. Perhaps the greatest architectural achievements of this age, however, were English and American. These were the iron bridges, truss bridges, and suspension bridges that are among the wonders of the time. On the Continent, major bridges were built, but the finished product was not that different from Roman or Renaissance bridges. As with the canals, the main continental development (introduction of caissons, open at first at the top above the waterline) preceded the beginning of the Industrial Revolution.

The first important iron bridge, built near the end of the 18th century over the river Severn in England, still stands. About the same time in the US wooden truss and suspension bridges began to be built. A major suspension bridge built by Thomas Telford in Wales in 1825 combined these ideas. Before the end of the period, the graceful Brooklyn Bridge in New York City reflected the state of the art.

Bridge and tunnel construction also led to significant studies of forces and the strength of materials. Pioneering work by French engineers, especially those connected to the Ecole des Ponts et Chaussées (School for Bridges and Highways), founded in 1747 and reorganized in 1793, led to better understanding of the mathematical side of engineering.

Communications.
At the beginning of the 18th century, the only way to communicate with anyone farther away than line of sight was to write a letter by hand and arrange to have it carried by a messenger. At the end of the 19th century, not only could you send a telegram across oceans and continents and converse over a telephone for shorter, but still substantial distances, but you could also use the government mails to send a typed message or a reproducible recording. You could even mail a photograph, for photography developed during this same period.

Surprisingly, the concepts of color photography and of television were implicit almost from the earliest days of

From earliest civilized times people built canals for irrigation. Anyone who has ever portaged even a small canoe from one stream or lake to another has envisioned a canal replacing his or her trail, so the canal idea was extended from providing water for crops to providing a highway for boats. The technology of a short, level canal between two bodies of water at the same height is as simple as can be—all you need is people to dig. In China, where finding enough people has never been a problem, transportation canals became important as early as the 3rd century BC. In the West, early transportation canals were developed primarily to shorten sea routes, but Chinese canals connected river systems and bound people together. Terrain was one of the reasons Chinese canals got a head start on European ones. Many of the obvious places that needed connecting in Europe were separated by rocky hills or even mountains. Where it was clear that a canal could help shipping, as across a narrow peninsula or between nearby bodies of water, Westerners built or at least planned to build canals.

Intervening changes in land height and different water levels posed problems. Canals could be and were built in sections; boats were hauled up artificial rapids from one level section to another or even on land around barriers. In the 10th century in China, the canal lock, or pound lock, was developed to permit better connections. Such locks enabled the Chinese Grand Canal to rise from sea level to 42 m (138 ft) above. Several hundred years later the idea of the lock was taken up by the West at a time when many Chinese influences began to be felt in Europe, although by that time locks and canals were no longer being built in China.

With the coming of locks, certainly known in Italy in the 15th century, European canal building began to enter its great period of expansion. France began to build major canals in the 17th century, while England and Northern Europe became more active in the 18th. In France, the grand project of connecting the two French coasts (Atlantic and Mediterranean) continued intermittently for more than 150 years before becoming a reality in 1681. Other major French canals were built around this time as well. Like most canals that were to follow, the French canals were too narrow for much sailing. Instead, ships were towed through the canals by animals, such as mules, that walked on paths along the banks.

The French canals inspired the English to do likewise. Notably the duke of Bridgewater, Francis Egerton, having seen French canals while on the Grand Tour, hired surveyor James Brindley, who had previously planned a canal but never built one, to construct a canal between coal mines on his estate at Worsley and the booming industrial center at Manchester. The Bridgewater (Worsley) Canal opened for traffic in 1761, and for some historians this date marks the beginning of the Industrial Revolution, for the cheap coal, available year-round, fueled all the factories of Manchester. Brindley was much sought after as a canal designer by various commercial enterprises. By 1792 there were no less than 30 different canal schemes competing for capital financing in England. The craze spread again from England back to the Continent, inspiring canal projects in Scandinavia, Russia, the Low Countries, the German duchies, and even the country of England's inspiration, France. Not only were new canals built, but also existing rivers were converted into canals by building walls and leveling river bottoms.

By the end of the 18th century or the beginning of the 19th, canal building had spread far beyond Europe. Canals were a feature of life in the US and Canada in the early 19th century. Egypt, with an early history of canal building that had lapsed two millennia earlier, linked Alexandria with the navigable Nile using a canal built by 350,000 peasants, working largely with their bare hands.

Although the railroad was to replace the canal for inland commercial hauling in the 19th century, there were still two great ship canals needed to carry goods beyond the capacity of rails. Ships carrying goods or passengers from Europe or the East Coast of North America, the two main commercial centers of the world of the 19th century, had to pass around either Africa or South America to reach the lucrative and populous Asian continent. Fairly short canals, albeit through difficult terrain, could shorten such journeys by thousands of kilometers. The solution for Africa was the Suez Canal, between the Mediterranean and Red seas, which opened in 1869, although it was not commercially viable until the British deepened and widened it late in the 19th century. The Panama Canal connected the Atlantic and Pacific oceans in 1914. Both great canals were among the most colossal engineering enterprises of their time. A glance at the map of the world shows that these two canals offered the only available opportunities to improve linkages so dramatically with relatively short waterways. Although canals will continue to be built for special purposes, the great age of canal building—on Earth at least—is over.

Although the age is past, canals delight people still. Devoid of serious shipping, remaining inland canals boast pleasure boats and even special canal excursions. A few canals—the Panama, Suez, and St. Lawrence Seaway—continue to have great economic presence, but for the most part canals reflect a more gracious period for shipping that passed well over a hundred years ago.

The mechanization of farming

Before the Industrial Revolution, the physical operations connected with farming had not changed much for about 4000 years. People in most parts of the world still turned over ground with plows pulled by animals, broadcast or sowed seed by hand, removed weeds with hoes or by pulling them out of the ground (if weeds were removed at all), cut grain with sickles or scythes and harvested most other crops by hand, removed grain from straw by beating it with devices called flails or by trampling the harvest with animal or human feet, separated grain from chaff (winnowed) by tossing the mixture in the air on a windy day, and ground grain afterward by rubbing it between rocks. In Europe, the last step had been mechanized since Roman times, while some of the other steps, notably sowing and winnowing, had been mechanized in China, also around the 1st century.

Chinese methods for farming were generally more advanced in terms of crop rotation, fertilizer, pest control, and mechanization than those in Europe before the Industrial Revolution. Some of the Chinese ideas, albeit not the actual Chinese devices, reached Europe as the China trade picked up in the 17th century. At the beginning of the 18th century, English inventors and farmers began to reinvent the devices and methods. In some cases, as in crop rotation or adding ground up rock to poor soil, earlier or localized European practices also were revived or spread, perhaps because growing population was making everyone conscious of the need to improve the food supply. Toward the end of the 18th century, Thomas Malthus was making the difficulty of feeding a growing population a serious intellectual issue.

Even so, innovation in farming usually spread slowly. By the end of the 18th century, Jethro Tull's seed drill, invented at the start of the century, was joined by the first threshing machines. Each new device took a generation or two to go from patents to popularity. It was not until the 1830s that full mechanization of farming got under way. Then all sorts of devices were developed to replace manual planting, cultivating, harvesting, and threshing.

Early farm machines were drawn by horses or oxen, unless, like the thresher, they sat in the field and were fed crops harvested with a different kind of machine. This last difference was eventually resolved when combined harvester-threshers, or combines, became popular. When the only power sources available were muscle, water, and wind, not much could be done except with animal power because waterwheels and powerful windmills could not be moved from field to field. A few advanced thinkers tried to move land machinery with sails, but this proved impractical.

Even when steam power was invented, the early steam engines were too heavy to move themselves over soil, although they could paddle boats and pull cars over rails. Some farmers plowed by putting a stationary machine engine at one end of a field and using it and cables to draw plows across the field. Some self-propelled tractors—*tractor* means puller—were made by combining heavy steam engines with treads, which did not sink into the fields as much as wheels did. By the end of the 19th century, however, lighter tractors with internal combustion engines were introduced. Still, it was not until halfway through the 20th century that more farms in the developed world used tractors instead of farm animals for pulling plows and machinery.

Although the main story of the mechanization of farming concerns grain farming, other farming operations also changed in the 19th century. Removing seeds from cotton, the equivalent of threshing grain, was among the first operations to be accomplished by engine, shortened from *engine* to *gin* by the workers. Animal husbandry is hard to mechanize, but milking machines and mechanical cream separators and churns came into being in the 1870s. A hundred years later, many of the operations of feeding and otherwise caring for animals were finally mechanized.

It was not until well into the second part of the 20th century that machines were developed for picking sensitive fruits or vegetables, such as tomatoes; and different machines had to be invented for each kind of fruit or vegetable. In many cases the plants that bore the fruit had to be specially bred for machine picking.

Some parts of farming have still not been mechanized. Despite new fruit-picking machinery, much is still picked by hand, although it may be graded by machine. Other operations, such as tree grafting and pruning, may be facilitated by machines, but essentially remain manual labor.

chemical photography and the earliest experiments with sending messages over wires. Inventors were caught up in the progressive spirit of the age and thought that almost anything was possible—although Edison was genuinely surprised by his discovery of sound recording. This was more unexpected than such fairly obvious applications of electromagnetism as the telegraph (practically everyone who learned about electricity seemed to think of a way it could be used to send signals) or such slowly developing technology as photography.

Energy. The revolution in energy was perhaps the most astounding to people at the time. At the beginning of the Industrial Revolution, energy was drawn solely from natural sources—muscles, water, and wind. By the end of this period, steam engines were prevalent; furthermore, the 1870s produced virtually modern forms of the internal combustion engine and the electric engine. The steam engine, powered mostly by coal, used a familiar source of energy. The main fuel for internal combustion engines (after some experimentation with other forms) was a new chemical, gasoline.

The development of current electricity was even more dramatic. To think that the force that attracted straw to amber that had been rubbed, the force that produced surprising "shocks" when a charged body was touched, the force just identified as present in lighting—that force!—could be produced in steady streams and transmitted by wires over considerable distances, was truly astonishing. Chemical means of producing an electric current were first introduced at the very beginning of the 19th century, and various electric batteries have been made since. With a source of current for experiments, scientists began to discover that currents could also be produced mechanically. At the end of the Industrial Revolution most people had begun to see that electricity was the wave of the future (James Clerk Maxwell explained its wave nature in 1873); it was already being applied in communication and lighting. The Electric Age was starting.

Food and Agriculture. Ancient peoples had learned that food could be preserved by salting or pickling (or adding some other chemical preservative, such as alcohol or honey, which in many ways is the equivalent of pickling), by drying, or, where natural ice was available, by cooling. Variations of these methods were all that existed until the 19th century, when canning in glass jars or metal cans was invented. Canning quickly became one of the main ways to preserve food, as it still it today. For one thing, canning does not alter the taste and texture of food as much as pickling and drying do.

Artificial refrigeration, although not so revolutionary an idea as canning, was even more important in practice, for nearly all perishable foods can be refrigerated for longer shelf life with virtually no deterioration in taste, texture, or nutrition for days or even weeks. Together, canning and refrigeration were truly revolutionary, for they directly improved the health of people all over the world. Although urban sanitation control is often cited as the main improvement in health and longevity of the past few centuries, certainly as important as immunization and antibiotics, the food preservation revolution (which has continued with frozen foods, freeze-dried foods, new forms of sterile packaging, and even food irradiation) merits the same recognition.

Although the Chinese had developed several ways of fighting insect pests, practices in the West lagged behind. As chemistry began to develop, some new chemicals and some old ones began to be used in systematic ways as pesticides. Most of these were poisons to vertebrates as well as insects, but the first glimmers of pest-specific poisons were seen during this period. Perhaps because farmers were aware of the dangers of 19th-century pesticides to themselves and to other animals, use of pesticides was not so extensive as to pose a serious environmental hazard at this time.

Materials. At the beginning of the Industrial Revolution there was wrought iron and precious steel and no other hard metal of significance in Europe, although China had known the secret of cast iron for centuries. By the end, cheap steel was the metal of choice for machinery and building. Wrought iron, although made much more cheaply and easily by the end of the 18th century, was almost all reserved for specialty purposes and cast iron was used only where its brittleness was not a deciding issue. The Iron Age began near the start of history, and in some senses it ended in the middle of the 19th century with a flood of cheap steel.

All this iron and steel would have been impossible without the substitution of coke for increasingly scarce charcoal. In the middle of the 18th century, a lucky combination of a source of low-phosphorus coal (phosphorus makes iron alloys brittle) and a strong foundry tradition enabled Abraham Darby to be the first to succeed in making a commercial success of coke-fired cast iron. Consequently, the demand for coke led throughout this period to better and better methods of producing that commodity. As an important consequence, coal gas was developed from coke production as the first good source of outdoor urban lighting and coal tar became the basis of much of the organic chemical industry of the late 19th and early 20th centuries.

Organic chemicals based on carbon were not the whole story, however. This was the period when the first major inorganic chemicals (other than metals) moved out of the laboratory and into industry. These included sulfuric acid (1746), sodium hydroxide, sodium carbonate, and oxymuriatic acid (1783—later identified as chlorine), useful rubber products (erasers, fabrics, and, in 1839, vulcanized rubber), nitric acid (1838), and sodium bicarbonate (1861). The chemical industry, essentially nonexistent at the start of the Industrial Revolution, was a major factor in society by its conclusion.

Although we think primarily of the coal-based chemicals and the inorganics during this period, a separate revolution was beginning, based on plant and animal materials, mainly wood. The plastics industry got under way with celluloid, like paper based on cellulose from plants. Finally, there was an array of new explosives (nitrocellulose, nitroglycerin, and dynamite) that were based to some degree on plant (cellulose) or animal (glycerin) products.

Medical Technology. Most authorities attribute today's longer life span primarily to improved sanitation. The main developments in the Industrial Revolution were the realization that contaminated water carries disease

Except for technical improvements made by John Smeaton, engineers in the middle of the 18th century built steam engines much the same way Thomas Newcomen designed his first engine.

Unlike Smeaton, who as a designer of machines naturally was led to the development of steam engines, the instrument maker James Watt started working with steam engines quite by chance. Watt had set himself up as the manufacturer of scientific instruments, and one of his first clients, the University of Glasgow, employed him to repair some of their scientific instruments. While repairing these instruments, he also worked with scientists, among them Joseph Black, and gained invaluable experience in physics. The university asked Watt to repair a model of a Newcomen engine. One of the first things Watt noticed was that the machine was extremely voracious for steam. Furthermore, most steam and heat were lost during the condensation of steam by a jet of water sprayed into the cylinder.

Watt reasoned that by keeping the cylinder hot at all times, which meant carrying out the condensation of steam in a separate vessel, he would be able to conserve steam and reduce fuel consumption. In 1765 he built a simple model of his steam engine with a separate condenser and found out that by avoiding the alternating heating and cooling of the cylinder he could save a lot of fuel. Watt also insulated the cylinder itself, reducing the loss of steam inside the cylinder by condensation on cold cylinder walls. Watt obtained a patent for "a new method of lessening the consumption of steam and fuel in fire engines" in 1769.

Later Watt realized another advantage of the separate condenser: The cylinder could be made double-acting. In a double-acting cylinder, steam is alternately directed above and under the piston, doubling the work that can be supplied by the piston.

Joseph Black had also brought Watt into contact with the industrial pioneer John Roebuck, and Watt and Roebuck formed a partnership. The partners started building an experimental beam engine (a steam engine whose action is to raise and lower a long lever, or beam) near Roebuck's residence, but the project did not progress. Roebuck had financial difficulties and Watt had to keep other commitments to earn his living. Roebuck went bankrupt in 1773 and Matthew Boulton, who had shown interest in Watt's machine since 1769, bought Roebuck's two-thirds share in the enterprise. Boulton proposed to build a factory to produce Watt engines "for all the world." In 1775 Boulton succeeded in extending Watt's patent, which would have expired before any machines could have been produced. In 1776, Boulton and Watt started work on the two first separate-condenser engines ordered, one by a coal mine for a pump and one by the ironmaster John Wilkinson, who needed a machine for pumping bellows for a blast furnace. Wilkinson had also just patented a new boring mill for cylinders. Use of Wilkinson's borer allowed the manufacture of cylinders for the steam engines to much higher specifications. The first two Watt engines were immediate successes, consuming only a quarter of the fuel of a Newcomen engine of similar power.

As soon as Watt developed his engine, Boulton proposed building a machine that could supply rotary motion. The idea of using a connecting rod and crank was not apparent because the stroke length of the beam machines used to vary considerably, and it was not understood that a flywheel, crankshaft, and connecting rod would solve this problem entirely. In any case, the crank used in conjunction with a steam engine was patented by another machine builder. Watt was forced to develop another method, and he chose to use the "Sun and planet" gear to obtain rotary motion. The piston rod was rigidly connected to the beam via a system of rods forming a parallelogram; the arrangement came to be called "parallel motion," although the important result was rotary motion.

Although the reciprocating motion supplied by the first steam engines was suitable for pumping water or powering bellows, it was the adaptation of those engines to rotary motion that made the Industrial Revolution possible. Previously, factories had to be set up near rivers and streams that could deliver waterpower, often far from cities and the necessary work force. Rotary steam power changed this situation completely. Factories were built in cities, near transport channels, such as rivers and canals, and near abundant work forces.

and that cleanliness is essential to surgery. Near the end of the period, the germ theory of disease explained both of these observations. Other progressive steps, such as the use of citrus fruit to prevent scurvy and vaccination to prevent smallpox, were found to work, but the rationale for them awaited scientific developments of the 20th century. Mental illness began to be looked at somewhat scientifically and treatments included hypnosis, new to science although probably known to shamans and traditional healers.

New tools for diagnosis included the ones most commonly used in a routine examination today: thermometers for body temperature, a stethoscope and thumping to listen to the heart and lungs, and early devices for measuring blood pressure. Dentists acquired the drill and became better at making and installing false teeth when the drill and various fillings, some invented during this period, failed to suffice. Familiar operations, such as the appendectomy and tracheotomy were introduced.

In many ways, however, the greatest immediate benefit

during the period was the introduction of general anesthesia, which made many of the other innovations possible.

Tools and Devices. While many of the changes we associate with the Industrial Revolution concern energy, materials, or transportation systems, none would have been possible without the development of new tools for production and operation. The advent of machine tools, including devices for precision boring, greatly improved lathes, the steam hammer, and sets of machine tools for specific purposes, helped make everything else possible (see "Machine tools," pp 190 and 192). Military needs were behind some of these advances. Even computation was gradually mechanized, starting with the first regular production of mechanical calculators in significant numbers and winding up with William Thomson's mechanical computer for solving differential equations (although not as advanced as Charles Babbage's prototype computer designs, it was actually built and used).

Although we think of this as a time that began with waterpower and ended with the steam engine and electric motor, not all the significant devices used engines for power. Hydraulics were employed for some and there were a number of devices, some of which we still use, based on compressed air.

With industrial progress came the need to measure both the environment and the output of the new devices. Over a dozen new measurement devices were developed, ranging from water and gas meters to at least five ways to measure electricity. Along with measurement, automatic safety devices came into use, including governors and safety valves. Despite all the measuring and safety innovations, however, steam engines and gas had an unhealthy way of blowing up throughout most of the 19th century.

Deliberate blowing up was considered a sign of progress, however. In addition to the new explosives (see Materials above), cannons and cannon manufacture were greatly improved. The revolver was developed, but well before either the self-contained cartridge or the familiar bullet-shaped projectile.

The fabric industry drove a lot of the invention and industry of the Industrial Age throughout the period, starting with the flying shuttle and followed by such colorfully named devices as the jenny and the mule. The whole production process of some fabrics was gradually mechanized, from carding and combing wool to spinning all sorts of fibers through weaving or knitting. The first sewing machines even appeared. And when all else failed to hold fabrics together, pin-making machines produced straight pins and even the safety pin. Although we think of England as the great fabric manufacturer, the development of automatic control

systems, which later influenced early computers, took place in France.

Transportation. As with other areas, transportation showed remarkable progress on all fronts, except perhaps for highways, still muddy ruts for most of the period (although gravel roads were introduced by McAdam). People and goods traveled largely by water, including on canals (see "The age of canals," p 159) as well as natural waterways, on rails, in horse-drawn carriages, and in carriages drawn by the steam-powered locomotives. The more adventurous even flew through the air in balloons and gliders.

The steam engine soon became small enough that inventors could envision it on vehicles that it moved—self-propulsion. Ships were most suitable—already adapted to carrying heavy loads and being propelled by the wind. Steamships of various types proliferated in Europe and the US, propelled by engines that shoved ships through the water by paddles or paddle wheels before the screw propeller was invented. Despite the importance of steamships for river travel—they could go upstream or down and did not have to rely on less certain winds even than at sea—economics kept freight shipment by canal and by sail in the open ocean strong.

The balloon had been invented in 1783, so it became a target for self-propulsion by steam, especially after the propeller was invented. Balloons, however, never became a major form of transportation. Another minor form of transport during this period—submarines and diving bells—went down instead of up. Effective submarines, however, had to await the development of a power source that did not depend on air.

Putting steam engines on carriages was the other logical step. This worked best when the carriages were on iron rails (even small steam engines are heavy, their fuel is heavy, and the water in their boilers is heavy). Steam-powered railroads were faster by far than the canals, and cheaper to build. Soon they had not only taken away virtually all passenger traffic, but also all but the bulkiest freight.

One modest means of propulsion made steady improvement from quite unpromising beginnings throughout this period. The bicycle gradually took the form of the one of today. In much of the world it is even today a more important means of personal transportation than the automobile.

At the very end of the Industrial Age, a new power source was about to revolutionize transportation. The internal combustion engine would not only power the car, truck, and tractor, it would also turn flight from an occasional adventure into what many today regard as a necessity.

1733 The Schemnitz Mining Academy (in Hungary) is the first technical college in the world; *See also* 1747 GEN

1734

1735

Jacques Vaucanson

Early in life Jacques Vaucanson became interested in the mechanical aspects of the human body, aiming to construct an artificial man, the final automaton. At the age of eighteen he completed a set of automatons that became instantaneously famous. The most intriguing automaton was a mechanical flute player. In this project Vaucanson was the first to use flexible tubing made of rubber, which had been introduced to France from South America a few years earlier by Charles Marie de la Condamine.

Exhibiting his automatons brought Vaucanson to the attention of the king, and in 1739 he was appointed to improve France's national silk industry, which could not compete with the more efficient silk manufacturers from the Piedmont (northern Italy). Vaucanson not only recommended new working methods, but he also improved the machines used in silk manufacture. He understood that automatic machines can reduce human error, and fully automated the semiautomatic weaving machines designed by Basile Bouchon in 1725 and Jacques de Falcon in 1728. Vaucanson's designs eliminated the need for a worker to feed perforated cards or paper strips into the machines. His machine, equipped with a perforated cylinder, was a direct precursor of the automatic loom of Jacquard.

Attempts to implement Vaucanson's plans for improving silk manufacturing led to one of the first social conflicts of the Industrial Revolution: Silk workers went on strike and revolted in Lyons, France, in 1744.

In addition to improving machines that made textiles (production machines), Vaucanson was also one of the earliest to understand that the quality and precision of the machine tools used to make production machines are of great importance. Tooling machines in Vaucanson's time were constructed entirely of wood and were not very precise. In 1751 Vaucanson introduced an iron frame for these machines, making them more rigid and less prone to inaccuracy. That year he also developed a tool holder that slid on a track made from two iron beams, giving the tool a firm position from which to work.

1736

1737 | Bernard Forest de Bélidor [b Catalonia, Spain, c 1693, d Paris, 1761] publishes the first volume of his four-volume *Architecture hydraulique*, completed in 1753 | Pierre Simon Fournier [b Paris, Sep 15 1712, d Paris, Oct 8 1768] introduces the point system for measuring type sizes; *See also* 1764 COM

1738 Jacques Vaucanson [b Grenoble, France, Feb 24 1709, d Paris, Nov 21 1782] completes an automaton consisting of a flutist that plays a tune on a real flute; a mechanism moves the lips and fingers and pumps air through the mouth | Charles Dangeau de Labelye develops the caisson, a device essential to building bridges and underwater tunnels, for a bridge over the Thames at Westminster, England | William Warburton [b Newark, England, Dec 24 1698, d Gloucestershire, England, Jun 7 1779] argues that writing developed from pictographs, which over the course of time became simplified into script; *See also* 1680–89 COM

1733

Haemostaticks by Stephen Hales relates his experiments on blood pressure with notes on the mechanical relation of the pressure to medical conditions, the velocity of blood flow, and the capacity of the different blood vessels; *See also* 1847 MED

John Kay (b Bury, England, Jul 16, 1704, d 1764] invents the flying shuttle loom; *See also* 1745 TOO

1734

Emanuel Swedenborg [b Stockholm, Sweden, Jan 29 1688, d London, Mar 29 1772] writes *Regnum subterraneum*, giving an account of mining and smelting techniques

1735

Abraham Darby II [b England, 1711, d 1763] introduces the use of coke for producing cast iron, one of the key elements of the Industrial Revolution; *See also* 1700–09 MAT

Antonio de Ulloa is the first to call European attention to platinum as a previously unrecognized element; *See also* 1550–59 MAT

John Harrison, in response to an award put up by the British Board of Longitude in 1714, builds his first marine chronometer, known as Number One; each of his prototype chronometers is given a similar designation, with Number Four being the most famous; *See also* 1710–19 TRA; 1759 TRA

1736

Claudius Aymand performs the first successful operation for appendicitis; *See also* 1822 MED

1737

1738

Spinning using rollers is introduced; *See also* 1530–39 TOO; 1764 TOO

1739

The Scottish goldsmith William Ged [b 1690, d 1749] invents stereotype printing; instead of printing from the type directly, pages are printed from a plate made of lead, which is obtained by pouring lead in a mold made from papier mâché of the composed type; the technique is not used in France until the 1790s; *See also* 1793 COM

1740

The western towers of Westminster Abbey in London, designed by Christopher Wren, are completed after his death

1742

Jacques Vaucanson compares industrial methods in silk manufacture in the Piedmont (in Italy) and in France; attempts to implement new manufacturing rules based on Vaucanson's findings lead to strikes and social unrest in Lyon; *See also* 1810 GEN

1743

Jean-Paul de Gua de Malves and John Mills begin a French version of Ephraim Chambers's English *Cyclopedia; See also* 1720–29 GEN; 1747 GEN

1744

The glow of an electric discharge, as demonstrated by Francis Hauksbee, is proposed to provide a safer light in mines; *See also* 1700–09 GEN

Benjamin Franklin invents the Franklin stove

1739

John Clayton describes his experiments in producing a close analog to natural gas by distillation of coal; this product later becomes known as coal gas; *See also* 1660–69 MAT; 1850 MAT

In Geneva, Switzerland, cannons are built by the process of casting a solid cylinder of metal and boring out the inside of the cylinder to make a chamber; *See also* 1540–49 TOO; 1774 TOO

1740

Dutch engineers develop a cannon with a shorter tube that can easily be loaded by hand; the English name for cannon of this type is howitzer, a corruption of the German name

1742

Benjamin Huntsman [b Lincolnshire, England, 1704, d 1776] introduces the crucible process for molten steel about this time; the cast steel is the hardest then known and is at first rejected by English manufacturers as too hard, although it becomes acceptable in England after the French demonstrate its benefits

Microscope made easy by Henry Baker [b London, May 8 1698, d London, Nov 25 1774] introduces the construction and use of the microscope to the layman; *See also* 1670–79 TOO; 1792 TOO

English mathematician and engineer Benjamin Robins [b Bath, England, 1707, d Madras, Spain, 1751] helps found the science of ballistics in his *New principles of gunnery*; *See also* 1530–39 GEN

The Newry River in Ireland is connected with Lough Neagh by a canal; *See also* 1690–99 TRA; 1745 TRA

1743

1744

Leonard Euler [b Basel, Switzerland, Apr 15 1707, d St. Petersburg, Russia, Sep 18 1783] succeeds in calculating the length of a rod that will buckle under its own weight when stood on one end; *See also* 1798 MAT

Serson is the first to attempt to use a gyroscope for steering at sea, but the experiment fails and he loses his life when the ship founders; *See also* 1852 TRA

1745

1746

In Jan the Leiden jar—the first practical way to store static electricity—is invented; the invention is made at about the same time by both Pieter van Musschenbroek and by Ewald Georg von Kleist; van Musschenbroek gets an electric shock when he first uses it, suggesting a connection between electricity and lightning; *See also* 1775 ENE

1747

The first civil engineering school, the *Ecole des ponts et chaussées* (School for bridges and highways), is established in France; *See also* 1733 GEN; 1771 TOO

Denis Diderot [b Langres, France, Oct 5 1713, d Paris, Jul 31 1784] becomes editor of what will become the *Encyclopédie;* he is assisted by Jean Le Rond d'Alembert [b Paris, Nov 16 1717, d Paris, Oct 29 1783], and replaces Jean-Paul de Gua de Malves; originating as a French adaptation of Ephraim Chambers's *Cyclopedia,* the new version is completely reworked and expanded; it is completed in 1772; *See also* 1743 GEN; 1751 GEN

Benjamin Franklin describes in a letter his discovery that a pointed conductor can draw off electric charge from a charged body; this discovery is the basis for the lightning rod even before Franklin proves that lightning is a form of electricity; *See also* 1744 ENE; 1749 GEN

1748

L'homme machine (Man the machine) by Julien Offroy de la Mettrie [b St. Malo, France, Dec 25, 1709, d Berlin (in Germany), Nov 11 1751] depicts humans as machines, without freedom or will

William Watson [b London, Apr 3 1715, d London, May 10 1787] observes "rays" of electricity in a tube; *See also* 1878 COM

FOOD & AGRICULTURE	MATERIALS	MEDICAL TECHNOLOGY	TOOLS & DEVICES	TRANSPORTATION	
	Gowin Knight [b Corringham, England, 1713, d Jun 8 1772] invents in England a method for preparing magnetized steel; *See also* 1749 TOO		Jacques Vaucanson invents the self-acting, tape-controlled loom for weaving silk; *See also* 1733 TOO; 1775 TOO	A canal joins the Elbe and Oder rivers (in Germany); *See also* 1742 TRA; 1755 TRA	**1745**
	John Roebuck [b Sheffield, England, 1718, d Jul 17 1794] introduces the lead-chamber process for the manufacture of sulfuric acid; *See also* 1783 MAT	The compressive bandage to stop bleeding from wounds is introduced; *See also* 1869 MED			**1746**
Andreas Marggraf [b Berlin, Mar 3 1709, d Berlin, Aug 7 1782] discovers sugar in beets, laying the foundation for Europe's sugar beet industry; *See also* 1500–09 FOO; 1802 FOO		*Primae lineae physiologiae* by Albrecht von Haller [b Berne, Switzerland, Oct 16 1708, d Berne, Dec 17 1777] is the first textbook on physiology			**1747**
	John Wilkinson [b Cumberland, England, 1728, d Jul 14 1808] establishes his ironworks in Bilston, England, where, after much experimentation, he develops an improved blast furnace; *See also* 1340–49 MAT; 1776 MAT		Daniel Bourn develops a carding machine; *See also* 1789 TOO Pierre Le Roy [b Paris, 1717, d 1785] invents a clock in which an impulse is transmitted to the escapement when the pendulum swings to one side but not when it swings to the other, making the escapement independent of the amplitude; *See also* 1765 TOO		**1748**

1749 Benjamin Franklin installs a lightning rod on his home in Philadelphia, mainly to test his theory that lightning is electricity that is attracted to sharp points; *See also* 1752 GEN

1750 Leipzig bookseller Johann Heinrich Zedler [b Breslau (Wroclaw, Poland), 1706, d Leipzig (in Germany), 1760] produces the 64-volume German encyclopedia *Universal lexicon*—the first volume was published in 1732; this is followed in 1754 by a supplement of 4 volumes; *See also* 1720–29 GEN; 1751 GEN

1751 The Ecole Supérieure de Guerre (Military Academy) is founded in Paris; *See also* 1794 GEN

On Jun 28 Denis Diderot and Jean Le Rond d'Alembert publish the first volume of their *Encyclopédie, ou dictionnaire raisonné des sciences, des arts, et des métiers* (Encyclopedia, or rational dictionary of sciences, arts, and trades); *See also* 1747 GEN; 1765 GEN

Benjamin Franklin describes electricity as a single fluid and distinguishes between positive and negative electricity in *Experiments and observations on electricity*; *See also* 1749 GEN; 1752 GEN

1752 Following suggestions from Benjamin Franklin and others, a 12-m (40-ft) iron rod is erected in a field at Marly near Paris; there it succeeds in attracting lightning

In Jun Benjamin Franklin performs his famous kite experiment; it shows that lightning is a form of electricity; *See also* 1749 GEN

Newgate Prison in London is equipped with an air ventilation system powered by a windmill (called a ventilator); it is designed by biologist, chemist, and inventor Stephen Hales

John Smeaton [b Leeds, England, Jun 8 1724, d Austhorpe, England, Oct 28 1792] invents the fantail, or fly, a small windmill at right angles to the main windmill that is used to keep the main sails facing the wind; *See also* 1759 ENE

1753 A supplement to Ephraim Chambers's *Cyclopedia,* largely prepared by Chambers before his death, but edited by botanist John Hill, is published; at seven volumes, the supplement is longer than either the first or second editions (1728 and 1738 respectively); *See also* 1720–29 GEN; 1788 GEN

In Scotland, Charles Morrison proposes in *Scots Magazine* the construction of a telegraph consisting of 26 electrical lines, each corresponding to a letter of the alphabet; individual letters would be indicated by the movement of a light object repelled when a current passes through one of the wires; *See also* 1774 COM

		John Canton [b Stroud, England, Jul 31 1718] develops a method for making an artificial magnet; *See also* 1745 MAT	Philip Vaughan patents radial ball bearings for carriage axles; *See also* 1490–99 TOO; 1862 TOO	**1749**
	Although cast iron is slowly replacing wrought iron in Europe, it still amounts only to 5 percent of all iron used			**1750**
		The first institution to treat mental patients is opened in London; *See also* 1764 MED		**1751**
		William Smellie [b England, 1697, d 1763] writes *Treatise on midwifery*, offering the first scientific approach to obstetrics	Duhamel du Monceau [b Paris, 1700, d Paris, Aug 13 1782] publishes his *Eléments d'architecture navale* (Elements of naval architecture)	**1752**
Zinnias and marigolds, both native to Mexico, are first imported into England; *See also* 1790 FOO		*Treatise on scurvy* by James Lind [b Edinburgh, Scotland, Oct 4 1716, d Hampshire, England, Jul 13 1794] establishes the curative effect of lemon juice on scurvy; *See also* 1795 MED		**1753**

1754

1755

The second Eddystone Light, built in 1709 after the first was swept away in a storm, is destroyed by fire; it was 28 m (92 ft) tall and built of wood; *See also* 1700–09 ARC; 1759 ARC

Lisbon, Portugal, after its destruction by an earthquake, is rebuilt using a rectangular layout of streets similar to that of midtown Manhattan in New York City

1756

1757

1758   A commission in England sets standards, known as the imperial standards, for length, capacity, weight, and other measures

Cast iron (part 2)

In Europe there was no apparent single discovery of cast iron and, unlike many innovations at the end of the Middle Ages, cast iron as a concept seems not to have reached Europe from China. Instead, forge designs were gradually improved to produce hotter and hotter fires. Some forges built in Austria in the late 13th or early 14th century were hot enough to melt the iron they produced. Improvements on these became early blast furnaces.

The European adoption of cast iron meant that availability of ore no longer determined the amount of metal that would be produced. Instead, the limiting factor soon became the availability of wood for charcoal. England, for example, had lost most of its forests by 1600, in large part because of iron smelting.

Coal, another product that had a much longer history in China than in the West, seemed to many to be the solution to the problem. Starting late in the 16th century in England, where coal was already a common fuel, iron makers attempted to use coal for iron instead of charcoal. Most of them failed, although some claimed success. Early in the 18th century, however, Abraham Darby in western England found out how to make high-quality cast iron using coke, coal that has been heated to drive off in gaseous form various impurities.

The grandson of Abraham Darby, also named Abraham Darby, demonstrated that cast iron made from coke could be used even in large bridges. Furthermore, the ease of making molds and specific shapes from cast iron led to the many different parts that were needed for the Industrial Revolution.

1754

The first iron-rolling mill is started at Fareham, England; *See also* 1790 MAT

John Smeaton [b Yorkshire, England, Jun 8 1724, d Yorkshire, Oct 28 1792] specifies that cast iron be used for the main shafts of windmills, influencing manufacturers to start employing cast iron; *See also* 1750 MAT; 1762 MAT

The University of Halle (in Germany) graduates the first female medical doctor

1755

French playwright Pierre-Augustin Caron de Beaumarchais [b Pierre-Augustin Caron, Paris, 1732, d Paris, May 18 1799] invents a clock that can be wound by turning a ring on the clock's face; previous clocks had to be opened to get to a mechanism for rewinding; *See also* 1770 TOO

Liverpool, England, merchants finance the digging of the short Sankey Brook Canal, which joins the river Mersey to a coalfield; this is the earliest recorded transportation canal in England; *See also* 1745 TRA; 1761 TRA

1756

The first cotton velvets are made at Bolton, England; *See also* 1845 MAT

Philipp Pfaff [b 1715, d 1767] publishes *Abhandlung von den Zähnen*, giving the first description of cast models for false teeth; *See also* 1771 MED

Charles A. Coulomb invents the diving bell; *See also* 1759 MAT; 1841 TRA

1757

John Wilkinson patents a hydraulic machine that uses waterpower to drive a bellows

Captain John Campbell [b Kirkcudbrightshire, Scotland, c 1720, d London, Dec 16 1790] improves the octant version of a sextant invented by John Hadley by extending the arc to a sixth part of a circle, the true sextant; *See also* 1730–32 TOO

1758

Jedediah Strutt [b Derbyshire, England, Jul 28 1726] invents the ribbing machine for the manufacture of stockings; *See also* 1580–89 TOO; 1816 TOO

1759

John Smeaton in Oct builds the first successful version of the Eddystone Lighthouse; his techniques are copied in most lighthouses built subsequently; the light itself still consists of tallow candles; the building lasts until it is replaced by a higher structure after more than a hundred years' use; *See also* 1755 ARC; 1877 ARC

John Smeaton publishes the results of his research on the performance of windmills in *Experimental inquiry concerning the natural power of water and wind;* he concludes that windmills cannot compete with the powerful steam engines; *See also* 1758 ENE; 1854 ENE

1760

Faber, a company in Nuremberg (in Germany), starts the commercial production of pencils; *See also* 1795 COM

1761

Originally started in 1675 by Jacques Buot, publication begins on *Descriptions des arts et métiers, faites ou approuvées par Messieurs de l'Académie des Sciences avec figures* (Description of the arts and trades made or approved by the members of the Academy of Sciences with illustrations), a huge compilation in 76 volumes on technology; the project is completed in 1789

1762

1759

John Smeaton reintroduces underwater mortar in building the Eddystone Light; *See also* 100 BC MAT; 1824 MAT

John Harrison completes Number Four, the marine chronometer that will eventually win the British Board of Longitude's prize for a practical way to find longitude at sea; *See also* 1735 TRA; 1761 TRA

1760

1761

Leopold Auenbrugger [b Graz, Austria, Nov 19 1722, d Vienna, Austria, May 18 1809] uses his musical knowledge to develop the technique of percussion for diagnosis of chest disorders; he publishes his findings in *Inventum novum*; *See also* 1826 MED

Giovanni Morgagni [b Forli (in Italy), Feb 25 1682, d Padua, Dec 5 1771] writes *De sedibus et causis morborum per anatomen indagatis* (On the causes of diseases), the first important work in pathological anatomy; *See also* 1774 MED

The Bridgewater Canal, designed by James Brindley [b Thornsett, England, 1716, d Turnhurst, England, Sep 30 1772] and built by the Duke of Bridgewater to carry coal from mines in Worsley to Manchester, opens on Jul 17; it is the first English canal to include an aqueduct portion in which the canal crosses over a river, the Irwell at Barton; *See also* 1755 TRA; 1766 TRA

With John Harrison's marine chronometer Number Four aboard and his son William Harrison to take the readings, the HMS *Deptford* sails off toward the West Indies to test whether Harrison's method can be used to find longitude at sea; *See also* 1759 TRA; 1765 TRA

1762

John Roebuck converts cast iron into malleable iron by action of pit fire and artificial blast at the Carron Iron-works in Stirlingshire, Scotland; *See also* 1754 MAT; 1784 MAT

Samuel Klingenstierna [b Linköping, Sweden, Aug 18 1697, d Stockholm, Sweden, Oct 26 1765] wins the prize of the Russian Academy of Science for the best method of constructing optical instruments free of chromatic abberation; *See also* 1830 TOO

The Bridgewater Canal is extended about 40 km (25 mi) to the Mersey River; *See also* 1761 TRA; 1773 TRA

1763 William Hutchinson fits out a lighthouse in Liverpool with bowl-shaped reflectors of tin and looking-glass mirrors as large as 3.65 m (12 ft) in diameter; *See also* 1720–29 GEN

Ivan Polsunov [b 1728, d 1766] designs a double-cylinder atmospheric steam engine that supplies rotary motion; the machine is completed in 1766 and drives the bellows at the Kolivano-Voskresensky silver mine at Barnaul, Russia; it works for a few months and then breaks down; *See also* 1730–32 ENE; 1781 ENE

1764

Pierre Trésaguet [b 1716, d 1796] develops a base for highways that is made of a low arch perpendicular to the direction of travel; the arch is built from flat stones set vertically; the base is later adopted by Thomas Telford and becomes known as a Telford base

Pierre Fournier's *Manuel typographique* is the first book on engraving and type founding; *See also* 1737 COM

1765 The final volume of Denis Diderot's *Encyclopédie* is published, although the illustrations will not be complete for another 7 years; *See also* 1751 GEN; 1772 COM

A steam engine fed by four steam boilers is built by William Brown; three boilers supply steam while the fourth is kept as a spare; it is the first example of intentional redundancy in technology; *See also* 1776 ENE

James Watt [b Greenock, Scotland, Jan 19 1736, d Heathfield, England, Aug 19 1819] builds a model of a steam engine in which the condenser is separated from the cylinder so that the cylinder does not lose heat each time the steam condenses, resulting in a power source six times as effective as the Newcomen engine; the first such full-scale engines are installed in 1776; *See also* 1710–19 ENE; 1769 ENE

1766

FOOD & AGRICULTURE	MATERIALS	MEDICAL TECHNOLOGY	TOOLS & DEVICES	TRANSPORTATION	
There are indications that ground tobacco leaves are used in France to control aphids; *See also* 1800 FOO	Josiah Wedgwood [b Staffordshire, England, Jul 12 1730, d Staffordshire, Jan 3 1795] patents the cream-colored earthenware that becomes the standard domestic pottery of England; *See also* 1782 MAT	The first American medical society is founded in New London CT	Joseph Oxley patents the use of a ratchet mechanism for creating rotary motion from a reciprocating steam engine; the system is not very efficient and is not commonly used; *See also* 1842 TOO		**1763**
		Robert Whytt [b Edinburgh, Scotland, Sep 6 1714, d Apr 15 1766] publishes *Observations on nervous, hypochondriacal, or hysteric diseases*, one of the first important textbooks on neurology; *See also* 1751 MED; 1840 MED	James Hargreaves [b Blackburn, England, 1720, d Nottingham, England, Apr 22 1778] introduces the spinning jenny, patented in 1770; his first model is able to spin 8 threads at once, although later models reach 120 threads; *See also* 1738 TOO; 1769 TOO		**1764**
Lazzaro Spallanzani [b Scandiano (in Italy), Jan 12 1729, d Pavia (in Italy), Feb 11 1799] suggests preserving food by sealing it in containers that do not permit air to penetrate; *See also* 1795 FOO	Rails made from cast iron are used in English mines; *See also* 1754 MAT

A mining academy is established in Freiberg (in Germany); *See also* 1734 MAT; 1775 MAT | John Morgan [b Philadelphia PA, Jun 10 1735, d Philadelphia, Oct 15 1789] founds the first medical school in America at the College of Pennsylvania | Thomas Mudge [b Exeter, England, 1720, d 1794] develops the detached-lever (free-anchor) escapement for balance-wheel clocks, an improvement over the cylinder escapement invented by George Graham because friction between the anchor and gear is reduced; this type of escapement continues in general use; *See also* 1748 TOO | John Harrison receives the first half of the prize offered by the British Board of Longitude by building a chronometer accurate within one-tenth of a second per day and demonstrating that it can be used to find the longitude accurately on a voyage to the West Indies; *See also* 1761 TRA; 1772 TRA | **1765** |
| | | Horace-Bénédict de Saussure [b Geneva, Switzerland, Feb 17 1740, d Geneva, Jan 22 1799] invents the electrometer, a device for measuring electric potential by means of the attraction or repulsion of charged bodies; *See also* 1786 TOO | | On Jul 26 the 224.6-km- (139.5-mi-) long Grand Trunk Canal officially opens; designed by James Brindley and John Smeaton, it joins the Bridgewater Canal with the river Severn; *See also* 1761 TRA; 1773 TRA | **1766** |

1767

1768

1769

1770

James Watt

Many people have grown up with a mental picture of young James Watt staring at his mother's teakettle, wondering if the pressure of the steam that raised the lid up and down could somehow be harnessed to do work. A truer picture would show the equivalent of a technician at a university who finds that one of his tasks is to repair the university's model of the Newcomen steam engine, a device that had been in use about half a century solely as a form of pump. Watt had been thinking of steam engines before he worked on the Newcomen model, but working on the model crystallized his thoughts. Within a few months he had not only made the repairs, but also developed a series of improvements that would greatly reduce the amount of fuel needed to make the device work. Watt's break came when a business-minded Birmingham, England, manufacturer, Matthew Boulton, picked up a two-thirds share in Watt's invention and also got Parliament in 1769 to grant Watt a patent on the new version of the engine, which Boulton then arranged to have extended through the rest of the 18th century.

The team of Boulton and Watt formed a company that quickly became one of the main driving forces of Britain's Industrial Revolution. Boulton worried about business matters, which he conducted in ways that would make a robber baron envious, and Watt continually improved the steam engine. Even by the time Boulton & Watt was formed, Watt had developed nearly a half-dozen improvements on Newcomen's engine, although it was still basically a pump. Boulton & Watt's first production engines achieved what Watt originally set out to do, which was to pump water from mines using a quarter of the fuel needed to lift the same amount with a Newcomen engine. Mines in England quickly replaced their Newcomens with Watts, which paid for themselves even under the severe financial conditions imposed by Boulton, which included a royalty to be paid to Boulton & Watt on the amount of fuel savings obtained by the new engines.

As Boulton moved strongly to protect Watt's earlier patents, Watt himself continued to patent new ideas. The most important of the improvements on the engine itself was the use of steam to push the piston against a vacuum in two directions, rather than just one as in previous engines. He also developed new methods of transmitting power from the pistons to rotary motion and ways to control the engine and to monitor its activity. Although his engines got better and safer, he consciously avoided one obvious improvement, the use of high-pressure steam. It is not clear whether he feared that the engines would blow up (as many early high-pressure engines did) or whether he just did not think the improvement was necessary.

Around the time Watt turned 65, his basic steam engine patents ran out. Watt and Boulton retired from Boulton & Watt, although their sons continued the famous partnership. Watt lived nearly 20 years more. While Boulton eventually went back into business, Watt stayed clear of steam engines, but could not resist inventing. Boulton's firm, before the advent of the Watt engine, had engaged in manufacturing devices for copying, and Watt worked on improvements along those lines in his advanced years. This was not a new interest, as his first copying patent, for production of documents that could be mechanically reproduced, had been filed while Watt worried over steam-engine development.

Watt's last inventions were devoted to the problem of copying sculpture, which he succeeded in doing as he had succeeded in every other inventive endeavor he attempted.

John Stewart (or possibly Robert Rainey) is the first to use a steam engine directly, without the intermediary of a waterwheel, to power machinery; earlier uses of steam to power machinery used the engine as a pump to raise water, thereby powering a conventional waterwheel when water flowed back into a reservoir; *See also* 1789 ENE

James Watt is granted a patent for his separate condenser steam engine, "a new method of lessening the consumption of steam and fuel in fire engines"; instead of creating a vacuum inside the cylinder by condensing steam with a spray of water, the steam is condensed in a separate vessel; the process results in a considerable saving of fuel because the steam cylinder remains hot; *See also* 1765 ENE; 1776 ENE

		MEDICAL TECHNOLOGY	TOOLS & DEVICES	
		Physician George Baker [b Devonshire, England, 1722, d 1809] is knighted for having proved that lead-lined cider presses cause paralysis, stomach aches, and other symptoms; *See also* 1768 MED		**1767**
		Benjamin Franklin points out to George Baker that lead still heads have been banned in MA because the lead in the rum causes "dry belly ache and the loss of the use of . . . limbs"; *See also* 1767 MED	Antoine Baumé [b Senlis, France, Feb 26 1728, d Paris, Oct 15 1804] invents the graduated hydrometer, with what is now known as the Baumé scale for specific gravities of liquids; *See also* 1824 TOO	**1768**
		Richard Arkwright [b Preston, England, Dec 23 1732, d Derbyshire, England, Aug 3 1792] receives a patent for the hydraulic, or water-frame, spinning machine, which is especially good for spinning warps; as it is too large and expensive to use in small areas, its invention leads to one of the key changes in Britain's Industrial Revolution, the introduction of centralized factories instead of decentralized cottage industries; *See also* 1764 TOO; 1771 GEN	Joseph Cugnot [b Poid, France, Sep 25 1725], a military engineer, builds a steam carriage that can carry four people at speeds up to 3.6 km per hour (2.25 mph), the first true automobile; *See also* 1670–79 TRA; 1784 TRA	**1769**
		Jacques Vaucanson develops the first Western chain drive, which is used in silk reeling and throwing mills; this is about 800 years after the development of the chain drive in China; *See also* 980 CE TOO; 1869 TRA Perrelet constructs a watch with automatic winding; *See also* 1755 TOO	R.D. Edgeworth invents in Ireland the tracked vehicle, which uses individual segmented wheels instead of the more familiar tread around a pair of wheels; *See also* 1825 TRA	**1770**

1771

Richard Arkwright becomes the founder of the modern factory system when he builds a cotton spinning factory at Cromford in rural Derbyshire, England, that uses his water frame as its principal machine; the Cromford factory operates continuously with hundreds of workers on shifts; *See also* 1769 TOO

The first edition of the *Encyclopaedia Brittanica* is published in three volumes; editing has been variously attributed to William Smellie [b 1740, d Jun 24 1795], printer Colin Macfaarquhar, and engraver Andrew Bell; *See also* 1751 GEN; 1796 GEN

1772

French captain Nicholas Marion du Fresne and his crew are the second European group to land on Tasmania, where they encounter and kill several natives; as Tasmania is settled, starting in 1803, Europeans learn that Tasmanians have only stone and simple wood tools that are not ground or hafted, no domestic animals, no bone tools, no sewn clothing, no spear throwers, and no boomerangs; Tasmanians do not fish or eat fish even though they live on an island; although it would appear that they colonized Tasmania before the Neolithic revolution, archaeological evidence shows that Tasmanians abandoned bone technology and fishing about 3500 years BP; *See also* 1640–49 GEN

Jean-Rodolphe Perronet [b 1708, d 1794], director of the Ecole des Ponts et Chaussées (School of Bridges and Highways), builds the Pont de Neuilly bridge over the Seine, below Paris; it is copied in 1827 by Thomas Telford in a bridge over the river Severn at Gloucester

The *Encyclopédie* of Denis Diderot and collaborators is issued in seventeen volumes of text, completed in 1765, and eleven of illustrations; a 7-volume supplement appears between 1776 and 1780; *See also* 1765 GEN

John Smeaton builds an atmospheric steam engine that is two to three times more efficient than the Newcomen engine because of better machined cylinder walls, resulting in a better seal between piston and cylinder walls; *See also* 1710–19 ENE

The atmospheric steam engine

Thomas Newcomen and Thomas Savery both thought of using steam to pump water from mines in 1698, the year that Savery applied for his patent for the Miner's Friend. Water, limiting the depth of mines in Britain, was then a major problem. With pumping water as the aim, steam engines of the first half of the 18th century did not provide rotary motion at all. Steam in a Savery or a Newcomen engine is at the same pressure as the atmosphere and then cooled to produce a partial vacuum that raises a piston in a cylinder. Such engines are termed "atmospheric."

For 60 years, many atmospheric engines were built that did not differ greatly from Newcomen's design, which was more efficient than Savery's. The versatile engineer John Smeaton, known for his scientific approach to practical engineering problems, started systematically investigating a number of Newcomen engines in England. In 1765 Smeaton built an experimental model engine near his home. He also studied in detail 15 of the approximately 100 existing atmospheric engines in England, and so determined the optimal proportions for atmospheric engines of different sizes. In addition, he improved the execution of engine parts, especially the piston and cylinder, and succeeded in doubling the performance of his engines.

Around 1760 engineers became interested in using steam engines for hauling coal from deep mines, and Smeaton completed in 1777 his "water coal-gin." An atmospheric engine pumped water into an elevated cistern that supplied water to an overshot waterwheel that could run in both directions. The same idea, using a steam pump to power a waterwheel, had been introduced a decade earlier. But it was the improvements in the Newcomen engine made by James Watt, who not only created machines that supplied rotary motion, but also increased vastly their performance, that made the steam engine the driving force of the Industrial Revolution. The first two of the Watt engines were already at work when Smeaton's water coal-gin was built.

Pierre Woulfe prepares picric acid; it is used principally as a yellow dye for a hundred years before it is realized that the substance is also a high explosive; *See also* 1871 MAT

The natural history of the human teeth by John Hunter [b East Kilbridge, Scotland, Feb 13, 1728, d London, Oct 16 1793] lays the foundations of dental anatomy and pathology; *See also* 1530–39 MED; 1790 MED

Watchmaker Ferdinand Berthoud [b Plancemont, Switzerland, Mar 19 1727, d Jun 20 1807] publishes his *Traité des horloges marines*

The Smeatonian Club for engineers, named after John Smeaton, is founded in London; *See also* 1747 GEN; 1828 GEN

1771

Joseph Priestley discovers that the volume of air decreases when an electric spark passes through it, but he does not determine why this happens; *See also* 1774 MAT

King George III of England intervenes to get the British Board of Longitude to give John Harrison the second half of the prize money he won for devising the marine chronometer; *See also* 1765 TRA

1772

The French describe technology

As early as 1675, Louis XIV instructed the French Académie des Sciences (Academy of Science) to create a treatise on all the machines used by artisans and manufacturers. Jacques Buot started work on this project, but died before he could accomplish anything. The *Descriptions des arts et métiers, faites ou approuvées par Messieurs de l'Académie des Sciences avec figures* (Description of the arts and trades made or approved by the members of the Académie des Sciences with illustrations) was not published until 1761. The first volume was *L'art du charbonnier* (The art of the coal miner). Etching of plates to illustrate the volumes, however, had been started much earlier, around the end of the 17th century. Consequently, once actual publication began, the *Descriptions des arts et metiers* was completed in twenty volumes by 1781.

The *Encyclopédie, ou dictionnaire raisonné des science, des arts, et des métiers* published by Denis Diderot and Jean de la Rond D'Alembert, took twice as long from first volume to supplement. But the *Encyclopédie* and not the *Description* became the most influential work of the Enlightenment, mainly because of the introduction of new ideas and views in the *Encyclopédie*, many opposing the existing order imposed by church and state. The authors shocked in other ways as well, as in Voltaire's learned article on fornication in Volume VII (1757).

By 1759 publication was stopped on orders of the government, but various means were found to proceed anyway. According to one story, told by Voltaire, the final permission to publish resulted from Madame de Pompadour remarking that she did not understand how her rouge or silk stockings were made. The duc de la Vallière said that the answers to her questions could be found in the outlawed *Encyclopédie*. When this had been demonstrated to Madame and to Louis XV, the king lifted the ban and publication was hurriedly completed before he could change his mind.

One of the aims of the *Encyclopédie* editors and authors, known as the *encyclopédistes*, was to give a complete picture of technology as it existed at the time. The *encyclopédistes* were convinced that technical knowledge had to be treated the same way as other forms of knowledge, thus making technology part of culture. In this Diderot defended the artisan and the manual worker, affirming that the expertise of the artisan is as valuable as that of the artist. The efforts of the *encyclopédistes* resulted in a very large collection of plates of machines and tools, and a large number of articles on the different arts and crafts, making the *Encyclopédie* the most complete source of information about the state of technology during the first half of the 18th century.

1773

Military engineer and scientist Charles *Augustin* Coulomb [b Angoulême, France, Jun 14 1736, d Paris, Aug 23, 1806] publishes *Essais sur une application des règles de maximus et minimus à quelques problèmes de statique relatifs à l'architecture* (Essay on an application of the rules of maxima and minima to some static problems in architecture); it provides the foundation of structural analysis for building construction; *See also* 1720–29 ARC

1774

George Louis Lesage [b Geneva, Switzerland, Jun 13 1724, d Geneva, Nov 9 1803] builds an electric telegraph with 26 electrical lines, one for each letter of the alphabet, following the design of Scottish inventor Charles Morrison; *See also* 1753 COM; 1796 COM

1775

Alessandro Volta [b Como (in Italy), Feb 18 1745, d Como, Mar 5 1827] describes his *eletrofore perpetuo* (electrophorus), a device for producing and storing a charge of static electricity; this device replaces the Leiden jar and eventually leads to modern electrical condensers; *See also* 1746 ENE; 1800 ENE

Pierre-Simon Girard invents a water turbine; *See also* 1798 MAT

				1773
				The Bridgewater Canal, designed by James Brindley, which opened in 1761, is finally completed with construction of the last locks; *See also* 1761 TRA; 1777 TRA

				1774
Henry Cavendish repeats Joseph Priestley's observation of sparks through air and concludes correctly that the cause of the decrease in volume noted by Priestley is the formation of nitric oxides; when electric power becomes less expensive, this "spark method" becomes a major way to manufacture nitrogen compounds; the spark method is also the principal way that nitrogen compounds are formed in nature; *See also* 1772 MAT; 1859 MAT	*Anatomy of the human gravid uterus* by William Hunter [b Lanarkshire, Scotland, May 23 1718, d London, Mar 30 1783] is his greatest work in anatomy; it contains 24 masterpieces of anatomical illustration Franz Mesmer [b Iznang (in Germany), May 23 1734, d Meersburg, (in Germany), Mar 5 1815] uses hypnotism to aid in curing disease; *See also* 1841 MED	John Wilkinson patents a precision boring mill; the machine, used for boring cylinders of steam engines and cannons, contains a rigid bar that leads the boring head through the cylinder; it is not self-centering; this setup results in a high degree of precision; *See also* 1739 TOO Hahn builds a machine that performs calculations with a precision of twelve digits; the machine is commercially successful, and is manufactured until 1820; *See also* 1690–99 TOO; 1820 TOO	

				1775
Farmers in colonial America report that serious soil erosion is resulting in poor harvests	A mining academy is established in Clausthal (in Germany); *See also* 1765 MAT	Sir Percival Potts suggests that chimney sweeps in London develop cancers of the scrotum and of the nasal cavity as a result of exposure to soot, giving the first indication that environmental factors can cause cancer William Withering [b Wellington, England, Mar 1741, d Birmingham, England, Oct 6 1799] introduces digitalis to cure the dropsy associated with heart disease; *See also* 1785 MED	Jacques Vaucanson develops an automatic weaving loom in which the index fingers (which lift and lower the warp threads) are controlled by a perforated rotating cylinder; *See also* 1745 TOO; 1801 TOO	David Bushnell [b Saybrook CT, 1742?, d Warrenton GA, 1824] invents a hand-operated, one-man submarine called the American Turtle; it is the first device to employ the propeller for motive power in water, using separate propellers for horizontal and vertical movements; *See also* 1620–29 TRA; 1777 TOO

1776 Adam Smith [b Kirkcaldy, Scotland, Jun 5 1723, d Edinburgh, Jul 17 1790], in his book *Inquiry into the nature and causes of the wealth of nations,* defines the origin of mechanization and invention of machines as a result of the division of labor; *See also* 1821 GEN

William Blakey patents in the Netherlands a high-pressure water-tube boiler; it fails to work, however, because the copper tubes do not sustain the steam pressure; *See also* 1765 ENE

The first two of Watt's steam engines are installed; *See also* 1769 ENE; 1781 ENE

1777 Johann Beckmann [b 1739, d 1811] publishes *Guide to technology,* which defines technology as the systematic study of technical procedures; *See also* 1806 GEN

The first buildings in France since Roman times are heated by a warm water central heating system; *See also* 100 BC ARC; 1784 ENE

1778 A lightning rod is installed on the two-century-old Genoa lighthouse, which at 61 m (200 ft) is a frequent lightning victim despite a nearby statue of St. Christopher intended to draw off the bolts

Joseph Bramah [b Stainborough, England, Apr 13 1748, d London, Dec 9 1814] patents an improved toilet with flushing water and a bowl encased in wood; *See also* 1580–89 ARC

1776

John Wilkinson uses a steam engine to create the blast of air in a blast furnace, increasing its efficiency dramatically; by the end of the century there are 24 steam-driven blast furnaces in England; *See also* 1748 MAT

Soldiers in the American Revolution have a decided advantage over the British because of their use of patched balls as charges in their rifles; a ball is "patched" when wrapped in greased buckskin or linen to give it a greater diameter while in the barrel, the patch being left behind when the lead shot emerges, so that the greater muzzle velocity is not slowed too much by air resistance; the patched ball was introduced in the American colonies earlier in the 18th century; *See also* 1700–09 TOO

1777

David Bushnell invents the torpedo; *See also* 1775 TRA

Augustin Coulomb invents the torsion balance independently of John Michell [b England, 1724, d Thornhill, England, 1793], who developed the device somewhat earlier; *See also* 1773 ARC; 1779 ARC

In Italy, the Paderno Canal is completed; *See also* 1773 TRA

The Grand Trunk Canal, officially opened in 1766 and combining the Trent and Mersey Canal with the Birmingham Canal to the Severn, is finally complete; the canal, primarily the design of James Brindley, who dies before it is complete, passes through five tunnels, including a famous one-way tunnel at Harecastle that is 2633 m (2880 yd) long; *See also* 1766 TRA; 1827 TRA

1778

Robert Barron invents a type of lock that resists picking; *See also* 1784 TOO

	GENERAL	ARCHITECTURE & CONSTRUCTION	COMMUNICATION	ENERGY
1779		Augustin Coulomb proposes using compressed air and caissons for underwater construction; *See also* 1680–89 TOO; 1781 ARC The third of the three famous men named Abraham Darby [b England, 1750, d 1791], son of the second, starting in 1775, builds the famous Iron Bridge over the river Severn at Coalbrookdale; it is the first really large iron structure—60 m (196 ft) long and built entirely of cast iron; *See also* 1750 MAT		
1780	Johann Beckmann starts his five-volume work *Beiträge zur Geschichte der Erfindungen* (Contributions to the history of inventions)	Victor Louis use an iron frame for building the theater at Bordeaux in France	The fountain pen is invented; *See also* 1884 COM James Watt invents a form of copying in which a special ink is used for making an original that then transfers to a sheet of paper pressed into it; adaptations of this method, using a purple dye, were common in schools before the introduction of dry copying	James Pickard patents a steam engine
1781		Augustin Coulomb writes on the construction of windmills; *See also* 1779 ARC; 1784 MAT		Jonathan Carter Hornblower [b 1753, d 1815] patents the compound steam engine invented by his father Jonathan Hornblower [b 1717, d 1789]; steam enters one high-pressure cylinder and subsequently expands in a second, larger cylinder; because of the low steam pressure used at the time, the machine is less efficient than the Watt engine; in 1799, English courts declare the Hornblower engine an infringement on James Watt's patents; *See also* 1776 ENE; 1804 ENE James Watt patents a way to change the power from a steam engine from back-and-forth motion to rotary motion by using the "Sun and planet" drive; Watt is forced to develop this method because the crankshaft method is protected by a patent; *See also* 1776 ENE; 1782 ENE

Joseph-Michel and Jacques-Etienne Montgolfier

When people before the Montgolfiers thought of human flight, they pictured birds. Leonardo da Vinci was one of the few who proposed a different concept, and that was the helicopter. But early in the 17th century the barometer was invented by Evangelista Torricelli, and Blaise Pascal used it to show that not only does air have weight, but also air density decreases with altitude. Soon the Jesuit monk Francesco de Lana realized that an evacuated globe should float on the heavier air and a Dominican friar, Joseph Galien, pictured capturing less dense air from the top of a mountain and dragging it downhill until it reached a level that would cause a craft attached to it to float. These impractical schemes embodied the idea that flight could be accomplished by the quiet mechanism that buoys a ship in water instead of by frantic activity.

It is not clear that any of this speculation ever reached the Montgolfier brothers, whose base was Annonay, France, south of Lyon and west of Grenoble. The Montgolfier family business there was paper manufacture. Access to paper in large quantities surely helped the Montgolfiers develop their first balloons. All accounts credit older brother Joseph with the original idea, but differ on whether it was inspired by watching bits of burning paper fly into the sky or by reading about experiments that described different gases with different weights. The brothers are thought never to have understood that a hot-air balloon floats because warm air is less dense than cold air. Instead, they believed that burning produced a gas different from air that was also lighter than air. In any case, Joseph began in 1783 with a small paper balloon filled with hot air by holding it over a fire. A stronger silk balloon, similarly inflated, rose to about 21 m (70 ft). By Jun 4 of that year the brothers had made a sphere 11 m (36 ft) in diameter of linen lined with their family's paper to retain the gas. In their public demonstration in Annonay, the balloon rose to a considerable height and floated in the air for 10 minutes, coming down about 2.4 km (1.5 mi) away from the launch site.

News of the success of the hot-air balloon quickly spread to Paris, where chemist Jacques Charles more than matched it with the first hydrogen balloon, which stayed in the air more than four times as long as the Annonay balloon and traveled ten times farther. The Montgolfiers had friends at Court, however, and their next experiment, carrying a sheep, a rooster, and a duck, was conducted before the king at Versailles. Using their paper-making connections, they obtained the materials for the balloon from a Parisian wallpaper house. One result was that the tradition of hot-air balloons covered with bright designs began with the Versailles balloon; it persists to this day.

Before the end of 1783, several humans had made their first flights in hot-air balloons, sometimes called *Montgolfières*, and also in Charles's hydrogen-filled devices. Charles himself took part in the first hydrogen test flight, but it was not until Jan 10 1784 that a Montgolfier went up—Joseph was part of a successful flight in Lyon that included six other passengers. That was the only trip for Joseph, and Jacques never did fly.

1779

Samuel Crompton [b Firwood, England, Dec 3 1753, d Bolton, England, Jun 26 1827] develops the spinning mule, a cross between the spinning jenny and the water-frame spinning machine that is capable of spinning either warps or wefts; *See also* 1769 TOO; 1825 TOO

The first versions of the bicycle appear in Paris; *See also* 1790 TRA

1780

1781

To build his telescopic equipment, William Herschel [b Hanover (in Germany), Nov 15 1738, d Slough, England, Aug 25 1822] uses a small brass-and-steel lathe that he operates with a foot pedal; a flywheel regulates and maintains the speed; this equipment is state-of-the-art for its time

The Marquis de Jouffroy d'Abbans [b Roces sur Rognan, France, 1751, d 1832] designs (and later builds) the *Pyroschape*, a steam-engine-powered paddleboat; it is tested successfully on the Saône River near Lyons in 1783; *See also* 1670–79 TRA; 1783 TRA

1782 Lazare Carnot [b Burgundy, France, May 13 1753, d Magdeburg, Germany, Aug 2 1823], in his *Essai sur les machines en général* (Essay on machines in general), uses scientific arguments to discuss the efficiency of machines

James Watt patents a double-acting steam engine: steam is admitted alternately on both sides of the piston, making the engine more efficient; he also patents his principle of cutting off steam before the piston completes its stroke; *See also* 1781 ENE; 1783 ENE

1783

Valentin Haüy [b St. Just, France, Feb 28 1743, d Paris, Jun 3 1822] introduces in France an alphabet for blind people; it consists of simplified italic characters embossed in paper; *See also* 1784 COM

James Watt completes construction of his first rotary-motion steam engine; the piston rod is connected to a "Sun and planet" gear that translates the back-and-forth action into rotary motion; steam is admitted to the cylinder during both strokes of the piston by a set of valves operated by a system of levers known as parallel motion; *See also* 1782 ENE; 1784 ENE

Machine tools: The lathe

Machine tools are devices used to build parts of machines; but usually the concept is interpreted more narrowly to denote tools that cut or drill, press or shear, or otherwise shape hardened materials into specific forms. Although casting is often one of the most important steps in producing a part, casting per se does not use machine tools; the mold may, however, have been shaped by machine tools. A classic description of how a sculptor works says that the sculptor takes a block of marble and removes everything that does not look like the statue. Most machine tools work the same way, removing metal or ceramic a little at a time until the part the designer had in mind is all that is left.

The first machine tool was among the first machines invented; the bow drill goes back at least to Neolithic times. But the main ancestor of machine tools was not a machine tool itself. The potter's wheel, dating from about 6000 years ago, was originally used to shape soft material only. It could not count as a machine tool because clay is not hard until after firing. By Classical Greek times, however, the potter's wheel had become a part-time machine tool, helping smooth imperfections from fired pottery.

Although records are scarce, that application of the potter's wheel apparently evolved imperceptibly into the principal machine tool used today, the lathe. A lathe is a device for rotating a hard object, originally wood and later mostly metal, so that the object can be shaped by a stationary cutter. We know that early lathes had been developed because we find parts such as chair legs that have clearly been produced by "turning" on a lathe. But it is not until the 15th century, when Leonardo da Vinci drew a lathe, that we learn anything about the device apart from its products. Leonardo's lathe is thought not to have been an invention, but a then-common type of wood lathe with a treadle and a spring pole.

During the 16th century, the art of lathe making advanced, and the lathe was adapted for making screws. Although early screws were used as fasteners, the application of screws for delicate adjustments was far more important. This use of screw adjustors ensured the accuracy of the machine parts produced. In addition to screw making, various adaptations of lathes throughout the 17th century using cams and patterns enabled the wood lathe to cut complicated figures, not just circles with different radii.

The 18th century began with a metal-cutting lathe built in 1701, but its builder clearly stated that his was not the first such lathe, but one of very few. By the end of the century, however, metal lathes were much more common. Jacques Vaucanson was among the first builders of the heavy, industrial lathe, which appeared in France before it did in England. A surviving lathe by Vaucanson was apparently built between 1770 and 1780. The English soon surpassed the French, however. English lathes by Henry Maudslay, built in the early 19th century, set the standard for the time. Maudslay turned out in his shop not only the best lathes of the time, but also the best machine tool manufacturers. Almost all the machine builders of the Industrial Revolution in Britain can trace their heritage directly or indirectly to Maudslay's shop.

1782

Josiah Wedgwood invents the pyrometer for checking the temperature in furnaces used to fire pottery; *See also* 1763 MAT

1783

The first plow-making factory opens in England; *See also* 1803 FOO

American Oliver Evans [b Newport DE, Sep 13 1775, d New York City, Apr 15 1819] invents a totally mechanized flour mill, the first ever to allow a continuous flow of wheat to be ground into flour; *See also* 1660–69 FOO; 1810 ENE

Chemist Nicolas Leblanc [b Issoudun, France, Dec 6 1742, d St. Denis, France, Jan 16 1806] wins a prize put forth by the French government for finding a practical way of making sodium hydroxide and sodium carbonate from salt, although the money is never paid; among other benefits, the Leblanc process makes possible the large-scale manufacture of soap; *See also* 1746 MAT; 1810 MAT

Oxidized hydrochloric acid (then known as oxymuriatic acid, since the unrecognized element chlorine was called muriatic acid) is introduced for bleaching of fabrics; this is the beginning of bleaching with chlorine; *See also* 1799 MAT

Thomas Bell develops cylinder printing for fabrics; *See also* 1790 TOO

Horace Benedict de Saussure [b Geneva, Switzerland, Feb 17 1740, d Geneva, Jan 22 1799] invents a hygrometer based on a hair that elongates when the weather becomes humid

L.S. Lenormand, influenced by accounts from China, becomes the first Westerner to use a parachute, the name of which he coins; *See also* 1793 TRA

The brothers Joseph-Michael [b Vidalon-les-Annonay, France, Aug 26 1740, d Balaruc-les-Bains, Jun 26 1810] and Jacques-Etienne [b Vidalon-les-Annonay, Jan 6 1745, d Serrières, Aug 1 1799] Montgolfier demonstrate the hot-air balloon at Annonay on Jun 5; *See also* 1785 TRA

Physicist Jacques-Alexandre Charles [b Beaugency, France, Nov 12 1746, d Paris, Apr 7 1823] builds the first hydrogen balloon on Aug 27; later in the year he begins a series of balloon ascents

Jean-François Pilâtre de Rozier [b Metz, France, Mar 30 1756, d Boulogne-sur-Mer, France, Jun 15 1785] and François Laurent, Marquis d'Arlandes, make a 25-minute flight on Nov 21 in a hot-air balloon designed by the Montgolfier brothers, becoming the first human beings to fly; *See also* 1784 TRA

Jacques-Constantin Perier operates a steamship on the Seine in France; *See also* 1787 TRA

1784

Aimé Argand [b Geneva, Switzerland, 1755, d 1803] patents the first really bright lamp, which burns oil in a wick between two concentric tubes, producing a circular flame with two sources of oxygen; a glass chimney provides constant airflow as it protects the flame from irregular breezes; *See also* 1800 GEN

Dutch pharmacist Johan Peter Minckelers is the first to use coal gas for lighting; this use is followed in the next year or two by experiments in coal-gas lighting by the English and Germans; *See also* 1786 GEN

Henry Shrapnel [b Bradford-on-Avon, England, Jun 3 1761, d Southampton, Mar 13 1842] invents the shrapnel shell

The first mail coach in England leaves Bristol for London and covers the distance in 17 hours; mail is transported by coach until 1846, when the bulk of mail begins to travel by train; *See also* 1550–59 COM; 1830 COM

Valentin Haüy prints his first book in the alphabet he has developed for the blind; *See also* 1783 COM; 1824 COM

James Watt uses steam pipes to heat his office, employing the first use of steam heat; *See also* 100 BC ARC; 1830 ENE

1785

William Murdock builds a model of a high-pressure steam engine; however, he does not gain approval for it from James Watt, who continues building atmospheric-pressure engines; *See also* 1799 ENE

Systems of machine tools

Henry Maudslay's reputation and the place of machine tools in industry were both set during a seminal operation at the Portsmouth Naval Yard at the start of the 19th century. Using a system set up by Marc Brunel and Samuel Bentham, the production of pulleys (known to shipbuilders as blocks) was completely mechanized. Although this operation had been partly mechanized as early as 1780, the new system used a special tool for each step in the operation. The arrangement was copied so extensively in the US and blended so well with earlier US developments of standardized parts that the whole process came to be called the American system.

The lathe is the basic machine tool. Variations on the lathe, such as the boring machine, grinder, milling machine, and planer, all of which use rotary motion to remove unwanted material, are generally considered separately. These tools, which advanced in the mid-19th century, were also essential to the growth of industry. The cannon-boring machine, adapted to making strong cylinders for steam engines, was a key development.

Standardized parts and mass production are worthless without excellent measurement standards, as early standardizers found, often to their dismay. Concurrent with and necessary to the development of good machine tools was the development of better ways to measure parts. Joseph Whitworth, from Maudslay's shop, was one of the first to recognize this. He developed various standards, including the bench micrometer, that improved accuracy from "can't fit a well-worn penny between the parts" early in the 19th century to one ten-thousandth of an inch or better by the middle of the 19th century. Whitworth also introduced plug-and-ring gauges, which led to the first "go-no-go" gauges—if the part was right it would fit into the larger or "go" gauge but not quite fit in the "no-go" smaller gauge. By the end of the century gauge blocks that could be combined to measure with accuracies of a millionth of an inch were available. These tools were used to measure other tools that were sometimes in turn used to measure others; although there was some loss in precision along the way, the tools used in the shop could always be recalibrated as necessary by the experts. As a result, standardized parts could become fully interchangeable all over the world.

1784

Andrew Meikle [b Scotland, 1719, d Houston Mill, near Dunbar, Scotland, Nov 27 1818] invents the threshing machine, designed to separate grain from straw without manual flailing or trampling on the harvested wheat or rye; its basic design is still used today; *See also* 1834 FOO

Henry Cort [b England, 1740, d 1800] patents the puddling process, also known as the reverberatory furnace, for making wrought iron from melted iron instead of hammering heated iron; the process, which greatly lowers the price of wrought iron by increasing the amount turned out by a single worker by a factor of 20 to 30, is one of the key inventions of the Industrial Revolution; *See also* 1762 MAT; 1843 MAT

Augustin Coulomb writes on the torsional elasticity of metal and silk fibers; *See also* 1781 ARC; 1822 MAT

Benjamin Franklin invents bifocal eyeglasses, mounting half lenses for near and for distant vision in the same frames; *See also* 1450–59 MED

Joseph Bramah in England invents an improved type of lock; a prize of 200 guineas is promised to anyone succeeding in picking the lock, which remains unpicked until 1851, when an American visitor spends 51 hours to achieve the task; *See also* 1778 TOO; 1818 TOO

Vincent Lunardi [b Jan 11 1759, d 1806] makes the first balloon ascent in England, using a balloon filled with hydrogen; *See also* 1783 TRA; 1785 TRA

William Murdock [b Auchinleck, Scotland, Aug 21 1754, d Birmingham, England, Nov 15 1839], an employee of steam-engine manufacturer Boulton & Watt, builds a working model of a steam-powered carriage; *See also* 1769 TRA; 1789 TRA

James Watt covers the possibility of a steam-powered carriage in one of his patent applications, but does not pursue the idea; *See also* 1769 TRA

1785

William Withering's *Account of the foxglove* reports on his discovery of the use of digitalis in the treatment of heart disease; *See also* 1775 MED

Edmund Cartwright [b Marnham, England, Apr 24 1743, d Hastings, England, Oct 30 1823] invents and patents a power loom; he uses the machine in his own factory in 1787; the machine, however, is imperfect, and power looms are used widely only after they have been improved in the 19th century; *See also* 1775 TOO; 1801 TOO

Jean-Pierre Blanchard [b Les Andelys, France, Jul 4 1753, d Paris, Mar 7 1809] and Dr J. Jeffries make the first balloon crossing of the English Channel on Jan 7; *See also* 1784 TRA; 1794 TRA

Jean-François Pilâtre de Rozier and P.A. Romain are killed on Jun 15 trying to cross the English Channel by balloon; they are the first casualties of flight; *See also* 1784 TRA

The Eider Canal, ordered dug by Danish King Christian VII in 1777, using three locks, the Eider River, and the Eider Lakes, connects Kiel Bay on the Baltic with the North Sea; this is the prototype ship canal for international traffic, as well as the predecessor to the level Kiel Canal (Nord-Orstee-Kanal); *See also* 1777 TRA; 1895 TRA

1785
cont

1786 Johann Georg Pickel is the first to experiment with coal gas for lighting in Germany; *See also* 1784 GEN; 1792 GEN

Jean-Rodolphe Perronet revives the Italian segmental-arch bridge of the Renaissance in the Pont Sainte-Maxence over the Oise north of Paris; *See also* 1772 ARC; 1791 ARC

1787

American patriot writer Thomas Paine [b Thetford, England, 1737, d New York NY, Jun 8 1809], brings a model of an iron bridge he has designed to England; the bridge had originally been planned for the Schuylkill near Philadelphia PA, but is reconsidered for the river Wear at Sunderland, England; *See also* 1779 ARC; 1796 ARC

Color and chemistry

Dyes are thought to predate clothing. The earliest indication of human interest in color is the preponderance of green and pink pebble tools produced by the Oldowan toolmakers. The Terra Amata *Homo erectus* site on the French River, dated a million and a half years after the time of Oldowan tools, is littered with pieces of the mineral pigment ocher in its yellow, brown, red, and purple forms.

A couple of hundred thousand years later, Neandertals dyed dead relatives with red ocher before burial, suggesting that the Neandertals painted, dyed, or otherwise decorated their own bodies while alive. The root of the madder plant was used for dying cloth used to wrap some Egyptian mummies. By early Classical times, the Phoenicians had discovered that a Mediterranean shellfish produces a deep purple dye; in one of the early marketing successes, they arranged for the dye to be used only in fabrics worn by rulers— royal purple. Naturally, everyone wanted it. Roman sources report that people in Great Britain used the plant woad to dye themselves blue. In the 16th century, Europeans learned that Native Americans were fond of a bright red dye from a cactus fly; both dye and fly are called cochineal.

Minerals, plants, shellfish, and insects continued in use by households and fabric manufacturers without much scientific investigation. Cornelius Drebbel, in addition to submarines and engineering works, developed in the mid-17th century a way to improve cochineal scarlet with tin and set off one of the first international fashion statements since royal purple. Thereafter, both the English and the French pursued the matter of dyestuffs and issued various handbooks and studies.

Dyes became big business in the mid-19th century with fast (nonfading) ones from coal tar. Most dyeing development was done by Germans (some while in England), and led to the great German work with organic chemistry. As colors like William Perkin's mauve and magenta, a French imitation of German fuchsin, were introduced, they swept through the fashion world—as, more recently, shocking pink did.

Dyes work differently from pigments because dyes chemically bond with fibers. It is not surprising that dyes effective for wool, cotton, and flax also bond with parts of living creatures. German medical workers, notably Paul Ehrlich, borrowed dyes and chemistry from the fabric industry to begin the era of chemical medicine. Their artificial remedies replaced and surpassed the herbal ones just as a few decades earlier artificial coloring replaced ancient herbs.

The continuing search for fiber

Although Western tradition ranks clothing among the basic needs along with food and shelter, the members of some human societies in the recent past wore no clothing, and many societies today wear very little. Climate is not the deciding factor for nakedness, as societies known to use the least amount of clothing, the Fuegan and Tasmanian of 200 years ago, lived on islands with notoriously bad weather. In Europe, however, clothing is known from very early times. Small statues depict clothing as early as 15,000 BC. This clothing is thought to be of animal skins, but linen cloth made by weaving existed at least as early as 6000 BC. In any case, animal skins are not the right shape or size for clothing and need to be sewn together with fiber of some sort. While thin strips of the same skin have been used as fibers, the Ice Man of 3300 BC used animal sinews in some places and grass in others.

Although the earliest recorded evidence of the use of fiber involves flax—linen is fiber from the flax plant—ambiguous evidence suggests that wool from sheep or goats was employed for cloth by that time as well. Sheep and goats were among the first domesticated animals; in the New World, llamas and alpacas, which also produce wool, were also domesticated early.

In warmer climates of both the Old and New worlds, cotton became the most common plant fiber from early times. By the 18th century CE, cotton was shipped all over the world. Introduction of new ways of processing cotton was a hallmark of the Industrial Revolution.

Humans are not the only animals to use fiber. Birds have long used animal hair and plant fibers for nest building. They also anticipated humans in the discovery of the utility of silk, a chemical fiber produced by spiders and caterpillars. Most birds prefer spider silk, some of which forms a micro-Velcro and some of which has adhesives attached, but people in China discovered that the caterpillar of the silkworm moth produces the longest silk fibers in nature, some 900 m (3000 ft) long. Shiny silk was seen from the first as superior to animal hair or plant fibers. During the Chinese's virtual monopoly of hundreds of years, trade in silk became the strand that held eastern Asia to the West along the famous Silk Road.

Silk begins as a liquid that is extruded through a tiny opening; it hardens into a solid in the air. As early as the 17th century inventors recognized that this process could be imitated by a machine if the right liquid were found. Some early attempts used gums or other natural plant products. Chemists in the 19th century perceived that cellulose from the breakdown of wood or some variation on it would do the trick. At least a half dozen different versions of such artificial silks emerged, the best of which we now call rayon. In the 20th century, the same idea was used with a different class of chemicals, the polymers. In bulk, the polymers are plastics, but as fibers they are nylon, Dacron, and polyester. For many purposes, the artificial fibers are better (wrinkle-free cloth is often cited as the greatest boon), but natural fibers continue to excel for others.

1785 cont The Hudson sloop *Experiment*, captained by Stewart Dean, sets sail from New York City on Dec 18 on a voyage to China, becoming the first American ship to make a direct trip to China and back, returning safely in Apr 1786; See also 1843 TRA

William Jessop develops the first metal rails for cars with flanged wheels to travel upon; See also 1797 TRA

1786 Abraham Bennet [b England, 1750, d 1799] invents the gold-leaf electroscope; it is an instrument that indicates the presence of electric charge by the mutual repulsion of two thin gold leaves; See also 1766 TOO; 1820 TOO

Matthew Boulton [b Birmingham, England, Sep 3 1728, d Birmingham, Aug 18 1809] applies steam power to operate machines that stamp coins; in 1790 he obtains a patent for this technology

1787 James Watt adapts the centrifugal governor for controlling the speed of steam engines; the device, already in use in flour mills, consists of two balls mounted on two arms; it is set in rotation by the engine; when the speed increases, the balls fly outward because of centrifugal force and close a butterfly valve that controls the admission of steam to the cylinder; See also 1783 ENE; 1790 ENE

The first steamboat in the Americas is demonstrated by John Fitch [b East Windsor CT, Jan 21 1743, d Bardstown KY, Jul 2 1798] to the Constitutional Convention on the Delaware River at Philadelphia PA; the steam engine drives paddles rather than the more familiar wheel; See also 1783 TRA

John Wilkinson casts the first iron barge for use in carrying armaments he is manufacturing down the Severn River in England

1788

Abraham Rees completes a project of combining the supplement with the original volumes of Chambers's *Cyclopedia* while adding more than 4000 new entries; the result is published alphabetically over a 10-year period and amounts to 418 short volumes in an inexpensive format; *See also* 1753 GEN

Sir John Soane [b John Swan, Whitchurch, England, Sep 10 1753, d London, Jan 20 1837] designs and rebuilds the Bank of England building, the first Greek Revival building in England, setting a style for both Britain and North America; Soane's design is closely copied from a round temple at Tivoli (near Rome), itself a Greek imitation by Roman architects; Continental architects also build Classic Revival buildings; *See also* 1819 ARC

1789

Ernst Chladni [b Wittenberg, Saxony, Nov 30 1756, d Breslau, Silesia, Apr 3 1827] invents the euphonium, a musical instrument in the tuba family

The first cotton factory driven by steam is opened in Manchester, England; *See also* 1767 ENE; 1801 ENE

1790

The US Congress introduces its first patent law, pursuant to Article I Section 8 of the Constitution, which became effective in 1789; *See also* 1620–29 GEN

James Watt builds a pressure gauge, or "indicator," that displays the pressure inside the cylinder of a steam engine; *See also* 1787 TOO; 1796 ENE

1788

French surgeon Pierre-Joseph Desault [b Magny-Vernois, France, Feb 6 1744, d Paris, Jun 1 1795] is appointed surgeon-major to the Hôtel Dieu, where he becomes famous for teaching improved surgical techniques and for bettering instruments used in surgery

1789

Edmund Cartwright develops a wool-combing machine that replaces more than 20 hand-combers; *See also* 1748 TOO; 1850 TOO

The Sapperton Tunnel in England, over 3 km (2 mi) long, links the highest navigable part of the Thames River to the Stroudwater Canal; *See also* 1777 TRA; 1805 TRA

Oliver Evans [b Newport DE, Sep 13 1755, d New York City, Apr 15 1819] takes out the first US patent for a steam-propelled land vehicle; *See also* 1784 TRA; 1801 TRA

1790

Dahlias, native to Mexico, are first imported into England; *See also* 1753 FOO

The first rolling mill driven by steam is opened in England; *See also* 1754 MAT; 1795 MAT

Porcelain begins about this time to be used as a baseplate material for dentures, replacing ivory, bone, and gold, all of which are hard to make fit correctly; *See also* 1756 MED; 1860 MED

American dentist Josiah Flagg [b c 1763, d 1816] invents the dentist's chair; *See also* 1832 MED

George Washington's dentist, John Greenwood [b Boston MA, May 17 1760, d New York City, Nov 16 1819], invents the dental drill; *See also* 1771 MED

William Nicholson [b London, 1753, d Bloomsbury, England, May 21 1815] patents a cylinder press; *See also* 1783 TOO

Count de Sivrac in France builds a céleri-ifère (roughly "quick strike"), a bicycle propelled by the feet, essentially an adult version of the 18th-century toy called the hobbyhorse; only a court plaything at first, an improved version becomes popular in Paris around 1816; *See also* 1779 TRA; 1816 TRA

John Fitch uses his steamboat on the Delaware River to establish commercial service between Philadelphia PA and Burlington NJ, about 30 km (20 mi) north; the boat, which travels about 6.4 km (4 mi) per hour, is not a financial success; *See also* 1787 TRA; 1797 TRA

	GENERAL	ARCHITECTURE & CONSTRUCTION	COMMUNICATION	ENERGY
1791	Slaves on the French-held island of Saint Domingue (Haiti and the Dominican Republic) begin in Aug what becomes the only completely successful revolt of African slaves against their European owners in history; *See also* 1830 GEN	Jean-Rodolphe Perronet builds the graceful Pont de la Concorde in Paris; *See also* 1786 ARC		John Barber patents a simple gas turbine: air, compressed by heating, is directed at a paddle wheel; however, its efficiency is very low; *See also* 1827 ENE
1792	Evidence suggests that manual operations are being timed with a stopwatch at the Derby china works in England with an eye toward speeding up production Inventor William Murdock introduces coal-gas lighting to England, using it to light his home in Redruth, Cornwall; *See also* 1786 GEN; 1799 GEN	Timothy Palmer builds the first modern truss bridge; made from timber, it spans the Merrimack River near Merrimac MA; *See also* 1570–79 ARC; 1805 ARC	The optical semaphore signaling system invented by Claude Chappe [b Brûlon, France, Dec 25 1763, d Paris, Jan 23 1805] and his brother Ignace [b 1760, d 1829] to send each other messages while at school is officially adopted by the French legislature; *See also* 1793 COM	Oliver Evans starts work on an internally fired steam boiler; he also publishes *Young millwright* and *The miller's guide*; *See also* 1776 ENE
1793			The first official semaphore telegram is sent on Aug 15 to announce the French victory in the war with Austria via Claude Chappe's network of semaphore stations; Chappe coins the term "telegraph" for his system of transmitting messages; *See also* 1792 COM Louis-Etienne Herhan, Firmin Gillot, and Nicolas-Marie Gatteaux develop between 1793 and 1797 a method for pressing the plates used in stereotype printing; *See also* 1720–29 COM	
1794	The Ecole Polytechnique opens in Paris in Dec; it will produce the scientific elite of France during the early years of the 19th century; *See also* 1751 GEN			

1791

1792

Edmund Cartwright invents the corderlier, a machine for making rope; *See also* 15,000 BC TOO

John Prince is the first to build a microscope in the US; *See also* 1742 TOO

1793

In Apr Eli Whitney [b Westboro MA, Dec 8 1765, d New Haven CT, Jan 8 1825] invents the cotton gin; it consists of a hand-cranked wooden cylinder with spikes that pass through the bars of a grid: the cotton passes through while the seeds are expelled

In Britain, the Board of Agriculture is formed to oversee "improvement societies" that have sprung up to educate British farmers on better agricultural practices

John Hunter is the first to ligate the femoral artery for treatment of aneurysm; *See also* 1771 MED

Mrs. Samuel Slater is the first woman granted a patent in the US; she invents cotton sewing thread; this invention allows her husband to flourish economically

Thomas Young [b Milverton, England, Jun 13 1773, d London, May 10 1829] is the first to apply engineering concepts to a biological problem: how the eye focuses

The Canal du Centre in France joins Digoin on the Loire River (which empties into the English Channel) with Chalon-sur-Saône on the Saône River (which flows into the Mediterranean); *See also* 1785 TRA; 1832 TRA

Jean-Pierre Blanchard makes the first successful parachute jump from a balloon; *See also* 1783 TRA

1794

Daniel Rutherford [b Edinburgh, Scotland, Nov 3 1749, d Edinburgh, Nov 15 1819] designs the first maximum-minimum thermometer; *See also* 1730–32 TOO

One of the first US canals, the Dismal Swamp Canal from Chesapeake Bay to Albemarle Sound, is completed; *See also* 1800 TRA

On Jun 2 Jean Marie-Joseph Coutelle and an ordnance officer of the French Army rise above the battlefield in the balloon *Entreprenant* to conduct the first aerial surveillance; as a result, the French win the Battle of Fleurus; *See also* 1785 TRA; 1858 TRA

1795

N.J. Conté develops a method of making pencil lead by mixing ground graphite into clay, forming the mixture into thin rods and firing it; this method has been used ever since; *See also* 1760 COM; 1822 COM

1796

The German encyclopedia now known as *Der Grosse Brockhaus* publishes the first volume of its first edition under the title *Konversations-Lexicon* and the editorship of Gotthelf Renatus Loäöbel; in 1808 Friedrich Arnold Brockhaus [b Dormund (in Germany), May 4 1772, d Leipzig (Germany) Aug 20 1823] purchases the rights to the encyclopedia, completes the first edition (1811), edits the second edition, and leaves the project to his sons, who carry on throughout the 19th century; *See also* 1771 COM; 1828 GEN

Rowland Burdon builds a cast-iron bridge over the river Wear that has a much greater span than the Darby bridge over the Severn; some attribute its basic design to Thomas Paine's model iron bridge; *See also* 1787 ARC

James Finley of PA builds the first suspension bridge in the US, using chains; although Native Americans had constructed suspension bridges in the Andes hundreds of years earlier, the Finley bridge is the first modern suspension bridge; *See also* 1809 ARC

Francisco Salva installs an electric telegraph linking Madrid to Aranjuez, Spain, a distance of 50 km (30 mi); the link consists of 44 wires that allow 22 characters to be transmitted; the signals are generated by electrostatic machines and detected by people holding the ends of the wires; later Salva uses the occurrence of sparks to detect signals; *See also* 1774 COM; 1809 COM

John Southern improves the "indicator" developed by James Watt; a pencil is attached to a pressure gauge connected to the cylinder of a steam engine; it records the variations of pressure on a tablet that follows the motion of the piston rod; the pencil describes a closed curve whose surface is a measure of the power produced by the machine; *See also* 1790 ENE

1797

Edmund Cartwright invents an engine that runs on alcohol

1795

FOOD & AGRICULTURE

The average sheep sold for mutton in London's Smithfield Market weighs more than double the typical 18-kg (40-lb) sheep of 1710 as a result of selective breeding for weight

Napoleon offers a prize for a practical method of preserving foods; the prize is won in 1809 by Nicolas Appert [b Châlons-sur-Marne, France, Oct 23 1752, d Massay, France, Jun 3 1841]; See also 1765 FOO; 1804 FOO

MATERIALS

Rolled iron plate becomes available commercially about this time; See also 1790 MAT

MEDICAL TECHNOLOGY

Physician Sir Gilbert Blane [b Blanefield, Scotland, Aug 29 1749, d London, Jun 26 1834] makes the use of lime juice to prevent scurvy mandatory in the British navy, as proposed by James Lind; this is the origin of the nickname "limey" for a British sailor; See also 1753 MED

TOOLS & DEVICES

Joseph Bramah invents an hydraulic press: a system consisting of two pistons, one with a wide diameter and one with a narrow, both filled with a fluid and connected; when pressure is applied to the small piston, the large piston exerts a greater force

TRANSPORTATION

Samuel Morey [b 1763?, d 1843] patents the use of a steam engine to propel a boat; See also 1790 TRA; 1826 TRA

1796

Johann Tobias Lowitz [b Göttingen (in Germany), 1757, d 1804] prepares pure ethyl alcohol

Physician Edward Jenner [b Berkeley, England, May 17 1749, d Berkeley, Jan 26 1823] gives the first inoculation against smallpox by infecting a boy with cowpox (vaccinia virus); See also 1720–29 MED; 1797 MED

Christoph W. Hufeland [b Langensalza (in Germany), Aug 12 1762, d Berlin, Aug 25 1836] publishes *Macrobiotics, or The art to prolong one's life*

Joseph Montgolfier invents the hydraulic ram, a system for lifting water to great heights using the kinetic energy of a small waterfall; the system is reintroduced in France during World War II

1797

Josiah Spode II adds bone ash to hard-paste porcelain in England, producing the first standard English bone china

It is discovered that diamond is a form of carbon; See also 1866 MAT

The Royal Society rejects Edward Jenner's inoculation technique for smallpox; See also 1796 MED

Engineer Henry Maudslay [b Woolwich, England, Aug 22 1771, d Lambeth, England, Feb 14 1831] perfects the slide rest for lathes, called one of the great inventions of history; it permits an operator to use the lathe without holding the metal-cutting tools in his hands; See also 1781 TOO; 1800 TOO

Edmund Cartwright patents a metal packing for piston rods; See also 1816 ENE

Some English roadways in Shropshire are converted to iron rails, along which wagons are drawn by horses; See also 1785 TRA; 1801 TRA

John Fitch applies the screw propeller to a steamboat; although both Benjamin Franklin and Thomas Jefferson recall seeing a boat propelled with a screw on the Seine at Paris, it is not clear that steam was also used; See also 1787 TRA; 1836 TRA

1798

COMMUNICATION

Aloys Senefelder [b Prague (in the Czech Republic), Nov 6 1771, d Munich, Germany, 1834] invents lithography; having made a note with a wax-based ink on a stone plate because paper was not at hand, he has the idea to treat the stone plate with acid, causing the surface protected by the ink to stand out in relief and making it possible to use it for block printing; *See also* 1808 COM

1799

GENERAL

The Royal Institution of Science is founded in England; its aim is furthering technology by means of science

Wilhelm Lampadius [b Hehlen, (in Germany), Aug 8 1772, d Freiburg, Germany, Apr 13 1842] experiments with coal gas for lighting in the palace of the elector of Saxony at Dresden (in Germany); *See also* 1792 GEN; 1801 GEN

Phillipe Lebon [b Brachay, France, May 29 1767, d Paris, Dec 2 1804] receives a patent for his "Thermo-lampe," which uses gas distilled from wood for lighting; *See also* 1792 GEN; 1801 GEN

ENERGY

William Murdock builds a rotary steam engine based on a pump design by Pappenheim in 1636; sealing problems cause this engine to have a very low efficiency; Murdock also patents the slide valve, which will replace the drop valves used in steam engines up to that time; *See also* 1802 ENE

1800

GENERAL

Bertrand G. Carcel adds a clockwork pump to the Argand lamp, producing a steadier light; *See also* 1784 GEN

COMMUNICATION

Lord Stanhope builds a printing press made of iron instead of wood and starts using such presses in London; *See also* 1811 COM

ENERGY

British engineer Richard Trevithick [b Illogan, England, Apr 13 1771] builds the first full-sized steam engine to use high-pressure steam; *See also* 1785 ENE; 1802 ENE

Alessandro Volta announces his invention, made in 1796 or 1799, of the Voltaic pile; it consists of a stack of alternating zinc and silver disks separated by felt soaked in brine; it is the first source of a steady electric current; *See also* 1775 ENE; 1816 ENE

1798

Nicolas Louis Robert [b 1761, d 1828] invents a machine for the continuous production of paper; *See also* 1660–69 MAT; 1807 TOO

Pierre-Simon Girard's *Traité analytique de la résistance des solides* (Analytic treatise on resistance in solids) reports on Girard's experiments with bending of beams and also provides a history of structural mechanics and an introduction to strength of materials; *See also* 1744 MAT; 1807 MAT

The 2-km (1-mi) Sault Ste. Marie Canal joining lakes Huron and Superior is completed by the Northwest Fur Company; the small lock, with a depth of about 0.5 m (1.5 ft) is the first in North America; *See also* 1821 TRA

Robert Fulton [b Little Britain PA, Nov 14 1765, d New York City, Feb 24 1815] builds a four-person submarine, the *Nautilus*; *See also* 1775 TRA

1799

Smithson Tennant [b Selby, England, Nov 30 1761, d Boulogne, France, Feb 22 1815] invents bleaching powder, calcium hypochlorite; bleaching is also introduced in the manufacture of paper, which can now be made from colored rags as well as from white ones; *See also* 1783 MAT

George Cayley [b Scarborough, England, Dec 27 1773, d Brompton, England, Dec 15 1857] designs the first flying machine configured with fixed wings, control surfaces on the tail, and a means of propulsion; *See also* 1804 TRA

1800

Copper sulfate, hydrogen cyanide, lead arsenide, nicotine, petroleum, and turpentine are used in fighting plant diseases; *See also* 1763 FOO; 1825 FOO

A pasta machine for making macaroni is invented in Naples; previously, pasta of all shapes had been made by hand

Chlorine is used by William Cruikshank [b Edinburgh, Scotland, 1745, d Jun 27 1800] to purify water in England

Humphry Davy [b Penzance, England, Dec 17 1778, d Geneva, Switzerland, May 29 1829] discovers nitrous oxide ("laughing gas") on Apr 9 and suggests its use as an anesthetic; *See also* 1844 MED

Henry Maudslay builds the first lathe that enables an operator to make screws of any desired pitch or diameter; *See also* 1797 TOO

The Santee and Cooper Canal, started in 1792, is completed in SC; *See also* 1794 TRA; 1803 TRA

1801

Phillipe Lebon uses his wood-gas system to light the Hotel Seignelay in Paris; it is the first known instance of a whole building being lit with inflammable gas; *See also* 1799 GEN

William Murdock installs coal gas for lighting at the engine house of Boulton & Watt's plant near Birmingham, England, to celebrate the short Peace of Amiens between Britain and France; *See also* 1792 GEN; 1803 GEN

John Rennie [b Phantassie, Scotland, Jun 7 1761, d London, Oct 4 1821] uses steam engines for the construction of the London docks; *See also* 1789 ENE; 1810 ENE

1802

The US Military Academy at West Point NY is established as headquarters of the engineering corps of the US Army; it remains entirely a school for training engineers until 1866; *See also* 1751 GEN

Samuel Bentham conceives of mass producing pulley blocks for the British navy; he enlists the help of Marc Isambard Brunel [b Hacqueville, France, Apr 25 1769, knighted 1841, d London, Dec 12 1849], who aids in planning, and Henry Maudslay, who develops 44 separate machines for performing the operations; as a result of mass production, 10 workers produce as many pulley blocks in a year as previously had been made by 110 workers

Humphry Davy demonstrates that metal strips can be heated to incandescence by passing sufficiently strong an electric current through them; *See also* 1820 GEN

Thomas Wedgwood [b Staffordshire, England, May 14 1771, d Dorsetshire, Jul 10 1805] announces to the British Royal Institution that he has found a method for creating images on nitrate of silver using a camera obscura; *See also* 1720–29 COM; 1810 COM

Matthew Murray [b 1765, d 1826] builds a rotary-motion steam engine mounted on a "bedplate," a cast-iron foundation on which all the parts of the engine are mounted; the bedplate does away with the requirement of supporting the engine components by masonry or building structures, and becomes generally adopted a few years later; Murray also invents a valve slide that becomes standard equipment on subsequent steam engines; *See also* 1799 ENE; 1807 ENE

William Symington [b 1764, d 1831] builds a direct-action steam engine for the paddle tug *Charlotte Dundas;* in such an engine the piston rod is directly connected via a rod to the crankshaft

Richard Trevithick obtains a patent for both his high-pressure steam engine and his steam carriage; the title of the patent is "Steam engines—Improvements in the construction thereof and Application thereof for driving Carriages"; *See also* 1800 ENE

1801

Thomas Young discovers the cause of astigmatism; *See also* 1864 MED

Robert Hare [b Philadelphia PA, Jan 17 1781, d Philadelphia, May 15 1858] invents the hydrogen-oxygen blowpipe, the ancestor of modern-day welding torches

Joseph-Marie Jacquard [b Lyon, France, Jul 7 1752, d Oullins, France, Aug 7 1834] develops an early version of his loom, in which each decision must still be carried out by a human operator; *See also* 1775 TOO; 1805 TOO

The first freight cars are used on the Surrey Iron Railway in England; *See also* 1797 TRA

Richard Trevithick builds a full-scale steam-powered carriage, completed on Dec 24; it runs well for four days, but burns up when he carelessly allows all the water in the boiler to evaporate; *See also* 1789 TRA; 1804 TRA

1802

German chemist Karl Archard founds in Prussia the first factory for manufacturing beet sugar; *See also* 1747 FOO; 1822 FOO

German explorer Alexander von Humboldt [b Berlin (in Germany), Sep 14 1769, d Berlin, May 6 1859] returns from South America with samples of guano; these droppings of pelicans have been used by the Incas of Peru as fertilizer; guano becomes a widely used fertilizer until the development of artificial fertilizers; *See also* 1690–99 FOO; 1820 FOO

Nathaniel Bowditch [b Salem MA, Mar 26 1773, d Boston, Mar 16 1838] writes *New American practical navigator* and sets the standards for navigation of sailing ships; *See also* 1843 TRA

The West India Docks, artificial basins in the Thames at London, designed and built in less than 2 years by William Jessop, open in Aug; *See also* 1826 TRA

Edward Massey invents and patents a ship's "log" that uses a rotor and gears to power a dial showing the distance a ship has traveled; however, it must be hauled on board to be read; it comes into general use; *See also* 1710–19 TRA; 1846 TRA

John C. Stevens builds a propeller-driven steamboat; *See also* 1797 TRA; 1803 TRA

The steamboat *Charlotte Dundas,* built by William Symington, operates as a towboat on the Firth and Clyde Canal in Scotland, but does not prove profitable; *See also* 1797 TRA; 1803 TRA

1803 William Murdock lights his main factory with coal gas; this is the first building to be routinely lit in this way; *See also* 1801 GEN; 1806 GEN

Records show that only six steam engines are in use in the US

The railroad system

There is no really satisfactory overall word to describe the mode of transportation in which a powerful, self-propelled vehicle pulls one or more passive wheeled vehicles along a a pair of equidistant supports. *Railroad* or *railway* only describes the supports; *locomotive* names the self-propelled puller; and *train* indicates the line of vehicles being pulled. The lack of a unifying word is rooted in history: the components of what we call a railroad arose separately.

The support system, or railroad, may be used to mean the whole system because it really was the first component. Planked roads in and out of mines extend far back into history. From an illustration in Agricola's book on mining, we know that they had evolved into rails as early as 1556.

Miners learned that railroads are more efficient for several vehicles linked together. Such trains of cars were pulled by mules, oxen, donkeys, or horses in different mines. In the early 19th century, such railways had gone beyond the mines and were being used to haul goods to canals. In England more than twenty such rail lines, with trains pulled by animals, had been authorized early in the 19th century.

Steam engines large enough to pull themselves—locomotives—or themselves and a train were considered too heavy for dirt or even gravel or paved roads, although some that were tried on the roads were no heavier than large modern automobiles. Furthermore, the earliest steam engines were associated with mines, where people were used to railways. So the combination was fixed early in the 19th century: a locomotive on rails pulling a train.

Such locomotives were the first vehicles in which turning wheels supplied the power to move forward. In all previous vehicles, humans or other animals moved forward by walking, and the wheels were used mainly to reduce friction. Few thought that a machine could be built that was strong enough to move itself and any significant load by spinning its wheels. Railroad pioneer Richard Trevithick won a bet of 500 guineas by pulling ten tons of iron and over 70 passengers a distance of 14.5 km (9 mi) in southern Wales. He used a mine's rails for the feat, and the trip was one way. When the engine arrived at the mine, it was left there to power stationary machinery. Even after that, it was generally believed in England that a locomotive could only pull loads on level rails. No early English railroad lines ran up hills—instead, they went around them or cut through them.

1804 Frederick Albert Winsor [b Friedrich Albrecht Winzer Brunswick (in Germany), 1763, d Paris, May 11 1830] patents an oven for the manufacture of coal gas

Arthur Woolf [b Cornwall, England, Nov 4 1766, d Guernsey, England, Oct 26 1837] develops a compound steam engine; however, he miscalculates the dimensions of the high-pressure and low-pressure cylinders, making the machine fairly inefficient; the Woolf engine is largely a reinvention of the Hornblower steam engine; *See also* 1781 ENE; 1811 ENE

Oliver Evans builds a double-acting vertical high-pressure steam engine using a steam pressure of three atmospheres; he demonstrates it in Philadelphia PA; however, his machine does not attract the backing of industrialists; *See also* 1802 ENE

1803

A self-sharpening cast-iron plow is introduced in England; *See also* 1783 FOO; 1819 FOO

Engineer Marc Isambard Brunel develops and installs between this date and 1810 the first complete set of machine tools, at the Portsmouth, England, dockyards, for the mechanical production of ships' blocks (pulleys); the tools, built by Henry Maudslay, replace manual operations completely; *See also* 1800 TOO; 1810 TOO

Richard Trevithick, while improving the safety of high-pressure boilers by installing double safety valves, introduces a rivet made from the metal lead mounted at the top of a boiler furnace; if the boiler boils dry, the rivet melts and allows steam to escape to the furnace; *See also* 1787 TOO

The Middlesex Canal, started in 1794, is completed between Boston MA and the Merrimack River; *See also* 1800 TRA; 1825 TRA

Robert Fulton's steamboat makes a successful trip on the Seine in France; *See also* 1802 TRA; 1807 TRA

1804

Nicolas Appert invents canning foods as a means of preservation and opens the world's first canning factory; along the way, he also invents the bouillon cube; *See also* 1795 FOO; 1809 FOO

Albert D. Thaer [b Zelle, Germany, May 14 1752, d Moeglin, Germany, Oct 26 1828] publishes *Grundsätze der rationellen Landwirtschaft* (Foundations of rational agriculture), which explains the benefits of rotation of crops; *See also* 1640–49 FOO; 1834 FOO

The soybean, which becomes a major crop in the US Midwest after World War II, is first cultivated in the US; it had been known for over 3000 years in Asia; *See also* 1500 BC FOO

Richard Trevithick develops a steam locomotive that runs on iron rails and successfully hauls 10 tons of iron 16 km (10 miles); *See also* 1801 TRA; 1808 TRA

George Cayley develops his first instrument to measure air resistance on Dec 1; during this year he also begins to build a series of gliders that will determine the basic principles of aerodynamics and writes, but does not publish, his *Essay on the mechanic principles of aerial navigation*; *See also* 1799 TRA; 1808 TRA

GENERAL	ARCHITECTURE & CONSTRUCTION	COMMUNICATION	ENERGY
1805	Theodore Burr, starting in 1803, completes a combination truss-and-arch bridge over the Hudson River at Waterford NY; *See also* 1792 ARC; 1820 ARC		
1806 William Murdock lights the very large Philips and Lee factory at Salford, England, with coal gas; about the same time, Samuel Clegg [b Manchester, England, Mar 2 1781, d Middlesex, England, Jan 8 1861] uses coal gas to light a mill in Halifax, England; *See also* 1803 GEN; 1812 GEN Frederick Albert Winsor forms the world's first coal-gas lighting company; it is initially called the National Light and Heat Company, but changes its name later to the Gas Light and Coke Company; *See also* 1807 GEN		Joseph Bramah invents a printing machine for bank notes that automatically numbers them	
1807 Frederick Albert Winsor's National Light and Heat Company installs coal gas lights on Pall Mall in London, although this is not followed with other major installations for the next few years; *See also* 1812 GEN		William Hyde Wollaston [b Norfolk, England, Aug 6 1766, d London, Dec 22 1828] invents the camera lucida; it produces images of drawings on another sheet of paper, allowing them to be copied by tracing over the image or enlarged or reduced	George Cayley invents the calorific, or hot-air, engine, used extensively in 19th-century industry; *See also* 1833 ENE Matthew Murray designs a "portable" steam engine that becomes the prototype of marine engines for the next 40 years; *See also* 1802 ENE
1808		Strixner and Piloty in Munich produce the first color lithographs; *See also* 1798 COM	

FOOD & AGRICULTURE	MATERIALS	MEDICAL TECHNOLOGY	TOOLS & DEVICES	TRANSPORTATION	
	The British win the Battle of Trafalgar in part because inferior French cast-iron cannons tend to burst when they are fired		Joseph-Marie Jacquard develops a method of controlling the operation of a loom based on punched cards, an idea that will be used in early computers; *See also* 1801 TOO	The largest canal tunnel in England, the Blisworth on the Grand Union Canal, is completed; it is 930 m (3056 ft) long, 4.6 m (15 ft) wide, and 3.25 m (10.75 ft) high; *See also* 1789 TRA; 1827 TRA	**1805**
In Germany the first gymnasium (similar to a US high school) devoted to preparing students for agriculture opens				About 30,000 Congreve rockets, developed by Sir William Congreve [b Woolwich, England, May 20 1772, d Toulouse, France, May 16 1828], are used in the bombardment of Copenhagen	**1806**
	Clergyman Alexander John Forsyth [b Scotland, Dec 28 1769, d Jun 11 1843] invents percussion powder, which leads to a way to ignite the charge in a cannon without using a match or burning stick; *See also* 1833 MAT Thomas Young introduces the concept now known as Young's modulus, defined as stress divided by strain, to measure elasticity or flexibility of a material; *See also* 1798 MAT; 1826 MAT		Henry [b France, 1766, d 1854] and Sealy [b France, d 1847] Fourdrinier invent and develop in England the first machine that makes paper by a continuous process instead of one sheet at a time; *See also* 1798 MAT	Robert Fulton's *North River Steam Boat* is tested in the East River off New York City on Aug 9; later renamed the *Clermont*, it makes its maiden voyage from Manhattan to Clermont NY on Aug 17; although not the first steamboat, it is the first practical and economical one; *See also* 1803 TRA; 1811 TRA	**1807**
	To demonstrate the importance of iron to the Industrial Age, John Wilkinson is buried, as he had previously arranged, in an iron coffin			Richard Trevithick builds a circular passenger railway on Gower Street in North London; *See also* 1804 TRA George Cayley builds a glider and becomes the first person to fly in a machine heavier than air; *See also* 1804 TRA; 1877 TRA	**1808**

1809

GENERAL

Humphry Davy uses a high-powered battery to induce a bright light between two strips of charcoal 10 cm (4 in.) apart, creating the first arc light; arc lights come into general use in the 1870s; *See also* 1845 GEN

ARCHITECTURE & CONSTRUCTION

The first Western suspension bridge capable of carrying vehicles, with a span of 74 m (244 ft), is built across the Merrimack River in MA; *See also* 1796 ARC; 1822 ARC

COMMUNICATION

Joseph Bramah develops and patents a machine that cuts a goose quill into a dozen or so parts that can be slipped into a holder, where each part functions as a pen; previously the entire goose quill was used as a pen; this device is successfully marketed and familiarizes the public with pens that combine a point (nib) and a holder; *See also* 600 COM; 1822 COM

Samuel Soemmering [b Thorn, Prussia (Germany), Jan 25 1755, d Frankfurt, Germany, Mar 2 1830] develops a multiwire telegraph; *See also* 1796 COM; 1812 COM

1810

GENERAL

The Cotton Spinners' Association of Manchester, England, organizes one of the first strikes of the Industrial Revolution; *See also* 1742 GEN; 1812 GEN

COMMUNICATION

Thomas *Johann* Seebeck [b Revel, Estonia Apr 9 1770, d Berlin, Germany, Dec 10, 1831] writes to Goethe that silver chloride exposed to light of a particular color tends to take on the color of the incident light; *See also* 1802 COM; 1891 COM

ENERGY

A steam engine built for operating a winch at a coal pit at Farme Colliery in England becomes one of the few steam engines from this time to remain in use in the 20th century; it is decommissioned in 1915

In Pittsburgh PA, American inventor Oliver Evans builds the first mechanized flour mill in the US to be powered by steam; *See also* 1783 FOO; 1834 FOO

In England, James Watt builds a demonstration flour mill powered by steam at Albion Mills; *See also* 1790 ENE; 1834 FOO

1811

GENERAL

Freiberg, a city in Saxony (Germany), is the first to be illuminated by gas light; *See also* 350 CE GEN; 1812 GEN

ARCHITECTURE & CONSTRUCTION

The Standedge Tunnel between Marsden and Diggle in England is opened for traffic on Apr 4 after 15 years of digging; at 4984 m (5451 yd), it is the longest tunnel excavated to this time; *See also* 1843 ARC

COMMUNICATION

Frederic Koenig builds in London the first mechanical press; it prints 3000 sheets per hour; *See also* 1800 COM; 1814 COM

ENERGY

Arthur Woolf compares the performance of his compound steam engine to the Watt engine and finds that the Woolf engine uses 50 percent less fuel; *See also* 1804 ENE

1809

Nicolas Appert, described by the *Courier de l'Europe* (Feb 10) as "making the seasons stand still" with his canning process, wins Napoleon's 12,000-franc award for inventing a method for preserving food with his invention of canning and steriliz-ing food in glass jars; the prize is awarded Jan 10 1810; *See also* 1804 FOO; 1810 FOO

George Cayley states the essential princi-ples required for any flying machine that weighs more than the air it displaces; *See also* 1808 TRA; 1843 TRA

1810

The mowing machine for hay is invented in the US; *See also* 1784 FOO

Nicolas Appert's *Le livre de tous les ménages, ou l'art de conserver pendant plusieurs années toutes les substances animales et végétales* (Art of preserving animal and vegetable substances for several years) explains heat sterilization of food; *See also* 1809 FOO; 1819 FOO

Pierre Durand patents food preservation in cans made of iron, but mysteriously fails to provide any way to open them; *See also* 1809 FOO; 1839 FOO

Augustin-Jean Fresnel [b Broglie, France, May 10 1788, d Paris, Jul 14 1827] develops a method of making soda (sodium carbonate) using limestone and com-mon salt; *See also* 1783 MAT

Friedrich Krupp [b 1787, d 1826] estab-lishes a steel plant at Essen (in Germany) that will be the foun-dation of Krupp enterprises

Organon of rational healing by Christian Friedrich *Samuel* Hahnemann [b Meis-sen (in Germany), Apr 10 1755, d Paris, Jul 2 1843] intro-duces homeopathy; *See also* 1811 MED

Marc Isambard Brunel applies his the-ories of mass produc-tion, previously exer-cised in the Portsmouth Ship-yards, to manufactur-ing footwear for the British army; *See also* 1803 TOO

In France, the 100-km- (60-mi-) long St. Quentin Canal, with three tunnels, opens; *See also* 1805 TRA

Thomas Walker pub-lishes *Treatise upon the art of flying by mechanical means*; *See also* 1808 TRA

1811

Samuel Hahnemann publishes a catalog of homeopathic drugs; *See also* 1810 MED

American John Hall patents a breech-load-ing rifle; *See also* 1776 TOO; 1820 TOO

Robert Fulton builds the *Chancellor Liv-ingston*, a ship 49 m (161 ft) long powered by a 60 hp steam engine; *See also* 1807 TRA; 1815 ENE

1812

The National Light and Heat Company changes its name to the Gas Light and Coke Company when Samuel Clegg joins the firm started by Frederick Albert Winsor; the company succeeds in lighting the city of London with coal gas over the next quarter century; *See also* 1811 GEN; 1816 GEN

Organized gangs in the North and Midlands region of England smash mechanized looms and other textile-manufacturing equipment; the gangs are called Luddites after (it is said) a Midland's apprentice named Ned Ludd who after a beating took a hammer to a loom; the British Parliament decrees that destroying machines will be punishable by execution; *See also* 1810 GEN; 1813 GEN

The German inventor Samuel Soemmering builds a telegraph that uses electrolysis for detecting electrically transmitted signals; 25 wires arrive in a tank; the ones that produce bubbles when a current is sent through them indicate the letters; *See also* 1809 COM

1813

In Jan the Luddite movement is smashed by English troops and magistrates with greatly expanded powers; seventeen of the Luddite leaders are hanged; *See also* 1812 GEN; 1819 GEN

1814

The Times of London is printed by a steam-driven cylinder press that prints 1000 sheets per hour; *See also* 1811 COM; 1847 COM

1815

English landowners influence Parliament to pass the Corn Laws, which restrict imports of grain to maintain high prices; manufacturers soon learn that they have to raise wages or their workers will be unable to afford food; *See also* 1846 GEN

Humphry Davy invents the safety lamp for use in coal mines

The first warship powered by steam is built in the US; *See also* 1811 TRA

1812

Maize, known in the US as corn, is hybridized successfully for the first time when two strains are crossbred; *See also* 1720–29 FOO

John Stevens of Hoboken NJ proposes a revolving gun turret for ships, but the idea is not developed for about 50 years; *See also* 1835 TOO

US railroads

The earliest railroads were in England. Although railroading on the English model gradually spread through Europe in the middle of the 19th century, the real advances in railroad travel came in the US, where greater distances fostered a different philosophy. As early as 1836 Americans tried to prove that a locomotive could pull a train up a moderately steep hill, so American railroad lines traveled over high hills and even over mountains, using switchbacks along with cuts and tunnels when needed. Long American locomotives bounced on quickly laid rails and had to be equipped with cowcatchers because the unprotected lines ran through open country. Stout British locomotives traveled smoothly on short, level lines that were built to give a smooth ride, because that was the only way British engineers thought the system would work. Fences were part of the plan. No cowcatchers for these protected railroad systems.

    The US system, different as it is, started almost as soon as the first British railways proved the concept. Less than 10 years after the first British common carrier, the state of Pennsylvania, worried about the opening of the Erie Canal in rival New York, started the Philadelphia and Columbia line. By the time the first railroad line opened in Greece, the coasts of the US were linked by rail. America had become the railroad leader of the world in terms of miles of track, freight hauled, number of operating locomotives, or any other measure.

1813

William Horrocks perfects the power loom by introducing a system for varying the speed of the batter; *See also* 1785 TOO; 1846 TOO

1814

The circular saw, developed in England, is introduced in the US; *See also* 1869 TOO

British engineer George Stephenson [b Wylam, England, Jun 9 1781, d Chesterfield, England, Aug 12 1848] introduces his first steam locomotive, the Blucher, on Jul 25; it is capable of hauling 30 tons at speeds faster than those possible from a horse-drawn system; *See also* 1808 TRA; 1825 TRA

1815

The first gas meters, known as "wet meters," are introduced for measuring consumption of natural or coal gas; they are not superseded by "dry" meters until about a hundred years later

Scottish engineer John Loudon McAdam [b Ayr, Scotland, Sep 21 1756) invents paving roads with crushed rock; although the name macadam is in his honor, he did not use tar or asphalt as in modern macadamized roads; *See also* 1819 TRA

1816

Gas lighting is introduced in Baltimore, the first city in the US to have its own plant for manufacturing coal gas for street illumination; *See also* 1812 GEN

Johann Maelzel patents the metronome

Joseph Nicéphore Niepce [b Chalôns, France, Mar 7 1765, d Chalôns, Jul 5 1833] begins experiments that lead to development of first photographs; *See also* 1802 COM; 1822 COM

William Wollaston improves the design of the Voltaic cell by adding an apparatus to remove the copper and zinc plates from the acid when the cell is not in use; this greatly extends the life of the cell; *See also* 1800 ENE; 1836 ENE

John Barton develops segmental wedge packing containing springs for piston rods of steam engines; it becomes the prototype of later packings; *See also* 1797 TOO

1817

Robert Stirling patents a hot-air motor that uses the expansion of heated air for driving a piston in a cylinder; there is little interest in the motor until about 1970, when companies such as Philips start investigating hot-air motors; *See also* 1807 ENE; 1985 ENE

1818

In England, the Institution of Civil Engineers is established to further "the art of directing the great Sources of Power in Nature for the use and convenience of man. . ."; *See also* 1828 GEN

Aloys Senefelder adapts his method of lithography to color printing; *See also* 1798 COM; 1851 COM

1816

Théophile *René* Laënnec [b Quimper, France, Feb 17 1781, d Brittany, France, Aug 13 1826] invents the stethoscope; *See also* 1819 MED

Marc Isambard Brunel invents and patents the round stocking frame, an improved mechanized knitting machine that can knit tubular fabric; *See also* 1758 TOO; 1864 TOO

David Brewster [b Jedburgh, Scotland, Dec 11 1781, knighted 1831, d Allerly, Scotland, Feb 10 1868] invents the kaleidoscope; *See also* 1847 COM

Joseph Nicéphore Niepce improves the célerifère, which is renamed the vélocifère; although still not steerable and pushed by the feet, this ancestor of the bicycle becomes popular in Paris; *See also* 1790 TRA; 1818 TRA

1817

Richard Roberts [b Wales, 1789, d Manchester, England, 1864] builds one of the earliest planes for smoothing metal; other early planes from the period 1810 to 1820 were constructed by Matthew Murray [1765-1826] and Thomas Fox [1789-1859]; *See also* 1800 TOO; 1818 TOO

1818

Jean-Baptiste Dumas [b Alais, France, Jul 14 1800, d Cannes, France, Apr 1884] treats goiter with iodine

Eli Whitney invents the milling machine, which consists of a power-driven horizontal table that passes at a right angle under a rotating cutter; this is used by Whitney to make oddly shaped parts for muskets, but can also be used for making parts for many other devices; Whitney does not patent the machine, but freely shares it with government manufacturers of muskets; the milling machine was more important for US industrial development than the more famous cotton gin—even English machinists import or copy the machines; *See also* 1817 TOO; 1821 TOO

Jeremiah Chubb invents a type of lock that resists picking; *See also* 1784 TOO; 1848 TOO

Drais von Sauerbronn invents the *Draisienne*, the first type of primitive bicycle that can be steered; it has a padded seat; *See also* 1816 TRA; 1839 TRA

The paddle steamer *Savannah* becomes the first steamship to cross the Atlantic (in 27 days, 100 hours), although it is propelled by its sails 87 percent of the trip; it uses steam power for only 85 hours because of the small amount of coal the ship can carry; *See also* 1807 TRA; 1838 TRA

1819

A crowd of over 50,000 British workers, women, and children gather in St. Peter's fields outside of Manchester to hear radical speaker Henry "Orator" Hunt call for political reform; they are charged by local militia who ride horses and wield sabers; 11 are killed, including 2 women, and 400 are injured; *See also* 1813 GEN

The Second Bank of the United States in Philadelphia PA is the first Greek Revival structure in the US; it resembles a temple with Doric columns; *See also* 1788 ARC; 1823 ARC

Hans Christiaan Oersted [b Rudkoebing, Denmark, Aug 14 1777, d Copenhagen, Mar 9 1851] accidentally discovers that a magnetized needle is deflected by a nearby electric current; his findings are published in 1820

1820

Warren De la Rue encloses a platinum coil in an evacuated glass tube and passes electricity through it in the first recorded attempt to produce an incandescent lamp; *See also* 1840 GEN

American architect Ithiel Town invents the lattice truss for bridges; it uses two long horizontal beams laced together with a number of diagonals in a lattice pattern; *See also* 1805 ARC

Eli Whitney

The realization that Eli Whitney was among the most important Americans at the start of the republic has come only gradually to historians, even historians of technology. Today it is clear that his most famous invention, the cotton gin, saved the US South economically while encouraging the slave and plantation systems that ultimately led to the Civil War. Yet it is likely that the gin, a simple device that Whitney made in a few days, would have been invented by someone else as well. Whitney's true worth came from the invention of the "American system" of manufacture and the creation of specialized tools. He also was the first American to utilize standardized parts, an idea probably originally owed to Leblanc, a French arms manufacturer. Whitney's development of cost accounting and establishment of stable government contracts originated methods still used for successful manufacture of all sorts.

Whitney showed inventiveness and intelligence from an early age, taking apart and reassembling his father's watch, for example. He used these skills to put himself through high school and Yale University. On gradua-

tion from Yale, deeply in debt (as circumstances were to keep him for most of his life), he had the pivotal encounter of his life. Taking a job as a tutor in far-off South Carolina, he arranged to travel with another young man, Phineas Miller, the manager of the considerable estate of the widow of General Nathanael Greene. All the evidence suggests that Whitney fell hopelessly in love with Catherine Greene, and remained so until her death. (Catherine, however, eventually married Miller.) At first, Whitney lingered at the Greene estate instead of taking his tutoring assignment. While there he learned of the problem of separating seeds from cotton fiber. Mrs. Greene said she was sure that Eli was clever enough to solve it, which turned out to be the case. Whitney formed a partnership with Miller to promote the invention, the cotton gin.

Miller, however, through bad policy came close to wrecking the business in a variety of ways. The gin was pirated, lawsuits were fought badly, and Miller lost money on land speculation as well. In the final analysis, Whitney never made any significant amount of money on his ubiquitous gin.

1819

In this year and 1820 food preservation factories that use the Appert process of canning in sterilized glass jars open in the US; *See also* 1810 FOO; 1839 FOO

Jethro Wood [b US, Mar 16 1774, d Ledyard NY, Sep 18 1834] of Scipio NY is granted a patent on a cast-iron plow with an improved moldboard designed by Thomas Jefferson; this becomes the standard plow of the period; *See also* 1803 FOO; 1833 FOO

Traité de l'auscultation médiate (On listening to organ sounds with the aid of an instrument) by physician René Laënnec treats the use of the stethoscope for investigating the lungs, heart, and liver; *See also* 1816 MED; 1826 MED

William Brunton [b Dalkeith, Scotland, May 26 1777, d Cornwall, England, Oct 5 1851] patents his coking stoker for steam boilers; termed by him the "peristaltic" grate, it consists of a series of mobile grates that move the coal forward within the boiler

Joseph Francis [b Boston MA, Mar 12 1801, d Cooperstown NY, May 10 1893] develops an "unsinkable" lifeboat that uses hydrogen tanks for buoyancy and a self-bailing feature; *See also* 1838 TRA

Practical essay on roads by engineer John McAdam describes his invention of road paving with crushed rock; *See also* 1815 TRA

Eli Whitney (Continued)

Nevertheless, he had access to enough capital to enter another business, this time without Miller. Learning that the US Army needed muskets, he invented in 1798 the "American system" of using powered, specially designed machines to make interchangeable parts. The then-standard manufacturing system used one skilled maker to produce all the parts for an individual musket by hand. Whitney needed several years to design the machines, build the plant, train the workers, and produce the muskets. Although he continued to be in debt much of this time, ultimately he demonstrated that his method was far superior.

Even though the strength of Whitney's system seems obvious today, it was not so at first to his contemporaries. He was 9 years late in fulfilling his first major government contract.

Aside from the manufacturing system, the most important of Whitney's inventions was the milling machine, a modified lathe that turns out irregularly shaped parts. Without the milling machine, it is difficult to see how standardized parts could ever have been invented. Whitney also invented the tumbler mill near the end of his life, although he was too ill to make anything more than drawings and plans.

Whitney's illness, which resulted in his death before the age of 60, was an enlargement of the prostate. He invented or reinvented the flexible catheter to relieve himself and probably thus gave himself another couple of years of life as well as some cessation of pain. That other inventive American, Benjamin Franklin, may have made the same invention when faced with a similar problem in 1784.

1820

The cultivator, a mechanical device for removing weeds by turning over the top layer of soil (like a mechanical hoe), is invented in the US; *See also* 1810 FOO

Guano, introduced by Alexander von Humboldt, becomes a widely used fertilizer in Europe; *See also* 1802 FOO

Charles X. de Colmar builds a mechanical desk calculator, called "Arithmometre," based on the principles of the Leibniz computer of 1673; more than 1500 such calculators are sold during the next 30 years; *See also* 1774 TOO

John Hall is manufacturing breech-loading rifles at Harpers Ferry VA (WV); *See also* 1811 TOO; 1845 TOO

Johann Salomo Christoph Schweigger [b Erlangen (in Germany), Apr 8 1779, d Halle, Germany, Sep 6 1857], upon learning of Oersted's discovery of electromagnetism, uses the effect to build the first galvanometer, a tool for measuring the intensity and direction of electric current; *See also* 1786 TOO; 1850 TOO

John Birkinshaw builds wrought-iron rails; *See also* 1785 TRA

1821

Mexico is liberated from Spain; *See also* 1825 GEN

James Mill coins the word *capitalism* to describe the economic system of industrial Britain; *See also* 1776 GEN

Michael Faraday [b Newington, England, Sep 22 1791, d Hampton Court, England, Aug 25 1867] reports his discovery of electromagnetic rotation in the paper "On some new electro-magnetical motions, and on the theory of magnetism," essentially presenting an application of the electromagnetic effect published in 1820 by Oersted; technically, Faraday creates the first two motors powered by electricity, although the rotating needle is not a motor in the normal sense and could not be used to power anything; *See also* 1819 ENE

1822

Brazil successfully declares itself free of Portugal; *See also* 1825 GEN

C. Frederick Bushman invents the accordion in Berlin; however, some claim it is a slightly later Viennese invention; *See also* 1829 GEN

Englishman George Pocock invents the kite carriage, also known as the *charvolant;* it is a pair of maneuverable arch-top kites attached to a light carriage with double bridles so that the kites can be guided away from trees and church steeples

Guillaume Henry Dufour [b Constance, Switzerland, Sep 15 1787, d Jul 14 1875] and Marc Seguin [b Annonay, France, Apr 20 1786, d Annonay, Feb 24 1875] start construction of the first large suspension bridge in Europe; the bridge, erected in Geneva, consists of two sections of 40 m (131 ft) each; *See also* 1809 ARC; 1823 ARC

Joseph Nicéphore Niepce and Claude Niepce produce the first permanent photograph; *See also* 1816 COM; 1839 COM

S. Mordan and John Isaac Hawkins start the mechanized manufacture of pencils in the United States; they also improve Joseph Bramah's pens, made from parts of goose quills, by replacing the quills with horn or tortoiseshell points (nibs) tipped with jewels or gold; *See also* 1809 COM; 1830 COM

John Mitchell reportedly develops the first machine-made steel pens; *See also* 1809 COM; 1830 COM

William Church of NY invents a typesetting machine; *See also* 1838 COM

1821

Thomas Blanchard [b MA 1788, d 1864] builds a machine to carve wooden gunstocks for the Springfield (MA) Armory; it is based on a rotary cutter guided by a mechanical linkage that follows the movement of a tracing wheel over a given gunstock pattern; *See also* 1818 TOO

Johann Seebeck observes that two different metals joined at two different places that are kept at different temperatures will produce an electric current, called thermoelectricity; the Seebeck effect will later be exploited in the development of semiconductor devices; *See also* 1810 COM

Canal building continues in Canada with a modest bypass of the Lachine Rapids on the St. Lawrence River; *See also* 1798 TRA; 1829 TRA

1822

Anselme Payen [b Paris, Jan 6 1795, d Paris, May 12 1871] discovers that charcoal can be used to remove impurities from sugar; *See also* 1802 FOO

Augustin Cauchy [b Paris, Aug 21 1789, d Sceaux, France, May 23 1857] presents his theory of stress and strain to the French Academy of Sciences; it provides the first careful definition of stress as the load per unit area of the cross section of a material; *See also* 1807 MAT; 1826 MAT

William Beaumont [b Lebanon CT, Nov 21 1785, d St. Louis MO, Apr 25 1853] starts his experimental study of digestion in the exposed stomach of a wounded man; *See also* 1838 MED

The first successful hysterectomy is performed; *See also* 1736 MED; 1825 MED

Charles Babbage [b Teignmouth, England, Dec 26 1792, d London, Oct 18 1871] develops the Difference Engine, a machine for calculating values of logarithms and trigonometric functions; it does not work well owing to an inability to make parts that live up to Babbage's design; soon he will abandon the Difference Engine for a general-purpose computer; *See also* 1805 TOO; 1832 TOO

John Stanley patents his "fire feeder," an overfeed stoker for steam boilers that sprinkles coal over the grate at regular intervals; *See also* 1819 TOO; 1834 TOO

1823

Augustin-Jean Fresnel invents the Fresnel lens, used in lighthouses; for a time, lighthouses that use reflectors are said to use the English system, while those with lenses use the French system; but by the end of the 19th century, most lighthouses use a combination of lenses and reflectors; *See also* 1763 GEN

Jan Evangelista Purkinje [b Libochovice, Bohemia (the Czech Republic), Dec 17 1787, d Prague, Jul 28 1869] is the first to classify various types of fingerprints; *See also* 1680–89 MED; 1860 GEN

The Grand Pont crossing the Seine at Fribourg, Switzerland, 273 m (896 ft) long, is the most successful large suspension bridge to this time; it remains in service until 1923, unlike many other suspension bridges of the early 19th century, which collapsed in the face of violent windstorms; *See also* 1822 ARC; 1825 ARC

The British Museum, designed by Robert Smirke [b 1781, d 1867] is built in the Greek Revival fashion; it resembles a temple with Ionic columns; *See also* 1819 ARC

Firmin Gillot uses acid to make copper printing plates from photographs, essentially a lithographic process; *See also* 1798 COM; 1872 COM

Samuel Brown invents an internal combustion engine with separate combustion and working cylinders and uses it to power a vehicle; *See also* 1859 ENE

1824

Réflexions sur la puissance motrice du feu (On the motive power of fire) by Nicholas L.S. (Sadi) Carnot [b Paris, Jun 1 1796, d Paris, Aug 24 1832] shows that work is done as heat passes from a high temperature to a low one; defines work; hints at the second law of thermodynamics; and suggests internal combustion engines; *See also* 1782 GEN; 1850 GEN

Louis Braille [b Coupvray, France, Jan 4 1809, d Jan 6 1852] introduces a method of writing the alphabet that uses a pattern of raised dots made with a stylus; this comes to be used by blind people all over the world, with many books published using the system; *See also* 1784 COM

Claude Burdin [b Lépin (in Italy), May 18 1778, d Nov 12 1873] coins the term "turbine" for a power source that derives motion from a flowing fluid; the word stems from the Latin *turbo*, a spinning object; Benoit Fourneyron [b Bornin St.-France, 1801, d St. Etienne, France, Jul 8 1867], one of Burdin's students, develops the first modern turbine; *See also* 1827 ENE

1825

Largely through the efforts of Simon Bolívar, all Spanish possessions in South America are liberated over a 10-year period; the last, now known as Bolivia, becomes independent in Aug; *See also* 1822 GEN

Thomas Drummond [b England, Oct 10 1797] and Goldsworth Gurney [b England, 1798] invent limelight, an intense beam of light focused by a parabolic mirror and produced by burning lime in an alcohol flame enriched by additional oxygen

A suspension bridge over the Menai Strait in Wales, built by Thomas Telford, with a single span of 176 m (579 ft), inaugurates the age of modern bridge construction; *See also* 1823 ARC; 1831 ARC

John Ayrton develops the traumatrope, a persistence of vision device that produces moving images; *See also* 1832 COM

Jacques de Montgolfier designs a steam engine in which the piston, equal in length to its stroke, acts as the exhaust valve; this type of engine becomes known as the uniflow engine during the 20th century; various versions are developed; *See also* 1827 ENE; 1908 ENE

FOOD & AGRICULTURE	MATERIALS	MEDICAL TECHNOLOGY	TOOLS & DEVICES	TRANSPORTATION	
	Charles Macintosh [b Glasgow, Scotland, Dec 29 1766, d Jul 25 1843] dissolves rubber in low-boiling naphtha and invents the waterproof fabric that bears his name (usually spelled mackintosh); raincoats made from this fabric or other waterproof fabrics are sometimes called mackintoshes; *See also* 1839 MAT			Marc Isambard Brunel designs swing bridges for the port of Liverpool; *See also* 1826 TRA	**1823**
	Stearin, a colorless, odorless fat produced from glycogen and stearic acid, is introduced for use in candles; *See also* 1850 MAT				
	English mason and building contractor Joseph Aspdin patents Portland cement; it is obtained by a very high calcination temperature; the end product looks like stone derived from the island of Portland, hence its name; *See also* 1759 MAT	Henry Hickman [b England, 1800, d 1830] uses carbon dioxide on an animal as a general anesthetic; *See also* 1800 MED; 1842 MED	Joseph-Louis Gay-Lussac [b St. Léonard, France, Dec 6 1778, d Paris, May 9 1850] invents the alcoholometer, a device for measuring the proof of liquors that is still in use today; *See also* 1768 TOO		**1824**
Michael Faraday is the first to prepare benzene hexachloride, which in the 20th century will be recognized as the first of a family of insecticides that appear to be almost harmless to most vertebrates; *See also* 1800 FOO; 1874 FOO	John George Appolt invents the chamber gas-producing retort	Pierre Bretonneau [b St. Georges-sur-Cher, France, Apr 3 1778, d Passy, France, Feb 18 1862] successfully performs the first tracheotomy to restore breathing to a child suffering from croup; *See also* 1730–32 MED	The automatic spinning mule is introduced; *See also* 1779 TOO	The 586-km (364-mi) Erie Canal from Albany NY to Buffalo NY, started in 1817, is officially opened on Oct 26; *See also* 1803 TRA; 1848 TRA	**1825**
			Henry Burden [b 1791, d 1871] of Troy NY manufactures railroad spikes by machinery rather than by hand	George Cayley invents the use of full treads instead of wheels for vehicles; *See also* 1770 TRA; 1908 FOO	
				George Stephenson's Active, later renamed Locomotion No. 1, makes its initial trip on Sep 27; it is the first steam locomotive to carry regularly both passengers and freight; *See also* 1814 TRA; 1829 TRA	

1826 Charles Babbage publishes *On a method of expressing by signs the action of machinery,* in which he develops a way of describing machines in a mathematical code applicable to all machines

Joseph Niepce uses sunlight to harden asphalt to reproduce a photograph on asphalt, which can then be used as a printing plate; *See also* 1823 COM; 1855 COM

1827 *Die galvanische Kette, mathematisch bearbeitet* (The galvanic circuit investigated mathematically) by Georg Simon Ohm [b Erlangen (in Germany), Mar 16 1789] contains the first statement of what is eventually called Ohm's law—that the current of electricity is equal to the ratio of the voltage to the resistance

John Walker [b Stockton-on-Tees, England, 1771, d May 1 1859] invents the friction match, the first match that can be ignited by pulling it through sandpaper; its tip is coated with antimony sulfide and potassium chlorate; *See also* 1830 GEN

Benoit Fourneyron builds the first enclosed water turbine of modern design; it develops 6 horsepower; the first waterwheels, 2000 years earlier, were also turbines; *See also* 1824 ENE; 1831 ENE

Jacob Perkins patents a steam engine with a piston that acts as the exhaust valve; this is a second version of the uniflow engine, as this type becomes known during the 20th century; *See also* 1825 ENE; 1885 ENE

1828 The Institution of Civil Engineers is granted a royal charter in England; *See also* 1771 TOO; 1847 GEN

Francis Leiber [b (Germany) 1798, d 1872] begins publication in Boston of the first edition of the *Encyclopedia Americana,* largely based on Brockhaus's *Konversations-Lexicon; See also* 1796 GEN

1826

MATERIALS: Claude-Louis-Marie-Henri Navier [b France, 1785, d 1836] puts the concept of Young's modulus into modern mathematical form, as stress over strain; this is equivalent to the slope of the stress-strain curve; *See also* 1807 MAT; 1829 MAT

MEDICAL TECHNOLOGY: René Laënnec writes *De l'auscultation médiate et des maladies des poumons et du coeur* (On using sound to diagnose maladies of lungs and heart); it presents an extension of his work on the use of the stethoscope; *See also* 1819 MED

TOOLS & DEVICES: The Collins Company is formed in CT to mass produce axes using machinery to make the parts

Alan Wood's water-powered rolling mill near Wilmington DE, the Delaware Iron Works, begins to make shovels and spades in large quantities by machinery

TRANSPORTATION: Marc Isambard Brunel designs floating landing stages for the port of Liverpool; *See also* 1823 TRA

Samuel Morey starts building the *Aunt Salley*, believed to be the first boat to be powered by an internal combustion engine; sometime between 1832 and 1843, the *Aunt Salley* sinks or is sunk by the inventor; *See also* 1795 TRA

1827

MATERIALS: Friedrich Wöhler [b Eschersheim (in Germany), Jul 31 1800, d Göttingen, Germany, Sep 23 1882] develops a method of preparing aluminum in pure form; because of the complexities of the chemistry required, aluminum remains the most expensive metal of its time; some aluminum jewelry is made during the 19th century; *See also* 1833 MAT; 1886 MAT

MEDICAL TECHNOLOGY: Charles-Eduard-Ernest Delezenne [b Lille, France, Oct 4 1776, d Aug 26 1866] develops his technique of "just noticeable differences" to study hearing

TRANSPORTATION: Thomas Telford designs a second tunnel on the Trent and Mersey Canal at Harecastle, enabling two-way traffic; unlike the earlier tunnel designed by James Bridley, this one even has room for a towpath; *See also* 1777 TRA

1828

FOOD & AGRICULTURE: Perhaps the first design for a combine, a combined harvester and threshing machine, is one patented this year by Samuel Lane; however, combines do not become common until after 1920; *See also* 1784 FOO; 1838 FOO

MATERIALS: John Thorp [b Rehoboth MA, 1784, d Nov 15 1848] introduces ring spinning, a method for continuous spinning still in use

MEDICAL TECHNOLOGY: Amalgam, a mercury alloy, is introduced to dentistry, but it does not replace gold until about 1910, following experimental work by G.V. Black in 1895 that shows that mercury in amalgam is safe; *See also* 1771 MED

TRANSPORTATION: Marc Seguin [b Annonay, France, Apr 20 1786, d Annonay, Feb 24 1875] patents the multitubular fire-tube or locomotive-type boiler; Robert Stephenson develops a similar boiler in 1829 for his Rocket locomotive; *See also* 1776 ENE; 1844 ENE

1829

Gas lighting is introduced in Paris; *See also* 1816 GEN

Jacob Bigelow [b Sudbury MA, Feb 27 1786, d Boston, May 10 1879] coins the English term *technology* in *The elements of technology*

According to some sources, Cyril Damian invents the accordion in Vienna, although others claim an earlier invention in Berlin; *See also* 1822 GEN

Charles Wheatstone [b Gloucester, England, Feb 6 1802, knighted 1868, d Paris, Oct 19 1875] invents the concertina (melodion), a bellows-operated instrument like the accordion, but with the two keyboards on each end of the bellows and with each reed playing the same note on both pumping actions; *See also* 1822 GEN

William Austin Burt [b Worcester MA, Jun 13 1792, d Detroit MI, Aug 18 1858] patents an unwieldy but workable typewriter, the Typographer; the operator turns an arm to the correct letter and presses, inking the letter with a pad and pressing the letter on paper; *See also* 1710–19 COM

1830

Slavery in the British Empire is abolished; *See also* 1791 GEN

Charles Sauria and J.F. Kammerer discover how to make matches that light when struck, using yellow phosphorus, sulfur, and potassium chlorate; *See also* 1827 GEN; 1855 GEN

Trains start carrying mail between cities in England; previously, the bulk of mail was carried by coaches; *See also* 1784 COM; 1840 COM

James Perry develops an improved form of the stepped pen point, or nib, that is made flexible by cutting slits in the body of the point and adding a hole; this form of pen becomes commonplace for the next hundred years; *See also* 1822 COM; 1938 COM

Jacob Perkins invents a radiator for use with hot-water central heating; *See also* 1784 ENE

1829

Patrick Bell develops a grain reaping machine that is moderately successful; *See also* 1810 FOO; 1831 FOO

Siméon-Denis Poisson [b Pithiviers, France, Jun 21 1781, d Paris, Apr 25 1840] introduces the constant of proportionality between primary and secondary stress now known as Poisson's ratio; if the primary stress and strain are compressive, the secondary strain will be tensile; if the primary stress and strain are tensile, the secondary stress will be compressive; *See also* 1826 MAT

James Mill [b Fofarshire, Scotland, Apr 6 1773, d Jun 23 1836] publishes *Analysis of the phenomena of the human mind*, which tries to show that the mind is nothing more than a machine and that it has no creative function

Joseph Henry [b Albany NY, Dec 17 1797, d Washington DC, May 13, 1878] completes a 4-year study of the electromagnet; he is the first to use insulated windings, discovers the law of proportion between electromotive force and resistance of the magnet, and works out the effects of different types of windings

The first electromagnetically driven clock is made; *See also* 1840 TOO

The Welland Canal in Canada is constructed to bypass the Niagara River between lakes Erie and Ontario, which includes among its rapids the famed Niagara Falls; *See also* 1821 TRA; 1932 TRA

In Oct George Stephenson's steam locomotive The Rocket is chosen over three competitors to open the Liverpool to Manchester railway, which begins operations in 1830; this is considered the start of the railroad boom; *See also* 1825 TRA; 1830 TRA

1830

Physiologist Marshall Hall [b Basford, England, Feb 18 1790, d Brighton, England, Aug 11 1857] denounces bloodletting as a treatment for disease

Joseph Jackson Lister [b London, Jan 11 1786, d West Ham, England, Oct 24 1869] succeeds in developing an achromatic lens for the microscope about 70 years after the first achromatic telescope lenses are made; *See also* 1762 TOO; 1834 TOO

Barthélemy Thimonnier [b 1793, d 1859] invents a sewing machine and installs 80 of them in a workshop for military clothing; the following year his workshop is destroyed by Parisian tailors fearing unemployment; *See also* 1846 TOO

Edwin Beard Budding [b 1796, d 1846] invents the lawn mower, patented by Budding's partner John Ferrabee; *See also* 1902 TOO

George Stephenson's steam locomotive The Rocket, along with seven others built to the same design by Stephenson, is used for operation of the Liverpool to Manchester railway, which formally opens on Sep 15; *See also* 1829 TRA; 1832 TRA

1831

GENERAL

Jacob Perkins patents an air-conditioning system that uses cold air from underground tunnels; *See also* 1837 GEN

ARCHITECTURE & CONSTRUCTION

Isambard Kingdom Brunel [b Portsmouth, England, Apr 9 1806, d Westminster, England, Sep 7 1859] designs a suspension bridge over the river Avon at Clifton, England, but it is not built until 1864; *See also* 1825 ARC

A new London Bridge, designed by John Rennie after the first stone bridge at the site is demolished, is completed by his son Sir John Rennie; this bridge is later moved to Lake Havasu AZ, piece by piece; *See also* 1170–79 ARC

COMMUNICATION

Joseph Henry develops the first electromagnetic telegraph; he sends an electric charge through 1500 m (5000 ft) of wire, where an electromagnet produces a force on a suspended permanent magnet that swivels and rings a bell; *See also* 1812 COM; 1832 COM

ENERGY

Joseph Henry discovers the principle of electromagnetic induction, but fails to publish his result; he also describes an electric motor; *See also* 1832 ENE

On Aug 29 Michael Faraday independently discovers Joseph Henry's earlier unpublished finding that electricity can be induced by changes in a magnetic field (electromagnetic induction); *See also* 1821 ENE; 1832 ENE

William Avery builds and sells steam turbines consisting of two jets mounted on a rotating hollow shaft; steam flows through the jets and the shaft rotates because of the reactive force of the steam; 50 of these turbines find applications in workshops; *See also* 60 CE TOO; 1849 ENE

1832

GENERAL

Charles Babbage's *On the economy of machinery and manufactures* takes stock of the changes brought about in Europe by the Industrial Revolution

ARCHITECTURE & CONSTRUCTION

Joseph Paxton [b Woburn, England, Aug 3 1801, d Sydenham, England, Jun 8 1865], while remodeling a greenhouse roof, is the first to use the concept of ridge-and-furrow construction; the idea for this system is owed to John Claudius Loudon; the bases of the V-shaped furrows and the vertices of their ridges each act as beams; the result is stronger than a roof of the same thickness that does not use this system; *See also* 1841 ARC

COMMUNICATION

Joseph Plateau [b Brussels, Belgium, Oct 14 1801, d Ghent, Belgium, Sep 15 1883], who has studied the retinal retention of the eye, develops the phenakistoscope; it consists of two disks mounted on a single axis, one carrying several images of an object or person in motion that can be observed through slits in the other disk; by rotating the disks and observing the images through the slits, the illusion of motion is created; *See also* 1825 COM

Baron Schilling links the summer palace of the czar in St. Petersburg to his winter palace using a telegraph with rotating magnetized needles; *See also* 1831 COM; 1833 COM

Charles Wheatstone invents the stereoscope; *See also* 1847 COM

ENERGY

Joseph Henry discovers self-induction, or inductance, the second current in a coil through which one current is passing that is induced by the first current; he publishes a paper about a spark that can be produced by this effect, but the effect is not described until 1835, when Michael Faraday independently discovers it; because of Henry's priority, the unit of inductance is known as the henry; note that *inductance*, discussed here, is not the same as *induction*, discovered independently by Henry and Faraday in 1831; *See also* 1831 ENE; 1835 ENE

Hippolyte Pixii demonstrates a hand-driven "magneto-electric machine," a generator in which a horseshoe magnet rotates in front of two coils; this is believed to be the first continuously acting current generator to produce alternating current (Faraday's generator of 1831 produced direct current); *See also* 1831 ENE; 1838 ENE

1831

Cyrus Hall McCormick [b Walnut Grove VA, Feb 15 1809, d May 13 1884] demonstrates his first version of the reaper in Jul; *See also* 1829 FOO; 1833 FOO

Pierre-Jean Robiquet [b Rennes, France, Jan 13 1780, d Paris, Apr 29 1840] and Jean Jacques Colin [b Riom, France, Dec 16 1784, d Mar 9 1865] discover the red dye alizarin; *See also* 1771 MAT; 1845 MAT

Chemist and physician Samuel Guthrie [b Brimfield MA, 1782, d Sackets Harbor NY, Oct 19 1848] discovers chloroform; *See also* 1846 MED

William Bickford patents the safety fuse, which consists of twelve strands of spun yarn with black powder funneled into the center

Walter Hancock introduces a steam-powered, ten-seat bus that travels between London and Stratford; *See also* 1801 TRA; 1835 TRA

1832

Dentists' chairs that recline are constructed, but manufactured versions do not become available until late in the 1860s; *See also* 1790 MED

The Warburton Anatomy Act legalizes the sale of bodies for dissection in England, ending the practice of body snatching and sometimes murder to provide bodies

Codeine is discovered

Charles Babbage conceives of the first computer, the Analytical Engine; it is a mechanical calculating machine driven by an external set of instructions, or program; although strikingly modern in concept, the computer is never built in workable form; *See also* 1822 TOO

The first rope-drilled artesian well, using a method of drilling originally developed by the Chinese 800 years before, is drilled in Europe; at Grenelle, France, near Paris, it is 547 m (1795 ft) deep; *See also* 1010–19 TOO; 1858 TOO

John Ireland Howe [b Ridgefield CT, 1793], using the shop of printing-press manufacturer Richard Hoe, builds a working model of a machine that can make straight pins in a single mechanized operation, which he patents; *See also* 1835 TOO

The Swedes, after 22 years of work, complete the Thomas Telford-designed Göta Canal, which completely crosses Sweden, from the Baltic to the Skagerrak; it is improved in 1877 and 1887; *See also* 1793 TRA

Franz Anton von Gerstner designs the first railroad passenger cars; they look much like the coaches then in use, except for the wheels, but the earlier coaches are drawn by horses; *See also* 1830 TRA; 1840 TRA

1833

ARCHITECTURE & CONSTRUCTION

COMMUNICATION

In Oct William Henry Fox Talbot [b Dorsetshire, England, Feb 11 1800, d Lacock Abbey, Sep 17 1877] independently conceives of photography while experimenting with the camera obscura; *See also* 1822 COM; 1834 COM

Karl Friedrich Gauss [b Braunschweig (in Germany), Apr 30 1777, d Göttingen, Germany, Feb 23 1855] and Wilhelm Weber [b Wittenberg (in Germany), Oct 24 1804, d Göttingen, Jun 23 1891] build an electric telegraph that operates over a distance of 2 km (1.25 mi); it uses a "mirror" galvanometer as a receiving device; *See also* 1832 COM

Xavier Progin builds a typewriter in which letters are mounted on independent hammers and inked by an ink pad; it is the first typewriter with a manual keyboard; he also develops a typesetting machine; *See also* 1829 COM; 1843 COM

ENERGY

John Ericsson [b Varmland, Sweden, Jul 31 1803, d NY, Mar 8 1889] builds a heat engine that uses superheated air as the driving force and that produces 5 horsepower; he publishes a description of his machine, *The caloric engine; See also* 1817 ENE

1834

GENERAL

Workers in Great Britain try to organize a federation of unions, the Grand National Consolidated Trades Union, but the leaders are transported to Australia and the federation fails; *See also* 1819 GEN; 1838 GEN

In Aug John Scott Russell [b Parkhead, Scotland, May 8 1808, d Isle of Wight, Jun 8 1882] observes a peculiar wave caused by a barge in the Union Canal near Edinburgh, Scotland; unlike ordinary water waves, this single 9-m- (30-ft-) long, 0.5-m- (1.5-ft-) high wave does not quickly dissipate, but maintains a constant speed of about 13 to 14 km/hr (8 to 9 mph) for several kilometers before Russell, who is following it on horseback beside the canal, loses it; the event is the first recorded observation of a soliton, a type of wave that will be applied to communications about 150 years later

COMMUNICATION

In Jan William Henry Fox Talbot begins experiments with silver nitrate that will eventually lead to the development of photography; *See also* 1833 COM; 1835 COM

William George Horner develops the zoetrope, a motion-picture device that is an improvement on Plateau's phenakistoscope; he introduces it to the US in 1867; *See also* 1832 COM

ENERGY

E.M. Clarke produces a commercial electromagnetic generator; *See also* 1832 ENE

1833

American inventor Obed Hussey (b ME 1792, killed by a train Aug 4 1860) patents a reaping machine based on reciprocating knives cutting against stationary guards; *See also* 1831 FOO

John Lane, a Chicago blacksmith, produces a plow made from wood covered with steel strips cut from a saw blade; it proves to be better for plowing Midwestern prairie soils than the cast-iron plows that are effective in the eastern US; *See also* 1819 FOO; 1837 FOO

French chemist Henri Braconnor discovers a flammable nitric acid ester or starch; this discovery leads to the first new explosives since gunpowder ("black powder"), the family of nitrogen explosives; *See also* 1807 MAT; 1838 MAT

Michael Faraday produces aluminum by electrolysis; *See also* 1827 MAT; 1854 MAT

1834

Jean Boussingault [b Paris, Feb 2 1802, d Paris, May 12 1887] discovers nitrogen fixation in plants, illuminating the main reason for the usefulness of crop rotation; *See also* 1804 FOO

US inventors John and Hiram Pitts develop the first efficient threshing machine; *See also* 1784 FOO

In Switzerland, Jacob Sulzberger improves the design of roller mills, which had not previously been widely accepted by millers; *See also* 1660–69 FOO

Johann Bodmer [b Zurich, Switzerland, 1786, d 1864] patents the traveling-grate stoker; it is improved on by John Juckes in 1841; the improved version is known as the chain-grate stoker; it becomes the most widely used stoker for steam boilers; *See also* 1822 TOO

Peter Barlow [b Norwich, England, Oct 1776, d Mar 1 1862] describes lenses with variable magnification; *See also* 1830 TOO

Swedish engineer Pehr Georg Scheutz and his son Edward build the first of several copies of the Difference Engine designed, but never built, by Charles Babbage; *See also* 1822 TOO

1835

During the summer William Henry Fox Talbot uses his new invention of photography to take experimental pictures of his home at Lacock Abbey; *See also* 1834 COM

Charles Havas founds in France the first press agency in the world, making use of the telegraph a few years later to collect news from abroad; *See also* 1848 COM

Joseph Henry develops the basic principles of the telegraph, which are put into more practical form 11 years later by Samuel F.B. Morse; these include the electric relay and the use of the Earth as a ground; *See also* 1831 COM; 1836 COM

Michael Faraday discovers independently of Joseph Henry the principle of self-induction, or inductance, and publishes it; *See also* 1832 ENE

Joseph Henry invents the electrical relay, which enables a current to travel long distances from its origin

1836

The arrival on Jan 1 of the paddle steamer *Jardine,* the first British steamship in Chinese waters, signals the beginning of British domination of China; *See also* 1842 GEN

French mathematician Jean-Victor Poncelet [b 1788, d 1867] publishes *Cours de mécanique appliquée aux machines* (A course in mechanics applied to machines); it introduces the use of mathematics and physical principles in the design of machines

Edward Davy [b Ottery, England, Jun 16 1806, d Jan 27 1885] discovers the electric relay independently of Joseph Henry and applies it to the telegraph; *See also* 1837 COM

John Frederic Daniell [b London, Mar 12 1790, d London, Mar 13 1845] invents the Daniell cell, the first reliable source of electric current, based on the interactions of copper and zinc; *See also* 1816 ENE; 1839 ENE

1835

Henry Burden of Troy NY begins to manufacture horseshoes by machinery, rather than making them by hand; *See also* 1825 TOO

Acetylene is recognized, although it will not be produced until Marcel Morren synthesizes it a quarter of a century later; *See also* 1859 MAT

William Gossage [b Burgh-in-the-Marsh, England, 1799, d Bowden, England, Apr 9 1877] invents a tower that absorbs hydrogen chloride, an important step in the development of the chemical industry; *See also* 1783 MAT

Samuel Colt [b Hartford CT, Jul 19 1814, d Hartford, Jan 10 1862] patents a revolver with a rotating breech; the breech turns in such a way that each time the trigger is pulled, a new bullet is placed in front of the barrel; Colt's invention was inspired by his observation of a helmsman turning a wheel that had a clutch that could lock it in several positions; *See also* 1812 TOO

John Ireland Howe, using a greatly improved version of his 1832 pin-making machine, founds the Howe Manufacturing Company, a successful manufacturer of straight pins in an age when pins are used not only for clothing, but also for most of the purposes for which paper clips are used today; *See also* 1832 TOO

James Nasmyth [b Edindurgh, Scotland, Aug 19 1808, d London, May 7 1890] develops the modification of a metal plane known as the shaper; *See also* 1817 TOO; 1842 TOO

Legislation in England promoted by horse-coach drivers inhibits further development of steam carriages by imposing prohibitive tolls; *See also* 1831 TRA

1836

John Stevens of England patents a screw propeller; the same year John Ericsson of Sweden patents a screw propeller for driving ships in England, eventually getting a fifth of the 20,000 pounds offered for improvements in steamship propulsion by the British Admiralty; *See also* 1797 TRA; 1839 TRA

1837

The Parliament in London begins to use an air-conditioning system devised by Jacob Perkins; *See also* 1831 GEN

Painter Samuel Finley Breese Morse [b Charlestown MA, Apr 27 1791, d New York City, Apr 2 1872] patents his version of the telegraph; the idea to use an electromagnet for transmitting signals comes upon him during a transatlantic trip when he sees a demonstration of one; *See also* 1835 COM; 1844 COM

Karl August Steinheil invents a telegraph system in which characters are printed on a paper ribbon; he is the first to use an induction machine as a current source; he also rediscovers that Earth can serve as a conductor, thus allowing a telegraph system linked by only one wire; the return wire is replaced by the Earth itself; however, the system is not commercially exploited; *See also* 1835 COM

Charles Wheatstone and William Fothergill Cooke [b Ealing, England, May 4 1806, d Surrey, England, Jun 25 1879] patent the first English telegraph, the "five-needle telegraph," a machine with five pointers that point to letters of the alphabet; they found the Electric Telegraph Company; their first telegraph line links Liverpool with Manchester in 1839; *See also* 1839 COM

1838

George Cayley founds the Polytechnic Institute in London for adult education

British cabinetmaker William Lovett and tailor Francis Place draw up the People's Charter calling for universal male suffrage and a secret ballot; the charter soon gains wide support from the working class, and supporters are termed Chartists; despite rallies, riots, and strikes, the Chartist movement fails to accomplish any of its goals and gradually disappears, with its last major rally in 1848; *See also* 1834 GEN

American David Bruce invents a machine for melting metal to make characters for typesetting; *See also* 1822 COM; 1840 COM

Louis-Jacques Daguerre [b Seine-et-Oise, France, Nov 18 1789, d Paris, Jul 12 1851] announces in Paris on Aug 10 his process for making daguerrotypes, silver images on copper plates; these are popular before photographs become common; *See also* 1822 COM; 1851 COM

Charles G. Page [b Salem MA, Jan 25 1812, d Washington DC, May 5 1868] is the first to build an electric generator with an electrically excited field magnet made from soft iron instead of a permanent magnet, as used by Faraday and Pixii; *See also* 1832 ENE; 1845 ENE

1837

John Deere [b Rutland VT, Feb 7 1804, d Moline IL, May 17 1886] of Grand Detour IL begins making plows of saw blade steel; they become the first popular horse-drawn plows; the company he founds remains one of the leading makers of farm equipment to this day; *See also* 1833 FOO; 1881 FOO

Moritz Hermann von Jacobi [b Potsdam (in Germany), Sep 21 1801, d St. Petersburg, Russia, Mar 10 1874] introduces electroplating, first developing silver plating in Russia, and then electroplating with nickel and chrome; *See also* 250 BC TOO

Isambard Kingdom Brunel builds the *Great Western,* the first ship to cross the Atlantic regularly on steam power, starting in 1838; he proves false the belief that a ship cannot carry enough coal for such a long voyage; *See also* 1818 TRA; 1838 TRA

1838

Justus von Liebig [b Darmstadt (in Germany), May 12 1803, d Munich, Apr 18 1873] develops a method for eliminating non-nutrient matter from preserved meat in order to reduce weight; *See also* 1819 FOO

Harris Moore and J. Hascall introduce their version of the combine, a combined thresher and harvester; *See also* 1828 FOO; 1935 FOO

Charles Frederick Kuhlmann [b Colmar, France, May 22 1803, d Lille, France, Jan 27 1881] introduces a new method for preparing nitric acid by the oxidation of ammonia in the presence of platinum as a catalyst; *See also* 1774 MAT; 1858 MAT

Théophile J. Pelouze develops nitrated paper as an explosive, a precursor of nitrocellulose; *See also* 1833 MAT; 1846 MAT

Italian chemist R. Piria is the first to isolate salicylic acid, which he obtains from willow bark; this step is an important precursor to the discovery of aspirin; *See also* 1853 MED

William Beaumont's *Experiments and observations on the gastric juice and the physiology of digestion* reports on his study of digestion *in vivo* and *in vitro* with a wounded man whose stomach remains partially accessible through a healed hole in the abdomen; *See also* 1822 MED

The steamship *Sirius* is the first to cross the Atlantic Ocean on steam power alone, taking 18 days and very nearly running out of coal before reaching NY from London, despite carrying 450 tons at the start; regular passage is instituted by 1840; *See also* 1818 TRA; 1837 TRA

Joseph Francis, inventor of the "unsinkable" lifeboat, patents a more practical design that uses a cable that attaches to shore when the lifeboat is used to aid a sinking vessel; *See also* 1819 TRA; 1845 TRA

1839

GENERAL

Edmond Becquerel discovers the photovoltaic effect, the first of several photoelectric effects discovered in the 19th century; he observes a voltage between two electrodes in an electrolyte produced by exposing one electrode to light; *See also* 1954 ENE

COMMUNICATION

The first commercial telegraph system starts operation in Great Britain; by 1850 the network extends over 5600 km (3500 mi); *See also* 1837 COM

On Jan 31 William Henry Fox Talbot describes his invention of photography to the Royal Society; *See also* 1835 COM

In Sep Fox Talbot, after an accidental discovery of the process, develops the first form of photographic negatives, called calotypes; he patents the process on Feb 8 1841; *See also* 1822 COM; 1851 COM

Sir John Herschel [b Slough, England, Mar 7 1792, d Collingwood, England, May 11 1871] demonstrates that the method used by Fox Talbot for slowing down the darkening of silver salts exposed to light could be greatly improved upon by substituting sodium thiosulphate (known as "hypo," short for sodium hyposulfite, an older name for the compound) for the sodium chloride (table salt) used by Talbot

ENERGY

William Robert Grove [b Swansea, Wales, Jul 11 1811, d London, Aug 1 1896] develops the first fuel cell, a device that produces electrical energy by combining hydrogen and oxygen; although theoretically an excellent way to produce electricity, fuel cells fail to become practical for most applications; *See also* 1800 ENE

William Robert Grove, in the same year as his fuel cell also develops a form of electric cell based on zinc and platinum electrodes in two forms of acid separated by a diaphragm; this cell nearly doubles the voltage of the Daniell cell; *See also* 1836 ENE; 1859 ENE

1840

GENERAL

Mémoire sur l'artillerie des anciens et sur celle du moyen âge (Ancient and medieval artillery) by Guillaume Henri Dufour is the most important work by the Swiss general, cartographer, and military writer

William Robert Grove succeeds in lighting an auditorium with impractical and very expensive incandescent lamps that use platinum coils in inverted glasses sealed with water; *See also* 1820 GEN; 1841 GEN

ARCHITECTURE & CONSTRUCTION

The British Houses of Parliament are constructed using the Gothic Revival style, popular in England for a century or so, but nearing the end of popular favor

The first iron-truss bridge in the US is built at Frankfort NY

COMMUNICATION

Prepaid postage is introduced in England with the "penny black" stamp; previously, the recipient had paid for receiving mail; *See also* 1830 COM

Scottish clockmaker Alexander Bain proposes a picture transmitter and receiver based on a pendulum apparatus; a metal version of the image to be sent is scanned by a pendulum, which sends electrical impulses over a wire when the pendulum is near the metal; at the receiving end, a matching pendulum transmits the electrical impulses to treated paper that changes color in response to an electric current; the entire mechanism at each end is controlled by clockwork; Bain's proposal is never developed as a practical device; *See also* 1847 COM

1839

Thomas Kensett and William Underwood switch their canning factories in the US from the use of jars to the use of metal containers; *See also* 1819 FOO

Charles Goodyear [b New Haven CT, Dec 29 1800, d New York, Jul 1 1860] discovers vulcanization of rubber (the addition of sulfur to rubber followed by processing the combination to make it stable when hot or cold) by "accident"; Goodyear had been seeking such a process for about 10 years when he spilled some of a sulfur-rubber mixture on a hot stove—the cooled rubber was the stable form he sought; sulfur produces crosslinks between rubber polymers; *See also* 1823 MAT; 1845 MAT

Isaac Babbitt [b Taunton MA, Jul 26 1799, d Somerville MA, May 26 1862] invents an antifriction alloy (called Babbitt metal) that becomes extensively used in bearings

James Nasmyth makes a drawing of a steam hammer on Nov 24, a design developed at the request of Isambard Kingdom Brunel, who needs a device to forge the paddlewheel shaft of his large steamships; the steam hammer is an extremely powerful tool for working iron, but can be adjusted to produce taps as light as those used in breaking an egg; *See also* 1842 TOO

On an order from the US Navy, John Ericsson outfits a British iron steamship with his screw propeller; the ship reaches New York City in May and Ericsson himself soon follows; the remainder of his ship-building career takes place in the US; *See also* 1836 TRA; 1844 TRA

Kirkpatrick MacMillan, in Scotland, develops a bicycle powered by treadles and driving rods; the first bicycle that does not require the rider's feet to touch the ground, it is called a velocipede; *See also* 1818 TRA; 1861 TRA

1840

Gottfried Keller, either upon observing children playing with a paste of sawdust and water or upon learning that wasps make paper from chewed up wood (or perhaps both), develops a method of preparing paper from ground wood; *See also* 1799 MAT

Following work by Justus von Liebig, mirrors coated with silver are introduced

Pathologischen Untersuchungen (Pathological investigations) by German pathologist and anatomist Friedrich Gustav Jakob Henle [b Fürth, Jul 19 1809, d Göttingen, May 13 1885] expresses his conviction that diseases are transmitted by living organisms, although he offers no hard evidence; *See also* 1876 MED

Italian physicist Giovanni Battista Amici [b Modena, Mar 23 1786] invents the oil-immersion microscope, one of his innovations in microscope building that results in instruments with an enlarging power of 6000 times; *See also* 1792 TOO

Alexander Bain [b Watten, Scotland, 1810, d 1877] develops a clock driven by electricity; *See also* 1829 TOO

The first railroad dining cars make their appearance in the US; *See also* 1832 TRA; 1868 TRA

British locomotive designer Daniel Gooch [b Bedington, Northumberland, Aug 24 1816, d Oct 15 1889] builds the Firefly, the most powerful locomotive of its time; it reaches average speeds of 80 km (50 mi) an hour; *See also* 1830 TRA; 1846 TRA

1840 cont

John Benjamin Dancer [b London, Oct 8 1812, d Nov 1887] develops a method for making photographs of microscopic objects with magnifications of up to twenty times; *See also* 1839 COM

John William Draper [b St. Helen's, England, May 5 1811, d Hastings NY, Jan 4 1882] takes what is today the oldest surviving photograph of a person; *See also* 1839 COM

Josef Petzval designs an improved camera lens for portraits, based on extensive calculations that had been performed by ten military calculators over several years; the lens is produced by the Viennese optician Voigtlaender; in 1856 Petzval also designs a lens for landscapes; *See also* 1860 COM

J.H. Young and A. Delcambre develop the first practical typesetting machine, using a design by Henry Bessemer; *See also* 1838 COM; 1884 COM

1841

Frederick de Moleyns obtains the first patent for an incandescent lamp, an evacuated glass containing powdered charcoal that bridges a gap between two platinum filaments; *See also* 1840 GEN; 1845 GEN

William Marshall builds a state-of-the-art flax mill at Leeds, England; the main factory floor covers more than 8000 m² (86,000 sq ft); rows of conical skylights provide natural lighting throughout the mill, while between the skylights sheep graze on the roof; the pillars that support the roof are pipes that conduct rainwater to the basement; steam engines in the basement not only power the machinery, but also provide the steam to heat the building and the hot water for baths for the employees; *See also* 1771 GEN

Joseph Paxton completes the Great Stove, a greenhouse (conservatory) at Chatsworth, England; this conservatory uses an arched ridge-and-furrow roof; *See also* 1832 ARC; 1849 ARC

Joseph Stephenson introduces the use of the steam hammer for driving piles into the ground (the pile driver), an important application in bridge construction; *See also* 1839 TOO

Henri Rossiter Worthington [b 1817, d 1880] patents a nonrotating pump for supplying feed water to boilers under pressure; the pump cylinder and the steam cylinder are directly in line with each other

Ada Lovelace

Ada, countess of Lovelace, was the only child of the poet Lord Byron and Anne Isabella Milbanke, who left Byron when her daughter was only a year old. The programming language ADA is named for her, although the countess has only a slender claim to the frequently used label of "first programmer."

Ada was a singularly attractive young woman when she met Charles Babbage, who was more than 20 years older. Babbage had already built a model of a hand-cranked machine that could calculate logarithms and was at work on a bigger project. Ada was greatly impressed by Babbage's ideas for machines that could calculate (and, some say, by Babbage himself). She utilized various of her talents, including considerable mathematical ability, to promote Babbage's planned steam-powered computer, the Analytical Engine, including writing a long series of "Observations on Mr Babbage's Analytical Engine" that remains one of the best sources of information about it and that demonstrates clearly that Ada understood the role of programming in relation to mechanical problem solving. She also translated what many consider the best contemporary description of how the machine would work, an account by Italian engineer L.F. Menabrea. And she really did write a program, one for calculating Bernoulli numbers—not a mean feat.

Ada and Babbage fell into mathematical error on at least one occasion. In an effort to finance construction of the machine, the pair developed a system for gambling on horse races. Like all such systems, this one failed and left them deeply in debt. Ada had to pawn jewels to escape creditors threatening blackmail.

Ada's health was never very good, and she died nearly 20 years before Babbage did. Three years after her death, the first of Babbage's calculators, the Difference Engine, was built in Sweden, where it worked as intended. It was not until after the development of electronic computers that any significant part of the Analytical Engine was constructed. It would have worked just as Ada always said that it would.

Although it is often written that English engineering of the time was not up to the demands of Babbage's device, this was not really the case; lack of funding was the obstacle. If Ada or Babbage had been richer, or if the gambling system had somehow worked, the first programmable computer would have been built during the 19th century, not at the end of World War II.

1841

John Augustus Roebling [b Mulhausen, Prussia, Jun 12 1806, d Brooklyn NY, Jul 22 1869] manufactures the first wire-rope cable in the US; this undertaking eventually brings him and his family into the business of building suspension bridges, although they also manufacture the cables for many suspension bridges built by others; *See also* 1845 ARC

The first beehive oven for making coke built in the US starts operating in PA; *See also* 1620–29 MAT; 1850 MAT

Charles Thomas Jackson [b Plymouth MA, Jun 21 1805, d Somerville MA, Aug 28 1880] discovers that ether is an anesthetic; *See also* 1842 MED

Surgeon James Braid [b Rylawhouse, Scotland, 1795, d Manchester, England, Mar 25 1860] renames mesmerism *hypnotism* and gives the practice some medical respectability by correctly explaining why it works; *See also* 1774 MED; 1872 MED

Joseph Whitworth [b Stockport, England, Dec 21 1803, d Monte Carlo, Monaco, Jan 22 1887] introduces the standard screw thread in *A uniform system of screw-threads*, a paper presented to the Institution of Civil Engineers; *See also* 1800 TOO; 1864 GEN

Painter John Goffe Rand patents the first squeezable tube, a device he has developed for keeping oil paints from becoming gummy when painting outdoors; *See also* 1891 TOO

James Buchanan Eads [b Laurenceburg IN, May 23 1820, d Nassau, Bahamas, Mar 8 1887] patents a new type of diving bell about this time and uses it for extensive salvage operations in the Mississippi River; *See also* 1756 TRA

1842

GENERAL

The Treaty of Nanjing, signed Aug 29, confirms British victory in the first Opium War with China; in addition to ceding Hong Kong to Great Britain, the treaty signals to other Western nations that China is weak and ripe for exploitation; soon the US, France, Belgium, and Sweden also establish beneficial trading agreements with China, all of which contribute to British trade as well; *See also* 1836 GEN; 1860 GEN

Julius Robert Mayer [b Heilbronn (in Germany), Nov 25 1814] is the first to state the law of conservation of energy, noting specifically that heat and mechanical energy are two aspects of the same thing; this is often called the first law of thermodynamics; *See also* 1824 GEN; 1850 GEN

ARCHITECTURE & CONSTRUCTION

Charles Ellet [b Penn's Manor PA, Jan 1 1810, d Cairo IL, Jun 21 1862] builds the first important US suspension bridge, over the Schuylkill River near Philadelphia PA; *See also* 1825 ARC; 1848 ARC

COMMUNICATION

Ernst *Werner* von Siemens [b Lenthe, Hanover, Germany, Dec 13 1816, d Berlin, Germany, Dec 6, 1892] invents and patents an electroplating process

ENERGY

William Howe invents the link-motion reversing gear for steam engines; it consists of two eccentrics, one for forward motion and one for backward, that alternately can be connected to the valve rod; this system was later used by Stephenson on locomotives; *See also* 1763 TOO; 1844 TOO

1843

ARCHITECTURE & CONSTRUCTION

Joseph Fowle develops the first drill powered by compressed air, for use in constructing tunnels; *See also* 1778 ARC

Marc Isambard Brunel builds the first tunnel under the Thames in London; it takes 18 years to complete the tunnel, which collapses several times, but it opens to pedestrians on Mar 25; the tunnel closes 20 years later, and reopens another 20 years later for trains of the London Underground system; *See also* 1811 ARC

COMMUNICATION

Ada Lovelace's *Notes*, which relate her experience with Charles Babbage's Analytical Engine, are published; *See also* 1826 GEN; 1853 COM

Charles Thurber [b E. Brookfield MA, Jan 2 1803, d Nashua NH, Nov 7 1886] of Worcester MA patents a form of typewriter in which the letters mounted on a wheel can be actuated by pressing directly on a letter; *See also* 1833 COM

1842

John Bennett Lawes [b Rothamsted, England, Dec 28 1814, knighted 1882, d Rothamsted, Aug 31 1900] patents superphosphate; in 1843, with aid from a student of chemist Justus von Liebig, he begins manufacturing the artificial fertilizer; *See also* 1820 FOO

The brothers M.A. and I.M. Cravath of Bloomington IL invent the disk plow about this time; *See also* 1837 FOO

The first use of ether in surgery is by Crawford Williamson Long [b Danielsville GA, Nov 1 1815, d Athens GA, Jun 16 1878] on Mar 30, but lack of publication allows credit for the discovery to go to William Morton in 1846; Long publishes his own results in 1849; *See also* 1841 MED; 1844 MED

James Nasmyth discovers that a competitor has copied and built his design for a steam hammer, which Nasmyth has not yet constructed; credit for the invention is generally assigned to Nasmyth, however (although there is also evidence that the first to envision such a tool was James Watt); *See also* 1839 TOO

Joseph Whitworth develops an improved planer that is able to cut in both directions; *See also* 1835 TOO

Thomas Woodward of Brooklyn NY patents a form of safety pin that lacks a spring to keep it closed; instead, it relies on the material being pinned being thick enough to hold the point of the pin in its shield; *See also* 500 BC TOO; 1849 TOO

Karl von Ghega (b 1802, d 1860) begins construction of the first railway over the Alps, linking Vienna and Trieste; the project is completed in 1854; *See also* 1830 TRA

W.S. Henson designs, continuing into 1843, the "aerial steam carriage," an early concept of a powered airplane based on the ideas put forth by George Cayley; *See also* 1809 TRA; 1871 TRA

1843

The steamship *Great Britain* is built largely from wrought iron produced by the puddling process; despite the fact that it lies on a beach unpainted for many years, it is still in good condition in 1988 and is salvaged; *See also* 1784 MAT

Oliver Wendell Holmes [b Cambridge MA, Aug 29 1809, d Boston MA, Oct 7 1894] advises doctors to prevent spreading puerperal fever (a common disease of mothers after childbirth at the time) by washing their hands and wearing clean clothes; *See also* 1847 MED

Emil Heinrich du Bois-Reymond [b Berlin, Prussia (Germany), Nov 7 1818, d 1896] demonstrates that electricity is used by the nervous system to communicate between different parts of the body; *See also* 1840 MED; 1861 MED

The aneroid barometer, which uses expansion and contraction of an evacuated metal tube to detect changes in air pressure, is invented; unlike previous barometers, the aneroid type indicates atmospheric pressure using a dial and pointer; *See also* 1640–49 TOO

The clipper ship is introduced in the US; this sailing ship is streamlined and designed for the fast transport of goods; *See also* 1785 TRA

Isambard Kingdom Brunel launches the *Great Britain*, the first steamship to cross the Atlantic driven by a screw propeller (in 1845); it has an iron hull and remains in service for 30 years; *See also* 1837 TRA; 1858 TRA

George Cayley designs the first aircraft designed to take off vertically, using four propellers, that would then swivel forward for level flight; *See also* 1804 TRA; 1853 TRA

1844

An experimental electric light is installed on the Place de la Concorde in Paris; *See also* 1841 GEN; 1845 GEN

Samuel Morse uses his telegraph system to send a famous message from Washington to Baltimore: "What hath God wrought?"; *See also* 1837 COM

Werner von Siemens develops a mechanical copying method based on reproducing printed matter with zinc plates; the part to be reproduced is raised into relief; *See also* 1842 COM

William Fairbairn [b Kelso, Scotland, Feb 19 1789, knighted 1869, d Moor Park, England, Aug 18 1874] and John Hetherington patent what comes to be called the Lancashire boiler; it consists of a cylindrical vessel through which passes a cylindrical flue; this type of boiler remains in use until the 20th century; *See also* 1828 ENE; 1856 ENE

1845

William Fairbairn introduces the riveting machine to the manufacture of steam boilers; he also designs and builds tubular steel railway bridges; *See also* 1841 GEN; 1846 GEN

American W.E. Staite patents in England an incandescent electric lamp; *See also* 1841 GEN; 1846 GEN

Thomas Wright obtains the first patent for an arc lamp; *See also* 1809 GEN; 1876 GEN

Isambard Kingdom Brunel builds the Hungerford suspension bridge over the Thames in London; in 1862, however, the Hungerford bridge is displaced by the Charing Cross railway bridge; *See also* 1825 ARC

Work begins on the library of Sainte Geneviève in Paris, designed by Henri P.F. Labrouste [b 1801, d 1875]; it is the first public building in which iron is used as part of the visible style, with iron vaults and columns; overall, the building combines the modern style, using uncovered iron, with more traditional features

John Augustus Roebling completes the first suspension aqueduct in the US; it carries the Pennsylvania State Canal across the Allegheny River near Pittsburgh PA; *See also* 1841 MAT; 1855 ARC

The first underwater telegraph cable is laid under the Hudson River between New York City and Fort Lee NJ; *See also* 1851 COM

The telegraph developed by Charles Wheatstone attracts publicity when it results in the arrest of the "Quaker murderer"; the suspect is spotted on board a London-bound train at Slough and, as a result of a telegram, is arrested when he arrives at Paddington Station; *See also* 1837 COM; 1910 COM

Charles Wheatstone develops an electric generator in which electricity is induced in coils by an electromagnet powered by a battery; later he will power these magnets with the current from the generator itself; *See also* 1838 ENE; 1848 ENE

1844

Horace Wells [b Hartford CT, Jan 21 1815, d New York City, Jan 24 1848] is the first to use nitrous oxide as an anesthetic in dentistry; *See also* 1800 MED

Charles Thomas Jackson suggests using ether to deaden pain to dentist William Thomas Green Morton; *See also* 1841 MED; 1846 MED

The Commission for Enquiring into the State of Large Towns establishes a connection between dirt and epidemic disease in England; *See also* 1800 MED; 1854 MED

About this time plaster of Paris begins to replace beeswax for making impressions of teeth; *See also* 1857 MED

Werner von Siemens develops a differential governor for steam engines; *See also* 1787 TOO

Egide Walschaërts invents a reversing gear for steam engines, using one eccentric; *See also* 1842 ENE

John Ericsson builds a propeller-driven, metal-hulled warship, the *Princeton*, which has its engines below the waterline for better protection; *See also* 1839 TRA; 1851 TRA

1845

A potato blight, which first appeared in Europe about 1840, devastates Ireland, which had come to depend on this American species as a staple; the blight is caused by a fungus; *See also* 1885 FOO

Peter Cooper, better known as the manufacturer of the locomotive *Tom Thumb*, invents the first gelatin dessert; *See also* 1897 FOO

Thomas Scragg invents a machine for making drainpipe that greatly facilitates land drainage in Britain

Thomas Hancock discovers ebonite by treating rubber with sulfur; it becomes widely used as an electrical insulator; *See also* 1839 MAT

August Wilhelm von Hofmann [b Giessen, Germany, Apr 8 1818, d Berlin, Germany, May 2 1892] develops a method for preparing aniline dye from benzene; *See also* 1831 MAT; 1856 MAT

John Mercer [b Dean, England, Feb 21 1791, d Nov 30 1866] discovers that cotton thread soaked in a solution of sodium hydroxide becomes heavier and shiny; this treatment is now called mercerization; *See also* 1756 MAT

The hypodermic syringe is introduced

An Italian manufacturer produces the first breech-loaded cannon of modern times, using a sliding-wedge design; this is quickly followed by other manufacturers using either the sliding wedge or an interrupted screw to prevent the explosion from forcing out the rear of the cannon; *See also* 1460–69 TOO

March of the Royal Arsenal surgery in England introduces the percussion tube, which ignites powder by pulling a rope that releases a hammer that explodes a tube of percussion powder, which then sets off the main charge of a cannon; *See also* 1807 MAT

The turret lathe is invented; *See also* 1800 TOO

Joseph Francis develops a wooden lifeboat that uses a cable that attaches to shore when the lifeboat aids a sinking vessel, but finds that it is not strong enough in actual practice; after much experimentation he patents a method of making boats out of corrugated iron that proves practical in many applications to building boats and other floating devices, such as docks and buoys; *See also* 1819 TRA; 1838 TRA

Robert William Thomson [b Stonehaven, Scotland, 1822, d Edinburgh, Scotland, Mar 8 1873] invents the rubber tire

1846

Under pressure from the Anti-Corn Law League, England's Corn Laws are repealed on Jun 15; *See also* 1815 GEN

John Draper patents an incandescent electric lamp with a platinum filament; *See also* 1845 GEN; 1850 GEN

Werner von Siemens improves the Wheatstone telegraph to make it self-acting by using "make-and-break" contacts; *See also* 1845 COM

Alexander Bain develops a method of sending telegraph messages by using punched paper tape, greatly improving on the speed of transmission; *See also* 1840 COM

Aimé Laussedat [b Moulins, France, Apr 19 1817, d Mar 19 1907] develops a photographic method, called photogrammetry, for measuring buildings and structures; it uses photographs taken from two different angles

---

The telegraph

Fires, smoke signals, and drums have been used since antiquity to transmit messages over long distances. The term *telegraph* was coined by Claude Chappe to describe such methods, a version of which was invented by him and his brothers to signal each other while in school. In 1793 Chappe introduced in France a form of this system for the transmission of messages based on stations with towers using a code to transmit signals by the position of crossarms. Today this kind of system is called a semaphore ("sign bearer"), not a telegraph ("far writer").

The idea of the electric telegraph was born when the first experimenters with electricity noticed that electric charges could travel through wires over distances. In 1753 in Scotland Charles Morrison described a system of 26 wires for transmitting the 26 letters of the alphabet. Electrostatic charges traveling through these wires deflected suspended pith balls at the receiving station. However, this was never developed as a practical system.

During the early 19th century, several scientists experimented with the transmission of messages through electric wires. At this time scientists had gained access to a steady, low-voltage source of electricity. Karl Friedrich Gauss and Wilhelm Weber transmitted signals over wires and detected them with sensitive galvanometers around 1833. In England Charles Wheatstone developed a telegraph with a five-needle galvanometer that indicated the transmitted letters. The Wheatstone telegraph actually came into use, linking Liverpool with Manchester in 1839. In Germany Karl Steinheil developed a telegraph that printed coded messages on a ribbon.

The electromagnet, a magnet that appeared when current was on and disappeared when it was off, was discovered in the 1820s. The American painter Samuel Morse first became acquainted with an electromagnet when it was shown to him by a young chemist he met on a transatlantic ship. Morse realized that a magnet turning on and off by transmission of a current from a distant source could be used to send messages. He soon enlisted America's greatest scientist of the time, Joseph Henry, to develop ways to cause an electromagnet to work at a distance. The electric telegraph became truly functional with the idea of using a code of dots and dashes to transmit the letters of the alphabet; this method was conceived by another collaborator, the American engineer Alfred Vail. Despite this technical help, Morse is given credit for the invention because he put together a practical system and got people to accept it.

1846

Louis Nicolas Menard [b Paris, Oct 19 1822, d Paris, Feb 9 1901] discovers collodion by dissolving nitrocellulose in a mixture of alcohol and ether; a number of other chemists independently discover virtually the same substance about this time, notably American J. Parkers Maynard (in 1848); *See also* 1855 MAT

Christian Schoenbein [b Metzingen, Swabia, Oct 18 1799, d Baden-Baden, Germany, Aug 29 1868] discovers nitrocellulose, an explosive, by soaking cotton in nitric acid; *See also* 1838 MAT; 1863 MAT

Ascanio Sobrero [b Casale (in Italy), Oct 12 1812, d Turin, Italy, May 26 1888] discovers the explosive nitroglycerin, which is later found to have medical uses as well, notably in controlling some forms of heart irregularities; *See also* 1838 MAT; 1863 MAT

William Thomas Morton [b Charlton City MA, Aug 9 1819, d New York City, Apr 2 1872] uses ether as an anesthetic during operations, as advised by Charles Jackson, who discovered that ether is an anesthetic; *See also* 1844 MED

Sir James Simpson [b Bathgate, Scotland, Jun 7 1811, d London, May 6 1870] discovers that chloroform is a better anesthetic than ether or nitrous oxide; he starts using chloroform during childbirth; his 1847 *Account of a new anesthetic agent* describes his discovery; *See also* 1831 MED

Erastus Brigham Bigelow [b Boylston MA, 1814, d Boston, Dec 6 1879] develops a power loom that produces carpets and tapestries at greater speed than previous looms and allows the incorporation of non-symmetrical patterns, such as flowers; *See also* 1813 TOO

Houillier, a French gunsmith, develops the first true one-piece cartridge for small arms, a charge of powder in a case with a percussion cap and a projectile, all in one waterproof package; an essential part of his invention is a mechanism for removing the case when the cartridge has been fired; *See also* 1807 MAT

Elias Howe [b Spencer MA, Jul 9 1819, d Brooklyn NY, Oct 3 1867] patents the lock-stitch sewing machine; *See also* 1830 TOO; 1851 TOO

Regular passenger service on steamships is established across the Atlantic Ocean between England and North America; packet service had preceded this for several years; *See also* 1838 TRA

Alexander Bain invents an improved form of Edward Massey's ship's "log"; *See also* 1802 TRA; 1861 TRA

Thomas Russel Crampton [b London, 1816, d 1888] designs a high-speed locomotive in which the driving axle is behind the boiler; 2 locomotives are put into service on the Liège-Namur line in Belgium, where they reach 97 km (60 mi) per hour; *See also* 1840 TRA

Daniel Gooch builds the Great Western, the most powerful locomotive of its time, with a boiler operating at 100 psi pressure and an average speed of 107 km (67 mi) per hour; *See also* 1840 TRA

The Gauge Act fixes the distance between rails in England; this gauge is adopted by all European nations except Spain and Russia; *See also* 1830 TRA

The telegraph (Continued)

Morse patented his telegraph in 1837 and officially inaugurated a link between Baltimore MD and Washington DC on May 14 1844 by transmitting the message "What hath God wrought." The message was transmitted by a telegraph key, a special switch that allows an electric current to be rapidly switched in and out; it was printed in the dot-dash code on ribbons of paper. In 1844 Vail determined that an operator could learn to hear the differences between dots and dashes; this became the preferred way to decipher messages.

Morse's telegraph quickly spread in the US, and later it superseded the existing systems of Wheatstone and Steinheil in Europe. In 1862, 240,000 km (150,000 mi) of telegraph cable covered the world, of which 77,000 km (48,000 mi) were in the US and 24,000 km (15,000 mi) in Great Britain. Europe and the US became linked by an underwater cable in 1866.

During the 20th century, the use of the telegraph declined, mainly because of lower prices for telephone and telex services. Also, wireless telegraphy, the first form of radio that used the same codes as ordinary telegraphy, was available. "Telegrams" increasingly were transmitted by telex or telephone instead of as actual telegrams. During the 1980s telegraph services disappeared altogether in most countries.

1847

The Institution of Mechanical Engineers is founded in England; *See also* 1828 GEN; 1851 GEN

Werner von Siemens suggests that gutta-percha be used to preserve electrical wiring from moisture, an idea that leads to the first underground and submarine telegraph cables

Frederick Bakewell invents a way of transmitting a picture painted on a conducting roller with shellac to another roller; this is not only an early form of facsimile machine, but also a precursor of television; Bakewell's replacement of the pendulum mechanism used by Alexander Bain with a rotating drum, and of Bain's metal figures with shellac, are great improvements, but synchronization of the drums at each end is a problem; Bakewell patents his version in 1848 but never commercializes it; *See also* 1840 COM; 1863 COM

Richard March Hoe [b Sep 12 1812] invents both the rotary and web printing presses; his rotary presses reach a capacity of 18,000 sheets printed on both sides per hour; their invention contributes to the rapid spread of large-circulation newspapers; the Philadelphia *Public Ledger* is the first newspaper to use Hoe's rotary press; *See also* 1814 COM; 1853 COM

Abel Niepce de Saint-Victor introduces the albuminized glass plate to photography

Sir David Brewster develops an improved stereoscope for viewing photographs taken from slightly different angles; each eye views one image, giving an impression of three-dimensional viewing when the two different images are combined into one by the brain; *See also* 1832 COM

1848

Charles Babbage's *Philosophy of manufacturers* supports Francis Bacon's ideas about the production of new knowledge by science and experiment; *See also* 1620–29 GEN

Charles Ellet builds the suspension bridge footpath over the Niagara River; *See also* 1842 ARC; 1849 ARC

The Associated Press is founded by six newspapers in New York City in order to share the costs of telegraphy; *See also* 1835 COM; 1849 COM

Jacob Brett improves the electric generator based on electrically excited soft iron field magnets by using direct current from the generator to keep the magnetism in the iron; it is still thought necessary to use an independent source of current to get the magnetic field started, however; *See also* 1838 ENE; 1857 ENE

1847

Hungarian physician Ignaz Philipp Semmelweiss [b Budapest, Jul 1 1818, d Vienna, Austria, Aug 13 1865] discovers that puerperal fever (childbed fever) is contagious; he has doctors working for him wash their hands in the hopes of reducing the number of cases of the disease in his hospital; *See also* 1843 MED; 1865 MED

Karl Friedrich Wilhelm Ludwig [b Witzenhausen (in Germany), Dec 29 1816, d Leipzig, Saxony, Apr 27 1895] develops a device that continuously records blood pressure, which he uses to show that the circulation of the blood is purely mechanical; no mysterious vital processes outside of ordinary physics need to be invoked; *See also* 1733 MED; 1863 MED

Matthew Townsand of Leicester, England, about this time invents the second form of knitting needle, the latch needle; *See also* 1589 TOO

1848

Linus Yale Jr. [b Salisbury NY, Apr 4 1821, d New York City, Dec 25 1868] patents a lock based on pins in a rotating cylinder, the most common type in use today and still known as the Yale lock; *See also* 1818 TOO

The Rideau Canal from Ottawa to Kingsbury, Canada, is opened, as is the first version of the St. Lawrence Seaway; *See also* 1829 TRA; 1959 TRA

The Illinois-Michigan Canal is completed, linking the Great Lakes to the Mississippi; *See also* 1825 TRA

1849

GENERAL

The first bombing raid from the air occurs when a pilotless Montgolfier balloon is used to drop bombs on Venice (Italy)

ARCHITECTURE & CONSTRUCTION

Charles Ellet builds a 303-m (1010-ft) suspension bridge over the Ohio, constructing one of the first major suspension bridges in the US; *See also* 1848 ARC; 1861 TOO

Joseph Paxton builds a greenhouse to contain the *Victoria regia* lily; the Victoria Regia Lily House is the first building to use ridge-and-furrow construction with a flat roof and the first to have a glass curtain wall hung from the girders; both features, the second in a modified form, are used in the Crystal Palace, which Paxton begins to build in 1850; *See also* 1841 ARC; 1851 ARC

COMMUNICATION

Paul Julius Baron von Reuter [b Kassel, Germany, Jul 21 1816, d Nice, France, Feb 25 1899] starts his famous press agency as a homing pigeon service for carrying stock prices between Brussels, Belgium, and Aachen, Germany; *See also* 1851 COM

ENERGY

George *Henry* Corliss [b Easton NY, Jun 2 1817, d Feb 21 1888] patents a steam engine that uses four valves instead of one; each end of the cylinder has spring-loaded inlet and exhaust valves, thus saving heat by regulating the amount of steam admitted; the Corliss valve engine is considered the most significant advance in steam-engine design since the work of James Watt; *See also* 1783 ENE; 1876 ENE

James Bicheno Francis [b Southleigh, England, May 18 1815, d Lowell MA, Sep 18 1892] develops the inward-flow turbine (also known as the Francis turbine), in which water flows inward along a radius and emerges near the shaft; this highly efficient turbine is still widely used for power generation; *See also* 1831 ENE

The advent of electricity

The Greeks and other ancient peoples knew that some substances attract others when one of them has been rubbed. Similarly, they must have encountered mild shocks. People still notice both today, as when clothing picks up lint without touching it or when a doorknob gives a slight shock on a dry winter day. Thales, often counted as the first scientist, investigated the attractive properties of rubbed amber in Ionia around 600 BC. He compared it to magnetism and noted that it was similar, but different. More than 2000 years later, around 1570, the English scientist William Gilbert also studied magnetism and the corresponding attraction of rubbed amber and various rubbed jewels. He modified the Greek and Latin terms for amber to produce the English word *electric* as a noun describing a material that behaved like amber. Nearly a hundred years later, in 1650, the term *electricity* was coined to refer to the force itself.

In the same year that *electricity* entered English, a German scientist, Otto von Guericke, was working with a method he had developed for making more electricity than could be made by rubbing a small piece of amber. He worked with a different electric, sulfur. He formed this electric into a ball, so it could be rubbed continuously by rotating it. Von Guericke was the first to observe light produced by electricity. Similar experiments produced light from electricity in England.

The electricity produced by von Guericke faded away soon after the ball stopped rotating. Shocks and lights were evanescent experiences. About a hundred years after the use of rotating balls, however, scientists in Leyden (Leiden, the Netherlands) learned that electricity could be stored in water in glass jars and conducted in and out with metal wires or nails. Modern versions are somewhat different in construction, but the main impact of the Leiden jar was the same. A large charge could be stored up over time and then released by touching the conductor. Some terrific shocks resulted, and the jars became popular in parlors as well as with scientists.

Still, no one knew what electricity was. It was not even clear whether there was one type or two (different electrics attracted or repelled different light substances, although the shocks seemed to be the same). Accidentally, a scientist in Italy, Luigi Galvani, found that severed frogs' legs twitched in response to electricity and that they also responded the same way when touching metal that was not charged. Another Italian, Alessando Volta, who had been experimenting with electricity, found that the reaction was chemical. He built an apparatus, which came to be known as the voltaic pile, that offered the first method of producing electricity in any quantity without rubbing. Improvements on the voltaic pile are the familiar batteries of today.

The first blossom in England of the gigantic (more than 30 cm, or 1 ft across) *Victoria regia* water lily is presented to Queen Victoria, for whom it is named; the South American lily, first noted to science in 1801 and rediscovered about 1830, is grown in a specially built house by gardener James Paxton on the estate of the duke of Devonshire; after opening twice, the flower sinks below the surface and produces edible seeds, known as *maiz del agua* (water maize)

The advent of electricity (Continued)

The main difference between electricity stored in a Leiden jar and that released by a battery is that chemically produced electricity does not all rush out at once. Instead, it flows like the current of a river. For the first time, electricity became available for periods of time, instead of in instants.

Within 20 years, the next key event occurred, also by accident. Hans Christian Oersted discovered the connection between electricity and magnetism, suspected since at least the time of Thales. An electric current could produce magnetic effects. In another 10 years the converse was shown, and magnets were being used to generate electric currents. With the development of powerful currents produced by magnetic generators, the stage was set for the use of electric power for light, for communication, and for production of motion.

James B. Francis builds the first Francis turbine; *See also* 1824 TOO

Compressed air is used in mining; *See also* 1878 MED

A Captain Minié of the French army combines the work of several innovators of the early 19th century to develop a cylindrical projectile for small arms that we now recognize as "bullet-shaped"; nicknamed the Minnie ball, versions are soon adopted by armies and hunters around the world

Eugène Bourdon [b Paris, Apr 8 1808, d Paris, Sep 29 1884] patents a "metallic manometer"; it is a pressure gauge consisting of a curved tube closed at one end that is linked with a pinion rack to a pointer; when pressure is applied, the tube straightens and moves the pointer; the Bourdon gauge is still the most used pressure gauge in industry

In a famous event in the annals of invention, Walter Hunt of New York City patents the safety pin, complete with spring closing, and assigns the rights to one Wm. or Jno. Richardson, the artist who made the patent drawings; it seems that Hunt owed money to the illustrator for past patent drawings and promised to assign him, in return for forgiveness of the debt, and an additional $400, the patent to any invention Hunt could make out of an old piece of wire; after about 3 hours of twisting the wire, Hunt comes up with the safety pin; *See also* 1842 TOO

A manually operated block-signaling system is introduced by the New York & Erie Railroad; when a train is on a certain section of track, no other train is allowed on that section; *See also* 1855 TRA

1849

1850

Über die bewegende Kraft der Wärme (On the driving power of heat) by Rudolf J.E. Clausius [b Köslin, Pomerania (in Poland), Jan 2 1822, d Bonn, Germany, Aug 24 1888], offers the first statement of the second law of thermodynamics, restated by Clausius in 1865 as "entropy always increases in a closed system," or energy in a closed system will change toward heat and disorder; the book supplies the theoretical foundation for steam engines

Edward G. Shepard is the first to make an incandescent lamp using charcoal, while Joseph Wilson Swan [b Sunderland, England, Oct 31 1828, knighted 1904, d Warlingham, England, May 27 1914] begins to work with carbon filaments made from paper; *See also* 1846 GEN; 1854 GEN

A suspension bridge in France fails and 200 soldiers lose their lives; *See also* 1877 ARC

The firm of Siemens & Halske lays the first submarine telegraph cable, from Dover, England, to Calais, France; however, the cable is cut by the anchor of a French fishing boat, and a new cable is laid the next year; *See also* 1847 GEN

Oliver T. Eddy of Baltimore MD patents a typewriter that uses a piano keyboard to input letters; it is the first typewriter to use an inked ribbon instead of a roller or pad to supply the ink; *See also* 1856 COM

John B. Fairbank patents the first typewriter to use a continuous-roll paper feed; *See also* 1856 COM

The telegraph network in England extends over 6000 km (4000 mi); *See also* 1839 COM

James Young [b Glasgow, Scotland, Jul 13 1811, d Edinburgh, May 14 1883] produces the first distilled coal oil (possibly in 1851); *See also* 1739 MAT; 1855 MAT

1851

In England the Amalgamated Society of Engineers is founded; it becomes the first British trade union to survive for more than a few years and is still in existence today; *See also* 1847 GEN; 1852 GEN

The Great Exhibition of the Works of Industry of All Nations is held in London, opening on May 1; it features the latest technical innovations in industry and promotes the application of science to technology; *See also* 1851 ARC

William Channing and Moses Farmer [b Boscawen NH, Feb 9 1820, d Chicago, May 25 1893] develop an electric fire alarm system; *See also* 1866 GEN

The Crystal Palace, designed and built by James Paxton, is opened on May 1 by Queen Victoria; a remarkable building of glass and iron, it is one of the highlights of the Great Exhibition in London; *See also* 1849 ARC; 1851 GEN

London and Paris are linked by a submarine telegraph cable; an earlier attempt at laying a cable between Calais and Dover failed when the cable was broken shortly after it had been laid; *See also* 1850 COM; 1857 COM

Frederick Scott Archer [b Bishop's Stortford, England, 1813, d London, May 2 1857] introduces the wet collodion process, also called the calotype wet-plate process, to photography; *See also* 1839 COM; 1864 COM

Paul Julius Baron von Reuter opens a telegraphic agency in London; he turns it into a press agency in 1858; *See also* 1849 COM

Georg Sigl in Vienna, Austria, develops a cylinder press for lithographic printing; *See also* 1798 COM

William Thomson [b Belfast, Ireland, Jun 26 1824, raised to peerage as Baron Kelvin of Largs 1892, d Ayr, Scotland, Dec 17 1907] describes the heat pump: heat is absorbed by the expansion of a working gas and given off at a higher temperature in a condenser; the air conditioner applies the heat-pump principle, since the pump heats in one direction and cools in the other; *See also* 1839 ENE; 1930 ENE

German physicist Heinrich Ruhmkorff [b 1803, d 1877] develops the induction coil; an iron core is surrounded by two coils, one with a low number of windings and one with a very high number; quickly and repeatedly interrupting a current in the first coil creates very high voltage in the second; such induction coils were used to produce the first artificial radio waves

1850

Cable plowing is introduced about this time; a stationary engine pulls a cable that draws a plow across a field; *See also* 1837 FOO

About his time the modern age of adhesives begins with the invention of rubber cement

The beehive method of making coke from coal in small quantities replaces earlier methods, even though the method wastes a great deal of coal, mainly because of the high-quality coke it produces; *See also* 1841 MAT; 1858 MAT

The hydrocarbon wax paraffin is developed and used in candles, in waxed paper, for sealing, and as a lubricant; *See also* 1823 MAT

Physician Carl Reinhold Wunderlich [b Sulz (in Germany), Aug 4 1815, d Leipzig, Germany, Sep 25 1877] about this time introduces the practice of taking accurate temperature with a thermometer as a regular part of diagnosis; *See also* 1730–32 TOO; 1866 MED

The power-driven combing machine for wool is introduced; *See also* 1789 TOO

The cylindrical grinder is introduced

William Armstrong [b Newcastle-on-Tyne, England, Nov 26 1810, d Rothbury, England, Dec 27 1900] develops an hydraulic accumulator, making hydraulic machinery independent of a water source

Macedonio Melloni [b Parma (in Italy) Apr 11 1798, d near Naples, Aug 11 1854] constructs a radiometer consisting of 100 thermocouples mounted in the form of a cube; *See also* 1820 TOO; 1878 TOO

1851

The first electrolytic process for the production of chlorine is patented in Great Britain by Charles Watt; *See also* 1835 MAT; 1858 MAT

Hermann von Helmholtz [b Potsdam, Prussia (in Germany), Aug 31 1821, d Berlin, Germany, Sep 8 1894] invents the ophthalmoscope independently of Charles Babbage's version of 1847; *See also* 1862 MED

Isaac Merrit Singer [b Oswego NY, Oct 27 1811, d Torquay, England, Jul 23 1875] patents the continuous-stitch sewing machine; *See also* 1846 TOO

J.T. King develops a washing machine with a rotating drum

William Siemens [b Karl Wilhelm Siemens, Lenthe, Germany, Apr 4 1823, knighted 1883, d London, Nov 18 1883] invents a water meter that is commercially successful; *See also* 1815 TOO

John Ericsson's ship *Ericsson* uses his radical design for a forerunner of the modern gas turbine, but Ericsson's engine is so heavy that the ship is slower than ships of the time with conventional steam engines; *See also* 1844 TRA

1852
The American Society of Civil Engineers is founded on Nov 5 in New York City; *See also* 1828 GEN

Aristide Boucicault opens the Bon Marché in Paris, the first true department store

John Ramsbottom patents a packing ring for locomotive cylinders

Intellectual and technological property

A patent is a capitalist device that like many great ideas works on a paradoxical principle: The way to spread the benefits of an invention is to restrict the number of people who can exploit it.

Other forms of intellectual property are protected in different ways. A scientific discovery, which is intended to be shared by everyone, is covered by informal agreement in the scientific community granting "ownership" with priority. No legal rights are deemed possible. A trademark or trade name can be registered and protected, but that is solely for the protection of the owner. Copyright is closer in idea to a patent, but is much more concerned with protecting structure and substance of thought than it is with providing a monopoly on an idea or a structure. Indeed, ideas are not copyrightable; only their expression and arrangement can be copyrighted. Ideas for inventions, however, are the basis of monopoly; and monopoly is the original purpose of the patent.

Capitalism and the middle class were babies of the Renaissance, but they did not become adult ideas until the Industrial Revolution. Patents have the same history, starting in Italy in the 15th and 16th centuries. Queen Elizabeth I may have been the first British monarch to issue monopolies, which included but were not limited to patents. By the time of James I, the business of royal monopolies had gotten out of hand, and a succession of efforts to control monopolies, which treated inventions differently from other monopolies, had the somewhat inadvertent effect of creating the first English Patent Law, although it was not codified as such until late in the 19th century, after the example of the Patent Law in the US.

By the time the US Constitution was being written, the Founding Fathers had a good philosophical grasp of intellectual property, and provisions regarding it were included: Article I, Section 8 gave Congress the power "To promote the progress of Science and Useful Arts, by securing for limited Times to Authors and Inventors the exclusive Right to their respective Writings and Discoveries." Almost as soon as the Constitution was ratified, Congress proceeded to set up a patent law to protect monopolies on inventions. By 1836 the US Patent Office was functioning effectively, and it soon became apparent that invention fared better under the US system than under any other. A monopoly for a limited time on a specific invention encouraged people to invent, knowing they could be protected, and enabled them to sell rights to others who had the capital or existing trade to manufacture and promote an invention.

In 1883 Great Britain consolidated its patent laws along US lines and the International Convention in Paris worked out a way to handle patents in its many signatory nations. The European Patent Organization of 1953 was among several agreements that were precursors to the European Community. Today there are effective ways for an inventor to file a patent once in one country and, with suitable payments and searches, have it accepted in nations around the world.

1853
Georg Scheutz builds a working copy of Charles Babbage's difference engine and gives a demonstration of it at a London exhibition; *See also* 1843 COM

Pierre Carpentier invents corrugated steel plate; it will find a wide application in construction; *See also* 1845 TRA

Josiah *Latimer* Clark [b Great Marlow, England, Mar 10 1822, d London, Oct 30 1898] builds a pneumatic system for sending telegrams between the Stock Exchange and the International Telegraph Company in London; documents are placed in cylinders that are sucked through a tube of 220 m (720 ft) long; *See also* 1858 COM

Richard March Hoe develops the Type revolving press, which prints 20,000 sheets per hour; *See also* 1847 COM; 1865 COM

The first attempts are made to use an electric generator to power an arc lamp; *See also* 1862 ENE

Kerosene is extracted from petroleum for the first time; *See also* 1855 ENE

1852

William Kelly [b Pittsburgh PA, Aug 21 1811, d Louisville KY, Feb 11 1888], working in a small town near Eddyville KY, and aided by four Chinese steel-making experts, invents a new process for making steel that anticipates the Bessemer process, invented only 4 years later; he fails to patent his ideas until after learning of the Bessemer process; *See also* 1856 MAT

Samuel Wetherhill develops an economic method for removing zinc oxide from ore

Karl Vierordt [b Germany, 1818, d 1884] makes the first accurate count of red blood cells, a measurement that later becomes an important tool for diagnosing anemia

Henri Giffard [b Paris, 1825, d 1882] builds and flies a 44-m- (144-ft-) long airship powered by a steam engine of 3 hp driving a three-blade propeller 3.3 m (11 ft) in diameter at 110 rpm; this is the first application of a propeller to flight; all powered lighter-than-air craft are termed "airships," while airships that can be steered are also termed "dirigibles," with both terms usually interchangeable; *See also* 1783 TRA; 1872 TRA

Elisha Graves Otis [b Halifax VT, Aug 3 1811, d Apr 8 1861] invents the first elevator that has a safety guard to keep the elevator from falling even when the cable is completely cut; *See also* 1854 TRA

Jean-Bernard-Léon Foucault [b Paris Sep 18 1819, d Paris, Feb 11 1868] uses a gyroscope to demonstrate the rotation of Earth, and predicts the use of the gyroscope as a compass; *See also* 1744 TRA; 1908 TRA

1853

According to legend, the potato chip is invented by chef George Crum at Moon Lodge in Saratoga Springs NY while he is trying to execute his version of a guest's memory of fried potatoes from France

Charles Frédéric Gerhardt obtains acetyl salicylic acid, a closer precursor of aspirin, from the bark of the silver birch; *See also* 1838 MED; 1859 MED

Jean Jacques Farcot develops a system for the continuous control of the speed of steam engines, a regulator that keeps speed steady for different speeds; *See also* 1863 TOO

Sir George Cayley's helicopter "toys," modeled on Chinese toy helicopters, are shown to go 27 m (90 ft) into the air; *See also* 1843 TRA; 1877 TRA

1854

Heinrich Göbel, a German watchmaker who emigrated to New York, constructs an incandescent electric lamp with a filament of carbonized bamboo placed in a glass vessel; in 1893 he wins a court case against Thomas Edison and receives credit as the inventor of the electric lamp; *See also* 1850 GEN; 1856 GEN

Commodore Perry forces Japan to trade with the US

George Boole [b Lincoln, England, Nov 2 1815, d Ballintemple, Ireland, Dec 8 1864] publishes *An investigation of the laws of thought, on which are founded the mathematical theories of logic and probabilities,* in which the first form of symbolic logic, known today as Boolean algebra, is developed; Boolean algebra will form the mathematical foundation for many applications in computer science a century later; *See also* 1660–69 COM; 1876 COM

Great Britain's postal services introduce stamps separated by perforated lines, dispensing with the need for scissors to separate them; *See also* 1840 COM

Georg Herman Babcock [b Unadilla Forks NY, Jun 17 1832, d Plainfield NJ, Dec 16 1893] and his father develop the polychromatic printing press; *See also* 1818 COM; 1858 COM

The total length of telegraphic networks in the world is 37,000 km (23,000 mi); in the US the telegraphic network covers 25,000 km (15,500 mi); *See also* 1850 COM; 1861 COM

Daniel Hallady produces the first US windmills, modeled on European devices but with wooden instead of cloth sails; *See also* 1759 ENE

1855

The safety match is invented in Sweden; such matches light only when struck on a chemically impregnated surface, while previous friction matches lit when struck on any rough surface; *See also* 1830 GEN

The International Exhibition is held in Paris

Robert Bunsen [b Göttingen (in Germany), Mar 31 1811, d Heidelberg, Germany, Aug 16 1899] begins using the gas burners that are named for him; however, the burners are developed by his technician, C. Desaga, and are similar to burners developed and used earlier by Michael Faraday

John Augustus Roebling completes the Niagara River suspension bridge for railroad travel; the 250-m (821-ft) span uses only one-fourth as much material as the 142-m (460-ft) span of the Britannia rail bridge over the Menai Strait in North Wales; *See also* 1825 ARC; 1869 ARC

In London, Giuseppe Devincenzi patents the first known electric writing machine, almost 50 years before the first successful electric typewriter; *See also* 1872 COM; 1902 COM

Frenchman Alphonse Poitevin invents a printing process called collotype that reproduces photographs much more accurately than most other methods; similar in some ways to offset lithography, collotype is difficult to use and most suitable for very short print runs; *See also* 1798 COM

William Thomson (Lord Kelvin) develops a theory of transmission of electrical signals through submarine cables that is applied to the first submarine telegraph cables; *See also* 1851 COM; 1857 COM

Benoit Fourneyron develops an improved water turbine; *See also* 1827 ENE

James Bicheno Francis publishes *The Lowell hydraulic experiments,* on the development of water turbines; *See also* 1849 ENE

Benjamin Silliman [b Trumbull CT, Aug 8 1779, d New Haven CT, Nov 24 1864] obtains several products by distilling oil; these include tar, naphthalene, gasoline, and solvents; *See also* 1853 ENE

1854

Henri-Etienne Sainte-Claire Deville [b St. Thomas, Virgin Islands, Mar 11 1818, d near Paris, Jul 1 1881] develops a new method of producing aluminum, and casts a 7-kg (15-lb) ingot; over the next 4 years the price of aluminum is reduced by a factor of 100, although it is still more expensive than steel; *See also* 1833 MAT; 1886 MAT

John Snow [b York, England, Mar 15 1813, d London, Jun 16 1858] shows that removing the pump handle of a well contaminated by sewage reduces the incidence of cholera in the vicinity of the well; *See also* 1844 MED; 1866 MED

Air is proposed as a lubricant for bearings; *See also* 1862 TOO

In May, Elisha Otis demonstrates his safe and workable freight elevator (lift) at New York's Crystal Palace; *See also* 1852 TRA; 1857 TRA

1855

French chemist George Audemars develops a process for making fibers from cellulose, but cannot produce the fiber in commercial quantities; this is accomplished by the same process about 30 years later; *See also* 1884 MAT

Alexander Parkes [b Birmingham, England, Dec 29 1813, d London, Jun 29 1890] obtains a patent in England on parkesine, a substance resembling collodion that he has developed by mixing anhydrous wood alcohol with gun cotton; *See also* 1846 MAT; 1862 MAT

F. Köller develops a type of steel that contains tungsten

Robert Arthur introduces the technique of filling teeth with cohesive gold foil, heating the foil to rid the surface of impurities and then cold-welding it in the cavity with pressure from instruments; *See also* 1450–59 MED

Alexander Thophilus Blakely patents in Britain a method for building cannon barrels from hoops of iron; Blakely demonstrates the superiority of this design mathematically and shows how it can be systematically implemented; *See also* 1774 TOO

James Clerk Maxwell [b Edinburgh, Scotland, Nov 13 1831, d Cambridge, England, Nov 5 1879] develops a new type of planometer, a device for measuring areas of plane figures

A version of the Babbage Difference Engine built by Swedish engineer Pehr Georg Scheutz and his son Edward is awarded a Gold Medal at the Paris Exposition; *See also* 1834 TOO

1856

French engineer C. de Chagny patents an incandescent lamp with a platinum filament for use by workers in mines; *See also* 1854 GEN; 1860 GEN

Alfred Ely Beach [b Springfield MA, Sep 1 1826, d Jan 1 1896] manufactures a typewriter that uses an inked ribbon; *See also* 1850 COM

Léon Scott de Martinville invents the phono-autograph, an instrument that creates on a rotating drum a trace that represents sound vibrations; the instrument is used in acoustical research and is an inspiration for the soon-to-be-invented phonograph; *See also* 1877 COM

Stephen Wilcox [b Westerly RI, Feb 12 1830, d Brooklyn NY, Dec 24 1872] patents a boiler with a design based on the safety water tube; *See also* 1844 ENE; 1859 ENE

The Crystal Palace

Certain structures stand out in the history of architecture as the inspiration for whole schools of design and construction. Many of these, such as the Abbey Church of St. Denis, progenitor of the Gothic cathedral, continue to inspire to this day and are still physically available to students of architecture. None, however, has at the same time been so influential and so fleeting as the Crystal Palace of Joseph Paxton. Conceived between Jun 7 and Jul 26 in 1850, it was opened to the public on May 1 1851 as the main building for the Great Exhibition of the Works of Industry of All Nations in London, an event often considered the first world's fair. After a highly successful run in Hyde Park at the Great Exhibition, the Crystal Palace was dismantled and reerected 2 years later in south London, where it housed various exhibitions and events for another 72 years before being destroyed by fire. Nearly all of its architectural impact, however, came during its short stay at Hyde Park.

Paxton was primarily a gardener for large estates who had become involved in greenhouse design and construction. The giant Crystal Palace was as long as eighteen football fields and as wide as eight, with another football-field wide (but nine times as long) addition on the side. Many of the ideas for its construction came during the construction of much smaller greenhouses during the preceding 20 years. The two principal earlier greenhouses were erected for the duke of Devonshire at Chatsworth; they are important for their own sakes as well as for their influence on the Crystal Palace. The first, known as

the Great Stove, was built between 1836 and 1841. Although a mere conservatory, the Great Stove actually had a span wider than that of London's familiar giant railway stations, which were built about the same time. The second Chatsworth building was a greenhouse designed to house a single flower, the giant *Victoria regia* water lily, which had been rediscovered by Europeans in 1832 and which the duke wished to bring to flower in England to present to the queen, for whom the plant had been named. (He succeeded in 1849, when Paxton's Victoria Regia Lily House was built; the plant produced 126 blooms during the following year.)

The Victoria Regia Lily House was, in some ways, a miniature Crystal Palace. Its main architectural features included a flat roof of ridge-and-furrow construction and curtain-wall construction. The ridge-and-furrow roof, a series of sharp peaks and valleys, has not been much imitated, although the basic principles involved, such as a design that produces virtual beams, are still influential. Curtain walls, which were not used in their pure form in the Crystal Palace, became a hallmark of modern office construction, especially in the work of Mies van der Rohe. The curtain walls of the Victoria Regia Lily House were glass panes that hung from cantilevered girders. The key to the curtain-wall system, introduced to modern office buildings in 1918, is that the outer skin of the building needs no supporting member. In the Crystal Palace the glass walls were hung but also supported by braces and columns.

Gail Borden [b Norwich NY, Nov 9 1801, d Borden TX, Jan 11 1874] patents his method for condensing milk using heat and a vacuum pan; during the Civil War his method gains widespread interest; *See also* 1885 FOO

Sir William Henry Perkin [b London, Mar 12 1838, d Sudbury, England, Jul 14 1907] synthesizes the first artificial (aniline) dye, mauve, and starts such a fashion craze that the next few years in England are known as the Mauve Age; *See also* 1845 MAT; 1858 MAT

English inventor Henry Bessemer [b Charlton, England, Jan 19 1813, d London, Mar 15 1898] introduces the Bessemer process for producing inexpensive steel; the process is based on his 1855 patents for the basic process for blasting air into molten iron and coke to make the fire hotter and to remove carbon, and on the 1856 patents of Robert Mushet [b Coleford, England, Apr 8 1811, d Cheltenham, England, Jan 19 1891] for removing sulfur impurities from steel; Mushet failed to stamp his patent application properly and would have died in poverty had Bessemer not paid him a pension of 300 pounds per year during the last years of his life; *See also* 1852 MAT; 1876 MAT

Friedrich Siemens [b Menzendorf, Hanover, Germany, Dec 8, 1826, d Dresden, Germany, Apr 24, 1904] develops the regenerative furnace; it burns previously unburnt gases for greater efficiency and is the forerunner of the open-hearth steel process; *See also* 1852 MAT; 1868 MAT

Karl Friedrich Wilhelm Ludwig is the first to keep animal organs alive outside the body, which he does by pumping blood through them

O.E. Carlsund develops an improved rolling mill with three cylinders; *See also* 1826 TOO

Alois Negrelli von Moldelbe's plans for construction of the Suez Canal from the Mediterranean to the Red Sea are accepted; *See also* 640 CE COM; 1859 TRA

The Crystal Palace (Continued)

The other architectural features of the Crystal Palace that came to influence modern architecture include the first system of portal bracing to counteract strong winds, needed because of the great size of the building; and the use of prefabricated modular units in construction, required to erect such a large building in just 17 weeks. As the Crystal Palace was the preeminent building of the Industrial Revolution, it is appropriate that it was also the first large freestanding building to use an iron frame (although much of the structure, including the columns framing the glass walls, was made from wood). The braces connecting columns to horizontal beams were iron, as were cross-braces used to strengthen the walls. The use of several different materials together in this way also became a hallmark of modern architecture. Finally, the Crystal Palace was among the first buildings to utilize prestressed members, in this case cast-iron beams.

Many of the statistics connected with the Crystal Palace remain astonishing today. It used 400 tons of glass, each 125-cm-sq (49-in.-sq) pane produced by blowing a cylinder, cutting it, and flattening it by reheating. With its modular construction, a team of 80 workers was able to install 18,932 panes in a week, resulting in a total surface array of glass of 85,000 m² (900,000 sq ft). The glass was supported by 3300 cast-iron columns and held in place by 330 km (205 mi) of wood sash.

Similar buildings were erected in 1852 in Dublin, in 1853 in New York City, and in 1854 in Munich, all to house, as the original had, exhibitions of new technology. Later in the century, many of the first department stores in France, England, and the US took elements from the Crystal Palace in the way that they displayed their wares. Another series of imitators, some of which are still in use, were city shopping arcades or galleries, in many ways the predecessors of today's suburban malls. It was not until the 20th century, however, that design principles of the Crystal Palace came to be common in buildings designed to be occupied for work or for living quarters. The essence of all 19th-century versions of the idea was display.

A barrier to erecting livable or workable versions of the Crystal Palace was that the Victorians, while they admired its utility and speed of construction, failed to recognize the architectural beauty of the building. It was just too strange. Architecture was still thought of primarily in terms of churches and great houses, built largely of stone. An insubstantial thing of glass, iron, and wood might serve a purpose, but no one would want to have such a giant greenhouse as an office and certainly not as a home.

1857

Alexandre Becquerel [b Paris, Mar 24 1820, d Paris, May 11 1891] experiments with coating electric discharge tubes with luminescent materials, a process that eventually leads to fluorescent lamps; *See also* 1934 GEN

An attempt at laying a transatlantic telegraphic cable fails when the cable breaks at a depth of 1800 m (6000 ft) and cannot be recovered; *See also* 1851 COM; 1858 COM

Werner von Siemens designs the shuttle winding for the armature of the electric generator (dynamo), a major step toward the development of a method of obtaining electrical energy from a generator without using permanent magnets; the design and other research is published by Werner's brother William in 1867; *See also* 1848 ENE

1858

The Foreland lighthouse in Kent, England, is the first to be equipped with electrically powered arc lights; *See also* 1809 GEN

The British government takes over most of India directly and controls the rest indirectly; this is the beginning of the British raj (rule) in India

Samuel Kier produces kerosene from petroleum and uses it for lighting

The first successful transatlantic telegraphic cable is laid, although it breaks after 3 months; *See also* 1857 COM; 1866 COM

Light emitted by electrons hitting a screen is observed for the first time; this phenomenon augurs the cathode-ray tube, still used in many applications, such as television; *See also* 1878 COM

Louis Ducos du Hauron [b Langon, France, 1837, d Agen, France, 1920] patents a method of color printing based on photographs made on zinc plates through color filters; *See also* 1854 COM; 1861 COM

Cromwell F. Varley [b London, Apr 6 1828, d Bexleyheath, England, Sep 2 1883] improves the pneumatic system for sending telegrams between the Stock Exchange and the International Telegraph Company in London; compressed air and suction are both used as motive powers, allowing documents to travel in two directions; *See also* 1853 COM; 1859 COM

Breckon and Davis introduce an improved beehive oven; although coke is already being produced as a by-product of the manufacture of coal gas and tar, such coke is far inferior for iron smelting to that from the Breckon and Davis ovens; *See also* 1850 MAT

E. Coppée introduces a new design for a coke oven in which large amounts of coke can be made in a single operation; however, the oven still cannot produce high-quality coke and recover by-products at the same time; *See also* 1881 MAT

1857

Ferdinand Carré [b France, 1824, d 1900] builds a refrigeration machine based on the expansion of ammonia gas; the device can produce temperatures as low as -30°C (-22°F); *See also* 1851 ENE; 1873 FOO

E.J. Hughes obtains a patent for an artificial silk made from starch, gelatin, resin, tannin, fat, and other ingredients; *See also* 1855 MAT; 1884 MAT

American metallurgist William Kelly, learning of similar work by Henry Bessemer, patents the "air-boiling process" for making steel in the US; *See also* 1852 MAT

American Joseph Cayetty invents the modern form of toilet tissue; its use spreads outside the US only after the First World War; *See also* 590 CE MAT

About this time a special mixture of resins and soapstone known as impression compound begins to replace plaster of Paris for making impressions of teeth; *See also* 1844 MED

Jean-Bernard-Léon Foucault develops the modern technique for silvering glass to make mirrors for reflecting telescopes; *See also* 1840 MAT

Jules Lissajous [b Versailles, France, Mar 4 1822, d Plombières-les-Dijon, France, Jun 24 1880] invents a machine for making visible Lissajous figures, which are curves obtained by superimposing different simple harmonic curves at right angles to each other

A .B. Wilson improves the sewing machine developed by Isaac Singer; *See also* 1851 TOO

Excavation of the tunnels for the London Underground railroad (subway) starts, using a method of iron casings developed by Marc Isambard Brunel while building the first tunnel under the Thames; *See also* 1863 TRA

Elisha Otis installs the first safety-equipped passenger elevator in a five-story downtown New York City china and glassware store; *See also* 1854 TRA; 1860 TRA

1858

Ezra Warner of Waterbury CT invents the can opener, nearly 20 years after US canners switch from jars to metal cans and almost 50 years after the metal can is invented (in 1810); previously, metal cans were opened with general-purpose tools, such as a hammer and chisel; *See also* 1839 FOO

August Wilhelm von Hofmann makes the artificial dye magenta (often called fuchsin, from its resemblence to the color of the flower of the fuchsia) from coal tar; it is quickly copied by French dyers, who name it the following year after the French defeat of the Italians at the Battle of Magenta; by 1860 it is being made in Germany as well; *See also* 1856 MAT

Henry Deacon [b London, Jul 30 1822, d Widnes, England, Jul 23 1876] introduces a process for making chlorine from hydrochloric acid and oxygen using a temperature of 400°C (750°F) and a copper chloride catalyst absorbed on pumice stone; *See also* 1851 MAT; 1861 MAT

The first practical powered dental drill, driven by a foot pedal, is constructed; *See also* 1790 MED; 1869 MED

Franciscus Cornelis Donders [b Tilburg, Netherlands, May 27 1818, d Utrecht, Netherlands, Mar 24 1889] discovers that farsightedness can be caused by too shallow eyeballs; *See also* 1801 MED

Théophile Guibal [b Toulouse, France, 1813, d Paris, 1888] invents the propeller fan, first used to refresh air in mines

Germain Sommeiller invents the pneumatic drill, which is used for the excavation of the Mont Cenis Tunnel; *See also* 1832 TOO; 1870 TRA

The first aerial photograph is taken from the balloon *Nadir* as it flies over Paris; *See also* 1794 TRA

Isambard Kingdom Brunel launches the *Great Eastern* on Jan 31; driven with both paddles and a propeller, with a displacement of 22,500 tons it remains the largest ship in service during the 19th century—210 m (693 ft) long and capable of carrying 4000 passengers; in 1866 it serves in laying the first transatlantic cable, but then is retired from active service, although exhibited to the public; it is scrapped in 1888; *See also* 1843 TRA

1859

GENERAL

William John Rankine [b Edinburgh, Scotland, Jul 5 1820, d Glasgow, Scotland, Dec 24 1872] writes *Manual of the steam engine and other prime movers*, introducing engineers to thermodynamics and the Rankine cycle; he also coins most of the modern terms used in the field today; *See also* 1824 GEN; 1850 GEN

COMMUNICATION

R.S. Culley and R. Sabine introduce a radial pneumatic mail system in London for the British Post Office; it carries telegrams from a central telegraph office to branches; *See also* 1858 COM; 1868 COM

John Benjamin Dancer demonstrates microdot photography; he succeeds in photographing pages of books on slides of 1/1600 part of the surface of an inch; *See also* 1840 COM

ENERGY

The first deliberate oil well produced as a commercial venture operates in Titusville PA; it is drilled by Pennsylvania Rock Oil Company employee Edwin L. "Colonel" Drake [b Greenville NY, Mar 29 1819, d Bethlehem PA, Nov 8 1880]; the well is 23 m (75 ft) deep and delivers 1600 L (422 gal) of oil per day; *See also* 1860 ENE

Joseph Harrison patents a sectional or "honeycomb" boiler; it is assembled from communicating cast-iron globes and can be built in different sizes; *See also* 1856 ENE; 1867 ENE

Jean-Joseph-Etienne Lenoir [b Mussy-la-Ville, Belgium, Jan 12 1822, d Varenne-St. Hilaire, France, Aug 4 1900] develops the first important internal combustion engine, using coal gas as a fuel; however, it is very inefficient; *See also* 1823 ENE

Gaston Planté [b Orthez, France, Apr 22 1834, d Paris, May 21 1889] invents the storage battery, which produces electricity from a chemical reaction and which can be recharged again and again; *See also* 1839 ENE; 1868 ENE

1860

GENERAL

Spiked shoes for sports are introduced in England about this time

Sir William James Herschel, working in India, begins to use fingerprints as a way of identifying government prisoners, but this concept is unknown elsewhere; *See also* 1823 GEN; 1880 GEN

John T. Way demonstrates that light can be produced by sending electricity through mercury vapor; this is an application of the Geissler tube, developed by Heinrich Geissler about 10 years earlier; *See also* 1892 GEN

ARCHITECTURE & CONSTRUCTION

True portal bracing, initiated in part by Joseph Paxton in the Crystal Palace, is used in the Royal Navy Boat Store; girders are riveted to columns throughout their whole depth in this building, which is completed this year; *See also* 1851 ARC

COMMUNICATION

C.C. Harrison and J. Schnitzler introduce the biconvex photographic objective lens, reducing distortion of the image; *See also* 1840 COM; 1866 COM

Charles Wheatstone invents a printing telegraph; *See also* 1844 COM

ENERGY

A patent for "cracking," a method of obtaining high-boiling fractions of petroleum, is issued to Downer's Kerosene Oil Company in Boston; *See also* 1859 ENE

Jean-Joseph-Etienne Lenoir builds a two-stroke internal combustion engine with a single cylinder and electric ignition; *See also* 1859 ENE; 1862 ENE

1859

Marcel Morren observes that an electric spark created with carbon electrodes in a hydrogen atmosphere produces an unknown gas, which he terms carbonized hydrogen; we now know this gas as acetylene; *See also* 1772 MAT; 1897 MAT

German chemist Albert Niemann isolates cocaine from the leaves of *Erythroxylon coca*; the stimulant properties of cocaine had long been known to natives of South America, and Europeans learned of them after their conquest of the Inca Empire; *See also* 1884 MED

Hermann Kolbe synthesizes salicylic acid, the active principle of aspirin, from inorganic chemicals; although effective in reducing pain and fever, it is poorly tolerated by the stomach; *See also* 1853 MED; 1893 MED

Ferdinand de Lesseps [b Versailles, France, Nov 19, 1805, d La Chenaie, France, Dec 7 1894] turns over the first shovelful of earth to start construction of the Suez Canal from the Red Sea to the Mediterranean; built according the plans of Alois Negrelli von Mold-elbe, the canal is completed 10 years later, on Aug 15 1869; *See also* 1856 TRA; 1869 TRA

1860

In France Leonce-Eugène Grassin-Baledans patents the use of twisted strands of sheet metal for fencing and to protect trees; this material is considered the precursor of barbed wire; *See also* 1868 FOO

Vulcanized rubber begins about this time to be used as a baseplate material for dentures, replacing porcelain; *See also* 1790 MED

Ellen Demorest begins marketing paper dress patterns, for which a new market exists because of the availability of sewing machines; *See also* 1863 GEN

Elisha Otis patents a steam-driven elevator; *See also* 1857 TRA

1861

GENERAL

The Massachusetts Institute of Technology (MIT) is founded

COMMUNICATION

James Clerk Maxwell, assisted by photographer T. Sutton, demonstrates in London the first color reproduction using a photographic process; he projects three separate images of the Scottish flag taken through red, green, and violet filters and matches them to form a single picture, using light of the same three colors to produce the color image; *See also* 1873 COM; 1891 COM

German schoolteacher Philipp Reis transmits musical tones over 100 m (300 ft) with a device he terms a telephone; although Reis later claims credit for the invention of the telephone, Alexander Graham Bell's successful device works on a slightly different principle; it is not known whether Reis ever tried to transmit speech; *See also* 1876 COM

A heliograph system is used in the US to send a signal over a distance of 145 km (90 mi); *See also* 1792 COM; 1890 COM

New York and San Francisco are connected by a telegraph line; *See also* 1854 COM; 1872 COM

ENERGY

Antonio Pacinotti [b Pisa (in Italy), Jun 17 1841, d Pisa, Mar 25 1912] builds a generator consisting of a ring-shaped solenoid rotating between magnets; the current is delivered by two trailing contacts on each side of the rotating ring

1862

GENERAL

Abraham Lincoln's Emancipation Proclamation, announced Jul 23, is that the slaves in the states rebelling against the US will be freed on the following Jan 1; this step leads to the complete freeing of any remaining US slaves following the Civil War, when antislavery amendments are made to the Constitution; *See also* 1830 GEN

ENERGY

A lighthouse in the Straits of Dover uses an arc lamp powered by an electric generator, demonstrating the first practical application of the generator; *See also* 1853 ENE; 1878 ENE

John T. Allen demonstrates at the London International Exhibition a high-speed horizontal steam engine; its speed is double that of other engines of its time, and the engine sets an example for the trend toward high-speed engines; *See also* 1849 ENE; 1866 ENE

Adolphe-Eugène Beau de Rochas [b Degne, France, 1815, d 1893] patents the four-stroke internal combustion engine; *See also* 1860 ENE; 1867 ENE

1861

Ernest Solvay [b Rebecq, Belgium, Apr 16 1838, d Brussels, Belgium, May 26 1922] discovers a process for making sodium bicarbonate from salt water, ammonia, and carbon dioxide; this process is much more economical than previous ones for developing sodium bicarbonate; regular production begins in 1863; See also 1810 MAT

Irish telegrapher Joseph May observes that sunlight affects the electrical resistance of instruments made from selenium; inventors soon use this discovery to produce crude and impractical television devices; See also 1876 MAT

Pierre-Paul Broca [b France, Jun 28 1824, d Paris, Jul 9 1880] demonstrates that a particular region in the brain (Broca's area) is connected to a particular faculty, in this case the faculty of speech, by discovering a lesion in the brain during an autopsy on a man that could not speak intelligibly; See also 1843 MED; 1868 MED

Joseph R. Brown, of Brown and Sharpe, invents the universal milling machine to make the grooves in twist drills used for making musket parts; the milling machine can also do all kinds of spiral and gear cutting; See also 1818 TOO; 1862 TOO

Charles Ellet builds a steam-powered ram used by the Union forces against the Confederate army on the Mississippi River in the US Civil War; in Jun 1862 Ellet is fatally wounded when he personally leads a fleet of five rams in the Battle of Memphis

On Oct 1 ship designer John Ericsson lays the keel on the *Monitor*, which becomes the first iron-clad ship on the Union side of the US Civil War to see battle; several ships on the Mississippi River are also armed by the Union at this time with 6-cm (2.5-in.) iron plate, but the *Monitor* is the first to be designed from the ground up as an ironclad; See also 1851 TRA; 1862 TRA

Pierre Michaux in France introduces a bicycle propelled by pedals attached to the front wheel and used as cranks; this version of the bicycle is often called the "bone-shaker," since it has no springs and, at first, uses steel tires; See also 1839 TRA; 1869 TRA

Thomas Walker invents the "har-poon," or frictionless ship's "log"; See also 1846 TRA; 1878 TRA

1862

Alexander Parkes displays at an international exhibition in London many items made from what comes to be known as celluloid, a plastic on which he had been working throughout the 1850s; See also 1855 MAT; 1868 MAT

Alfred Bernhard Nobel [b Stockholm, Sweden, Oct 21 1833, d San Remo, Italy, Dec 10 1896] begins experimenting with nitroglycerin; this results in his invention of a patentable percussive detonator in 1863; See also 1867 MAT

Alexander Pagenstecher introduces the use of yellow mercury oxide salve as an ophthalmological ointment; See also 1851 MED; 1864 MED

Pierre Michaux patents the ball bearing in France; ball bearings are first used in 1879 on bicycles; See also 1854 TOO

W.B. Bement introduces the first vertical milling machine, which makes it easy to cut dies used in die stamping; this leads to die stamping as an important and inexpensive manufacturing system; See also 1861 TOO

In Nov Richard Jordan Gatling [b Maney's Neck NC, Sep 12 1818, d New York City, Feb 26 1903] develops the first machine gun

In the US Civil War, on Mar 9 the North's ironclad *Monitor* defeats but does not sink the Confederate ironclad *Merrimac*, a previously scuttled warship renamed the *Virginia* after getting its 10-cm (4-in.) iron plate; this occurs in the first naval engagement between ironclad ships; on Mar 8 the *Merrimac* sunk one and bullied two Union wooden-clad ships; See also 1861 TRA

1863

Ebenezer Butterick [b Sterling MA, May 29 1826, d New York City, Mar 31 1903] and Eleanor Butterick patent the paper dress pattern; *See also* 1860 TOO

French physicist Giovanni Caselli patents the pantelegraph, his apparatus for the transmission of facsimile, in the US; developed between 1857 and 1865, the pantelegraph is based on earlier ideas of Alexander Bain and Frederick Bakewell and consists of two synchronized pendulums to which metallic pointers are attached; one detects a message handwritten on a metal plate with nonconducting ink; the pantelegraph operates between Paris and Lyon from May 16 1865 to 1870, with another branch between Paris and Marseilles starting in 1867; war ends the service, which is never restored; *See also* 1847 COM

Samuel Van Syckel builds an 11-km (7-mi) pipeline with a 5-cm (2-in.) diameter between his oil production plant and a railway station in PA; it is the first known pipeline; *See also* 1860 ENE; 1872 ENE

1864

American manufacturer William Sellars develops a standard screw thread that, especially in the US, replaces the English standard of Joseph Whitworth; *See also* 1841 TOO

B.J. Sayce and W.B. Bolton introduce silver bromide emulsion to photography; *See also* 1851 COM; 1871 COM

1865

The Jennings Pedestal Vase Company installs the first known pay toilets, using Joseph Bramah's design, outside the Royal Exchange in London; the price is one penny, which remains immune to inflation for the next hundred years; *See also* 1778 ARC

Joseph Hinks introduces the duplex burner oil lamp

William Bullock [b Greenville NY, 1813, d Philadelphia PA, Apr 12 1867] produces a rotary press that can print on a roll of paper; it is used for printing newspapers; *See also* 1853 COM

The forerunner of the International Telegraph Union (ITU) is established; twenty countries agree to cooperate in telegraph communications

1863

Louis Pasteur [b Dôle, Jura, France, Dec 27 1822, d near Paris, Sep 28 1895] discovers the microorganism that sours wine and turns it into vinegar; he introduces heating to kill such microorganisms, a process that comes to be known as pasteurization

Pierre-Emile Martin [b Bourges, France, Aug 18 1824, d Fourchambault, France, May 24 1915] succeeds in producing steel by heating a combination of steel scrap and iron ore with gas burners; *See also* 1856 MAT

German chemist J. Wilbrand invents the explosive trinitrotoluene, better known as TNT; *See also* 1846 MAT

Alfred Krupp founds a physical laboratory for testing steel

Johann Friedrich Adolf von Baeyer [b Berlin, Germany, Oct 31 1835, d Starnberg, Germany, Aug 20 1917] develops the first barbiturate, which he supposedly names for his girlfriend Barbara

Physiologist Etienne-Jules Marey [b Beaune, France, Mar 5 1830, d Paris, May 15 1904] invents the sphygmograph, the predecessor of the sphygmomanometer used to measure blood pressure today; *See also* 1847 MED

Jean Jacques Farcot develops a mechanism for controlling the rudder of a ship from the bridge by way of a hydraulic linkage; he terms this type of device, in which the linkage is an intermediate control, a "servomechanism"; *See also* 1853 TOO

Jean-Joseph-Etienne Lenoir builds the first "horseless carriage" that uses an internal combustion engine for power (previous versions used steam engines); it can reach a speed of 5 km (3 mi) per hour; *See also* 1789 TRA; 1877 ENE

The first underground railways in London start operation; *See also* 1857 TRA; 1868 TRA

Gabriel de la Landelle combines the Latin roots *avis* (bird) and *actio* (action) to obtain the new English word *aviation* to describe things connected with flight; *See also* 1889 TRA

1864

Franciscus Cornelis Donders [b Netherlands, 1818, d 1889] discovers that astigmatism is caused by an uneven curvature of the lens or cornea of the eye; *See also* 1801 MED

S.C. Barnum invents the rubber dam to help dentists control saliva while working on teeth; *See also* 1860 MED

Brown and Sharpe in Providence RI develops the first commercial grinding machine; *See also* 1900 TOO

William Cotton patents a knitting machine for hosiery that produces complete stockings sewn up the back; this becomes the basic type of knitting machine; *See also* 1816 TOO

Norwegian whaling engineer Sven Foyn [b Tonsberg, Norway, Jul 9 1809, d Ramdal, Norway, Nov 29 1894) invents a gun for firing harpoons with explosive heads; this gun enables hunting of the elusive fin, sei, and blue whales

George M. Pullman [b Brocton NY, Mar 3 1831, d Chicago, Oct 19 1897], working with Ben Field, builds the first of his line of sleeping cars famous for exceptional comfort; these cars become known as Pullman cars after the Pullman Palace Car Company is founded in 1867; *See also* 1840 TRA; 1868 TRA

1865

Edmund La Croix designs the first modern version of a middlings purifier, a device that removes bran from wheat flour; his device combines vibrating screens, similar to those proposed by Ramelli in 1588, and a current of air to lift the bran; *See also* 1580–89 FOO

Alexander Parkes produces celluloid from experiments with nitrated cellulose; *See also* 1862 MAT; 1868 MAT

John D. Rockefeller [b Richford NY, Jul 8 1839, d Ormond Beach FL, May 23 1937] founds the first large oil refinery; *See also* 1859 ENE

Joseph Baron Lister [b Upton, England, Apr 5 1827, d Walmer, England, Feb 10 1912] introduces phenol as a disinfectant in surgery, reducing the surgical death rate from 45 to 15 percent; *See also* 1847 MED

Hermann P. Sprengel [b Schillerslage, Germany, Aug 29 1834, d London, Jan 14 1906] develops a pump that creates a vacuum sufficient for experiments with electric lamps; *See also* 1871 TOO

The novel *From the Earth to the moon* by Jules Verne [b Nantes, France, Feb 8 1828, d Amiens, France, Mar 24 1905] depicts three scientists and a journalist being shot to the Moon from a great cannon situated at Cape Canaveral FL; *See also* 1969 TRA

1866

French medical doctor François Carlier invents the chemical fire extinguisher; the extinguishing agent is a mixture of sodium bicarbonate and sulfuric acid; this mixture forms a film that cuts off the fire's air supply; *See also* 1851 GEN; 1905 GEN

Cyrus West Field [b Stockbridge MA, Nov 30 1819, d NY, Jul 12 1892] succeeds in laying a telegraph cable across the Atlantic Ocean; *See also* 1858 COM

Mahlon Loomis [b Oppenheim NY, Jul 21 1826] sends telegraph messages over radio waves between two mountains in WV using aerials held in the air by kites; *See also* 1883 COM

Adolph Steinheil and **John Dallmeyer** [b Loxton, Westphalia, Germany, Sep 6 1830, d New Zealand, Dec 30 1883] develop independently a photographic objective consisting of exactly symmetric double lenses with the diaphragm placed in between them, thus decreasing halo and spherical aberration; *See also* 1860 COM

Alexander Graham Bell [b Edinburgh, Scotland, Mar 3 1847, d Nova Scotia, Aug 2 1922] incorrectly assumes that sound travels through electric wires; *See also* 1875 COM

Charles Brown [b 1827, d 1905] and **Heinrich Sulzer** build a steam engine using superheated steam; because superheated steam does not condense on cylinder walls, its use increases the efficiency of steam engines; *See also* 1862 ENE; 1871 ENE

1867

Charles Sanders Peirce [b Cambridge MA Sep 10 1839, d Milford PA, Apr 19 1914] observes that Boolean algebra can be used to describe switching circuits, where ON and OFF replace TRUE and FALSE

Charles Warren [b Bangor, Wales, Feb 7 1840, d Weston-super-Mare, England, Jan 21 1927] discovers the vertical well in Jerusalem now known as Warren's Shaft; later research suggests that it is a natural sinkhole that ancient Israelites converted into a well; *See also* 1980 FOO

The basis for what will become the first successful typewriter, the Remington Model I, is designed by Christopher *Latham* Sholes [b 1819, d 1890], Carlos Glidden, and Samuel W. Soulé and patented in 1868; *See also* 1710–19 COM; 1872 COM

George Herman Babcock and **Stephen Wilcox** patent the Babcock Wilcox boiler, in which the steam-generating tubes are placed in vertical rows over the grate and are connected to separation drums in which water is separated from the steam and recycled back; this boiler becomes the standard method for steam generation; *See also* 1859 ENE

Nikolaus August Otto [b Holzhausen, Germany, Jun 10 1832, d Cologne, Germany, Jan 26 1891] and **Eugen Langen** [b Cologne, Germany, 1833, d 1895] develop an internal combustion engine that is an improved version of Lenoir's engine; *See also* 1862 ENE; 1870 ENE

1866

The first known diamond to be found in southern Africa is picked up by a fifteen-year-old boy strolling along the Orange River, leading to the Kimberly region's being recognized as the world's greatest diamond-producing area; *See also* 1797 MAT

William Budd [b North Tawton, England, Sep 14 1811, d Clevedon, England, Jan 1880] demonstrates in Bristol, England, that limiting the contamination of a town's water supply can stop a cholera epidemic; *See also* 1854 MED

Physician Sir Thomas Clifford Allbutt [b Dewsbury, England, Jul 20 1836, d Cambridge, England, Feb 22 1925] develops the clinical thermometer; previously, thermometers used in medicine were very long and took about 20 minutes to determine temperature; *See also* 1850 MED

US inventor J. Ousterhoudt develops a metal can that can be opened with a key that rolls up the top of the container (familiar to many as a method commonly used for canned fish products)

Engineer Robert Whitehead [b Lancashire, England, Jan 3 1823, d Beckett, England, Nov 14 1905] invents the torpedo

The British Aeronautical Society is founded

1867

Alphonso Dabb of Elizabethport NJ receives on Apr 2 the first US patent for a barbed fence attachment; later in the year, two other inventors receive patents for single-stranded wire fitted with barbs; *See also* 1860 FOO; 1868 FOO

Alfred Bernhard Nobel patents dynamite in Great Britain; the explosive is based on nitroglycerin but mixed with other compounds to make it safer; Nobel is granted a US patent in 1868; *See also* 1862 MAT; 1875 MAT

Joseph Monier [b 1823, d 1906], a gardener at the Versailles Palace, patents reinforced concrete and its use in concrete containers; *See also* 1824 MAT; 1877 MAT

Emeline Brigham patents a pessary, described on the application as a womb supporter, but clearly intended for (then illegal) contraception; *See also* 1876 MED

The friction tube, which uses percussion powder that is set off by drawing a rough bar through the powder to ignite a cannon's charge, is invented; *See also* 1845 TOO

American engineer George Westinghouse [b Central Bridge NY, Oct 6 1846, d New York City, Mar 12 1914] solves a major problem of the railways by inventing the air brake

The automatic block signaling system for railroads is introduced in the US; *See also* 1871 TRA

Léon François Edoux introduces the elevator driven by an expanding column operating under water pressure; such an elevator, traveling over 160 m (535 ft), is installed in 1889 in the Eiffel Tower; *See also* 1860 TRA

1867
cont

Cromwell F. Varley discovers that an electric generator does not need to be started with an outside source of magnetism, but can use Earth's magnetic field to induce enough magnetism to get started; Sir Charles Wheatstone discovers the same principle independently, and William von Siemens also announces discovery of the concept; Varley patents the idea, although it was probably originally discovered by Werner von Siemens, perhaps in 1866; *See also* 1857 ENE

Zénobe Théophile Gramme [b Jehay-Bedegnée, Belgium, Apr 4 1826, d Bois-Colombes, France, Jan 20 1901] invents the ring armature and builds the first commercially practical generator for producing alternating current; *See also* 1857 ENE; 1871 ENE

1868

James Clerk Maxwell, in an article on the governor, specifies the concept of "feed back," that is, the information sent back to a system to effect a correction or change

A pneumatic mail system is installed in Paris; it operates for over a hundred years and the tubes eventually extend to more than 400 km (250 mi); *See also* 1859 COM

Georges Leclanché [b Paris 1839, d 1882] invents the zinc-carbon battery, a precursor to the dry cell and the familiar flashlight battery; *See also* 1859 ENE

Perpetual motion (part 2): An obsession

During the 19th century, with the advent of electricity, inventors planned and sometimes built machines in which an electric motor drove a generator that, in turn, supplied power to the electric motor. The goal was perpetual motion and a "free" source of energy. A large number of other designs, some very intricate and based on hydraulic, pneumatic, and magnetic forces, appeared at that time with the same purpose.

None of these machines fulfilled their promise. Some seemed to work, but invariably they were found to be hoaxes. Also during the 19th century, scientists discovered why perpetual motion machines would not work when they learned the law of conservation of energy.

Until then, the quest for free energy became comparable to the UFO obsession during the 20th century. Several inventors cashed in by building complicated machines and charging admission for demonstrations or receiving funding for their inventions. A famous case was that of Charles Redheffer, who, at the beginning of the 19th century, displayed a perpetual motion machine in his house in Philadelphia PA, charging admission fees and receiving funding. Redheffer was soon found out because a young boy noticed that the teeth on a cogwheel were worn on the wrong side, showing that the machine was driven by an external power source.

Redheffer promptly moved his show to New York City, where he was unmasked by the famous engineer Robert Fulton, who, upon seeing the machine, exclaimed "Why, this is a crank motion!" Fulton had noticed an irregularity in the rotation of the machine, the telltale sign of a manually driven crank.

Perpetual motion (part 2): An obsession (Continued)

But some perpetual motion machines were bona fide attempts at extracting energy from the forces of nature. The Zeromotor was a famous project of a British professor, John Gamgee. His idea was based on replacing water in a steam engine with ammonia, a liquid that boils at −33.5°C (−31.9°F). He reasoned that the ambient heat would cause the ammonia to boil. Even at a temperature as low as 0°C (32°F), boiling ammonia turns to a gas that increases in volume enough to produce 4 atmospheres of pressure, sufficient to drive a piston in a cylinder. During expansion in the cylinder, Gamgee believed, the ammonia gas would condense and the resulting liquid could be pumped back to the "boiler."

In practice, the gas did not liquefy, and the Zeromotor did not work. What Gamgee had overlooked was that the ammonia gas would condense only if the cylinder had cooled below the boiling point of ammonia, back below to −33.5°C. Unfortunately, the energy required to cool the cylinder would be greater than the machine could supply. Just as in all other perpetual motion machines, there was a hitch. As in most such cases, the hitch was the second law of thermodynamics, a law discovered by the French engineer Sadi Carnot in 1824.

Understanding of the laws of thermodynamics made it clear to scientists that a machine that can supply power from nothing is a physical impossibility. Notwithstanding, intrepid inventors are still building perpetual motion machines today.

1867 cont

British marine engineer William Froude [b 1810, d Simon's Town, South Africa, May 6 1879] starts experimenting with model ships to study the hydrodynamical properties of hulls; he develops the Froude number, one of the basic concepts of fluid dynamics

1868

Michael Kelly of New York City becomes the first to patent double-stranded barbed wire, although his method fails to provide a way to keep the barbs in place; *See also* 1867 FOO; 1874 FOO

Hyppolyte Mergé-Mouriés wins a competition for a substitute for butter with the first blended margarine, an oil-in-water emulsion of fats

James Oliver [b Whitehaugh, Scotland, Aug 28 1823, d 1908] in South Bend IN develops a process for producing chilled cast iron that revolutionizes plow making; *See also* 1837 FOO

William H. Perkin synthesizes coumarin, a scent and flavoring used in foods until 1954, when it is discovered to cause liver poisoning

William Siemens and his brother Freidrich develop the open hearth process for making steel that supersedes the Bessemer process; by 1950 about 85 percent of plain steel is made by the open hearth process, but thereafter newer methods begin to take over; *See also* 1856 MAT

Daniel Spill invents the synthetic plastic trademarked Xylonite, a mixture of pyroxylin (a lower-nitrated cellulose nitrate), alcohol, and ether; it is considered a precursor to the similar plastic trademarked Celluloid that is invented about the same time; *See also* 1865 MAT; 1869 MAT

Sir Francis Galton shows that mental abilities of human beings form a normal distribution, lying along the familiar bell-shaped curve; *See also* 1861 MED

Thomas Alva Edison [b Milan OH, Feb 11 1847, d West Orange NJ, Oct 18 1931] patents his first invention, a device to record votes in Congress developed with the intent of speeding up proceedings; Congress informs Edison that slow votes have advantages over fast ones and does not buy his invention; *See also* 1870 TOO

Charles Henry Gould invents the wire stitcher for use in binding magazines; it works from an uncut piece of wire, cuts and inserts the wire in the fold of a magazine, and folds the wire ends over; such a stitcher, the predecessor of the modern stapler, can stitch only parallel to a fold and cannot insert its "staples" close together; *See also* 1877 TOO

The first underground railways (subways) in New York City start operation; *See also* 1863 TRA

The Pullman Palace Car Company begins making a dining car developed by George M. Pullman; *See also* 1864 TRA; 1887 TRA

John Stringfellow [b 1799, d 1883] designs the triplane and builds a model; when airplanes are invented in the 20th century, the Stringfellow design is remembered and leads directly to the biplane

1869

Edward Everett Hale [b Boston, Apr 3 1822, d Roxbury MA, Jun 10 1909] publishes *The brick moon* about an artificial satellite placed in orbit as a navigational aid, reviving an idea not much discussed since the time of Isaac Newton; *See also* 1680–89 TRA

Harriet Morrison Irwin is the first woman in the US to patent an architectural innovation; it is for a design of a hexagonal building, which she builds in Charlotte NC as her home

John Augustus Roebling designs and begins work on the Brooklyn Bridge between Brooklyn and Manhattan NY, but dies as a result of an accident; when the bridge is completed by his son and his son's wife, it becomes the longest suspension bridge of its time; *See also* 1855 ARC; 1883 ARC

Louis Ducos du Hauron describes in *Les couleurs en photographie, solution du problème* (Colors in photography, the solution to the problem) the method he has developed for recording a color image on a single photographic plate and projecting it with a single projector; the method requires breaking the light from the projector and the image on the plate into tiny dots that are handled so as to produce separate dots of color on the screen, which the eye then combines; he also states the additive and subtractive ways that different colors can be obtained; *See also* 1861 COM; 1873 COM

Independently of Louis Ducos du Hauron, Charles Cros [b Fabrezan, France, 1842, d 1888] discovers the subtractive principle used in color photography; *See also* 1861 COM; 1873 COM

The Post Office in Great Britain obtains the exclusive right to transmit telegrams; *See also* 1854 COM

Zénobe Gramme builds the first commercially practical generator for producing direct current; *See also* 1867 ENE

1870

James Clerk Maxwell proposes that all measurement standards be based on atoms because all atoms of a substance are thought to be exactly alike (a view modified by later physicists); Maxwell's idea is not put into practice until more than a hundred years later, largely because physicists do not know enough about atoms to use them in this way until the 1960s

Mary Potts invents a hot iron for pressing clothes that has a cool handgrip

Pope Paul V builds the Acqua Marcia-Pia aqueduct, the first new aqueduct in Italy since the Aqua Alexandrina of 226, thus, the first new Italian aqueduct in 1644 years; *See also* 230 CE ARC

Julius Hock is the first to build an internal combustion engine that works on Lenoir's principle, but uses liquid gasoline as fuel; *See also* 1867 ENE; 1877 ENE

1869

According to Foster Manufacturing, the commercial toothpick is invented by Charles Foster in his basement in Boston

John Wesley Hyatt [b Starkey NY, Nov 28 1837, d Short Hills NJ, May 10 1920] develops the substance trademarked in the US as Celluloid, probably not quite independently of Alexander Parkes, who had produced the same substance or something close to it a few years earlier, or of Daniel Spill, who made a similar plastic; it is the first commercial artificial plastic; Hyatt uses his new plastic to win a $10,000 prize for a substitute for ivory in billiard balls, and later commercializes the product as a base for film in photography; essentially, Celluloid is the same cellulose nitrate as guncotton, but less nitrated and therefore not explosive, although still quite flammable; *See also* 1865 MAT; 1909 MAT

German surgeon J. Friedrich A. von Esmarch [b 1823, d 1908] demonstrates the use of a prepared first-aid bandage on the battlefield; *See also* 1746 MED; 1877 MED

The first electrically powered dental drill is constructed; *See also* 1790 MED; 1874 MED

A large band saw is developed and patented by Jacob R. Hoffman; it soon becomes the standard device for heavy sawing in the US; *See also* 1814 TOO

On May 10 the Union Pacific and Central Pacific railroads link up at Promontory Point UT, creating the first transcontinental rail system in North America

The firm of A. Guilmet and Meyer develops a bicycle in which the power from the pedals is transmitted to the back wheel by a chain; J.F. Tretz also uses a chain drive to propel the large front wheel of the popular bicycle, called a "high wheeler," so that people with short legs are not at a disadvantage; *See also* 1861 TRA; 1876 TRA

Ferdinand de Lesseps completes construction of the Suez Canal, linking the Mediterranean and Red seas; *See also* 1859 TRA; 1880 TRA

1870

American photographer Thomas Adams invents chewing gum; it is intended as a substitute for chewing tobacco

William Lyman of W Meridan CT patents a can opener that uses a wheel for continuous operation, improving on the lever-and-chisel variety that preceded it; the can must be pierced in the center of the lid and the opener must be adjusted to fit each size can; *See also* 1858 FOO; 1925 FOO

Between 1867 and 1870 Alexander L. Holly introduces the manufacture of steel by the Bessemer process into the US; *See also* 1856 MAT

Thomas Alva Edison improves his "stock ticker" or "gold indicator" of 1869 by making it possible to adjust remote indicators to match a home station; he sells his device to a Wall Street firm for a reported $40,000, a fortune at the time; *See also* 1868 TOO

During this decade Fred Kimble of IL invents choke boring for shotguns, a method of constricting the barrel for part of its length so that the shot forms a tighter pattern on emerging

Margaret Knight patents a machine that glues and folds paper bags; its design is still in use today

Germain Sommeiller's Mont Cenis Tunnel through the Alps, the first major railroad tunnel, is completed; started in 1857, it is 14 km (8.7 mi) long; pneumatic drills and dynamite are used in its construction; *See also* 1858 TOO

1871

COMMUNICATION

Richard L. Maddox introduces the dry plate (bromide on gelatine) to photography; *See also* 1864 COM; 1884 COM

ENERGY

Peter Brotherhood [b 1838, d 1902] patents a radial three-cylinder steam engine; the cylinders are placed at angles of 120 degrees and the connecting rods are linked to a common crank pin; since its invention, the Brotherhood engine has been used as an air motor for torpedoes and as an air compressor; *See also* 1866 ENE

Zénobe Gramme presents his version of the dynamo to the Academy of Sciences in France; *See also* 1867 ENE

1872

GENERAL

The Siemens and Halske research laboratory is founded in Germany; *See also* 1876 GEN

Russian physician Alexandre de Lodyguine develops an incandescent lamp that uses graphite for a filament within a glass globe filled with nitrogen; although about 200 of these are installed by the Russian Admiralty in St. Petersburg, their unreliability and cost make them impractical, as was the case with all incandescent lamps before those of Thomas Edison and Joseph Swan in 1879; *See also* 1860 GEN; 1879 GEN

Jane Wells patents the Baby Jumper

COMMUNICATION

Thomas Edison patents an electric typewriter, the prototype of the teletype machines that will later be used by press agencies; *See also* 1855 COM; 1902 COM

Latham Sholes develops an improved form of the typewriter of 1867, which is then manufactured by Philo Remington of Ilion NY; this typewriter introduces the qwerty keyboard that is now standard; it goes on sale in 1874; *See also* 1867 COM; 1878 COM

Charles Gillot (son of Firmin Gillot, the first to print photographs using lithography) is the first to apply the Poitevin collotype process in regular printing; as a result, he is considered by some to be the inventor of photogravure; *See also* 1855 COM

Elisha Gray [b Barnesville OH, Aug 2 1835, d Newtonville MA, Jan 21 1901] forms the Western Electric Manufacturing Company, later the Western Electric Company, a pioneering communications firm; *See also* 1876 COM

England and Australia are linked by telegraph via submarine cables; *See also* 1861 COM

ENERGY

The first long-distance fuel pipeline is completed; located in Rochester NY and made entirely of wood, it carries natural gas over a distance of 40 km (25 mi); *See also* 1863 ENE

George B. Brayton builds a two-stroke internal combustion engine; exhibited in 1876 in Philadelphia PA, it inspires Nikolaus Otto to build his engine; however, Otto's engine uses the four-stroke cycle devised by Beau de Rochas; *See also* 1867 ENE; 1877 ENE

1871

German chemist Hermann Sprengel shows that the yellow dye picric acid can be detonated and is an effective explosive; *See also* 1771 MAT; 1885 MAT

American inventor Ives W. McGaffey constructs an industrial vacuum cleaner powered by a steam engine; *See also* 1865 TOO; 1902 TOO

Otto Lilienthal [b Auklam, (in Germany), May 23 1848, d Rhinow, Germany, Aug 10 1896] begins to study bird flight and the construction of kites with the idea of building a heavier-than-air flying machine; *See also* 1842 TRA; 1877 TRA

Franklin Pope [b Great Barrington MA, Dec 2 1840, d Great Barrington, Oct 13 1895] develops a centrally controlled signaling system for the Boston & Lowell Railroad; *See also* 1867 TRA

1872

Amanda Theodosia Jones [b E. Bloomfield NY, Oct 19 1835] patents a vacuum process for preserving food; *See also* 1809 FOO

Physician Jean Martin Charcot [b Paris, Nov 29 1825, d Nievre, France, Aug 1893] uses hypnosis as part of his treatment for therapy; in 1885 Sigmund Freud is a student of Charcot's and learns this use of hypnotism from him; *See also* 1841 MED

Paul Haenlein [b Cologne, Germany, Oct 17 1835, d Mainz, Germany, Jan 27 1905] builds the first airship to be powered by an internal combustion engine; it obtains its fuel from the gas in the balloon; *See also* 1852 TRA; 1883 TRA

The first oil tanker is built

Amédée Bollé [b LeMans, France 1844, d 1917] builds a steam-driven car that reaches a speed of 40 km (25 mi) per hour; *See also* 1835 TRA

1873

Hermann Vogel [b Dober-lug, Germany, Mar 26 1834, d Berlin, Dec 17 1898] improves the sensitivity of photographic collodion plates to green light by treating them with an aniline dye; subsequently, he improves the sensitivity of photographic plates to other colors, thus making it possible to reproduce truer colors; *See also* 1861 COM; 1891 COM

Hippolyte Fontaine [b Dijon, France, 1833, d Paris, 1917] demonstrates at the Exhibition of Vienna that the electric generator developed by Zénobe Gramme can also serve as a motor; he shows that a large generator attached to a small one will turn the small one with power from the larger one; *See also* 1871 ENE

1874

James Buchanan Eads [b Lawrenceburg IN, May 23 1820, d Nassau, Bahama Islands, Mar 8 1874] builds a two-level highway and rail bridge over the Mississippi at St. Louis MO; the first US bridge to use steel in its construction, the tubular-arch bridge still stands; *See also* 1779 ARC

Thomas Edison develops a method for transmitting simultaneously two messages over one telegraph line or four messages in each direction over the common arrangement of a pair of telegraph wires; thus the invention, which secures Edison's early reputation as an inventor, is called the perfected quadruplex telegraph; *See also* 1872 COM

Emile Baudot [b Haute-Marne, France, 1845, d 1903] introduces a binary code that uses five bits to represent characters; this code remains in international use until 1930

Alexander Graham Bell patents a multiple or harmonic telegraph that can simultaneously transmit two or more musical tones, similar to a more complex harmonic telegraph developed by Elisha Gray; it is a failed attempt to develop a telephone; *See also* 1876 COM

1873

Fred Hatch attempts to build the first modern grain silo in McHenry County IL; similar airtight storage bins for fodder had been built by both the Romans and the Chinese, but they are no longer in use by the 19th century; *See also* 50 CE FOO

The Australian wool merchant T.S. Mort and the French engineer E. Nicolle build the first refrigeration plant for meat in Australia; *See also* 1857 FOO; 1877 FOO

Willoughby Smith states that anecdotal reports of the effects of light on selenium have some basis in fact; conductivity of selenium measurably increases on illumination; this becomes known as the photoconductive effect; *See also* 1861 MAT; 1876 MAT

Charles Thomson, while on a round-the-world expedition on HMS *Challenger,* discovers manganese nodules about 300 km (185 mi) southwest of the Canary Islands

Karl Paul Gottfried von Linde [b Berndorf, Germany, Jun 11 1842, d Munich, Germany, Nov 16 1934] begins work on the first movable refrigerator, using methyl ether as a coolant; *See also* 1857 FOO; 1876 TOO

Hermann Sprengel introduces a mercury pump that is able to create a vacuum sufficient for the production of electric filament lamps; *See also* 1871 TOO

Christopher M. Spencer [b 1833, d 1922] builds a machine tool that automatically feeds in the piece to be cut, moves the workpiece, and changes tools as needed; it is the precursor of a type of automatic machine tool called the Hartford automats

1874

Joseph Farwell Glidden [b Charleston NH, Jan 18 1813, d Oct 9 1906] of De Kalb IA patents double-stranded barbed wire with the barbs twisted in place along the wire so they cannot slip; the Glidden invention, called "the Winner," becomes the main type of barbed wire used thereafter; two other De Kalb entrepreneurs patent their own varieties of barbed wire; *See also* 1868 FOO; 1915 TRA

Othmar Zeidler prepares DDT, but does not discover its insecticidal properties; *See also* 1825 FOO; 1939 FOO

The role of fluorides in preventing dental decay is discovered

The "destructor," a machine designed to incinerate waste, is introduced in Nottingham, England; *See also* 1885 GEN

Martin Wiberg of Stockholm, Sweden, builds an improved version of Babbage's Difference Engine; *See also* 1855 TOO

1875

On May 20 the Treaty of the Meter, establishing the International Bureau of Weights and Measures, with headquarters at Sèvres, France, is signed in Paris; the treaty provides for a prototype meter and kilogram on which scientific and other measures are to be based; *See also* 1889 GEN; 1967 GEN

Work is completed on the Opera House in Paris, designed by Charles Garnier [b 1825, d 1898]; the building demonstrates that various traditional styles can be combined in an artistic and pleasing way

Long after an unsuccessful attempt by Roman Emperor Claudius to drain Lacus Fucinus (Lago Fucino) in Italy, Swiss and French engineers working for Prince Alessandro Torlonia rebuild and extend the tunnel from the lake to the Liri River, starting in 1852; this time the project works and the lake is converted to land; in 1951 the Italian government settles 8000 families on the former lake; *See also* 50 CE ARC

British anthropologist E.R. Jones identifies tiny marks cut into bones by Neolithic cultures as tallies used to count game or otherwise used in hunting

Alexander Bell transmits sounds over electric cables; *See also* 1874 COM; 1876 COM

Austrian printer Karl Klietsch develops etching on metal via a photographic method called heliogravure; the resulting plate can be used for printing images; *See also* 1872 COM; 1895 COM

George R. Carey proposes a form of "television" system in which the transmitter consists of an array of light-sensitive selenium cells that switch an array of electric lights on or off as an object moves in front of the cells; *See also* 1861 MAT; 1873 MAT

1876

Edison's laboratory at Menlo Park NJ marks the start of modern industrial laboratories in the US; *See also* 1872 GEN

Paul Jablochkoff [b Serdobsk, Russia, Sep 14 1847, d Russia, Mar 19 1894] develops the "electric candle," an improved version of the arc light that will burn for 2 hours with no mechanical adjustment because its two carbon rods are separated by porcelain clay that vaporizes as the arc burns; alternating current is used to ensure that each rod vaporizes by the same amount while the lamp is in operation; *See also* 1845 GEN; 1878 GEN

William Thomson, later Lord Kelvin, shows that machines can be programmed for all sorts of mathematical problems; *See also* 1843 COM

Alexander Graham Bell patents the telephone; the microphone and receiver are identical: both consist of a magnet surrounded by a solenoid placed close to an iron membrane; vibrations in the membrane induce currents in the solenoid; these currents cause the membrane at the receiving end to vibrate; the first words are transmitted inadvertently by Bell to his assistant in another room: "Mister Watson, come here, I want you."; *See also* 1875 COM

On Feb 14, 2 hours after Alexander Graham Bell files for a patent on his telephone, Elisha Gray files a notice with the Patent Office for a device that would allow the transmission of voice and sound (indeed, experts think that Gray's version would have worked better than the device Bell patented); after many years of litigation, the US courts declare Bell the inventor; *See also* 1872 COM; 1895 COM

Henry Corliss builds a large, 2500-horsepower engine for the Centennial Exposition in Philadelphia PA; the beam vertical engine with two vertical cylinders, revolving at 360 rpm, is the largest engine built at the time; *See also* 1849 ENE

1875

Swiss chocolate maker F.L. Cailler, following a suggestion from his son-in-law, Daniel Peter, begins to manufacture milk chocolate; previously, milk was mixed into chocolate by the consumer just before making a chocolate-and-sugar drink

Alfred Nobel develops blasting gelatin, a combination of nitroglycerin and nitrocellulose, which he patents in 1876; *See also* 1867 MAT; 1888 MAT

1876

Fred Harvey, commissioner of the Chicago, Burlington and Quincey railway line, invents the cafeteria for quick food service at railroad stops; *See also* 1902 FOO

Sidney Thomas [b London, Apr 16 1850, d Paris, Feb 1 1885] and his cousin Percy Gilchrist [b Lyme Regis, England, Dec 27 1851, d Dec 15 1935] develop a method for using the Bessemer process with other than low-phosphorus ores and pig iron; the Thomas-Gilchrist process involves lining a converter with a basic material such as lime and adding lime to the mixture of ores, coke, and pig iron; this removes phosphorus oxides that would otherwise result in brittle, useless steel; *See also* 1856 MAT

William G. Adams [b Laneast, England, Feb 6 1836, d Apr 10 1915] and R.E. Day conduct an experimental study of the photovoltaic properties of selenium and discover the effect in selenium barrier-layer cells; *See also* 1873 MAT; 1930 ELE

Robert Koch [b Klausthal-Zellerfeld, Hanover, Germany, Dec 11 1843, d Baden-Baden, Germany, May 27 1910] discovers that the microorganism responsible for cattle anthrax can be grown in culture; *See also* 1840 MED

Lydia E. Pinkham [b Lynn MA, Feb 9 1819, d Lynn, May 17 1883] starts advertising her patent medicine for female reproductive disorders, the main active ingredient of which is alcohol; *See also* 1867 MED

Karl von Linde develops the first practical refrigerator, replacing methyl ether as a coolant with ammonia; previously Ferdinand Carré had used ammonia for the same purpose; *See also* 1873 TOO; 1877 FOO

Werner Siemens develops a practical selenium cell; when illuminated, its electrical resistance drops; *See also* 1873 MAT

William Thomson, later Lord Kelvin, designs a mechanical computer, called *Tidal Harmonic Analyser*, for solving differential equations; his ideas will be developed further by Vannevar Bush in the 1920s; *See also* 1927 ELE

H.J. Lawson develops the first "safety" bicycle, in which the power from the pedals is transmitted to the back wheel by a covered chain; wheels of equal size are used, minimizing the danger from falls that led some cyclists in the early 1870s to switch to the tricycle; different sized gears connected by the chain give the mechanical advantage; *See also* 1869 TRA; 1885 TRA

1877

A strike against the Baltimore and Ohio railroad constitutes the first US national labor action; riots that ensue between railroad police and strikers are halted by federal troops, an action considered a victory for capital over labor; *See also* 1851 GEN

On Aug 15 Thomas Alva Edison writes to T.B.A. David, president of the Central District and Printing Telegraph Company in Pittsburgh, on the subject of introducing telephone service: "Friend David, I don't think we shall need a call bell as Hello! can be heard 10 to 20 feet away. . . ."; Alexander Graham Bell envisioned answering the telephone "Ahoy," so it seems likely that Edison invented the US custom of saying "Hello"; in Britain, people answer early telephone calls with "Are you there?"; *See also* 1876 COM

After 5 years of construction, the bridge over the Tay in central Scotland is completed; it consists of 85 sections of which the longest is 75 m (220 ft); on Dec 28 1879, while a train is crossing the bridge, it collapses, killing the 300 passengers aboard; *See also* 1850 ARC; 1883 TRA

Work on the present-day Eddystone Light is started; although the construction techniques developed by John Smeaton for the third Eddystone Light are used, the base is a solid cylinder that breaks up the waves much better than the tapering design Smeaton and others had employed; tapered lighthouses actually direct waves and spray upward toward the light itself and toward the weaker portions of the structure; *See also* 1759 ARC

Charles Cros, independently of Thomas Edison, proposes using the phono-autograph system invented by Scott de Martinville to reproduce sound, although he does not actually build a device; the Cros conception precedes Edison's device by a few months, resulting in a French claim to the invention; *See also* 1856 COM

Between Nov 29 and Dec 6 a workman named John Kruesi follows a sketch made by Thomas Edison and builds the first phonograph, in which a needle attached to a diaphragm traces a groove on tinfoil placed around a rotating cylinder; the recording is replayed by the same needle and diaphragm; the first recording is of Edison reciting "Mary had a little lamb"; after various minor improvements and adjustments, Edison prepares a patent application and demonstrates the device in the offices of *Scientific American* on Dec 7; the application is filed on Dec 15 and the invention announced in the Dec 22 issue of the magazine

Nikolaus Otto develops the four-cycle internal combustion engine, the basis of the most common type of engine today; *See also* 1870 ENE

Zénobe Gramme derives an electric alternator from his dynamo; it produces alternating current and does not need brushes—a source of electrical loss and mechanical wear; however, it is only after the introduction of the transformer a few years later that alternating current becomes important; *See also* 1832 ENE; 1884 ENE

1878

Charles F. Brush in Sep develops an improved carbon arc lamp, which he demonstrates in a Cleveland department store; *See also* 1876 GEN; 1879 GEN

Musician David Edward Hughes [b London, May 16 1831, d London, Jan 22 1900] discovers that the pressure of sound waves can be used to change electrical resistance in carbon lying loosely between electrical terminals; Hughes uses this effect to create the first microphone

Hiram Stevens Maxim [b Sangerville ME, Feb 5 1840, knighted 1901, d Streatham, England, Nov 24 1916] invents a graphite-rod incandescent lamp that eventually loses in commercial competition to Edison's lamp; *See also* 1872 GEN; 1879 GEN

On Jan 28 the first commercial telephone exchange opens in New Haven CT; *See also* 1876 COM

The Remington No. 2 typewriter introduces the shift bar; *See also* 1872 COM

William Crookes [b London, Jun 17 1832, d London, Apr 4 1919], in an address to the Royal Society on Nov 30, describes his experiments in passing electric discharges through an evacuated glass tube, an early form of the cathode-ray tube; *See also* 1895 MED

Frederick E. Ives (b Litchfield CT, Feb 17 1856, d Philadelphia PA, May 27 1937] invents the first halftone process for printing photographs; *See also* 1881 COM

Charles F. Brush [b Euclid OH, Mar 17 1849, d Cleveland OH, Jun 15 1929] builds and patents an electric generator to go with the arc lamp he has also invented; *See also* 1862 ENE

James Wimshurst [b Poplar, England, Apr 13 1832, d Clapham, England, Jan 3 1903] develops an efficient electrostatic generator that uses two counter-rotating plates; *See also* 1775 ENE

1877

Eleanor A. Ormerod [b Gloucestershire, England, May 11 1828, d St. Albans, England, Jul 19 1901] starts publication of the *Annual report of observations of injurious insects,* pamphlets that reach print runs of 170,000 and are consulted by agriculturalists all over the world

Charles Tellier [b Amiens, France, 1828, d Paris, 1913] starts the transport of meat cooled to 0°C (32°F) from Argentina to France aboard the ship *Paraguay,* equipped with refrigeration by Ferdinand Carré; *See also* 1873 FOO

Joseph Monier patents the use of reinforced concrete for the manufacture of railroad sleepers; *See also* 1867 MAT; 1892 MAT

Louis Pasteur notes that some bacteria die when cultured with certain other bacteria, indicating that one bacterium gives off substances that kill another; 60 years pass before this observation is put to use, when René Jules Dubos discovers the first antibiotics produced by a bacterium; *See also* 1939 MED

J. Friedrich A. von Esmarch introduces the antiseptic bandage; *See also* 1869 MED

A form of wire stitcher is developed that uses precut, U-shaped staples (similar to staples already in use by carpenters) instead of wire cut from a continuous piece; the staples must be inserted one at a time into the machine; *See also* 1868 TOO; 1894 TOO

Italian inventor Enrico Forlanini builds a helicopter with two counter-rotating propellers driven by a steam engine; he achieves a height of 15 m (45 ft) for a flight of 1 minute's duration; *See also* 1853 TRA; 1907 TRA

Otto Lilienthal about this time begins to build and test a series of gliders that he launches from the top of a hill and steers by moving his legs; *See also* 1871 TRA

John Thornycroft [b Rome, Feb 1 1843, d Isle of Wight, Jun 28 1928] patents an air-cushion vehicle, but he does not succeed in building a working prototype; *See also* 1955 TRA

1878

Anna Baldwin invents the suction milking machine; the first successful application of the device is made in 1918 by Carl Gustav Laval

Bar soap with air bubbles trapped in the mixture so that the soap floats in water is discovered by accident when a batch of a new soap, called White Soap, is left in the stirring machine too long; the soap that floats is renamed Ivory

Physiologist Paul Bert [b Auxerre, France, Oct 17 1833, d Hanoi (in Vietnam), Nov 11, 1886] announces that dissolved nitrogen in the blood of people working under pressurized air causes the disease commonly known as the bends or caisson disease; he proposes, correctly, that lowering air pressure by stages will prevent the disease; *See also* 1779 ARC

Samuel Pierpont Langley [b Roxbury MA, Aug 22 1834, d Aiken SC, Feb 22 1906] builds a radiometer consisting of two metallic grids connected to a Wheatstone bridge; when one grid is exposed to the radiation to be measured, the increase in temperature changes its resistance, which is then measured by the bridge; *See also* 1850 TOO

Thomas Walker invents the cherub ship's "log," which does not have to be hauled out of the water to be read as his harpoon log did; improved models of this log are still in use; *See also* 1861 TRA; 1911 TRA

OVERVIEW

The Electric Age

Many inventions that mark the technological development of the 20th century were put in place during the last two decades of the 19th century: the telephone, electric generator, electric light, steam turbine, cheaper processes for making steel and aluminum, automatic machine tools, and, most important in its effect on society, the internal combustion engine.

One important consequence of these inventions was that technology did not remain confined to large enterprises only, as was true for most of the 19th century. Through most of the 19th century ordinary people came into very little in contact with technology and lived the same way as during the earlier centuries. Around the end of the 19th century, mainly through the impetus of electricity, technology started to become an important part of everyday life in the industrialized world. First electric light and the telephone entered the home. The widespread availability of electricity opened the way to a host of kitchen and other appliances. The vacuum cleaner and electric iron soon became common, followed during the 1920s and 1930s by the refrigerator and washing machine.

A second factor that contributed to the spread of technology was the enormous growth of the automobile industry. Automobile use spread the most quickly in the US, where there were 2.5 million registered motor vehicles in 1915. Automobile manufacture drove the economy of the US and other developed countries, paving the way for sharp increases in materials manufacture and adding to the prosperity of many small manufacturers who made parts for automobiles as well as related devices or who made machine tools.

Some newly emerging technologies were not immediately accepted, however; paradoxically, the automobile and the truck are good examples. When Karl Benz built his first automobiles in the 1880s, there was generally no perception of a need for such vehicles—the public became interested in automobiles only about 20 years later. Trucks underwent the same fate. Developed in the 1890s, it was only after they had proved their usefulness during World War I that they became fully accepted in the US, although a study at Cornell University in 1900 had shown that their operation was 25 to 40 percent cheaper than that of the horse-drawn wagon.

The same delay in acceptance occurred with farm tractors. Although tractors equipped with internal combustion engines were available before World War I, farmers, especially in Europe, used them reluctantly, and many relied on horses until the 1960s. One reason for the slow acceptance of the tractor was that it went against farmers' tradition of self-sufficiency: Farmers could breed their own horses and feed them with their own crops. Horses and other livestock also produced manure to fertilize the farmers' fields. The tractor made farmers more dependent on outside sources for fuel, replacement parts, and fertilizer.

The spread of technology increased the demand for energy during the first decades of the 20th century. The oil industry quickly converted to the production of gasoline, introducing new methods, such as cracking petroleum to produce specialized hydrocarbons. Coal remained the main source of energy, however, and the rapidly spreading electric power stations became its largest consumers.

Electricity: A revolution in technology

Although electricity had been known since the 18th century—it was then more a subject of curiosity and entertainment—the widespread introduction in the workplace during the 1880s caused a profound change of technology. Electricity was the first practical way to transport energy over long distances between central power stations and homes or factories. It also allowed the construction of numerous new devices that were impractical by mechanical means only. The electric magnet, whether in electric motors, electric relays, or electric mechanisms of other kinds, became the core element of a new technology that quickly developed: electrotechnology. Thermostats, servomechanisms, and a large number of automatic systems became based on electricity. The use of electricity allowed the construction of machines that in pure mechanical form would be impracticable because of friction and inertia. Many formerly mechanical devices became much smaller, more efficient, and faster through the introduction of electric components. The transition from purely mechanical devices to electromechanical devices is comparable to the transition, after World War II, from electromechanical devices to purely electronic devices.

The introduction of the electric motor made possible the development of underground railways. Diesel engines are not suited for use underground because of their need for oxygen and even more because of problems caused by the products of combustion they produce. Electrotechnology also allowed the development of automatic switching systems and safety systems in railways, a development that with mechanical devices alone would have been impossible.

Electric power, still supplied by small power stations at the turn of the century, became more centralized because of the availability of large steam turbines and electric generators. The first electric power stations were built close to consumers to minimize the losses caused by the resistance of electric wires. The transformer was a very important invention: It permitted the transformation of the voltage of generators to high values so that electric energy could be transported over much longer distances with very little loss. The establishment of high-voltage distribution systems of electricity ultimately freed industry from geographical constraints. Earlier, industry would establish itself where energy was available, either waterpower or coal. Now it could establish itself wherever there was electric power, or almost everywhere.

Although factories in which a central steam engine powered the majority of machines through a system of belt transmissions existed until the end of the 1960s (the important Bowler machine shop in Bath, England, used this method until it went bankrupt in 1969), industry converted from the early 1900s on to machines driven by independent electric motors. For example, when the hydraulic power station at Niagara Falls was completed in 1886, it attracted several new industries using electric motors to drive their machines. The electric motor, cheap and easy to operate, also stimulated the creation of numerous small workshops with a few machines each. Electricity-producing companies encouraged this development by offering cheap industrial rates. In the early days of electric power generation, electricity served only for lighting, so demand was high during the evening hours and very low during the day. The increased use of electric motors by workshops and industry evened this imbalance.

Science and technology

A new type of laboratory emerged during the 1880s, a laboratory for testing safety, especially the safety of railway systems. Such laboratories appeared first in Germany, but also in the US. Besides testing, scientists became involved in the development of new products.

The necessity for engineers to rely on science and scientific research became apparent in many other areas. German chemical industrial labs were followed by laboratories for electricity and electronics. Although Thomas Edison was able to experiment with the electric bulb ignoring Ohm's law, and Zénobe Gramme built his electric generator knowing little of physics, early developments in radio communication and electronics required a much better understanding of the physics at the base of these devices.

Large companies, including General Electric, Du Pont, and Bell Telephone established research laboratories. Their aim was to develop or even invent new products. Researchers at several of these laboratories were given free rein and often ventured into basic research. Some of these industrial laboratories developed into cradles of experimental and fundamental science, on a par with the best university laboratories.

Major advances

Communications. The telephone had a profound influence on society, bringing instant communication to the home. Broadcasting also played an important role in changing the society of the 20th century.

The telephone as well as the early radio communication devices were electromechanical devices rather than electronic ones, although the crystal detector, developed in 1901, can be viewed as the first electronic device in the modern sense. The discovery of the vacuum tube increased substantially the technical possibilities of telephone and radio communication, mainly because vacuum tubes can amplify weak signals or function as oscillators or as rectifiers. Vacuum tube amplifiers allowed telephone signals to travel over thousands of miles, and radio communications to span continents.

The development of communications using electricity and electromagnetic waves is probably the chief technological factor that influenced society in the 20th century. Radio, and later television, also played an important role in society's acceptance of new inventions during the 20th century, spreading ideas swiftly through even illiterate populations. The development of sound recording and broadcasting also had a profound impact on culture. Just as printing made literature accessible to everyone, now music came into the lives of all people.

Motion pictures, after groundwork in both the US and France, and made practical by the invention of celluloid film, developed quickly and became an important form of entertainment. Sound motion pictures became a reality during the mid-1920s and color pictures appeared in the 1930s. This rapid progress in less than 40 years culminated in the most famous movie ever made, *Gone With the Wind*, a color sound film of 1939 that had the largest number of viewers in the history of film until very recent times.

Often overlooked because of the revolutionary development of broadcasting, printing and typesetting advances in this period changed the way that books, magazines, and newspapers were created. Among the advances were the ways in which printers reproduced photographs, especially color photographs, the ways in which type was set (even the Linotype, which seems almost primitive today, was first introduced at the beginning of this period), and the use of offset printing. The creation of publications, as well as private and business correspondence, was also revolutionized by rapid advances in the mechanical typewriter. The electric typewriter did not achieve widespread use until after World War II, however.

Food and shelter. Handling of food at both the production end and in the kitchen changed dramatically during this period. The main agent of change in production was the steady improvement of the tractor and its increased use on farms in Europe, but especially in North America, where a tradition of large fields had developed to take advantage of mechanical planters and reapers. Even in the US, however, farms were rebuilt by removing stone walls or brush hedges to make for still larger fields that took better advantage of the power of the tractor. Other mechanized farm equipment was also improved, but with few dramatic breakthroughs—a lot of the development was plagiarized from the growth of the automobile and truck industries. The self-propelled combine was one noticeable step forward, however.

Although one thinks of mechanization of farming in terms of planting and harvesting crops, herding was also partly mechanized during this period, primarily in the way milk came to be extracted from cattle with mechanical milkers and in the way milk was separated and stored after milking.

At the end of this period, signs of the next great change in agriculture become apparent with the recognition of potent insecticides and selective herbicides.

In the kitchen, the electrification of small appliances was the main change. With electrification, brand new appliances, such as the blender, also were invented. Although freeze-drying and freezing were introduced in this period, they did not affect the way people lived until after World War II. One significant change in eating habits at home came with the introduction of prepared breakfast cereals; previously, breakfast cereals, when they were eaten at all, consisted of cooked grains, such as oatmeal. A sign of eating habits to come took place outside of the home: Cafeterias and restaurants where the patrons did not sit down were the precursors of the fast food establishments of today.

Two new important elements entered architecture and construction during the last decades of the 19th century. The Bessemer process allowed the manufacture of cheaper steel, suitable for construction. The Eiffel tower demonstrated that the use of iron and steel was not restricted to bridges, but also could be used in vertical buildings. Steel became an important component of skyscrapers, many-storied office buildings that began to be built in Chicago IL before the turn of the 20th century.

Reinforced concrete is made possible by the lucky coincidence that steel and Portland cement have the same thermal expansion coefficient, so changes in tem-

It is often argued that science and technology develop independently, and that new technologies only rarely are directly derived from scientific developments. There are enough examples of recent technological developments, such as the transistor or the laser, to contradict this. The development of radio communications is an early example of the direct application of physics.

It was an international collaboration. The theoretical basis of radio waves was laid by British physicist James Clerk Maxwell, who gave a thorough mathematical description of electromagnetic waves in 1873. Maxwell showed that light is an electromagnetic wave, but that electromagnetic waves with a much longer wavelength than that of light—now known as radio waves—also exist. Ten years later Irish physicist George Francis Fitzgerald suggested the way in which such waves could be produced. German physicist Heinrich Hertz was the first to deliberately produce such waves, showing that they are reflected and refracted in the same way as light.

In 1887 Hertz started his experiments, creating radio waves by producing sparks with a high-voltage induction coil, a device not very different from the coil that delivers the high-voltage pulses to the spark plugs in an automobile engine. As "receiver" he used a device consisting of two small spheres attached to the ends of an open loop of wire. When he brought the loop in the beam of electromagnetic radiation produced by the spark generator, he observed small sparks that jumped between the two spheres.

The coherer, invented independently a half-dozen years later by British physicist Oliver Lodge and French scientist Edouard Branly, was a more sensitive detector for radio waves than the receiving loop of Hertz. It consisted of a small tube filled with iron filings connected to electric wires at each end. In the presence of an electromagnetic wave, the iron filings lined up (as Lodge expressed it, became "coherent") and the coherer became conducting. Branly could detect radio waves produced by a spark generator up to a distance of 30 m (100 ft), and Lodge was able to double that feat.

Around the same time, Italian engineer Guglielmo Marconi also began to generate and detect radio waves, but from the beginning he concentrated on producing useful effects, such as ringing a bell from a distance. Marconi also went somewhat further than the physicists by experimenting with antennas, for which a physical theory was still nonexistent.

Marconi became a key figure in the development of radio—his contribution to radio as a communication tool can be compared to the role of Steven Jobs in the development of the personal computer. Marconi was the first to set out deliberately to use electromagnetic waves for communication purposes. By also using coherers and experimenting with antennas he could quickly increase the distance between receiver and transmitter, resulting in the bridging of the Atlantic by radio transmission in 1901.

Mainly under the impetus of Marconi, radio communication systems developed quickly. The early transmitters were all of the spark-generator type, producing electromagnetic radiation in a pulselike manner. Such transmitters produced radiation in a wide spectrum, with the consequence that nearby transmitters easily interfered with each other. Although tuned circuits, first introduced by Oliver Lodge and Marconi, were an improvement, the greatest breakthrough in radio communication was the vacuum-tube oscillator. Such an oscillator could produce a continuous radio wave occupying only a narrow frequency range.

Canadian inventor Reginald Aubrey Fessenden was producing voice transmission as early as 1904, and soon was making experimental voice broadcasts that could be received all up and down the Atlantic coast of the US. The introduction of the vacuum tube in 1913 also made possible a better way to modulate radio waves. Fessenden and other inventors, such as the American Edwin H. Armstrong, soon produced much more powerful transmitters and improved receivers based on vacuum tubes, although some early programs continued to use spark-generated signals until the early 1920s.

perature do not cause serious internal stresses that would lead to cracks or other flaws. Reinforced concrete became a very important building material. Concrete gave architects much more freedom in their designs. For example, the homes designed by Frank Lloyd Wright and Le Corbusier made ample use of reinforced concrete, freeing them from the constraints of the traditional building materials.

Electronics and computers. Not everyone agrees as to what distinguishes an electronic device from an electric one. Many define a circuit as "electronic" if it incorporates components that do not respond linearly to changes in current or voltage. Vacuum tubes fulfill this requirement, as does the crystal detector (its resistance depends on the polarity of the applied voltage), but according to this definition, the coherer, invented by Edouard Branly and used by Marconi in 1897 in his radio receiver, is also an electronic device. It consists of a tiny glass tube filled with iron filings. Normally, the resistance across the filings is high, but it drops drastically on detection of a radio wave.

The development of the vacuum tube (or *valve* in evocative British English) was the most important fac-

It is generally accepted that the telephone is the invention of Alexander Bell. Just as with the electric lamp and the telegraph, several people had the same idea at once, in this case of transmitting sound through wires during the 1860s. In Germany, the teacher Philip Reis, who first coined the term "telephone," experimented in 1861 with the transmission of musical sound through electric wires. His microphone was extremely simple; it consisted of a metal point in contact with a metal strip attached to a membrane. When the membrane vibrated because of sound waves, the electrical contact was interrupted with the same frequency as the sound wave. The receiver consisted of a long metal needle attached to a resonator and placed into a solenoid—a coil of wire used to generate a magnetic field when a current passes through the wire. When a varying current passed through the solenoid, the length of the metal needle changed because of magnetostriction, causing the resonator to vibrate. It is not known whether Reis transmitted speech with his telephone.

Alexander Bell was a professor in phonetics and interested in the rehabilitation of deaf people. He had read a book on the physiological theory of music written by the German physicist Hermann Helmholtz in 1863. In this book the physicist described how he created sound by the action of an electromagnet on a tuning fork. Bell started experimenting himself, replacing the tuning fork by metal strips of different length, of which some would vibrate when in tune with the alternating current in the magnet.

Bell then replaced the vibrating strips by a metal diaphragm, and found that it would vibrate at all sound frequencies and thus be able to reproduce speech. In 1876 Bell filed his patent for the telephone, slightly preceding another American inventor, Elisha Gray, who filed his patent 2 hours later. A legal battle between the two ensued, but the courts decided in Bell's favor.

In Bell's telephone, the microphone and receiver were identical. Each consisted of a magnet surrounded by a solenoid with many windings, and placed close to an iron membrane. When the membrane picked up sound it caused changes in the magnetic field of the magnet, which in turn generated currents in the solenoid that followed closely the sound vibrations. These currents caused fluctuations in the magnetic field of the magnet in the receiver, causing the membrane to vibrate in turn, thus reproducing the sound wave.

Bell's telephone was usable only over distances of a few miles because the currents generated by the transmitter were very weak. In 1877 David Edward Hughes developed the more powerful carbon microphone, but it was Thomas Edison who patented in 1878 a telephone system that used such a microphone. In the carbon microphone a membrane is in contact with carbon granules that are enclosed between two electrical contacts. When the membrane vibrates, it disturbs the carbon granules, causing the electrical resistance between the two contacts to vary accordingly. Connecting such a microphone in series with a battery produces a much stronger signal

tor in the evolution of electronic technology. Diode vacuum tubes served for the detection of radio waves, and triodes became widely used for amplifying radio and audio frequencies. In a triode, a small change in the current in the middle grid controls the passage of a much larger current between the two outer grids, permitting amplification of the current that flows through the middle grid. The development of tetrodes and pentodes (tubes with one and two extra grids) increased their suitability for a range of applications. During the 1920s and 1930s, their use expanded into television, sound recording and sound film, scientific instruments, radar, and, at the end of World War II, early electronic computers. In computers vacuum tubes became part of logical circuits, such as flip-flops (circuits that can occupy two electric states, such as a toggle switch).

Energy. Coal remained the chief source of energy throughout the end of the 19th and the first half of the 20th century. For example, in 1900, 92 percent of energy in the world was supplied by coal and only 3.5 percent by petroleum. Coal mining became mechanized

with the introduction, starting in the 1860s, of coal-cutting machines that operated underground and giant excavation machinery used in open-pit mines. Coal remained the main energy source for steam power, railroad locomotives, steamships, the steel industry, and gasworks. Most of the electric power stations ran on coal, but an increasing part of electricity was generated hydraulically, especially in mountainous countries such as Italy or Canada. Gas produced from coal served mainly for lighting until the arrival of electricity, when it came into use for cooking and heating.

Toward the end of this period, vehicles powered with diesel engines began to become commonplace. In the US, General Motors was able to use its near monopoly to replace electric-powered tram lines in cities with diesel-powered buses. High-speed passenger trains also began to use diesel locomotives. Other heavy equipment, such as earth-moving equipment, gradually shifted from steam power to diesel power as well.

Materials. Chemistry, as compared with other sciences, made an enormous leap forward during the second half

than that of Bell's microphone. Edison also introduced the induction coil, a type of transformer that increased the voltage of the electric sound signals so they could travel farther.

The first commercial telephone exchange was established in 1878 in New Haven CT; it had 21 subscribers. In the early exchanges, all telephone connections had to be made manually by plugging in and out connectors in boards. Very quickly, the telephone spread in the US. In 1880, 54,000 customers rented telephones from Bell. In 1884 Boston and New York City were linked by telephone wires; in 1892, New York and Chicago; and in 1915 the first link between the East and West Coast was established. A new invention, the vacuum-tube amplifier, made such long telephone links possible.

Another important innovation contributing to the rapid spread of the telephone was the automatic exchange, invented in 1892 by Almon B. Strowger. Strowger was an undertaker in Kansas City. The wife of a competitor worked at the telephone exchange as an operator. Strowger suspected her of directing any calls concerning funeral arrangements to her husband. His automatic exchange solved that problem, and many others as well.

The automatic exchange was based on switches actuated by magnets that received pulses from push buttons placed on telephones. Soon the dial telephone automated the production of pulses. Strowger's automatic exchange quickly became adopted by telephone services all over the world and remained in use well into the 1970s, when such exchanges were replaced by electronic semiconductor systems.

Several other inventions improved telephone services. One of them was the introduction of multiplexing—transmission of several telephone conversations over one wire by using several frequency bands. The first multiplexed telephone signals were transmitted by "twisted" wires. The introduction of coaxial cable, more suitable for the transmission of high-frequency signals, made it possible to transmit several thousands of telephone conversations simultaneously through one cable.

Shortwave radio was first used for transatlantic telephone conversations. Surprisingly, the first telephone cables crossing the Atlantic Ocean were not laid until 1956. They provided 24 channels only, so getting a transatlantic call through was an event. Within 10 years, however, communication satellites with much greater signal capacity were available and transatlantic calling became almost routine.

Soon optical cables also greatly extended the capacity of telephone networks. These, along with other improvements such as the change from mechanical to electronic exchanges, have made telephone lines suitable for the transmission of fax messages and high-speed computer data. As the 20th-century was coming to a close, implementation of high-volume, high-speed electro-optical networks became a reality, although the effects of this development are yet to be known fully.

---

of the 19th century. Organic chemistry had opened up a Pandora's box of new products, especially dyes. Although the first artificial dyes were developed in England, the major development of the organic chemistry-based industry took place in Germany. Because of its colonies, a rich source for natural dyes, industrialists in England showed little interest in synthetic dyes.

From 1850 on, Germany became practically the sole manufacturer of dyestuffs. Because of fierce competition between dye-producing companies, it became important to come up continuously with new dyestuffs; this was the reason that industrial research laboratories appeared in Germany during the 1850s. In the 1880s large firms, such as the Badische Anilin und Soda-Fabrik (BASF), introduced modern industrial research methods, based on research methods at German universities. The result was that linen and cotton could be dyed in a much wider variety of colors than previously possible, and people started wearing clothes with brighter and different colors.

Probably the most important result of the flowering of the chemical industry during the end of the 19th century was the development of many new materials. Several of these materials contributed to the development of other technical fields. Celluloid, because of its transparency and flexibility, became the chosen material for the film industry. Bakelite, derived from phenol and formaldehyde, because of its properties as an electric insulator and because it is easily molded, became the material of choice for many electric and radio components, such as switches, control knobs, and vacuum-tube sockets.

Chemists, especially in Germany, set out to develop substitutes for organic materials such as rubber or fibers because of their high cost or unavailability. Scientists at German industrial laboratories discovered many plastic materials during the 19th century, but these products became commercially available only much later, in the first half of the 20th century.

The introduction of new artificial fibers had a large impact on the clothing and fashion industries. In the 1920s, rayon, also known as artificial silk, started replacing cotton in clothing, making low-cost fashionable clothing widely available. The most spectacular of new man-made fibers was nylon, discovered in 1938 by

W.H. Carothers. Nylon first became famous as the material of choice for women's stockings, but later it was used in applications such as rope and carpeting because of its intrinsic toughness.

During the first decades of the 20th century, reinforced concrete became the building material for bridges, factory buildings, military fortifications, roads, and even homes.

Medical technology. More than any other discipline, medicine profited from many discoveries in the life sciences, chemistry, and physics during this period. In physics, the discovery of X rays was probably the most important finding, not only for setting broken bones, but in the improvement of diagnostic medicine generally.

Another important finding based on physics, by Dutch physiologist Willem Einthoven, was the tiny electric currents generated by the heart muscle. By studying a graph of heartbeat-produced voltages, physicians can find abnormalities of the heart that would otherwise be difficult to detect.

Near the end of the 19th century, France and Germany entered a fierce competition for identification of the pathological agents that cause disease. The discovery of large numbers of microbes and small protozoans that cause disease was followed by the introduction of better hygienic measures for avoiding contamination. The search for the agent causing syphilis was started on a grand scale in both Germany and France. It resulted in the discovery of the microbe that causes the disease in 1906 and the subsequent Bordet-Wassermann test for syphilis. Scientists also discovered drugs that could destroy specific disease-causing microorganisms.

Alexander Fleming uncovered the most powerful agent for killing bacteria in 1928, a mold of the strain *Penicillium,* the source of the chemical penicillin. Because of the difficulty of obtaining penicillin in sufficient quantity, it became widely available only during World War II, after a major research program in the US found ways to grow the *Penicillium* mold in vats as yeast was grown. Streptomycin also became available during the war for use against tuberculosis. The availability of effective antibiotics, and a better understanding of the transmission mechanisms of disease, has caused many diseases to be nearly eradicated.

For a long time viruses, as the causative agent of diseases, remained undetected. However, in 1908 Karl Landsteiner isolated the poliomyelitis virus. The electron microscope, developed during the 1930s, made it possible to examine viruses directly. However, viruses were found to be untouched by the antibiotics that were so deadly to bacteria; such antibiotics interfere with development pathways in bacteria that are not shared by either humans or viruses. An improved understanding of the immune system became the only path to fighting viral diseases. By the end of the 19th century, several researchers, including Louis Pasteur, Robert Koch, Emil Behring, and Shibasuro Kitasato, understood the role of the immune system in fighting bacteria and had developed several vaccines. Viral vaccines became available later; the first one, against influenza, was developed in 1945.

In 1910 Peyton Rous showed that viruses play a role in cancer when he discovered chickens that suffered from certain sarcomas caused by viruses. Until the end of the 19th century, cancer often went undiagnosed, but throughout the 20th century, the incidence of all types of cancer became increasingly noted. However, the cause of cancer remained little understood, although scientists began to recognize that cancer could be caused by chemicals and by radioactivity. Surgery was the main method of treatment in the first half of the 20th century, even though the effect of X rays in destroying tumors had been discovered at the beginning of the century.

Tools and devices. During the early part of the 19th century, machines, railway cars, and engines were mostly built entirely by hand by skilled workers, using a variety of tools. Machines were used in special applications, for example, the boring of steam cylinders, mainly because machines allowed the creation of cylinders of much greater precision. The gradual introduction of several types of tooling machines not only accelerated the manufacture of metal parts and other devices, but also improved the precision with which these devices were built. The manufacture of machine tools, sometimes called "the industry of industry," became the backbone of commerce. Lathes, milling machines, and boring machines served for the manufacture of all kinds of objects, but also for the creation of better tooling machines themselves.

The first automatic tooling machines appeared around the 1870s. These machines could manufacture large numbers of simple metal parts, such as screws, nuts, and bolts. American Christopher M. Spencer developed one of the first automatic machine tools in 1873, the precursor of the so-called Hartford automats. The first automatic machine tools were still driven via belts by steam power, but later machines incorporated electric motors, making them much easier to operate. During this century engineers created many of the automatic machines that could manufacture more complicated parts, such as gears or rifle parts.

The cutting and grinding tools of these machines consisted of carbon steel that had to be much harder than the material of the workpiece. The cutting speed attainable with these tools was between 5 and 10 m (15 and

30 ft) per minute. The invention of the so-called high-speed steel allowed cutting speeds that were much higher; by 1912 cutting speeds of 30 to 50 m (90 to 150 ft) per minute were common. During subsequent years, even harder and more heat-resistant alloys were developed, allowing the manufacture of metal parts in a fraction of the time required by the older carbon-steel tools.

Transportation. The development of the automobile is often viewed as the direct consequence of the development of the internal combustion engine. Interestingly enough, most automobiles built around the turn of the century were not powered by internal combustion engines. For example, in 1900, of the 4192 cars manufactured in the US, 1681 were equipped with a steam engine, 1575 with an electric motor, and 936 with an internal combustion engine. During subsequent years, the balance tipped in favor of the gasoline engine. However, during the beginning of the time of car manufacturing, it was not clear that the internal combustion engine would win out over other engine types. Even experiments with early airplanes showed that the steam engine could be turned into a lightweight power source. Clement Ader's first airplane, in 1890, the *Eole*, was equipped with a steam engine that weighed 15 kg per horsepower. His second airplane, the *Avion III*, completed in 1897, was equipped with two small steam engines weighing 3 kg per horsepower, something that could not be attained with an internal combustion engine at the time. For example, the airplane of the Wright brothers that flew in 1903 was powered by an internal combustion engine weighing 6 kg per horsepower.

In 1908 Henry Ford equipped his Model T automobile with a gasoline internal combustion engine; it was this model, of which there were 15 million built, that set the stage for the further development of the automobile. Several improvements to the internal combustion engine made it eventually more suitable for mobile applications; among those improvements were electric ignition and the carburetor. Using the high-energy, low-weight, low-volume fuel gasoline, a car powered by an internal combustion engine did not need to stop frequently to take in water or have its battery charged.

During the first decades of the 20th century, the gasoline engine underwent further improvements. During the early years of the automobile industry, engines were limited in power output because "knock" limited the compression ratio in the cylinder. Knock is caused by the spontaneous ignition of the gas mixture before the piston has reached its highest position, causing loss of power, damage to the piston, and overheating. During the mid-1920s, gasoline that was more resistant to knock became available, and engines were built with higher compression rates and higher performance capability.

The development of airplanes was even more tightly linked to the development of the internal combustion engine. Two factors were important; the weight of the engine and the amount of power the engine could deliver. The "Gnome" engine that equipped airplanes during World War I weighed 1.5 kg per horsepower. This ratio was reduced to 0.5 kg per horsepower for airplane engines used during World War II. Supercharging—that is, the compression of air before it enters the cylinders—substantially increased the power achievable with airplane engines. Engines of a few hundred horsepower were available during World War I, but much more powerful engines were available during World War II. A supercharged 36-L V–12 engine, delivering 2000 horsepower, powered the most advanced versions of the famous Spitfire, giving it a maximum speed of 730 km (450 mi) per hour.

The speed of the Spitfire was the limit that could be achieved with a propeller because the efficiency in delivering thrust of the propeller diminishes rapidly with higher speeds. During World War II, both in England and Germany, engineers developed the turbojet engine. The turbojet engine is very efficient at delivering thrust at high speeds; it made possible the construction of very fast airplanes during the years after World War II.

Besides engines, airplanes underwent several other improvements. Instead of wood and fabric, manufacturers turned to aluminum for building airplanes, giving them a much more aerodynamic shape. Also, retractable wheels diminished the resistance of air at high speeds. The most successful plane developed before World War II was the Douglas DC–3, also known as the *Dacota*. Eventually, 15,000 DC–3 airplanes were built, and it became the workhorse of American forces during the war. Many of these twin-engined planes are still flying today. The Clipper four-engined seaplane inaugurated transatlantic passenger service in 1939; however, this service was interrupted because of the outbreak of war.

In the 20th century, up to World War II, three coexisting technologies powered commercial and navy ships: sails, traditional steam engines, and the newly developed steam turbines. A few ships were also powered by diesel engines, which soon became the most popular choice for small- and medium-sized ships after World War II.

In the 1930s, a fierce competition started among England, France, and Germany for offering fast and comfortable passenger service between Europe and the US. Each of these countries built giant ships that could cross the Atlantic Ocean in four days—the *Normandie*, the *Queen Elizabeth* and the *Queen Mary* are examples.

1879

Thomas Alva Edison in the US and Joseph Wilson Swan in England each produce carbon-thread electric lamps that can burn for practical lengths of time; Edison's lamp uses a carbon fiber derived from cotton and burns on Oct 21 for at least 13½ hours, although there is some evidence that one of Edison's first lamps of this design burned for about 40 hours; the same year 115 lamps are installed in the paddle steamer *Columbia*; *See also* 1872 GEN; 1880 GEN

Charles F. Brush installs in a public square in Cleveland OH a dozen of the carbon arc lamps he has developed; *See also* 1878 GEN

William Sugg invents the Sugg-Argand lamp; he uses coal gas for fuel in an Argand-type lamp, producing an economical light of high intensity; *See also* 1784 GEN; 1887 GEN

George Eastman [b Waterville NY, Jul 12 1854, d Rochester NY, Mar 14 1932] patents a photographic emulsion coating machine, allowing him to start mass producing photographic dry plates; *See also* 1871 COM

London gets its first telephone system as Bell Telephone opens its first exchange on Coleman Street in the City; during the same month, Thomas Edison forms the Edison Telephone Company Ltd, which opens exchanges in Lombard and Queen Victoria streets; *See also* 1878 COM; 1892 COM

Elihu Thomson demonstrates the use of induction coils to step down voltage while increasing current, employing the basic principle of the electric transformer; Thomson operates the coils in a parallel circuit, the secret of the commercially successful transformer developed 6 years later; *See also* 1851 ENE; 1883 ELE

1880

Pierre Curie [b Paris, May 15 1859, d Paris, Apr 19 1906] discovers the piezoelectric effect; certain substances produce an electric current as a result of pressure on them; also, an electric current can change the dimensions of these substances slightly

Thomas Alva Edison discovers that bamboo produces a carbon fiber of superior quality, lighting his newly developed incandescent lamps for as long as 1200 hours; *See also* 1879 GEN; 1904 GEN

Henry Faulds writes to *Nature* from his university post in Japan to point out that since fingerprints are never identical, they can be used to identify people positively and perhaps aid in criminal investigations; this sparks a report by Sir William James Herschel on his similar use of fingerprints in India and propels the British system of fingerprint identification; *See also* 1860 GEN; 1901 GEN

The New York *Daily Graphic* of Mar 4 contains the first use of a screened photograph; photographs are screened because printing smears ink on continuous tones, which do not reproduce; a screen of tiny dots indicates shading by the density of the dots; *See also* 1878 COM; 1885 COM

Alexander Graham Bell patents the photophone; a phone circuit sets a mirror in vibration and reflected Sun rays are detected by a selenium detector, thus allowing the transmission of sound by light

Fifty-four thousand telephones, rented to customers by Bell, are in use in the US; *See also* 1879 COM; 1882 COM

Jacques-Arsène d'Arsonval [b La Borie, France, Jun 8 1851, d La Borie, Dec 1940] invents an improved form of the galvanometer, now known as the D'Arsonval galvanometer; *See also* 1820 TOO; 1884 ELE

George Bailey Brayton [b 1830, d 1892] proposes a gas turbine consisting of a combustion chamber in which air is pumped by a turbine-driven pump; because of the unavailability of adequate materials, the engine never becomes operational; *See also* 1791 ENE; 1903 ENE

1879

Carl Gustav Patrik de Laval [b Orsa, Sweden, May 9 1845, d Stockholm, Feb 2 1913] patents a centrifugal separator for making butter from milk; *See also* 1890 FOO

Ira Remsen [b New York City, Feb 10 1846, d Carmel CA, Mar 5 1927] and Constatin Fahlberg [b Tambow, Russia, Dec 22 1850, d Nassau/Laun, Aug 15 1910] discover the artificial sweetener saccharin while studying coal tar; *See also* 1965 FOO

The first skyscraper, the Leitner Building, is completed in Chicago IL; a wave of Chicago skyscraper building soon follows; *See also* 1883 FOO

The Wakefield Rattan Company in the US develops a method of making wicker furniture from the fibrous pit of the rattan stalk (a plant native to Indonesia and the Philippines long used in its natural state to make rattan furniture or split to make cane furniture); unlike traditional rattan furniture, the new form, which Wakefield calls reed furniture, can be stained, painted, and, when wet, bent into scrollwork; *See also* 1920 TOO

William Siemens invents an electric furnace of the arc type, which, with modifications and improvements, is the arc furnace used today; *See also* 1893 MAT

Louis Pasteur discovers by accident that weakened cholera bacteria fail to cause disease in chickens and that chickens previously infected with the weakened virus are immune to the normal form of the virus; this paves the way for the development of vaccines against many diseases, not just smallpox; *See also* 1796 MED; 1881 MED

Dayton OH saloon keeper James Ritty, having seen how ships' logs function, uses similar principles to develop and patent the first cash register, which he called Ritty's Incorruptible Cashier; it is not a commercial success; *See also* 1878 TRA; 1884 TOO

Werner von Siemens demonstrates an electric railway 500 m (1640 ft) long at a Berlin exhibition; 2 years later he builds the world's first public-service electric street railway at Lichterfelde, a suburb of Berlin; *See also* 1880 TRA

1880

The first refrigerated meat from Australia arrives in London aboard ships; *See also* 1877 FOO; 1881 FOO

Para red, the first azo dye, is discovered; although it dyes cellulose fabrics a brilliant red, it is not very fast; a series of improvements on para red over the next 75 years form the azo family of increasingly fast and colorful dyes; *See also* 1912 MAT

Louis Pasteur's "On the extension of the germ theory to the etiology of certain common diseases" develops the germ theory of disease; Pasteur also demonstrates his findings on vaccination to the French Academy of Medicine; *See also* 1882 MED

Physician Josef Breuer [b Vienna, Austria, Jan 15 1842, d Vienna, Jun 20, 1925] treats a patient suffering from psychological disabilities by having her relate her fantasies, sometimes using hypnosis; this relieves her difficulties; Breuer tells Sigmund Freud of his experience, and Freud soon begins similar treatments for his patients; *See also* 1872 MED; 1885 MED

Major Rubin, a Swiss army officer, invents the jacketed bullet about this time; a lead core for weight is encased in a hardened metal "jacket" to protect the core from being blown apart by large charges or being deformed by air resistance or in the process of striking a target; *See also* 1846 TOO; 1884 TOO

After chairing an international conference on the idea in 1879, Ferdinand de Lesseps begins construction of a sea-level Panama Canal through Central America (locks are incorporated into the design in 1887); *See also* 1869 TRA; 1889 TRA

Thomas Alva Edison demonstrates an electric tramway in Menlo Park NJ (Edison NJ); *See also* 1879 TRA; 1888 TRA

About this time Otto Lilienthal builds a pair of connected gliders, one flying above the other, in hopes of getting more lifting power; *See also* 1877 TRA; 1896 TRA

1881

GENERAL

US President James A. Garfield, shot by an assassin on Jul 2, is given air-conditioning from a device that sucks in outside air, then passes the air over salted ice and into the sickroom; it lowers the temperature by 11°C (20°F) for a period of 58 days, using over 250 tons of ice; Garfield dies anyway; *See also* 1880 FOO; 1902 GEN

COMMUNICATIONS

Frederic E. Ives [b Litchfield CT, Feb 17 1856, d Philadelphia PA, May 27 1937] prints the first three-color halftone photographic reproduction; *See also* 1878 COM; 1885 COM

Etienne-Jules Marey develops a camera that takes many pictures of the same scene in a short period of time; it is a predecessor of the motion picture camera; Marey uses it to learn about animal locomotion; *See also* 1888 COM

ENERGY

Camille Faure improves the Planté lead-acid storage battery, making it essentially like the ones used to start automobiles today; *See also* 1868 ENE; 1900 ENE

Fernand Forest [b Clearmon-Ferrand, France, 1851, d Monaco, 1914] introduces an internal combustion engine in which the cylinder is cooled by air; air-cooled cylinders have since been used in motorcycle engines and some airplane and automobile engines

1882

GENERAL

Physicist Ernst Abbe [b Eisenbach, Germany, Jan 23 1840, d Jena, Germany, Jan 14 1905] founds the Zeiss Laboratory for Optical Technology in Jena, Germany; *See also* 1887 GEN

Percival Everitt patents in England a machine that supplies goods when you put a coin in it

Henry W. Seely invents the electric iron

COMMUNICATIONS

Lars Magnus Ericsson introduces the telephone nicknamed the "dachshund," the first telephone for which the handset combines the microphone and earpiece and rests on a cradle; *See also* 1880 COM

ELECTRONICS & COMPUTERS

Physicist Marcel Deprez [b France, 1843, d 1918] shows that electric energy can be transported over longer distances by increasing its voltage; *See also* 1891 ENE

ENERGY

The Pearl Street power station in New York City, built by Thomas Edison, brings electric lighting to the US on Sep 4; it consists of three 125-horsepower steam generators that supply current for 5000 lamps in 225 houses; earlier in the year Edison opened a similar station in London, England, the first power station ever built; *See also* 1890 ENE

The first hydroelectric power plant goes into operation in Appleton WI on Sep 30; a 107-cm (42-in.) waterwheel powers two direct-current generators that together provide about 25 kw of power; *See also* 1869 ENE; 1890 ENE

Two French engineers, Abel Pifre and Augustin Mouchot, demonstrate in Paris the use of solar energy; solar radiation, captured by a 2.2-m (7-ft) parabolic mirror, heats water in a small boiler, thus producing steam to drive a steam engine that in turn powers a printing press that prints 500 copies per hour of Pifre's newspaper, the *Soleil-Journal* (Sun newspaper); *See also* 1891 ENE

1881

Refrigerated railway cars are used in the US for transporting meat from the slaughterhouses in Chicago and Kansas City; *See also* 1880 FOO; 1918 FOO

John Froelich, an Iowa farm-machinery dealer, builds the first mechanically successful internal-combustion-engine tractor; *See also* 1837 FOO; 1901 FOO

The Society of Chemical Industry is founded in England; *See also* 1908 GEN

Henry Simon in Manchester, England, and separately Albert Hussner in Gelsenkirchen, Germany, develop a system for making high-quality coke while recovering coal gas, oil, and tar during the process; improvements on this method continue to be used today; *See also* 1858 ENE

Louis Pasteur develops the first artificially produced vaccine, against anthrax, a deadly disease that affects both animals and humans; on May 5 he successfully demonstrates that vaccination of sheep and cattle prevents their falling ill after injection with live bacteria; *See also* 1879 MED; 1885 MED

Alexander Graham Bell invents two types of metal detectors for locating bullets in the human body; one type, the induction balance, is used on President Garfield when he is assassinated

William Edward Ayrton [b London, Sep 14 1847, d London, Nov 8 1908] and John Perry invent a device that uses a band or rope around a pulley for measuring force (a dynamometer), and put forward the first electromechanical principles for using measurement of electrical resistance to obtain values for force measurements; as part of their work they also develop a brake system for railroad safety

Samuel Langley invents the bolometer, an ultrasensitive device for measuring temperature

The first electric streetcar is introduced in Berlin; *See also* 1879 TRA

Hermann Ganswindt [b Germany, Jun 12 1856, d Oct 25 1934] calculates the velocity a vehicle has to reach for escaping Earth's gravitational force; this is known as the escape velocity and varies for different heavenly bodies; for Earth its value is 11.3 km (7.2 mi) per second; *See also* 1891 TRA

1882

The largest apartment complex yet is built in New York City; it is the first to be larger than Pueblo Bonito, built by the Anasazi of western North America around 1100; *See also* 1100–09 ARC

J.M. Swen produces artificial silk from cellulose acetate; *See also* 1857 MAT; 1884 MAT

Robert Koch discovers the bacterium that causes tuberculosis, the first definite association of a germ with a specific human disease; *See also* 1880 MED; 1890 MED

Paul Ehrlich [b Strehlen, Germany, Mar 14 1854, d Bad Homburg, Germany, Aug 20 1915] introduces his diazo reaction for diagnosing typhoid fever

The modern form of the saliva ejector for draining a patient's mouth while dental work progresses is introduced; *See also* 1864 MED

The Saint-Gothard Tunnel (Paso del San Gottardo) in Switzerland is completed; it is 15 km (9.3 mi) long; the two excavation teams meet each other in the middle with an error of only 20 cm (8 in.) in the horizontal axis; *See also* 1870 TRA; 1927 TRA

The first attempt to build a tunnel beneath the English Channel is halted for political, not technical, reasons; *See also* 1990 TRA

1883

The International Convention for Patents, signed on Mar 20, is ratified and goes into effect in Jul 1884; it is subsequently revised and extended on Jun 2 1911 and Nov 6 1925; the most important provision states that if a patent is granted in one member country, subsequent patents in other countries are granted as of the date of the first patent; *See also* 1953 GEN

Consolidating rules that had developed since Parliament had taken over legislation of monopolies from King James I, a basic patent law for new inventions is passed by the British Parliament; *See also* 1620–29 GEN; 1907 GEN

George Francis Fitzgerald [b Dublin, Ireland, Aug 3 1851, d Dublin, Feb 21 1901] points out that James Clerk Maxwell's theory of electromagnetic waves indicates that such waves can be generated by periodically varying an electric current; later Heinrich Hertz demonstrates that this is true; radio waves are still generated in this way; *See also* 1866 COM; 1894 COM

Thomas Edison discovers the "Edison effect"; he introduces a metal plate in an evacuated incandescent lamp in an attempt to prevent it from turning black; it does not succeed in preventing the lamp from turning black, but Edison discovers that there is a current between the filament and a separate electrode, thus finding the basic principle of the operation of the vacuum tube; Edison, seeing no immediate application for his discovery, loses interest; *See also* 1904 ELE

Lucien Gaulard of France and John Dixon Gibbs of England together introduce an induction coil system used as an electric transformer; the coils are arranged in a series circuit; *See also* 1879 ELE; 1885 ELE

Stuart Perry [b Newport NY, Nov 2 1814, d Feb 9 1890] introduces the first steel-bladed windmill; *See also* 1854 ENE; 1927 ENE

1884

Ottmar Mergenthaler [b Hachtel, Germany, May 11 1854, d Baltimore MD, Oct 18 1899] invents the Linotype typesetting machine; the machine is equipped with a typewriterlike keyboard and produces "slugs" that consist of entire lines; *See also* 1840 COM; 1887 COM

Paul Nipkow [b Liauenberg, Germany, Aug 22 1860, d Berlin, Aug 24 1940] invents the scanning disk named after him, a precursor of television; the rotating disk has a hole through which a narrow beam of light from an object passes; its intensity is measured by a selenium photocell; *See also* 1926 COM

The ammeter is introduced in electrical engineering; *See also* 1880 ELE

Charles Algernon Parsons [b London, Jun 13 1854, d Kingston, Jamaica, Feb 11 1931] patents the first steam turbine generator for electric power; *See also* 1880 ENE; 1888 ENE

American carpenter Lester Allan Pelton [d 1908], as a result of observing a misplaced bucket on a waterwheel at a California gold mine, invents the Pelton wheel, a type of water turbine with blades shaped like divided buckets, so that the direction of a water jet directed on them is reversed entirely, thus using a maximum of the kinetic energy of the jet; *See also* 1880 ENE; 1887 ENE

1883

The ten-story Home Insurance Building is started in Chicago IL; designed by William Le Baron Jenney [b 1832, d 1907] with a combination cast-iron and wrought-iron skeleton covered by a brick exterior, it uses steel for the first time in a major way in a building; See also 1879 FOO; 1889 FOO

Metallurgist Robert Abbott Hadfield [b Sheffield, England, Nov 29 1858, knighted 1908, d London, Sep 30 1940] patents manganese steel, a superhard alloy that is the first of the specialty alloy steels; See also 1855 MAT; 1888 MAT

Sydney Ringer [b Norwich, England, 1835, d Lastingham, England, Oct 14 1910] discovers that an isolated frog heart kept in a saline solution will beat longer if calcium and potassium are added to the solution; the combination is known today as Ringer's solution; Ringer also finds that other activities of cells require calcium

Antipyrene, a powder used to reduce fever and relieve pain, is synthesized

Francis Galton's Enquiries into human faculty introduces the term eugenics and suggests that human beings can be improved by selective breeding; See also 1901 MED

Inventor Gottlieb Wilhelm Daimler [b Schorndorf, Germany, Mar 17 1834, d Kannstatt, Württemberg, Germany, Mar 6 1900] develops the first of his high-speed internal combustion engines and uses it on a boat, the first true motorboat; See also 1877 ENE; 1909 TRA

Washington Augustus Roebling [b Saxonburg PA, May 26 1837, d Trenton NJ, Jul 21 1926] completes his father's design of the Brooklyn Bridge, although from 1872 on he is confined to his apartment as a result of caisson disease; his wife Emily Warren Roebling oversees construction; the completed bridge has a main span of 486 m (1595 ft) and is dedicated on May 24; See also 1869 ARC; 1931 TRA

Albert and Gaston Tissandier launch on Oct 8 the first airship to be powered with an electric engine; See also 1872 TRA; 1884 TRA

1884

A new law in CA prohibits using water to extract ores because of the pollution of agricultural land the waste water causes

The Washington Monument is completed; See also 1450 BC ARC

Louis-Marie-Hilaire Bernigaud, comte de Chardonnet [b Besançon, France, Mar 1 1839, d Paris, Mar 25 1924], begins to produce an artificial fiber made from cellulose, which comes to be known as rayon; See also 1855 MAT; 1889 MAT

P. Böttinger discovers the dye congo red (bisazo), the first dye that can be used on cellulose fiber without treating the fiber with another chemical first; See also 1856 MAT; 1897 MAT

Surgeon Carl Koller [b Schüttenhofen, Bohemia (Czech Republic), Dec 3 1857, d New York City, Mar 21 1944] uses cocaine as a local anesthetic; See also 1890 MED

Hiram Maxim [b Sangerville ME, Feb 5 1840, d Streatham, South London, Nov 24 1916] develops the Maxim machine gun, a single-barrel, beltfed, water-cooled weapon that fires 600 shots per minute; it is adopted by the British army in 1889 and thereafter by many other armies and navies; See also 1862 TOO; 1920 TOO

The airship La France, designed, built, and piloted by Charles Renard [b Damblain, France, 1847, d Paris, 1905] and Arthur C. Krebs [b 1847, d 1935], on Aug 9 becomes the first dirigible, an airship capable of being steered well enough to return to its launch point after a flight of about 8 km (5 mi); like the Tissander brothers' airship of the preceding year, La France is powered by an electric motor; See also 1883 TRA; 1897 TRA

1884 cont

George Eastman and William H. Walker introduce the use of roll film into photography; the negative images, obtained on paper, have to be mounted on glass for reproduction; *See also* 1864 COM; 1889 COM

Lewis E. Waterman [b 1837, d 1901] patents the modern type of fountain pen; *See also* 1938 COM

Telephone wires connect Boston to New York City; *See also* 1880 COM

Nikola Tesla [b Smiljan, Yugoslavia, Jul 10 1856, d New York City, Jan 7 1943] invents the electric alternator, an electric generator that produces alternating current; *See also* 1867 ENE; 1888 ENE

1885

The first garbage incinerator in the US is installed on Governors Island in New York Harbor; *See also* 1874 TOO; 1896 GEN

The Kitson lamp uses a platinum mantle and oil under pressure; *See also* 1887 GEN

Frederic E. Ives invents the modern method called halftone printing for screened photographs, a different and much improved method from the one he used earlier; results of the two methods are alike, but the new procedure is much less complicated; *See also* 1881 COM

Charles S. Tainter [b Watertown MA, Apr 25 1854, d Apr 20 1940] develops the Dictaphone, a machine for recording dictation; *See also* 1877 COM

George Eastman patents a machine for producing continuous photographic film; *See also* 1884 COM; 1889 COM

William Stanley [b Brooklyn NY, Nov 22 1858, d May 14 1916] greatly improves the electric transformer invented by Gaulard and Gibbs by returning to a parallel instead of a series construction and by replacing the iron wires with rings and plates; *See also* 1883 ELE

James Prescott Joule [b Salford, England, Dec 24 1818, d Sale, England, Oct 11 1889] builds an internal combustion engine based on the incomplete ideal Carnot cycles that uses a porous piston through which the exhaust escapes; this engine is on the line of development that eventually leads to the diesel; *See also* 1890 ENE

Leonart Jennett Todd patents a steam engine with a so-called "terminal exhaust" cylinder; it has two hot inlets and a common central cold exhaust; the cylinder acts as the exhaust valve; the engine is more efficient than previous steam engines because the cylinder is kept hot at both ends; this type of design later becomes known as the uniflow engine because steam flows in one direction through the cylinder; *See also* 1827 ENERGY; 1908 ENERGY

1884 cont

MATERIALS

French chemist Paul Vielle develops the first smokeless gunpowder that is reliable enough for military purposes; previous smokeless gunpowders were useful only in shotguns; *See also* 1889 MAT

Carl Zeiss [b Weimar, Germany, Sep 11 1816, d Jena, Germany, Dec 3 1888] develops a type of glass that resists heat; the resistance is conferred by adding boric acid to the silicon-dioxide mixture used to make glass; *See also* 1915 MAT

TOOLS & DEVICES

John Patterson buys the rights to Ritty's Incorruptible Cashier and forms the National Cash Register Company; an aggressive sales campaign results in sales of a million cash registers by 1911; *See also* 1879 TOO

TRANSPORTATION

The 65-km (40-mi) Danube-Black Sea Canal, built by Romania, is opened, providing a shortcut around the Danube delta that saves 240 km (150 mi); *See also* 1848 TRA

Frank Julian Sprague [b Milford CT, Jul 25 1857, d Oct 25 1934] develops a direct-current motor for electric locomotives; *See also* 1881 TRA; 1895 TRA

1885

FOOD & SHELTER

John B. Meyenberg of Highland IL introduces canned evaporated milk

Bordeaux mixture (a combination of copper sulfate and hydrated lime dissolved in water) is discovered by French horticulturist P.M.A. Millardet to control a serious fungus disease of grapes; invented by wheat farmers to prevent a fungus disease of seed, Bordeaux mixture is blue-green from copper sulfate and had been applied to the grapes in Bordeaux as a dye to discourage grape thieves; Bordeaux mixture is quickly discovered to control other fungal diseases, such as late potato blight (cause of the Irish famine of 1845-1846); it is still used today; *See also* 1845 FOO; 1896 FOO

MATERIALS

French chemist Eugène Turpin discovers how to load artillery shells with picric acid, which explodes on impact (a "bursting charge"); *See also* 1871 MAT

MEDICAL TECHNOLOGY

Louis Pasteur develops a vaccine against hydrophobia (rabies) and uses it to save the life of Joseph Meister, a young boy bitten by a rabid dog; *See also* 1881 MED; 1890 MED

Austrian psychiatrist Sigmund Freud [b Freiberg, Germany, May 6 1856, d London, England, Sep 23 1939] studies hypnotism with Jean Martin Charcot, the beginning of his path toward the development of psychoanalysis; *See also* 1880 MED; 1893 MED

Swiss surgeon Emil Theodor Kocher [b Aug 25 1841, d Jul 27 1917] develops surgical removal of the thyroid as a cure for goiter about this time

Dr. Scott, an American dentist, patents, 75 years before they become common, an electric toothbrush; it does not become a commercial success because the device is too noisy and too expensive

TOOLS & DEVICES

Dorr Felt [b Beloit WI, Mar 18 1862, d Chicago IL, Aug 7 1930] develops the Comptometer, a key-driven adding and subtracting machine; *See also* 1889 TOO

TRANSPORTATION

Karl Benz builds a three-wheel automobile powered by a gasoline engine; *See also* 1877 ENE; 1889 TRA

The Rover Company, following the concepts of Harry J. Lawson, is the first to manufacture a vehicle that would be recognized today as an ordinary bicycle, except that it lacks air-filled rubber tires; *See also* 1876 TRA; 1888 TRA

Gottlieb Daimler installs one of his internal combustion engines on a bicycle, creating the world's first motorbike; *See also* 1883 TRA; 1901 TRA

The first transcontinental rail link across Canada is opened; *See also* 1869 TRA; 1904 TRA

1886

French criminalogist Alphonse Bertillon [b Paris, Apr 23 1853, d Paris, Feb 13 1914] develops the system of identification named for him; the Bertillon system uses various physical characteristics, especially head length to identify individuals, especially criminals; fingerprints are included as part of the system; *See also* 1880 GEN; 1901 GEN

Alexander Graham Bell uses wax cylinders for recording sound with a modified version of Edison's phonograph; *See also* 1877 COM; 1888 COM

R.S. Waring introduces coaxial cables for telephone lines to reduce interference from electric power lines

---

Louis Pasteur

Louis Pasteur's career began by solving the outstanding problem in chemistry of his day: Why did two substances that were the same chemically react differently to light? His work for his doctoral degree, when he was only 26, led to the understanding that crystals can exist in right-handed and left-handed forms, a finding with implications that are still being worked out today. In this, as in many other enterprises, luck in choice of materials helped Pasteur. Of course, it was not fortune alone that made him a great scientist. As he said himself, "Chance favors the prepared mind."

Pasteur became the dean of chemistry at the French University of Lille in 1854. In 1856 an industrialist asked Pasteur to investigate why beer and wine often turn sour when aging. Such leading chemists of the time, as Justus Liebig and Friedrich Wöhler, believed that fermentation was a chemical reaction. Pasteur showed that both fermentation and spoiling are caused by minute living organisms. Pasteur also developed a way to prevent wine spoiling, in a process now called pasteurization. Heating wine to about 50°C (120° F) kills microorganisms that might spoil it.

Pasteur proceeded to study interactions of microorganisms and other organic materials. He showed, for example, that boiled meat extract does not spoil unless exposed to dust particles in air. In 1865 Pasteur investigated a disease that was killing silkworms. He was not able to isolate the microorganism, but suggested killing the infected silkworms and starting new cultures free of infection.

Pasteur had recognized that infectious diseases are caused by microorganisms. This concept soon led others to introduce sterilization, disinfection, vaccines, and eventually antibiotics.

Pasteur himself developed several vaccines, including those for rabies and anthrax. He created weakened microorganisms by heating them. They could not cause the disease but evoked immunity to these germs in patients. Pasteur's work was crowned by the successful inoculation of a boy bitten by a rabid dog in 1885.

1887

Karl Auer [b Vienna, Austria, Sep 1 1858, becomes Baron von Welsbach 1901, d Treibach, Aug 4 1929] invents the mantle for gas lamps, a cotton net soaked in thorium oxide that emits an extremely white light when heated by a Bunsen burner (similar to the Coleman white-gas lamps often used by campers today); *See also* 1879 GEN; 1885 GEN

Mansfield Merriman [b Southington CT, Mar 27 1848, d New York City, Jun 7 1925] founds the first experimental laboratory for hydraulics; *See also* 1882 GEN; 1898 GEN

Western Union is the first to apply multiplexing on telephone lines by using very low frequencies for the transmission of telegraphic signals and other frequencies for the transmission of voice; *See also* 1874 COM

Tolbert Lanston [b 1844, d 1913] patents the Monotype typesetting machine; *See also* 1884 COM; 1949 COM

The effect of light on a spark gap, the first hint of the photoemissive effect, is discovered by Heinrich Hertz [b Hamburg, Germany, Feb 22 1857, d Bonn, Jan 1 1894], who observes that ultraviolet light affects the length of the spark; the photoemissive effect is the form of photoelectric effect in which electrons are emitted from a metal struck by light; it is sometimes confused with a different photoelectric effect, the change in conductivity caused by light, which is the photovoltaic effect; *See also* 1888 ELE

Carl Gustav Patrik de Laval develops a small steam turbine rotating at 42,000 revolutions per minute; *See also* 1884 ENE; 1888 ENE

1886

John S. Pemberton invents Coca-Cola

Thomas Crapper [b England, 1837, d Jan 27 1910] invents what amounts to a modern form of flush toilet, with the tank of reserve water a good height above the bowl and a chain and lever system that allows a charge of water to both expel and dilute the material in the bowl; he also includes a bend in the pipe leading to the sewer so that there is only clean water in the pipe after the fouled water passes on to the sewer; *See also* 1778 ARC; 1889 FOO

Gustave Eiffel [b Dijon, France, Dec 15 1832, d Paris, Dec 27 1923] starts building the tower named after him; it is assembled from 12,000 prefabricated metal parts and is built in 2 years without any worker's loss of life; *See also* 1889 FOO

The Statue of Liberty is unveiled in New York Harbor; *See also* 300 BC ARC

A major source of gold ore is found in southern Africa at the Witwatersrand gold reef in the Transvaal

Hamilton Young Castner [b Brooklyn NY, 1859, d 1899] invents the sodium process of aluminum production; *See also* 1854 MAT

Charles M. Hall [b Thompson OH, Dec 6 1863, d Daytona Beach FL, Dec 27 1914] and Paul-Louis-Toussaint Héroult [b Thury-Harcourt, Calvados, France, Apr 10 1863, d near Antibes, France, May 9 1914] independently discover an economical way to make aluminum from abundant alumina and electric power; *See also* 1854 MAT; 1887 MAT

Ernst von Bergmann introduces steam sterilization of surgical instruments in his Berlin clinic; *See also* 1890 MED

Inventor Elihu Thomson [b Manchester, England, Mar 29 1853, d Swampscott MA, Mar 13 1937] patents a welding method that uses heating through electrical resistance; *See also* 1898 MAT; 1901 TOO

1887

Anna Connelly patents the fire escape

The first commercial production of aluminum starts in Neuhausen, Switzerland, using electric energy obtained from hydraulic generators powered by the Rhine River falls; *See also* 1886 MAT; 1889 MAT

Hannibal W. Goodwin invents celluloid photographic film and develops a production process for it

F.A. Muller, a German glassblower, develops the first form of contact lens, which covers the whites of the eye as well as the cornea; *See also* 1784 MED

G.B. Grant introduces the first successful machine for cutting gears; he uses a cutting device called a hob that is fed across the revolving gear blank; gears are still made with the descendants of this machine and of the gear shaper of about a decade later; *See also* 1861 TOO; 1896 TOO

Gottlieb Daimler uses his internal combustion engine to power a four-wheeled vehicle, one of the first automobiles; *See also* 1885 TRA

Siemens & Halske introduce the electrically driven elevator at the industrial exhibition in Mannheim, Germany; *See also* 1867 TRA

The Pullman Palace Car Company begins making a vestibule car, with an enclosed area at one end; *See also* 1868 TRA

1888

Brazil becomes the last nation to outlaw slavery, finally ending the slave trade between Africa and the New World; *See also* 1862 GEN

Heinrich Hertz detects and produces radio waves for the first time; radio waves are called Hertzian waves until renamed by Guglielmo Marconi, who calls them radiotelegraphy waves; *See also* 1883 COM; 1894 COM

Thomas Edison and William Dickson make a sound "motion picture"; a phonograph controls a Kinetoscope, a cylinder with a series of photographs on it for synchronization of sound and image; the photographs are viewed through a magnifying lens and, as in later cinema and television, persistence of vision gives the illusion of motion; a description is filed with the US Patent Office on Oct 8; *See also* 1891 COM

Emile Berliner [b Hanover, Germany, May 20 1851, d Washington DC, Aug 3 1929] invents the flat disk form of the phonograph, an improvement over Thomas Alva Edison's cylinder system that is quickly adopted by the record industry; *See also* 1877 COM

In Jun, George Eastman introduces the first commercial paper roll-film camera, a box camera that he names the Kodak; it costs $25 and takes 100 pictures; this is a high price, and the original Kodak is something of a luxury item; the entire camera with film inside is returned for processing; *See also* 1889 COM

Oberlin Smith describes in an American journal a magnetic sound recording system, a precursor of today's magnetic tape recorder; *See also* 1893 COM

The release of negative electricity as a result of shining ultraviolet light on zinc, discovered by Wilhelm Hallwach [b Darmstadt, Germany, Jul 9 1859, d Dresden, Germany, Jun 20 1922], offers the first confirmation of the photoemissive effect discovered by Heinrich Hertz; *See also* 1887 ELE; 1899 ELE

Oliver Schallenberger invents an electric meter for measuring amounts of alternating current; *See also* 1884 ELE

Nikola Tesla invents an alternating current induction motor; *See also* 1884 ENE

Charles Van DePoele [b Lichtervelde, Belgium, Apr 27 1846, d Lynn MA, Mar 18 1892] introduces carbon contacts in electric motors instead of metallic brushes to drain current from commutators, reducing sparking substantially

The first Parsons steam turbine driving an electric generator is installed at Forth Banks in England; *See also* 1884 ENE; 1895 ENE

Carl Gustav Patrik de Laval invents the impulse turbine; *See also* 1887 ENE; 1890 ENE

Thomas Alva Edison

Thomas Alva Edison is the most successful and well known inventor of all time, with almost 1300 patents to his name. His most famous inventions are the incandescent lamp, the phonograph, and motion pictures, but he also contributed inventions for telegraph systems.

Others also invented the devices we associate with Edison, but he made them better and he got people to use them. His success began with hard work; for example, he performed thousands of experiments to find a suitable filament that would resist the intense heat that results when producing light. However, this approach did not always yield results. Edison also tried 8000 designs for a new storage battery, this time in vain.

Edison showed his enterprise early in life. As a youth, while working as a newspaper boy on a train, he set up a chemical laboratory on the train. He financed his experiments by running a printing press on the train and by operating a fruit and vegetable business. He picked up goods at station stops and sold them at the terminal stop, usually in a large town.

Another factor that made Edison a successful inventor was a keen interest in anything mechanical. As a teenager he worked as a telegraph operator and was repeatedly fired because of his constant tampering with the equipment.

Theophilus Van Kennel patents the revolving door on Aug 7 in Philadelphia PA

Ernst Abbe and Otto Schott develop barium glass that has a very high refraction index and that allows the construction of achromatic lenses

John Wesley Hyatt develops thin films of celluloid as a replacement for glass as the base for plates in photography; *See also* 1869 MAT; 1904 MAT

Chemist George W. Kahlbaum [b Berlin, Germany, Apr 8 1853, d Basel, Switzerland, Aug 28 1905] fabricates plastic bottles from metacrylate, in one of the first uses of plastic; *See also* 1975 MAT

William J. Keep [b Oberlin OH, Jun 3 1842, d Sep 30 1918] discovers that the various types of cast iron identified by workers in the industry are determined by the amount of silicon in the alloy; *See also* 1883 MAT

Alfred Nobel develops ballistite, the first smokeless blasting powder, but in Great Britain, where cordite (essentially the same thing) is invented in 1889, the patent is contested in 1894 and 1895 and the British courts rule against Nobel; *See also* 1875 MAT; 1889 MAT

William S. Burroughs [b Rochester NY, Jan 28 1855, d Citronelle AL, Sep 14 1898] patents an adding machine; *See also* 1885 TOO; 1889 TOO

Charles Vernon Boys [b Wing, Rutland, England, Mar 15 1855, d Andover, England, Mar 30 1944] develops a very sensitive radiometer; *See also* 1878 TOO

John Boyd Dunlop [b Ayrshire, Scotland, Feb 5 1840, d Dublin, Oct 23 1921] introduces air-filled rubber tires in England; *See also* 1885 TRA

Frank J. Sprague operates the first commercial American tramway on a 27 km (17-mi) line in Richmond VA; *See also* 1880 TRA; 1892 TRA

Thomas Alva Edison (Continued)

Edison's success is often attributed to his business sense, but this is open to question. His first major patent, a vote recorder for Congress, did not rouse interest—one congressman told Edison that legislators wanted to keep voting records vague. He then vowed to make only inventions for which there was demand. Although correct about the electric light, his efforts with electric automobiles were superseded by the internal combustion engine. In 1878 he suggested these uses for his phonograph: a dictating machine for letter writing, spoken books, teaching of speech, reproduction of music, archiving voices of famous people, music boxes and toys, speaking clocks, study of language, educational recordings, and transmission of recorded messages over the telephone. Not only is recording of music halfway down the list, but initially Edison resisted even the idea. Possibly, his partial deafness was a factor in his failure to recognize the importance of recorded music; in any case, all of his other ideas have come to pass, although nearly always using magnetic tape technology instead of his purely mechanical method.

There are other examples of Edison's inability to foresee the uses to which his inventions might be put. He fought against projection for motion pictures, and believed that more money was to be made on peep shows than in movie houses. Consequently, although Edison showed that motion pictures were possible and made some of the early films, credit for the cinema outside of the US is usually granted to the Lumière brothers of France, who showed motion pictures to small audiences from the first.

Where Edison's business acumen was more evident was in the development of manufacturing and distribution related to his inventions. The electric generator had been available for decades when Edison opened the first power plants in London and New York City. He correctly recognized that his light bulbs would be of no use unless electric power was available and that money was to be made not only by selling the bulbs, but also by supplying the electricity.

Edison was not much interested in science for its own sake. In 1883 he discovered the basic principle of the vacuum tube, still known as the Edison effect, but paid no attention to something for which he failed to see a use.

1889

As provided in the Treaty of the Meter, the International Bureau of Weights and Measures at Sèvres, France, establishes as prototypes the international meter and the international kilogram, physical standards wrought of a special platinum-iridium alloy and kept under lock and key; scientific and other measures are to be based on these prototypes under rigidly decreed conditions, such as temperature and position on Earth's surface; *See also* 1875 GEN

British inventor William Friese-Greene [b Bristol, Sep 7 1855, d London, May 5 1921] patents a camera that takes ten photographs per second on a paper roll film; filmed scenes at Hyde Park Corner are probably the first cinematic pictures ever projected (Edison's Kinetoscope did not involve projection); *See also* 1888 COM

George Eastman produces a transparent version of roll film that is much simpler to work with than his previous paper roll film; this soon leads to a roll-film camera that is easy to use and opens the way for popular photography; *See also* 1888 COM

On Nov 23 Louis Glass installs a phonograph with four listening tubes in his Palais Royale Saloon in San Francisco CA; for a nickel, a customer can put an ear to the tube and hear a song; this is the ancestor of the jukebox; *See also* 1888 COM; 1927 COM

Fernand Forest builds a four-cylinder internal combustion engine; *See also* 1885 TRA; 1890 ENE

"'Bout as high as a building ought to go . . ."

The skyscraper mentioned in a song from the musical *Oklahoma* was in Kansas City MO and was seven stories high. About the same time as the setting for the show, another midwestern city, Chicago IL, boasted the earliest buildings recognized as skyscrapers, two or three times as high.

The development of the skyscraper is often viewed as a consequence of the elevator. Elevators did permit buildings as tall as ten to sixteen stories, but lower floors were almost unusable. Concrete walls had to be 6 m (20 ft) thick to support the upper stories. The true key to the skyscraper was the steel frame that then supported the walls and floors. Even so, the tallest building to the end of the 19th century, the Park Row Building in New York City, was only 132 m (435 ft) high, half as tall as the Eiffel Tower, built around the same time. Also, early hydraulic elevators were only effective up to about twenty stories. Modern high-speed cable elevators, when introduced in 1900, greatly facilitated the construction of what we would view today as skyscrapers.

William Le Baron Jenney in Chicago first used a mixture of cast iron and steel to support a tall office building, the Home Insurance Building. Soon he and other Chicago architects were using all-steel frames. What they mounted on the frames, however, was often a tall version of a Classic Revival building. Louis Sullivan, the Chicago architect who designed the spare, modern looking buildings and who is best known today, was not a success during most of his career. His turn-of-the-century buildings that showed clearly the influence of the steel frameworks in vertical and horizontal lines were not acclaimed at the time, and Sullivan died in poverty in 1924 after years with no work. His influence lived on, however, as a result of his teaching.

In the 20th century, most of the much taller skyscrapers were in New York City, notably the Flatiron, Woolworth, Chrysler, and Empire State buildings. But Chicago also had its share of famous skyscrapers. Today Chicago still has the tallest building, the Sears Tower.

A hotel in Bernina, Switzerland, supplied with electricity from a waterwheel, installs an electric oven; *See also* 1890 FOO

Gustave Eiffel's tower in Paris, named after him, at 303 m (993 ft), is the tallest free-standing structure built to that time; *See also* 1886 FOO

The Rand McNally Building in Chicago IL is the first to use steel with no cast iron in its framework throughout; *See also* 1883 FOO; 1892 FOO

Machine Hall in Paris, built from glass panels over a frame of steel girders, permits a large interior space; *See also* 1851 ARC

The Auditorium Building in Chicago IL is one of the first major buildings to use forced-air ventilation

The "wash-down" toilet, similar to the modern flush toilet with a ball float in the tank, is invented; *See also* 1886 FOO

The Pittsburgh Reduction Company introduces the first aluminum kitchen utensils; *See also* 1887 MAT

Frederick Augustus Abel [b Woolwich, England, Jul 17 1827, d London, Sep 6 1902] and James Dewar invent cordite, smokeless gunpowder; *See also* 1888 MAT

The artificial fiber rayon, a form of cellulose manufactured by the comte de Chardonnet, wins the Grand Prix at the Paris Exposition; *See also* 1884 MAT; 1890 MAT

US chemist Francis Despard Dodge develops the alcohol citronellol with a rose or geranium odor and citronellal with various flower odors; these chemicals advance the artificial perfume industry; *See also* 1921 MAT

An air lock, called a hospital lock, is introduced to eliminate caisson disease (nicknamed the bends) as part of an abortive attempt to build a tunnel under the Hudson River at New York City; by providing a place for slow decompression for workers from the pressurized, underwater caissons, the hospital lock completely eliminates the disease, which previously had affected as many as one worker in four; *See also* 1779 ARC; 1907 MED

Baked porcelain inlays for lost teeth are introduced; *See also* 1890 MED; 1907 MED

Léon Bollée builds the Millionaire, the first mechanical calculator to include a built-in multiplication table; previous calculators by Leibniz and Colmar used repeated addition for multiplication, while Schickard's calculator was based on Napier's bones, which used the principle of the slide rule for multiplying; the Millionaire remains in production until 1935; *See also* 1820 TOO; 1892 TOO

Dorr Felt's Comptometer is equipped with a built-in printer; *See also* 1885 TOO

Karl Benz and Wilhelm Maybach [b Heilbronn, Germany, Feb 9 1846, d Dec 29 1929] build a four-wheeled vehicle propelled by an internal combustion engine; its design is based on the three-wheeled automobile built by Benz, but it is viewed as the prototype of the modern automobile; *See also* 1885 TRA; 1895 TRA

The Forth railway bridge over the Firth of Forth in Scotland, designed by John Fowler [b Sheffield, England, Jul 15 1817, d Bournemouth, England, Nov 20 1898], is built using over 50,000 tons of steel; it is the first use of steel in a large and important structure; the amount of painting it takes to keep the bridge from rust has been a stock joke in Britain ever since; the bridge is still in use; *See also* 1779 ARC; 1904 TRA

Ferdinand de Lesseps abandons the Panama Canal project when the first French company formed to build the canal goes bankrupt; *See also* 1880 TRA; 1894 TRA

Otto Lilienthal publishes *Der Vogelflug als Grundlage der Fliegekunst* (Bird flight as a basis for aviation), which becomes an influential book for aviation pioneers; *See also* 1877 TRA; 1895 TRA

1890

Sven Foyn introduces the factory ship for hunting and processing whales; *See also* 1864 TOO

William Kemmer becomes the first person to be executed in the electric chair, a device developed by Harold P. Brown and E.A. Kenneally that uses alternating current; Thomas Alva Edison had arranged for prisons to have alternating current for the electric chair as an element in his fight to have direct current adopted for household use

A heliograph system is used by the US Army to send signals from mountaintop to mountaintop over a distance of 350 km (215 mi); *See also* 1861 COM

Paul Rudolph develops the Protar Zeiss photographic lens; it is derived from the symmetrical Dahlmeyer lens, and consists of a pair of divergent and convergent lenses; *See also* 1899 COM

Edouard Branly [b Amiens, France, Oct 23 1844, d Paris, Mar 24 1940] invents the coherer, a small glass tube filled with iron filings; normally, the resistance of the iron filings is great, but in the presence of electromagnetic radiation, the resistance becomes very small; *See also* 1894 COM

Herman Hollerith [b Buffalo NY, Feb 39 1860, d Washington DC, Nov 17 1929] develops an electrically driven census system based on punched cards; the US uses this system in 1890, the Russians in 1895, and the Austrians in 1900; *See also* 1896 ELE

The first hydroelectric plant to produce alternating instead of direct current electricity is built on the Willamette River near Oregon City OR; *See also* 1882 ENE; 1894 ENE

Akroyd Stuart designs an internal combustion engine functioning at low pressure that uses kerosene as a fuel and that relies on the heat of the wall of the cylinder to vaporize the fuel and to ignite the air-fuel mixture; this is an important predecessor of the diesel engine; *See also* 1885 ENE; 1893 ENE

Carl Gustav Patrik de Laval builds a steam turbine consisting of a single disk rotating at 30,000 revolutions per minute; *See also* 1888 ENE

Alexander Graham Bell

Alexander Graham Bell's interest in speech therapy, which he inherited from his father and grandfather, led to the telephone. At age fifteen, Bell left his native Scotland and went to live in London with his grandfather. There he acquired knowledge of the mechanisms of speech and sound. After a short period of teaching in Bath, England, Bell and his parents moved to Canada. In 1871 Bell started teaching a phonetic system for the deaf devised by Bell's father, called "visible speech," to teachers of the deaf in Boston.

Bell had read a book written by Hermann Helmholz in 1863 on the physiological theory of music in which Helmholz described how sounds could be transmitted by electrical currents and magnets that caused metal reeds to vibrate. Bell started his own experiments, producing sound from metal reeds by sending rapidly varying currents through electric magnets. This work resulted in the "multiple telegraph," a telegraph system that could transmit several messages simultaneously over one wire by using different frequencies. Bell patented the multiple telegraph in 1875.

In 1876 he replaced the metal reeds by a metal membrane and found that he could reproduce any sound with it. The first transmission of a voice message happened inadvertently. His assistant, Thomas A. Watson, who was in an adjoining room, heard Bell say clearly through the telephone, "Mr. Watson, come here, I want you." A few months later Bell talked to his father over a telephone line of 13 km (8 mi). Around the end of the year, he had extended the length of this telephone line to 230 km (143 mi). Bell patented his telephone in 1876, and its commercial application spread quickly through the country.

In 1880 Bell invented the photophone. He used sound to modulate a light beam and a selenium crystal as receiver. The electrical resistance of selenium varies with the intensity of light falling on it. The photophone was the first wireless way of transmitting speech, but it was limited to line-of-sight uses. Today telephone signals are also transmitted by light, not through the air, but through optical fibers.

Bell also improved Thomas Edison's phonograph by using wax cylinders instead of cylinders of metal foil. He developed devices for detecting metal fragments in the human body. One type of metal detector was tried out on US President Garfield when he was shot in 1881. Bell also became interested in flight and experimented with several types of large kites.

The Carpenter Electric Heating Manufacturing Company in Saint Paul MN begins selling electric ovens; *See also* 1889 FOO; 1897 FOO

The Exchange Buffet in New York City introduces the practice of standing at a shelf while eating in a restaurant instead of sitting at a table; *See also* 1876 FOO; 1902 FOO

Stephen M. Babcock [b Bridgewater NY, Oct 22 1843, d Jul 2 1931] develops a test to measure the fat content of milk; *See also* 1879 FOO

Louis-Henri Despeissis patents a method for making an artificial cellulose fiber that begins with the cellulose dissolved in a solution of copper ammonium; this method was probably first developed by M.E. Schweitzer in 1857 and perhaps was known even earlier than that; Despeissis's patent leads to commercial production of rayon by a similar process starting in 1898; *See also* 1889 MAT; 1892 MAT

Surgeon William Halsted [b New York City, Sep 23 1852, d Baltimore MD, Sep 7 1922] introduces the practice of wearing sterilized rubber gloves during surgery at Johns Hopkins Hospital in Baltimore; *See also* 1886 MED; 1896 MED

Emil von Behring [b Deutsch-Eylau, Germany, Mar 3 1854, d Marburg, Germany, Mar 31 1917] develops a vaccine against tetanus and diphtheria and introduces the concepts of passive immunization and antitoxins; *See also* 1885 MED

About this time physicians in Australia discover that children are being poisoned by ingesting old paint that has turned to powder; *See also* 1768 MED; 1917 MED

Robert Koch announces the discovery of tuberculin as a cure for tuberculosis; *See also* 1882 MED

Cocaine is used as a local anesthetic in dentistry; *See also* 1884 MED; 1904 MED

Benzocaine is developed in Germany; it is a local anesthetic that is given the trade name Anesthesin; *See also* 1884 MED

The porcelain jacket crown to cover lost portions of teeth is introduced by C.H. Land; the dentist removes the enamel from the tooth and mounts a porcelain shell baked onto a platinum matrix over what is left of the tooth; *See also* 1889 MED; 1907 MED

Albert Charles Pain [b 1856, d 1929] patents a system of forced lubrication in machines; oil is conducted to joints and bearings through small canals; the system is now universally used in all kinds of machines; *See also* 1865 MAT

C. Northrop develops an automatic weaving machine; *See also* 1775 TOO; 1920 TOO

Clément Ader's *Eole* is the first full-size aircraft to leave the ground under its own power; driven by a 20-horsepower steam engine, it flies a distance of about 60 m (200 ft) carrying its inventor as pilot, although not very high off the ground at any time; Ader never has control of the craft and it crashes on landing, which is why the invention of the airplane is seldom credited to him; *See also* 1877 TRA; 1897 TRA

1891

Thomas Edison develops a system of taking motion pictures on the recently developed celluloid film of George Eastman, but his patent application of Aug 24 (intended to replace his description of the cylinder device filed with the Patent Office in 1888) fails to suggest projecting the images; the strip of film is to be viewed through a magnifying glass as the film is wound from one reel to another inside a box; *See also* 1881 COM; 1895 COM

French physicist Gabriel Jonas Lippmann [b Hollerich, Luxembourg, Aug 16 1845, d Atlantic Ocean, Jul 13 1921] presents a paper to the French Academy of Science in which he describes his color photography method that uses a thick emulsion over a mercury surface (used as a reflector) that captures the incident waves of light; the method is not very practical as the colors can only be seen at a right angle; *See also* 1810 COM; 1908 COM

The International Electrical Exhibition is held in Frankfurt, Germany

Nikola Tesla invents the Tesla coil, which produces high voltage at high frequency; *See also* 1882 ENE

The first long-distance high-voltage line for carrying electricity is completed between Lauffen and Frankfurt am Main, Germany; at 8000 volts it bridges a distance of 177 km (120 mi); *See also* 1882 ELE; 1931 ENE

Clarence M. Kemp of Baltimore MD patents the first commercial solar water heater; *See also* 1882 ENE; 1977 ENE

1892

The defeat of a five-month strike against the Homestead Steel Works, Andrew Carnegie's plant near Pittsburgh PA, effectively ends the labor movement in the US steel industry for a generation; *See also* 1877 GEN; 1894 GEN

Leon Arons [b Berlin, Germany, 1860, d 1919] designs the first full-fledged mercury-vapor lamp, but does not manufacture it for general use; *See also* 1901 GEN

Henri Moissan describes the use of acetylene gas for lighting; about the same time Thomas L. Willson also describes the production and use of acetylene to produce a brilliant white light; *See also* 1859 MAT; 1897 MAT

Almon B. Strowger, an undertaker, invents the first practical telephone switching system, in which the caller determines who the receiver will be without the aid of a human operator; according to legend, Strowger is driven to this invention when the switchboard-operating wife of his chief rival begins to misdirect calls for Strowger to his chief business rival; Strowger's invention leads to the dial telephone; *See also* 1876 COM

Thomas Oliver patents and produces the first practical typewriter in which the operator is able to see the words as they are typed; *See also* 1872 COM

New York City and Chicago, 1500 km (900 mi) apart, are linked by telephone; *See also* 1879 COM

1891

In May, Carnegie Hall in New York City, the first US auditorium designed for musical performance (by architect William Burnet Tuthill), opens, with 2247 seats and 65 boxes; Pytor Ilich Tchaikovsky conducts the opening program; *See also* 1895 GEN

Edward Goodrich Acheson [b Washington PA, Mar 9 1856, d New York City, Jul 6 1931], while trying to make artificial diamonds, accidentally discovers a process for making carborundum (silicon carbide), a material almost as hard as diamond; *See also* 1929 MAT

An antitoxin for diphtheria is tested for the first time on humans; *See also* 1901 MED

Paul Ehrlich uses methylene blue for treating malaria; *See also* 1660–69 MED; 1934 MED

George Redmayne Murray treats myxedema (the common form of hypothyroidism) with a hormone extracted from the thyroid gland

John Wesley Hyatt, best known for his invention of celluloid, develops and markets the Hyatt roller bearing, still used in modern machines; *See also* 1490–99 TOO

Reinhard Mannesmann [b May 13 1865, d 1922] patents a method of producing seamless tubes on special rolling mills, invented a few years earlier; *See also* 1856 TOO

The electric fan is invented

A. Blondel invents the oscilloscope; *See also* 1893 TOO

Karl Elsener devises the Swiss army knife

In a public lecture about spaceflight, eccentric German inventor Hermann Ganswindt proposes a spaceship that would be propelled by firing a cannon in the direction opposite to the one in which a person wishes to travel; *See also* 1881 TRA

1892

Marquis, a new strain of wheat, is formed by the crossing of a Canadian type with an Indian strand; it is now the most commonly cultivated variety; *See also* 1900 FOO

The 21-story Masonic Building in Chicago IL uses light, thin screens between columns and beams in a modified form of curtain wall construction; *See also* 1849 ARC; 1918 FOO

An artificial fiber, often called viscose rayon, made from cellulose zanthate by the team of Charles Frederick Cross [b Brentford, England, Dec 11 1855, d Hove, England, Apr 15 1935], Edward J. Bevan [b Birkenhead, England, 1856, d 1921], and C. Beadle using their "viscose process," is the foundation of much of the artificial fiber industry that follows; *See also* 1890 MAT

François Hennebique [b 1843, d 1921] improves reinforced concrete by introducing steel bars laid out into three rectangular frames at right angles to each other; the frame makes the material more resistant to bending, traction, and shearing forces; *See also* 1877 MAT

Robert Koch introduces filtration of water for controlling a cholera epidemic in Hamburg, Germany; *See also* 1866 MED

Biologist Dmitri Ivanovsky [b Gdov, Russia, Nov 9 1864, d USSR, Jun 20 1920] shows that viruses exist

James Dewar [b Kincardine, Scotland, Sep 20 1842, d London, Mar 27 1923] invents the Dewar flask or Dewar bottle, the first form of thermos bottle; the Dewar bottle (and the thermos bottle) is a double-walled flask with a vacuum between the walls; the walls are silvered to reflect heat radiation and the vacuum prevents heat passing through by conduction or convection— thus, cold materials in the flask stay cold, which was Dewar's goal, and hot materials stay hot as well; *See also* 1865 TOO; 1898 MAT

William S. Burroughs introduces an adding-subtracting machine with printer; *See also* 1888 TOO

The first underground railway (subway) and elevated train system (the el) in Chicago starts operation; *See also* 1868 TRA; 1896 TRA

Konstantin Tsiolkovsky [b Izheskaye, Russia, Sep 17 1857, d Kaluga, Russia, Sep 19 1935] starts construction of a number of wind tunnels for measuring air friction on moving vehicles; *See also* 1895 TRA

1893 Whitcomb L. Judson patents a "clasp locker or unlocker for shoes"; after various changes and improvements over the next 30 years, the "clasp locker" evolves into the slide fastener best known as the zipper; *See also* 1905 GEN

Australian Lawrence Hargrave [b England] invents the box kite

Valdemar Poulsen [b Copenhagen, Denmark, Nov 23 1869, d Copenhagen, Jul 1942] develops the first magnetic sound recorder, the telegraphon; the sound is recorded as magnetized regions on a steel wire that is wrapped around a cylinder; *See also* 1888 COM

Chemist Leo H. Baekeland [b Ghent, Belgium, Nov 14 1863, d Beacon NY, Feb 23 1944] introduces a photographic paper that is sensitive enough for printing by artificial light (Velox); this revolutionizes the process of making photographic prints; *See also* 1889 COM

Louis Ducos du Hauron invents a method of producing three-dimensional-appearing images in which the stereoscope uses images that are viewed by filters with a different color for each eye; *See also* 1847 COM

Rudolf Diesel [b Paris, Mar 18 1858, d English Channel, Sep 30 1913] describes an internal combustion engine, first built in 1892, that will be named after him; the ignition of the gas mixture in the cylinder is obtained by heating during the compression cycle; *See also* 1890 ENE; 1914 ENE

Wilhelm Maybach [b Heilbronn, Germany, Feb 9 1846, d Stuttgart, Germany, Dec 29 1929] develops the float-feed carburetor in which gasoline and air are mixed to form the explosive mixture entering the cylinder of a gasoline motor; Maybach's carburetor is used on the four-stroke Daimler engine in 1897 and most automobile motors thereafter

The perfect machine: The turbine

Turbines are devices that spin in the presence of a moving fluid. The difference between waterwheels or windmills and turbines is largely one of emphasis and degree. Also, the most useful turbines for many purposes are those that can be propelled with heat, turning heat energy into motion.

During the 18th and 19th centuries, much progress was made toward extracting the kinetic energy of flowing water by devising water turbines. Leonard Euler, applying fluid mechanics, developed a water turbine as early as 1750. During the 18th century several engineers, such as Benoit Fourneyron, succeeded in building water turbines that by giving the blades special shapes by far outstripped the conventional waterwheels. The term "turbine" was coined by Fourneyron's professor Claude Burdin; he derived the term from *turbo*, a spinning object.

A typical turbine based on heat is the steam turbine. The idea of a steam turbine is much older than the steam engine itself. Around 60 BC the Alexandrian Greek Hero (also known as Heron) used jets of steam to turn a kettle. In 1629 the Italian engineer Giovanni Branca depicted in his machine book *Le machine* a steam turbine in which a jet of steam is directed at the vanes of a wheel like a waterwheel. No doubt others observed that escaping steam is like the rushing wind and could be used to push mills.

When practical steam engines were built at the start of the 18th century, however, they moved a cylinder back and forth (reciprocating motion) instead of pushing a wheel around. Although the reciprocating steam engine had reached a high level of sophistication during the 19th century, it had several inconveniences. Such steam engines were bulky, had slow rotation speeds, and much energy was wasted by the machine itself in moving the heavy pistons back and forth. When first applied to drive electric generators, reciprocating steam engines proved difficult to maintain at a fixed rotation speed as the load on the generator changed.

Turbines are as simple as reciprocating engines are complex. Because they have essentially only one moving part, they are sometimes called the perfect engines, almost directly turning heat into rotary motion.

The first to build a steam turbine was the British engineer Charles Algernon Parsons. In 1884 he completed a small turbine that rotated at 18,000 revolutions per minute and that delivered 10 horsepower. The Swedish engineer Carl Gustav Patrik de Laval, experimenting with steam turbines, achieved greater power and higher rotation rates. In 1890 he built a turbine consisting of a 30-cm (12-in.) disk with 200 blades mounted on a flexible axis. The steam was admitted to the blades by special nozzles (Laval nozzles) that accelerated the steam to very high velocities, thus transferring the energy of the steam in the form of kinetic energy to the blades.

Léon Appert invents reinforced glass; it contains a wire mesh, making the glass less prone to scattering; *See also* 1905 MAT

Edward Drummond Libbey [b Chelsea MA, Apr 17 1854, d Nov 13 1925] produces fabric woven from glass fibers and silk; *See also* 1710–19 MAT

Henri Moissan [b Paris, Sep 28 1852, d Paris, Feb 20 1907] describes the use of an electric furnace for smelting metals; *See also* 1879 MAT; 1907 MAT

Henri Moissan announces that he has made artificial diamonds from carbon; later researchers learn that Moissan did not have enough heat or pressure to produce diamond; it is believed that someone perpetrated a hoax on him; *See also* 1797 MAT; 1955 MAT

Felix Hoffman, working for the Aldolf von Bäyer firm in Elberfeld, Germany, synthesizes aspirin (acetyl salicylate); it is made by a variant of the method used earlier by Hermann Kolbe to produce salicylic acid, but aspirin is not as unfriendly to the stomach lining as salicylic acid is; Hoffman's motivation is to develop a treatment for his father's rheumatoid arthritis; *See also* 1859 MED

Surgeon Daniel Williams [b Holidaysburg PA, Jan 28 1858, d Aug 4 1931] performs the first open-heart surgery on a patient injured by a knife wound; *See also* 1914 MED

Sigmund Freud's and Josef Breuer's collaboration in studying the psychic mechanism of hysterical phenomena becomes the foundation of psychoanalysis; *See also* 1880 MED

James J. Harvey patents Harvey's Pneumatic Dusting Machine, a sort of vacuum cleaner operated by two persons, one of them working a bellows and the other pushing the dust collector from place to place; *See also* 1902 TOO

William Du Bois Duddell [b England 1872, d Nov 4 1917] improves the oscilloscope and gives it the design it has today; *See also* 1891 TOO

George Moore builds a steam-driven automaton in the shape of man; it is powered by a 0.5 horsepower gas-fired boiler that enables it to walk at 14 km (9 mi) per hour; it also vents steam through a cigar in its mouth

A project under consideration from time to time for about 2500 years is finally completed when a shipping canal is dug across the isthmus of Cornith; *See also* 60 CE COM

The perfect machine: The turbine (Continued)

The design of steam turbines developed into a science during the end of the 19th century. Better materials allowed the construction of turbine blades that were resistant to corrosion. Charles Curtis developed the multistage turbine in which the blades and disks became progressively larger when the steam expanded. Parsons developed in 1894 the ship turbine engine. The slow-revolving turbine consisted of several sections of increasing diameter. High-pressure steam was admitted to the turbine and pressure differences in each section drove the turbine blades. The first ship to be equipped with such a steam turbine, the *Turbinia*, immediately established a speed record with 31 knots (35.7 mi per hour).

During the beginning of the 20th century, reciprocating steam engines were quickly replaced by steam turbines. Steam turbines can deliver much more power than reciprocating engines and need less maintenance. Steam turbines also supplanted marine steam engines on ships.

A similar evolution took place with internal combustion engines, mainly driven by the need for lightweight and powerful airplane engines. Most large modern airplanes are now powered by either turboprop or turbojet engines. These turbines are spun by the expansion of jet fuel instead of by the expansion of water into steam and gain additional rotary power from intake of air, which also expands.

1894

Carbon filaments based on squirted cellulose are introduced, they become the standard filaments for Edison and Swan incandescent lamps; *See also* 1879 GEN; 1904 GEN

American Daniel M. Cooper invents the time clock; *See also* 1892 GEN; 1896 GEN

John Patterson, sole owner of National Cash Register Company, recasts the factory as a paternal organization with free medical care, a cafeteria, and the first ever suggestion system to reward workers for ideas for improving operations; *See also* 1892 GEN

Electrical engineer Marchese Guglielmo Marconi [b Bologna, Italy, Apr 25 1874, d Rome, Jul 20 1937] builds his first radio equipment, a device that will ring a bell from 10 m (30 ft) away; *See also* 1888 COM; 1899 COM

Edouard Branly detects radio waves produced by a spark generator at a distance of 30 m (100 ft) with his coherer; the receiving circuit consists of a battery, a coherer, and a galvanometer; the galvanometer indicates a current as soon the spark generator starts working; *See also* 1890 ELE; 1901 COM

Oliver Lodge [b Staffordshire, England, Jun 12 1851, d Wiltshire, England, Aug 22 1940] demonstrates the transmission of radio signals over a distance of 60 m (180 ft) at the annual meeting of the British Association for the Advancement of Science; *See also* 1888 COM; 1897 COM

Eugene Porzolt invents a photographic method for the composition of type based on letters projected on a sensitive plate; the system, now known as photocomposition, becomes commercially available 50 years later; *See also* 1884 COM; 1949 COM

On the Willamette River at Oregon City OR, the first dam specifically built to produce power to drive a hydroelectric plant begins operation at a site that already has a hydroelectric plant that uses the existing fall of the river for energy; *See also* 1890 ENE; 1895 ENE

Guglielmo Marconi

Guglielmo Marconi is generally credited as the inventor of radio communication as such, although transmitters and receivers had been built before him by other experimentalists. In 1894, reading about the electromagnetic waves discovered by Heinrich Hertz, he realized that they could be used for the transmission of signals in the same way as electric wires carried electromagnetic waves.

Marconi started experimenting with radio waves with the help of the physicist Augusto Righi. Lacking formal training in physics, Marconi proceeded empirically. He used a transmitter consisting of a high-voltage coil and a spark gap connected to an antenna circuit. The receiver consisted of a coil connected to a coherer, a tube containing iron filings that became conductive on the reception of a radio signal.

By connecting an antenna and an Earth conductor to the transmitter and receiver, Marconi was able to transmit radio signals over increasing distances. He experimented with different kinds of antennas and soon succeeded in transmitting signals over 2.4 km (1.5 mi).

Marconi tried to interest the Italian government in his experiments, but the government showed no interest; so Marconi moved to England to continue his experiments there. In England he successfully applied for a patent for his system of radio transmission, and in 1897, with the help of wealthy relatives, he set up the Wireless Telegraph and Signal Company.

Continuing his experiments with antennas and increasing the power of the transmitter, Marconi transmitted signals over much larger distances: 240 km (150 mi) in 1900 and across the Atlantic Ocean in 1901.

The British army and navy became Marconi's first customers, buying systems developed by his company. By 1900 Marconi founded a subsidiary that leased radio communication systems, including radio operators, to ships and shore stations.

In 1909 Marconi received the Nobel Prize in physics, which he shared with German physicist Ferdinand Braun, who had developed a sparkless antenna circuit for transmitters.

Artificial emery (aluminum oxide, considered a variety of corundum) is produced; *See also* 1891 MAT; 1902 MAT

Charles Cross and Edward Bevan develop a dye of the type called azo; it stays fast better than other azo dyes because it is chemically linked to the cellulose-based artificial fiber; this is the first fiber-reactive dye, but the process of dyeing with it is too complicated to use commercially; *See also* 1880 MAT; 1953 MAT

Herman Frasch [b Germany, Dec 25 1851, d Paris, May 1 1914] begins to develop the method of removing sulfur from deep deposits by melting it with superheated water; *See also* 1902 MAT

In Greenland Robert E. Peary [b Cresson PA, May 6 1856, d Washington DC, Feb 20 1920] notes that natives make tools from copper broken from a large meteorite; the remaining 37 tons of native copper is now in the American Museum of Natural History in New York City; *See also* 8000 BC TOO

A stapler using precut, U-shaped staples is invented; the staples can be inserted one at a time into a magazine that then feeds them to the machine; *See also* 1877 TOO; 1896 TOO

The first electric automobiles appear on the market; the electric motor is powered by a rechargeable battery; *See also* 1881 ENE; 1900 ENE

A new French company charged with building the Panama Canal is formed, although the project is no longer backed by the French government after the US declares that such government backing would be against the Monroe Doctrine; *See also* 1889 TRA; 1898 TRA

The Manchester Ship Canal opens in England, giving the manufacturing center of Manchester access to the Atlantic Ocean

B.F.S. Baden-Powell uses kites to lift human beings into the air; *See also* 1871 TRA

Charles Parsons invents the ship turbine engine; *See also* 1888 ENE

John I. Thornycroft and Sydney W. Barnaby discover strong vibrations from the propellers of the first British destroyer and find the cause of it: erosion caused by bubbles, known as cavitation; *See also* 1844 TRA; 1920 ENE

1895 Physicist Wallace Clement Ware Sabine [b Richwood OH, Jun 13 1868, d Cambridge MA, Jan 10 1919] becomes the first acoustical engineer in response to a request to improve the sound capabilities of a lecture hall; he develops a method to measure the time of decay of pure tones in a room that is empty and a room in which absorbing materials are present; later he uses acoustic principles to design Boston's Symphony Hall; *See also* 1891 GEN

Auguste [b 1862, d 1954] and Louis Lumière [b Besançon, France, Oct 5 1864, d Bandol, France, Jun 6 1948] patent the *cinématographe,* a projection version of Thomas Edison's Kinetoscope; the first public showing of a motion picture using their process takes place in a café in Paris in Dec; the showing is preceded in the spring in the US by a crude projected version of a prizefight using modified Edison equipment; the Lumières use 16 frames per second and 35-mm film, both of which become early standards for the industry; *See also* 1891 COM; 1896 COM

Herman Castler patents the Mutoscope; inspired by Thomas Edison's first version of a Kinetoscope, the Mutoscope consists of a large number of cards mounted on a rotating drum; when viewed one by one in rapid succession, the cards create a moving image; coin-operated versions soon appear in "picture parlors"; *See also* 1888 COM

In England, Karl Klietsch combines forces with printer Samuel Fawcett to develop an improved form of machine photogravure on copper plates; the method, nicknamed "Rembrandt photogravure," becomes generally used in the early 20th century; *See also* 1872 COM

John T. Underwood [b London, Apr 12 1857, d Wianno MA, Jul 2 1937] begins to manufacture a typewriter patented in 1893 by Franz X. Wagner; like the Oliver typewriter of 1892, it allows the operator to see the words as they are typed; *See also* 1872 COM

The first commercially introduced ballpoint pen is produced, although patents for similar pens are filed throughout the 19th century; it will be another half century, however, before a version of the ballpoint becomes popular;

At Niagara Falls, hydraulic electric generators installed by the Westinghouse Electric Company start operation, producing an alternating current of 3750 kw at 25 Hz to supply the city of Buffalo NY, 35 km (22 mi) away, with electric power; *See also* 1894 ENE

Charles G. Curtis [b Boston, Apr 20 1860, d Central Islip NY, Mar 10 1953] patents a multistage steam turbine that operates by transforming both the impulse and the reaction forces caused by the steam into rotary motion; *See also* 1888 ENE; 1920 ENE

Charles William Post [b Springfield IL, Oct 26 1854, d Santa Barbara CA, May 9 1914] begins his first health food business by marketing Postum, described as a "cereal food coffee"; *See also* 1897 FOO

The Reliance Building in Chicago IL is the first office building to use glazed terra-cotta over a steel frame

Independently, Karl Paul Gottfried von Linde [b Berndorf, Germany, Jun 11 1842, d Munich, Germany, Nov 16 1934] and William Hampson [b Cheshire, England] develop a regenerative (feedback) method of mass producing liquid air; the same method can be used to liquefy any gas except neon, hydrogen, or helium; *See also* 1898 MAT; 1908 MAT

Wilhelm Konrad Roentgen [b Lennep, Germany, Mar 27 1845, d Munich, Germany, Feb 10 1923] discovers X rays on Nov 8; he finds that these rays pass through matter, a property that immediately leads to their application in diagnostic medicine; *See also* 1896 MED

By this time, the heliocoidal wire cutter for cutting stone is introduced at marble quarries; in addition to the wires, some of which could be a 2 km (1 mi) long, a cooling but abrasive slurry of sand in water is used to cut stone blocks; *See also* 1978 TOO

The German firm Benz builds a gasoline-powered bus that can transport six to eight passengers; *See also* 1831 TRA

Otto Lilienthal and his brother Gustav design and fly the first glider that can soar above the height of takeoff; *See also* 1889 TRA

Percy Pilchard [d 1900] starts experiments with kites and gliders in hopes of developing heavier-than-air craft; *See also* 1894 TRA; 1900 TRA

Russian physicist Konstantin Tsiolkovsky proposes that liquid-fueled rockets can be used to propel vehicles in space; *See also* 1892 TRA

The first electric locomotive starts service in the US; it has four motor-driven axles and runs on 550-volt DC current; *See also* 1884 TRA; 1912 TRA

The Kaiser-Wilhelm-Kanal, later renamed the Nord-Ostee-Kanal, but known in English as the Kiel Canal, opens, joining the North Sea via the Elbe and the Baltic Sea at Kiel Bay; it is claimed that the Kiel Canal is used by more ships than any other international maritime waterway; *See also* 1785 TRA; 1914 TRA

Looking into people (part 1)

Almost as soon as Wilhelm Konrad Roentgen had demonstrated in public that the bones inside his hand could be observed without removing the flesh, doctors began to use Roentgen's X rays in setting bones. Initially, bone was the only obvious tissue observed on X-ray photographs, although keen-eyed surgeons soon learned to read the shades of gray. Tumors and spots on the lungs were visible to educated physicians. Seeing the interior of the body with no apparent harm to the tissues was one of the wonders of the 20th century. Soon in every small town across the US and other industrialized nations X-ray machines were installed in shoe stores so that children and their parents could make sure that new shoes were not squeezing young bones. Dentists learned to find the details of previously hidden cavities.

X rays were miraculous, but only fairly hard tissue showed up. Within the first year of their use, clever doctors found fluids that were opaque to radiation, which could be swallowed or injected or otherwise inserted into a desired region of the body. Stomach ulcers and colon cancers could be observed after a glass of chalky barium solution or a barium enema. Other fluids were injected into the circulatory system to reveal the interior of the heart or brain. The other problem with X rays, as doctors learned, is that X rays cause much more damage than had been expected. The shoe store machines were withdrawn and doctors and dentists learned to leave the room while an X ray was being taken.

Radioactivity was discovered shortly after X rays. From the beginning, scientists knew that radioactivity also could be used to make photographs through opaque substances; indeed, that is how it was discovered. It was not until the 1930s, however, that doctors found ways to use radioactivity to see inside the body. Accelerators used for studying subatomic particles also produced various radioactive versions of familiar elements as well as some new elements. Radioactive iodine has been used since the 1940s to observe the thyroid, which concentrates iodine. Starting in 1960, the artificial radioactive element technetium (extremely rare in nature) became a popular way to observe many different organs. Like X rays, radioactivity damages tissue. Both are frequently used to treat tumors as well as in diagnosis.

Although X rays and one form of radioactivity, gamma rays, are both energetic electromagnetic waves, there is an essential difference in the way the two are used in diagnosis. X rays are generated outside the body on one side, and captured on the other side. Radioactive "tracers" are inserted into the body, where they generate radiation that is collected on the outside. Newer methods used in diagnosis also fall into one of the two categories. Sonograms and CT scans, for example, are generated outside the body, while PET scans and thermography use radiation generated within the body. Magnetic resonance imaging partakes of both techniques (see part 2 on p 400).

1896

Donald Cameron devises the first septic tank that promotes anaerobic fermentation; *See also* 1885 GEN; 1914 GEN

Engineer Frederick Winslow Taylor [b Philadelphia, PA, Mar 20 1856, d Philadelphia, Mar 21 1915] writes *The adjustment of wages to efficiency*; he is among the first to make time-and-motion studies of workers with the idea of eliminating unneeded steps; *See also* 1894 GEN

Eastman Kodak introduces the Brownie camera, priced at $1 in the US; millions buy the simple camera and take hundreds of millions of photographs; *See also* 1889 COM; 1923 COM

Thomas Armat of Washington DC develops a projector for motion pictures that permits the image to stay on the screen longer with a shorter interval between images; the Vitascope greatly reduces the flickering in the "flickers"; *See also* 1895 COM; 1900 COM

The telephone dialer is developed by an unknown inventor in Milwaukee WI; before this time, telephone connections were made only by giving the number of the person to be reached to a telephone operator or, if an automatic exchange had been installed, by pushing a button to produce the pulses for each number; *See also* 1903 COM

Herman Hollerith, after succcessfully applying his punched-card technique to the US census, founds Tabulating Machine Company, which later changes its name to International Business Machines, and still later to IBM; *See also* 1890 ELE; 1901 ELE

1897

Physicist Alexander Popov [b Bogoslavsky, Russia, Mar 16 1859, d St. Petersburg, Russia, Jan 13 1906] uses an antenna to transmit radio waves over a distance of 5 km (3 mi); *See also* 1894 COM; 1901 COM

Karl Ferdinand Braun [b Fulda, Germany, Jun 6 1850, d New York City, Apr 20 1918] develops a cathode-ray tube; it consists of an evacuated electron tube in which electrons, aimed by electromagnetic fields, form an image on a fluorescent screen; *See also* 1878 COM; 1911 COM

Joseph John Thomson [b Chetham Hill, England, Dec 18 1856, d Cambridge MA, Aug 30 1940] discovers the electron, the particle that makes up electric current and the first known particle that is smaller than an atom, in part because he has better vacuum pumps than previously available; he and, independently, E. Wiechert, determine the ratio of mass to charge of the particles by deflecting them by electric and magnetic fields

1896

Selective weed killers are used in France; *See also* 1885 FOO; 1940 FOO

The invention of the alloy invar by Charles Guillaume [b Fleurier, Switzerland, Feb 15 1861, d Sèvres, France, Jun 13 1938], a metal with a very small expansion or contraction with temperature, solves the long-standing problem of compensating for the way that temperature affects the operation of pendulum clocks; pendulums of invar instead of brass do not require compensating mechanisms; *See also* 1720–29 TOO

Johannes von Mikulisz-Radecki invents the gauze mask to be worn by surgeons when performing surgery; *See also* 1890 MED

Michael I. Pupin [b Idvor, Austria-Hungary, Oct 4 1858, d New York City, Mar 12 1935] of Columbia University takes the first diagnostic X-ray photograph in the US; Eddie McCarthy of Dartmouth NH has a broken arm set with the new diagnostic aid less than 3 months after the discovery of X rays by Roentgen; *See also* 1895 MED; 1897 MED

F.W. Fellows invents a machine that manufactures almost any kind of gear cheaply and rapidly; *See also* 1887 TOO

Grinding wheels using artificial emery and other forms of corundum go on the market in the US; *See also* 1894 MAT; 1900 TOO

Thomas Briggs founds the Boston Wire Stitcher Company to market wire stitching machines similar to those invented in 1868 by Englishman Charles Henry Gould; in 1948 the company becomes Bostich, Inc., a principal manufacturers of office staplers; *See also* 1868 TOO; 1914 TOO

Henry Ford [b Greenfield Village MI, Jul 30 1863, d Dearborn MI, Apr 7 1947] builds his first automobile; *See also* 1889 TRA

F.W. Lanchester is the first to develop a "live" back axle for a full-sized automobile; it is cut in the middle and the cut ends are connected through a gear box so that the rear wheels can rotate at different speeds when the vehicle turns

Samuel Langley tests his steam-driven flying machine on the Potomac, flying for 1.2 km (0.75 mi) before crashing; *See also* 1890 TRA; 1897 TRA

The first underground railways (subways) in Budapest start operation; *See also* 1892 TRA; 1898 TRA

1897

The electric coffee grinder is introduced; *See also* 1890 FOO; 1900 FOO

C.W. Post introduces his first breakfast cereal, called as it is today Grape Nuts, although it contains neither grapes nor nuts (Post erroneously thought that the manufacturing process resulted in grape sugar and that the cereal had a nutty taste); *See also* 1895 FOO; 1904 FOO

Pearl B. Wait manufactures the gelatin dessert invented in 1845 by Peter Cooper under the trade name Jell-O, a name coined by Wait's wife Mary; *See also* 1845 FOO

Johann Friedrich Adolf von Baeyer synthesizes indigo; his method will be adopted by Badische Anilin- & Soda-Fabrik (BASF), which starts industrial production of indigo in 1898; *See also* 1901 MAT

Georges Claude [b Paris, France, Sep 24 1870, d Saint-Cloud, France, May 23 1960] and A. Hess observe that acetylene dissolves remarkably well in acetone; since this time, the standard method for storing acetylene has been as so dissolved; *See also* 1859 MAT

Adolf Spittler discovers, probably by accident, casein plastics; *See also* 1869 MAT; 1906 MAT

Christiaan Eijkman [b Nijkerk, the Netherlands, Aug 11 1858, d Utrecht, Nov 5 1930] shows the relationship between the occurrence of beriberi and the consumption of polished rice; however, he does not attribute the disease to the absence of a vitamin in polished rice; the role of vitamins in preventing diseases is discovered in 1906; *See also* 1901 MED

Walter Bradford Cannon [b Prairie du Chien, WI, Oct 19 1871, d Franklin NH, Oct 1 1945] discovers that a bismuth compound can be used to make intestines visible to X rays; *See also* 1896 MED; 1927 MED

Almroth Wright introduces a vaccine against typhoid; *See also* 1890 MED; 1917 MED

Clément Ader, according to his own account, flies with his steam-powered airplane *Avion III* over a distance of 300 m (960 ft); most observers of the trial on Oct 14 say that *Avion III* did not fly at all or barely got off the ground; *See also* 1890 TRA; 1901 TRA

David Schwarz builds in Germany the first rigid-hulled airship, using an aluminum frame and sheeting; *See also* 1884 TRA; 1900 TRA

Charles Parsons's *Turbinia*, the first steamship to use steam turbines for power, demonstrates its superiority over conventional steamships by unexpectedly scooting through a formal naval review in front of Queen Victoria; *See also* 1894 TRA; 1907 TRA

1898

Karl Auer (Baron von Welsbach) introduces the osmium filament for light bulbs; it offers a longer bulb life because of the high melting point of the dense metal filament, but osmium is too rare and expensive for the lamps to be a commercial success; *See also* 1856 GEN; 1902 GEN

German physicist and mathematician Félix Klein [b Düsseldorf, Apr 25 1849, d Göttingen, Jun 22 1925] founds a department of technical physics at Göttingen University; the department is partially funded by industry; *See also* 1887 GEN; 1906 GEN

Nikola Tesla demonstrates wireless control of model ships in Madison Square Garden in New York City

George Washington Carver

George Washington Carver is remarkable for what he did not do as well as for what he accomplished. Although offered opportunities at higher salaries to work for Thomas Edison and Henry Ford, he remained at a low-paying post in Alabama, where he felt that he could contribute more to the lives of impoverished blacks. Furthermore, although he developed a large number of products that had commercial success, he gave all of his ideas to the public, without making any effort to enrich himself. He became famous in his own lifetime and was given a number of prestigious awards, but he maintained his simple life-style. What money he had when he died he willed to a foundation to continue his work in Alabama.

Carver was born into slavery on the Missouri farm of Moses Carver (whose last name he took after Emancipation). Rustlers captured young George along with his mother and others and took them to another state. George was ransomed back for a horse said to have been worth $300, but the rustlers sold his mother elsewhere. After Emancipation, there was a window of opportunity for black people in the US South for several decades, and Carver worked hard to get an education, eventually taking a masters degree in agriculture from Iowa State Agricultural College, which welcomed him onto the staff. Soon, however, another black educator of this time of Reconstruction opportunity, Booker T. Washington, invited Carver to come to his new school, Tuskegee Institute in Alabama. In the South, the opportunities for black advancement that had been available to Carver and Washington were fading fast, with voting rights restricted and segregation becoming the law of the land. In addition to these problems, blacks in Alabama had to farm in soil that had been depleted by decades of poor practices in growing cotton and tobacco.

Carver worked with local farmers to reform agricultural practices. He preached organic and regenerative farming before those labels had been invented. He taught that depleted soil could be renewed by working in animal manure or plant remains.

Carver also knew that one of the main nutrients plants need is nitrogen (see "Nitrogen: A matter of life and death," page 332). Certain plants, notably the legumes, have formed a symbiotic relationship with bacteria that can take nitrogen from the air. Carver encouraged Alabama farmers to plant legume crops to help restore the soil.

1899

Guglielmo Marconi establishes the first radio link between England and France, transmitting greetings to the French scientist Edouard Branly; *See also* 1894 COM; 1901 COM

Thomas Dallmeyer [b 1859, d 1906] develops a lens with variable focus; *See also* 1946 COM

Philipp Lenard [b Pressburg, Austria-Hungary, Jun 7 1862, d Messelhausen, Germany, 1947] in Germany and Joseph J. Thomson in England each demonstrate that the photoemissive effect of ultraviolet light on zinc is caused by the release of electrons; this effect is explained by Albert Einstein in 1905 in a famous paper that first introduces wave-particle duality to physics; *See also* 1888 ELE; 1900 ELE

Agricultural chemist George Washington Carver [b near Diamond Grove MO, c 1861, d Tuskegee AL, Jan 5 1943] publishes his first paper on agriculture from Tuskegee Institute in AL, "Feeding acorns to livestock"; as head of Tuskegee's Department of Agricultural Research, Carver works to develop techniques for regenerating land by growing sweet potatoes and peanuts; he also develops hundreds of new uses for both plants

The bacteria-killing effects of X rays are noted, although no scientific work on preservation of food with radiation is pursued until the 1940s; *See also* 1906 FOO

James Dewar succeeds in liquefying hydrogen; *See also* 1895 MAT

Johann *Hans* Goldschmidt [b Berlin, Germany, Jan 18 1861, d Baden-Baden, Germany, May 20 1923] develops thermite, a mixture of aluminum powder and iron or chromium that burns at a high temperature, leaving a residue of pure iron or chromium; its most common use is in welding; *See also* 1886 TOO; 1901 TOO

Engineers Frederick Winslow Taylor, better known for his work on time-and-motion studies, and Maunsel White develop a new method of tempering steel that permits metal-cutting operations at high speed; the hardening process involves alloying steel with small amounts of tungsten, chrome, vanadium, and molybdenum; the resulting "high-speed steel," or Taylor steel, keeps its cutting properties even when glowing hot, starting a revolution in the design of cutting tools; *See also* 1904 TOO

Herta Ayrton [b Portsea, England, Apr 25 1854, d Aug 26 1923], who invented a sphygmograph (an instrument for monitoring the human pulse), becomes the first woman to be elected to England's Institution of Electrical Engineers

The first underground railways (subways) in Vienna and Paris (the Metro) start operation; *See also* 1896 TRA

The second effort by the French to build a canal across Panama fails; *See also* 1894 TRA; 1904 MED

1898

George Washington Carver (Continued)

Poor farmers in Alabama needed plants that would grow in their depleted soil and hot climate. Two plants from South America that had become popular in Africa met both requirements. Furthermore, one of them, the peanut, is a legume. The other is the relative of the morning glory that we call the sweet potato. Both were easy to grow in Alabama and produced abundant harvests.

The problem was that there was not enough demand for peanuts and sweet potatoes to keep prices high enough to support the farmers. So long as all people did with them was eat them as snacks or starches with meals, the economy would not support very many farmers growing these plants. Carver realized that what was needed were industrial uses for peanuts and sweet potatoes. Maize (corn), for example, is not only animal feed, but also a source of corn oil, corn sugar, alcohol, and even raw materials for plastics. Very little of the corn seen growing so profusely in the US Midwest is intended for consumption as a vegetable.

The peanut is a particularly good candidate for a variety of uses, since it is about half oil and a quarter starch, both useful in developing other products. Peanut oil breaks down at a higher temperature than other common vegetable oils, which not only has advantages in cooking but also makes it suitable for industrial use. Both the peanut and the sweet potato are good animal fodder. Carver succeeded in developing hundreds of other uses for both plants, greatly expanding the market and with it the economy of Alabama.

Belgian driver Camille Jenatzky breaks the 100 km (62 mi) per hour land speed record with a car shaped as an artillery shell

1899

1900

Reginald Aubrey Fessenden [b Milton, Quebec, Canada, Oct 6 1866, d Bermuda, Jul 22 1932] discovers the radio-frequency alternator that produces AM radio signals, although several years of development of the device are still ahead; *See also* 1904 COM

Raoul Grimoin-Sanson projects a motion picture on a circular screen using ten projectors, the spectators are in the middle and see a moving image all around; a similar idea is revived in 1951 under the name Cinerama; *See also* 1951 COM

Constantin Perskyi coins the term "television"; *See also* 1884 COM

Johann Phillip Elster [b Bad Blankenburg, Germany, Dec 24 1854, d Bad Harzburg, Germany, Apr 6 1920] and Hans F. Geitel [b 1855, d 1923] devise the first practical photoelectric cell, although earlier models go back to 1896 and they have been working on the theory of photoemission since 1889; in a photoelectric cell, electrons are emitted when the cell is struck by light; *See also* 1888 ELE; 1905 ELE

The first offshore oil wells are drilled; *See also* 1859 ENE; 1923 ENE

Thomas Alva Edison invents the nickel-alkaline electric battery; however, it does not produce nearly as much power as Edison is hoping to achieve for use in electric automobiles; *See also* 1881 ENE; 1903 ENE

1901

Walter Nernst [b Breisen, Germany, Jun 25 1864, d Bad Muskau, Germany, Nov 18 1941] patents a form of incandescent lamp that uses a filament consisting of oxides of rare earths; it is introduced commercially but does not have wide use; the "Nernst glower" spurs activity toward the development of more practical filaments; *See also* 1879 GEN; 1904 GEN

Peter Cooper-Hewitt begins to market a mercury-vapor lamp, partially solving the main problem with the lamp, its spectrum that contains no red, by using a fluorescent screen that converts some of the excess blue, green, and yellow light into red; the first practical installation of the Cooper-Hewitt lamp is in 1903; *See also* 1892 GEN; 1933 GEN

Guglielmo Marconi receives the letter "S" in St. Johns, Newfoundland, that has been transmitted from England; it is the first transatlantic telegraphic radio transmission; *See also* 1899 COM

Reginald Aubrey Fessenden discovers the heterodyne principle: When two radio signals are mixed in a receiver, a third signal with a frequency equal to the difference in frequency of the two received signals is created; this principle is fundamental to the superheterodyne receiver developed by Edwin Armstrong; *See also* 1918 COM

Herman Hollerith develops the first numerical keyboard for punching cards for tabulating machines; *See also* 1896 ELE; 1911 ELE

1900

The electric meat grinder is introduced; *See also* 1897 FOO; 1909 FOO

Clément, a priest from Oran, Algeria, develops the hybrid of the bitter orange and the mandarin orange that is now known as the clementine; *See also* 1892 FOO

Johann August Brinell [b 1849, d 1925] develops the test named after him for estimating metal hardness; when a steel ball or diamond cone is pressed against a metal, the hardness of the metal is inversely proportional to the depth of impression

C.H. Norton of the Norton Company of Worcester MA builds the first powerful grinding machines with wide wheels and precision controls; the machines are used to grind almost all automobile crankshafts early in the 20th century; *See also* 1896 TOO; 1904 TOO

Nikola Tesla proposes the use of radio waves for the detection of moving objects; it is the first conception of radar; *See also* 1904 TOO

Emil Wiechert [b Tilsit, East Prussia, Dec 26 1861, d Göttingen, Germany, Mar 19 1928] invents the pendulum seismograph, essentially the type used today; *See also* 130 CE TOO

The first Browning revolvers are manufactured; *See also* 1884 TOO; 1920 TOO

In Germany, Count Ferdinand von Zeppelin [b Konstanz, Germany, Jul 8 1838, d Charlottenburg, Germany, Mar 8 1917] launches on Jul 2 the first of his line of rigid-framed, hydrogen-filled airships, which come to be known as zeppelins; these are built at a rapid pace during World War I, and after 1918 the main stock is taken away by Allied victors; by 1924 the Zeppelin company is back in operation in Germany; *See also* 1897 TRA; 1937 TRA

Enrico Forlanini builds a hydrofoil catamaran, a two-hulled ship, that reaches 80 km (50 mi) per hour by 1905; *See also* 1918 TRA

Percy Pilchard is killed in a glider accident; *See also* 1895 TRA

1901

Charles Hart and Charles Parr start building and marketing the Hart-Parr internal-combustion-engine tractor; it is the first commercially successful gasoline tractor, selling 200 by 1906; *See also* 1881 FOO; 1917 FOO

The first form of instant coffee is produced, although it is not successful; *See also* 1938 FOO

In Russia, a practical technique for artificial insemination of farm animals is introduced

The first synthetic vat dye, known either as indanthrone or as indanthrene blue, is developed when German chemist René Bohn improves synthetic indigo; *See also* 1897 MAT

Gerrit Grijns shows that beriberi is caused by the removal of a nutrient from rice during polishing; *See also* 1897 MED; 1912 MED

Adolf Windaus [b Berlin, Germany, Dec 25 1876, d Göttingen, Germany, Jul 9 1959] shows that the molecule of vitamin D can be affected by sunlight; *See also* 1897 MED

The journal *Biometrika*, in which psychologists support the idea of eugenics, is founded; *See also* 1883 MED

Welding using oxygen and acetylene is introduced; *See also* 1886 TOO; 1963 TOO

The Wright brothers, Wilbur [b Millville IN, Apr 16 1867, d Dayton OH, May 30 1912] and Orville [b Dayton, Aug 19 1871, d Dayton, Jan 30 1948] fly their first glider, reaching distances up to 183 m (600 ft); *See also* 1900 TRA; 1903 TRA

Motor-driven bicycles are introduced commercially; *See also* 1885 TRA

1901 cont

GENERAL

General Electric is the first company in the US to open a research laboratory; *See also* 1876 GEN; 1902 GEN

Sir Edward Richard Henry, who had used fingerprints in India, establishes a system of fingerprint classification for the British police that continues (with improvements) to be used in English-speaking countries; a different system, developed shortly before this time by Juan Vucetich in Argentina, is used in Spanish-speaking countries; *See also* 1880 GEN; 1987 GEN

COMMUNICATIONS

William Du Bois Duddell discovers that an electric arc can generate electromagnetic waves; however, he achieves frequencies no higher than 10,000 Hz and does not develop it for radio applications; Duddell also experiments with recording sound on photographic film

Donald Murray of New Zealand invents a new five-unit code that becomes the basis for telex starting around 1930; it is a method for communicating between two teletypewriters over telephone lines; *See also* 1874 COM

ELECTRONICS & COMPUTERS

Karl Braun uses a crystal detector for the detection of radio waves; *See also* 1897 ELE; 1906 ELE

1902

GENERAL

Willis H. Carrier invents the first air conditioner, although the name is first used in 1906 to describe a different device; *See also* 1881 GEN; 1903 GEN

Light bulbs with osmium filaments are produced commercially; *See also* 1898 GEN; 1905 GEN

Following the example of General Electric, Du Pont Company and Parke-Davis both establish research departments; *See also* 1901 GEN; 1911 GEN

The electric hair drier is introduced

The teddy bear is introduced

Edwin Binney of Easton PA combines paraffin, stearic acid, oil, and various pigments to create Crayola crayons

COMMUNICATIONS

German physicist Arthur Korn [b 1870] develops a method of transmitting photographs by first breaking them down into components and then reconstructing them at the other end using selenium cells; the resulting fax machine soon becomes indispensable for transmitting photographs of news events; *See also* 1907 COM

The first electric typewriter to be sold worldwide, the Blickensderfer Electric, is produced; it is invented by George Blickensderfer of Stamford CT; *See also* 1895 COM

Otto von Bronk [b Danzig, Germany, Feb 29 1872, d Berlin, Oct 5 1951] applies for a patent for color television in Germany; *See also* 1925 COM

ENERGY

Robert Bosch [b Sep 23 1861, d Stuttgart, Germany, Mar 12 1942] invents the spark plug; *See also* 1893 ENE

G. Honold develops high-voltage ignition for internal combustion engines based on electromagnetic induction; *See also* 1893 ENE

Hugh Longbourne Callendar [b Gloucestershire, England, Apr 18 1863, d London, Jan 23 1930] publishes *The properties of steam and thermodynamic theory of turbines*, which becomes a standard text on the topic; *See also* 1895 ENE; 1912 ENE

1901 cont

Emil von Behring of Germany wins the Nobel Prize in physiology or medicine for his discovery of diphtheria antitoxin; *See also* 1890 MED

Frank Hornby invents Meccano, a children's toy consisting of simple metal parts that can be assembled to construct mechanical structures, such as bridges and cranes; similar toys in in the US are known as Erector Sets

---

1902

John Kruger, imitating the Scandinavian smorgasbord as well as the Harvey railroad restaurants, uses the Spanish word *cafeteria* to describe his restaurant; *See also* 1890 FOO

Charles S. Bradley and D.R. Lovejoy start a small factory in Niagara Falls NY to exploit cheap electricity to make nitrogen chemicals directly from air; they use the "spark process" discovered in the 18th century by Henry Cavendish; *See also* 1774 MAT; 1903 MAT

Herman Frasch fully develops his method of removing sulfur from deep deposits by melting it with superheated water; *See also* 1894 MAT

Arthur D. Little [b Boston MA, Dec 15 1863, d Aug 1 1935], William H. Walker, and Harry S. Mork obtain the first US patent for an artificial fiber, a yarn based on rayon; *See also* 1884 MAT

Auguste Victor Verneuil patents a method for synthesizing crystals of corundum, which he colors red to make an artificial ruby; the process becomes a successful way to make jewel bearings for mass-produced watches and clocks; *See also* 1894 MAT

Millar Hutchinson invents in NY the first electrical hearing aid

After observing a machine that blows dust away in 1901, English engineer Hubert Cecil Booth develops the first successful vacuum cleaner, a complex and unwieldy machine driven by a 5-horsepower engine and mounted on a horse-drawn cart; its success is ensured after it is used to clean up from the rehearsal for the coronation of Edward VII, which leads to the purchase of one for Buckingham Palace and one for Windsor Castle; *See also* 1893 TOO

James Edward Ransome [b Ipswich, England, July 13 1839, d London, Jan 30 1905] manufactures the first successful motor-powered lawn mower; *See also* 1830 TOO

Richard Adolf Zsigmondy [b Vienna, Austria, Apr 1 1865, d Göttingen, Germany, Sep 23 1929] invents the ultramicroscope, for seeing small particles in a colloidal solution; *See also* 1903 TOO

The first practical French airship, *Le Jaune*, is launched by the Lebaudy brothers; *See also* 1897 TRA; 1919 TRA

Louis Renault [b Paris, Jan 12 1877] develops the drum brake; *See also* 1867 TRA; 1919 TRA

Frederick W. Lanchester invents in Great Britain the disk brake, which has better heat dissipation properties than the drum brake; *See also* 1867 TRA; 1919 TRA

J. Wilkinson designs the first air-conditioning system for an automobile, which is used on the 1902 Franklin; it is based on water cooling, as in a conventional automobile radiator

The British company Thorpe & Salter introduce the speedometer for automobiles; the first ones produced have a scale of 0 to 35 mi (0 to 55 km) per hour

1903

Willis Carrier builds the first whole-building air conditioner for a manufacturing plant in Brooklyn NY; *See also* 1881 GEN

Valdemar Poulsen patents an arc transmitter that generates continuous radio waves, producing a frequency of 100 kHz and receivable over 240 km (150 mi); *See also* 1900 COM

Mechanical telephone repeaters are installed by Herbert Shreeve [b Cambridge, England, Aug 1 1873, d 1942] between Amesbury, MA and Boston; a metallic diaphragm, activated by the incoming signal, is connected to a carbon transmitter; the sound quality, however, is very bad

Bell Telephone Company starts research in developing an automatic telephone switching system; *See also* 1896 COM; 1910 COM

Giovanni Conti builds in Larderello, Italy, the first geothermal electric power station, supplying power to four light bulbs; a 290-megawatt geothermal station is completed more than 50 years later, supplying electricity to Italian railroads

Thomas Alva Edison opens a plant to make an improved version of his nickel-alkaline electric battery; but when the battery goes on sale in 1904, it is so unreliable that Edison soon recalls it and halts manufacture while he goes back to the laboratory; *See also* 1900 ENE; 1908 ENE

Hans Holzwarth patents the explosion gas turbine; the gas mixture is introduced in a combustion chamber and ignited by a spark plug; the exhaust from the explosion drives a turbine; *See also* 1908 ENE; 1930 ENE

1904

Ultraviolet lamps are introduced

After nearly 15 years of development, the first commercial installation is made of electric gas-discharge lamps developed by Daniel McFarlan Moore; these use nitrogen for a yellow light and carbon dioxide for a pinkish-white light, but the gas pressure must be carefully regulated for the lamps to work properly; *See also* 1934 GEN

Willis R. Whitney develops a metalized heat-treated carbon filament that leads to the development of the tungsten filament; this filament continues in use to this day, since tungsten lasts longer because of its high melting point of 3400°C (6150°F); *See also* 1880 GEN; 1906 GEN

Physicist Reginald Aubrey Fessenden demonstrates the transmission of speech by the modulation of a radio wave generated by his high-frequency alternator; however, more satisfactory results are obtained a few years later with continuous-wave transmitters using vacuum tube oscillators; *See also* 1900 COM; 1906 COM

Ira W. Rubel accidentally invents offset printing about this time when he observes that a rubber pad can transmit an image from a lithographic plate to paper

L.C. Smith & Bros. introduces a replacement for its Smith Premier typewriter, one that has a shift to produce capital letters; the original Smith Premier (date uncertain) uses two keyboards, one for capital letters and another for lowercase; *See also* 1872 COM

John Ambrose Fleming [b Lancaster, England, Nov 29 1849, d Sidmouth, England, Apr 18 1945] files a patent for the first vacuum tube; also called the "Fleming valve," it is a diode that acts as a rectifier, a device that makes current flow in a single direction instead of alternating back and forth; hence, it changes alternating current (AC) to direct current (DC); *See also* 1883 ELE; 1907 ELE

1903

Ludwig Roselius, finds that coffee beans accidentally soaked in seawater can be used to produce acceptable flavor in coffee with 97 percent of the caffeine removed; he names his product Sanka for *sans caffeine*; *See also* 1901 FOO; 1908 FOO

The distinctive Flatiron Building (so named because of its triangular cross section) is constructed in New York City; at 69 m (226 ft), it is often considered the first skyscraper in New York City; *See also* 1879 FOO; 1913 FOO

The Ingals Building in Cincinnati OH is the first office building constructed with reinforced concrete; *See also* 1892 MAT; 1904 TOO

The destroyer H.M.S. *Wolf* is equipped with mechanical strain gauges attached at various points and set to sea to look for bad weather to learn which points suffer the most stress and strain

Olaf Kristian Birkeland [b Oslo, Norway, Dec 13 1867, d Tokyo, Japan, Jun 18 1917] and Samuel Eyde [b Arendal, Norway, Oct 29 1866, d Asgardstrand, Norway, Jun 21 1940] develop a method for producing nitric acid from atmospheric nitrogen by producing a reaction inside an electric arc; the method produces very small quantities and uses a lot of electricity; *See also* 1902 MAT; 1908 MAT

Surgeon Georg Perthes [b Germany, Sep 17 1869, d 1927] discovers that X rays inhibit the growth of tumors and proposes X-ray treatment for cancer; *See also* 1895 MED; 1941 MED

Dutch physiologist Willem Einthoven [b Semarang, Java, May 22 1860, d Leiden, the Netherlands, Sep 29 1927] develops in the Netherlands the string galvanometer, the forerunner of the electrocardiograph, used to measure tiny electrical currents produced by the heart; *See also* 1880 ELE; 1924 MED

The safety razor, first thought of by King C. Gillette in 1895 and under development since, goes on sale, selling a total of 51 razors and 168 blades during 1903; in 1904 sales of blades go up to 123,648; *See also* 1928 GEN

The spinthariscope, a magnified fluorescent surface that can detect individual subatomic particles, is invented; *See also* 1902 TOO

The first successful airplane is launched at Kitty Hawk NC by Wilbur and Orville Wright on Dec 17; powered by an internal combustion engine of 12 horsepower, its best flight of the day lasts 59 seconds; *See also* 1901 TRA; 1905 TRA

Konstantin Tsiolkovsky proposes that liquid oxygen be used for powering rocket engines for space travel; he also proposes the use of multistage rockets in which used up stages are discarded to decrease weight during escape from Earth's gravity; like Hermann Ganswindt in 1881, but independently, Tsiolkovsky calculates Earth's escape velocity; *See also* 1892 TRA; 1914 TRA

1904

C.W. Post calls his new brand of corn flakes "Elijah's Manna" at first, but finds that Christian ministers oppose it under that name, so he renames the cereal "Post Toasties"; *See also* 1897 FOO

Leon Guillet [b St. Nazaire, France, Jul 11 1873, d Paris, May 9 1946] develops the first stainless steels, but unaccountably fails to note that they resist corrosion; *See also* 1898 MAT; 1911 MAT

W.C. Parkin develops in France a nonflammable celluloid by adding a soluble metallic salt to ordinary celluloid; *See also* 1862 MAT

Albert Einhorn synthesizes procaine, also known as Novocaine, a local anesthetic that is the first usable substitute for cocaine in medicine; *See also* 1890 MED; 1906 MED

Dr. William Scholl [b La Porte IN, Jun 22 1882, d Mar 29 1968] sells the first of his newly patented arch supports, now known as the Foot-Eazer

Work resumes on the Panama Canal, this time by the US; US Army Colonel William C. Gorgas [b Mobile AL, Oct 3 1854, d London, Jul 3 1920] becomes chief sanitary officer for the project and introduces modern public health measures; *See also* 1898 TRA; 1914 TRA

The Heald Company introduces the piston-ring grinder; *See also* 1900 TOO; 1905 TOO

Small vacuum cleaners are introduced; unlike almost all later models, they are not powered by electricity; *See also* 1902 TOO

Reginald Aubrey Fessenden develops a depth finder (fathometer) that uses electromagnetic waves; *See also* 1900 TOO; 1938 TOO

Christian Hülsemeyer patents a detection system based on reflected radio waves, an early version of radar; *See also* 1900 TOO; 1931 TOO

The Trans-Siberian Railway, started in 1891, is completed in Russia; it links Moscow to Vladivostock; *See also* 1885 TRA

Louis Renault introduces the taxi, a car equipped with a meter for registering distance and speed

Benjamin Holt designs the modern tracked vehicle; *See also* 1908 FOO

A bridge with a 60-m (230-ft) reinforced concrete arch is built over the river Isar at Grünwald, Germany; it is designed by E. Mörsch; *See also* 1867 MAT; 1913 TRA

1905

GENERAL

Whitcomb L. Judson develops a new version of his "clasp locker" of 1893 that is called the hook-and-eye fastener, although its trade name is the punning C-Curity; *See also* 1893 GEN; 1913 GEN

Russian inventor Alexander Laurent develops a new agent for use in fire extinguishers; it is a mixture of sodium bicarbonate and aluminum sulfate; *See also* 1866 GEN

The firm of Siemens & Halske introduces the tantalum filament for light bulbs; it has a longer lifetime than carbon filaments because of the high melting point of the metal; around the same time various other metal filaments are tried, notably osmium and an alloy of osmium and tungsten, but these are not as successful as the tantalum filament; *See also* 1902 GEN

COMMUNICATIONS

Guglielmo Marconi invents the directional radio antenna; *See also* 1901 COM

ELECTRONICS & COMPUTERS

Albert Einstein [b Ulm, Germany, Mar 14 1879, d Princeton NJ, Apr 18 1955] explains why the photoemissive photoelectric effect, release of electrons as a result of light striking a metal, varies with the frequency of the light instead of with the intensity, invoking for the first time the idea that entities could be waves and particles at the same time; it is for this work that Einstein wins the Nobel Prize in physics in 1921; *See also* 1900 ELE

1906

GENERAL

General Electric Company patents a method for making a tungsten filament for use in incandescent lamps; *See also* 1904 GEN; 1907 GEN

Gustave Eiffel founds a laboratory for the study of aerodynamics; *See also* 1898 GEN

COMMUNICATIONS

On Dec 24 Reginald Aubrey Fessenden uses his high-frequency alternator to transmit music and voice via radio waves to ships at sea that he has equipped with receivers; the transmission is early AM radio—the amplitude (or intensity) of the radio waves varies according to an audio signal that is superimposed on it—although Fessenden calls it wireless telephony; Fessenden requests anyone who hears his signal to write to him, and he gets replies from as far away as the West Indies; *See also* 1904 COM

Arthur Korn transmits images over telegraph lines over a distance of 1600 km (1000 mi); *See also* 1902 COM; 1907 COM

The photostat is developed; a black-and-white photograph, a piece of art, or a printed page is made using a special camera and coated paper instead of film; one of the virtues of the photostat (or "stat") is that it can be exactly sized as it is made

ELECTRONICS & COMPUTERS

General Henry N.C. Dunwoody and G.W. Pickard each discover the rectifying properties of crystals, Dunwoody working with carborundum and Pickard using silicon; *See also* 1901 COM; 1947 ELE

1905

Safety glass is patented; it consists of two layers of glass and a layer of celluloid, and is widely used for automobile windows; *See also* 1893 MAT; 1919 MAT

Adolf von Baeyer of Germany wins the Nobel Prize in chemistry for his work on organic dyes; *See also* 1897 MAT

George Washington Crile [b Chile OH, Nov 11 1864, d Jan 7 1943] performs the first direct blood transfusion

Alexis Carrel [b Lyon, France, Jun 28 1873, d Paris, Nov 5 1944], working at the Rockefeller Institute in New York City, develops techniques for rejoining severed blood vessels, paving the way for organ transplantation; *See also* 1856 MED; 1963 MED

The first modern instance of the transplant of an animal organ into a human takes place, without success; *See also* 1963 MED

John B. Murphy [b Appleton WI, Dec 21 1857, d 1916] develops the first artificial joints for use in the hip of an arthritic patient; *See also* 1938 MED

Percy Bridgman [b Cambridge MA, Apr 21 1882, d Randolf NH, Aug 20 1961] develops the first devices that can produce pressures higher than 100 atmospheres; *See also* 1955 MAT

The Heald Company introduces the automobile-engine cylinder grinder; *See also* 1904 TOO; 1912 TO

The first German U-boat submarine is launched; *See also* 1798 TRA

Cameron B. Waterman patents the first commercially successful outboard motor for boats; *See also* 1883 TRA; 1909 TRA

Gabriel Voisin, Ernest Archdeacon, and Louis Blériot [b Cambrai, France, Jul 1 1872, d Paris, Aug 1 1936] start the first airplane factory in Billancourt, near Paris; *See also* 1903 TRA; 1908 TRA

1906

Jacques Arsène d'Arsonval [b Borie, France, Jun 8 1851, d Borie, Dec 31 1940] and George Bordas develop freeze-drying in Paris; *See also* 1898 FOO

Leo Baekeland, who has already made a fortune from his invention of the photographic paper Velox, discovers that a reaction between phenol and formaldehyde produces a synthetic resin; by 1909 he converts the resin into the plastic Bakelite; *See also* 1909 MATERIALS

Michael Semenovich Tsvett [b Asti, Italy, May 14 1872, d Voronezh (Russia) Jun 26 1919] invents chromatography as a means of separating plant pigments; it is popularized in the 1920s by Richard Willstätter [b Karlsruhe, Germany, Aug 18 1872, d Locarnon, Switzerland, Aug 3 1942]; *See also* 1938 MAT

Bacteriologist August von Wasserman [b Bamberg, Germany, Feb 21 1866, d Berlin, Mar 16 1925] begins to develop his famous test for syphilis when unexpected destruction of red blood cells appears in his work; initially very sensitive to the skills of the bacteriologist performing the test, it is gradually refined into a reliable test after several international "Wasserman conferences" arranged by the League of Nations and various improvements by workers in the field; *See also* 1908 MED; 1910 MED

Procaine is introduced as a local anesthetic in dentistry; *See also* 1904 MED

1907

The British Patent Act of 1907 includes provisions that require patentees to grant licenses when a patent is not "being worked" in the UK and revokes patents that are worked only outside the UK after a few years; *See also* 1883 GEN; 1919 GEN

The first commercial tungsten-filament incandescent lamps become available in the US; they are based on a pressed tungsten process developed by Alexander Just and Franz Hanaman; *See also* 1906 GEN; 1910 GEN

Arthur Korn telegraphs a photograph from Munich to Berlin, Germany, to inaugurate commercial tele-photography; *See also* 1902 COM

Auguste and Louis Lumière introduce the first commercial color photography system for use by amateur photographers; *See also* 1891 COM; 1912 COM

Reginald A. Fessenden and E.F.W. Alexanderson invent a high-frequency electric generator that produces radio waves with a frequency of 100 kHz; *See also* 1906 COM; 1908 COM

Boris Rosing works on a television system consisting of a mechanical camera and an electronic receiver

Lee De Forest [b Council Bluffs IA, Aug 26 1873, d Hollywood CA, Jun 30 1961] and R. von Lieben invent the amplifier vacuum tube (triode); it is based on the two-element vacuum tube invented by John Ambrose Fleming, but contains a third element, the grid, placed between the cathode and anode; the grid allows the modulation of the current through the valve with very small voltage changes; *See also* 1904 ELE; 1912 ELE

1908

The first form of electric razor is introduced; *See also* 1903 TOO; 1928 GEN

The American Institution of Chemical Engineers is founded; *See also* 1881 MAT

General Electric develops a 100 kHz, 2 kW alternator for radio communication purposes; *See also* 1900 COM

G. Smith and C. Urban develop a system for reproducing color pictures using green and red only; the color quality is too low for the system to be generally accepted; *See also* 1912 COM

French physicist Gabriel Jonas Lippmann wins the Nobel Prize in Physics for invention of the first method of color photography; *See also* 1891 COM

The international ampere is adopted by the International Conference on Electrical Units and Standards; a device called the current balance, based on a design by W.E. Ayrton and J. Viriamu Jones, is used at the National Physical Laboratory in England to measure the absolute ampere; the Weston cell is adopted as a standard voltage source; *See also* 1889 GEN; 1967 GEN

Thomas Alva Edison, after 8 years of experimentation, greatly improves his nickel-alkaline battery by adding lithium hydroxide to the electrolyte; although he does not know exactly why it works, the lithium solves most of the problems with the battery and the Edison Storage Battery Company resumes manufacture, halted in 1904, of batteries, primarily for use in electric-powered vehicles; *See also* 1903 ENE

Hans Holzwarth builds the first functional explosion gas turbine; *See also* 1903 ENE; 1930 ENE

Johann Stumpf starts designing uniflow steam engines; a large number of them are built in Germany and England; *See also* 1885 ENE

1907

Henri Moissan uses an electric furnace for the production of tungsten carbide, used in the manufacture of very hard steel; *See also* 1893 MAT

John Scott Haldane [b England, May 3 1860, d Bhubaneswar, India, Dec 1 1964] develops a method for deep-sea divers to rise to the surface safely; *See also* 1889 MED

Ross G. Harrison [b Germantown PA, Jan 13 1870, d New Haven CT, Sep 30 1959] demonstrates the *in vitro* growth of living animal tissue; *See also* 1876 MED; 1931 MED

W.V. Taggart introduces a practical method of casting gold inlays for teeth using wax impressions and lost-wax casting of molten gold; *See also* 1889 MED

The paint spraygun is invented

Leyner develops a water-powered drill for use in mines; *See also* 1908 TOO

The *Mauretania*, equipped with steam turbines and displacing 31,000 tons, is launched; it is the first ship to reach a speed of 25 knots; *See also* 1897 TRA

French bicycle dealer Paul Cornu builds the first helicopter that can take off vertically carrying a human; after a 20-second flight at 30 cm (1 ft), it breaks up on landing; *See also* 1877 TRA; 1908 TRA

1908

The Holt Company of California develops the first tractor to use moving treads, based on the work of Benjamin Holt; *See also* 1904 TRA; 1917 FOO

Melitta Bentz starts using circles of porous paper to prepare coffee by filtering the coffee through the paper; in 1909 she exhibits her filter coffeepot at the Leipzig trade fair; by 1912 she is manufacturing her own line of coffee filters; *See also* 1901 FOO

Jacques *Edwin* Brandenburger invents cellophane

Fritz Haber [b Breslau (Wroclaw, Poland), Dec 9 1868, d Basel, Switzerland, Jan 29 1934] invents in Germany the Haber process for extracting nitrogen from air in the form of ammonia; *See also* 1903 MAT; 1909 MAT

Heike Kamerlingh Onnes [b Groningen, Holland, Sep 21 1853, d Leiden, Holland, Feb 21 1926] succeeds in liquefying helium at a temperature only 4.2 °C (7.7 °F) above absolute zero; *See also* 1895 MAT

Alfred Wilm [b Niederschellendorf, Germany, Jun 15 1869, d Saalberg, Germany, Aug 6 1937] invents duraluminum, a strong, lightweight alloy of aluminum

The tuberculin test, a skin test for tuberculosis based on an immune reaction, is introduced; *See also* 1882 MED; 1906 MED

The Hughes Tool Company develops the steel-toothed rock-drilling bit, revolutionizing the oil industry by enabling drilling through hard rock for the first time *See also* 1907 TOO

Frederick G. Cottrell [b Oakland CA, Jan 10 1877, d Berkeley CA, Nov 16 1948] patents the electrostatic precipitator, which removes fly ash, dust, and mist particles from smoke from industrial smokestacks, a process that not only reduces air pollution but also allows some chemicals to be recovered and reused

Henry Ford starts building his Model T, powered by a 4-cylinder, 200-horsepower engine; *See also* 1896 TRA

Orville Wright makes the first airplane flight that lasts an hour; *See also* 1905 TRA; 1909 TRA

Elmer Ambrose Sperry [b Cortland County NY, Oct 12 1860, d Brooklyn NY, Jun 16 1930] patents gyrostabilization, a method for reducing roll in ships by tilting forward or backward the vertical axis of spinning gyroscopes; *See also* 1912 TRA

H. Anschütz-Kaempfe invents the gyroscopic compass; *See also* 1852 TRA; 1911 TRA

Louis Bréguet [b Paris, Jan 2 1880, d May 4 1955] builds the *Gyroplane*, a helicopter with four propellers; it lifts itself 4 m (13 ft) from the ground; *See also* 1907 TRA; 1923 TRA

1909

The first cigarette lighters are introduced; they use "flints" made from a magnesium-iron alloy named Auermetal after Karl Auer, Baron von Welsbach, who invented the alloy

The A.G. Spalding Company offers shoes coated in rubber as sport shoes, each weighing about 0.5 kg (1 lb); *See also* 1860 GEN

Guglielmo Marconi of Italy and Karl Ferdinand Braun of Germany win the Nobel Prize in physics for wireless telegraphy; *See also* 1901 COM

1910

William D. Coolidge [b Hudson MA, Oct 23 1873, d Schenectady NY, Feb 3 1975] develops a method for producing drawn tungsten filaments, a great improvement over the pressed filament previously used; he patents his method in 1913; *See also* 1907 GEN; 1913 GEN

Over 7 million telephones, most of them rented to customers by AT&T, are in use in the US; in addition, the first automatic telephone exchanges become available in the US; *See also* 1903 COM

Radio communications gain much publicity when the captain of the *Montrose* alerts Scotland Yard via radio that the escaping murderer Doctor Crippen is aboard his ship, leading to his capture; *See also* 1845 COM

Charles Proteus Steinmetz [b Breslau, Germany, Apr 9 1865, d Schenectady NY, Oct 26 1923] warns in *Future of electricity* about air pollution from burning coal and water pollution from uncontrolled sewage disposal into rivers

1911

The Bell System establishes a research branch, which in 1925 becomes the Bell Telephone Laboratories; *See also* 1902 GEN; 1925 GEN

A.A. Campbell Swinton gives the first description of a modern television system with an all-electronic scanning system using cathode-ray tubes in both the transmitter and the receiver; *See also* 1907 COM; 1919 COM

Panoramica film, which displays images with a similar aspect ratio as that of CinemaScope, is introduced; *See also* 1889 COM

Herman Hollerith merges his Tabulating Machine Company with two others to form the Computing Tabulating Recording Company, which becomes International Business Machines (IBM) in 1924; his tabulating machines, developed for the US census, are used in England for the first time for their census; *See also* 1896 ELE; 1935 COM

1909

FOOD & SHELTER

General Electric begins to market the world's first electric toaster; *See also* 1900 FOO

Peter Behrens [b 1868, d 1940] designs the AEG Turbine Factory in Berlin, Germany; it is the first factory to use modern curtain wall techniques, which are based on glass walls supported by a steel-and-concrete frame; *See also* 1892 FOO; 1918 FOO

MATERIALS

Leo Baekeland patents Bakelite, the first plastic that solidifies on heating and the first to be widely used; one day later British scientist James Swinburne files for a patent on the same substance; *See also* 1906 MAT; 1917 MAT

Karl Bosch [b Cologne, Germany, Aug 27 1874, d Heidelberg, Germany, Apr 26 1940] develops the Haber process; *See also* 1908 MAT; 1913 MAT

TRANSPORTATION

On Jul 25 Louis Blériot becomes the first human to fly across the English Channel in an airplane, the *Blériot XI*; the trip from Calais to Dover takes 37 minutes; *See also* 1905 TRA; 1927 TRA

American Ole Evinrude invents a lightweight outboard motor for boats, spurred by a need to cross a lake to buy ice cream for his fiancee; *See also* 1905 TRA

1910

MATERIALS

Rayon stockings for women are manufactured in Germany; *See also* 1884 MAT

MEDICAL TECHNOLOGY

Paul Ehrlich and Sahachiro Hata introduce salvarsan (arsphenamine), also known as 606, as a "magic bullet," or cure, for syphilis, launching modern chemotherapy; *See also* 1906 MED

Major Frank Woodbury of the US Army Medical Corps introduces the use of tincture of iodine as a disinfectant for wounds

TOOLS & DEVICES

Wolfgang Gaede [b Lehe, Germany, May 25 1878, d Munich, Germany, Jun 24 1945] develops the molecular vacuum pump; this pump creates a vacuum of 0.00001 mm of mercury and allows the production of improved vacuum tubes; *See also* 1873 TOO; 1913 GEN

Electric washing machines are introduced

TRANSPORTATION

Eugene Ely becomes the first person to take off in an airplane from the deck of a ship, showing that aircraft carriers are possible; *See also* 1922 TRA

1911

MATERIALS

William Burton [b Cleveland OH, Nov 25 1865, d Miami FL, Dec 29 1954] introduces thermal cracking for refining petroleum; *See also* 1855 MAT; 1913 MAT

Heike Kamerlingh Onnes discovers superconductivity, the ability of a substance to conduct electricity with no resistance; he finds the phenomenon in mercury that has been cooled to a temperature close to absolute zero; *See also* 1908 MAT; 1957 MAT

MEDICAL TECHNOLOGY

London doctor William Hill develops the first gastroscope, a tube that can be swallowed by a patient so that the doctor may look at the inside of the stomach through the tube

Physiologist Walter Bradford Cannon, in *Mechanical factors of digestion*, describes his use of bismuth compounds to make soft internal organs visible on X rays; *See also* 1897 MED; 1913 MED

TRANSPORTATION

On Jul 4 a Valkyrie B canard (pusher) airplane designed by Horatio Barber [b 1875, d 1964] delivers the first known air cargo in Great Britain, a carton of Osram lamps; *See also* 1914 TRA

Elmer Ambrose Sperry invents the gyrocompass; *See also* 1908 TRA; 1919 TRA

**1911
cont**

The Model T

Although Henry Ford cannot be credited with the invention of the automobile, his Model T and the innovations associated with it broke through the barriers that previously made motor cars high-priced toys. Ford paved the way for a world that, in industrialized nations at least, has been completely reshaped by high-speed, simple-to-use personal transportation.

The Model T was not Ford's first low-priced car; that was the Model N. (The Model K, earlier than the Model N, was an unsuccessful attempt at a high-priced vehicle.) The Model A, which followed the Model T, was a better and possibly a more successful automobile. Most of the very old Fords seen at antique car rallies are Model A's, not Model T's. But the Model T was the agent of change.

Henry Ford is remembered today more as an industrialist than as an engineer, but the Model T was essentially his own design. It featured a 20-horsepower, 4-cylinder gasoline-fueled internal combustion engine, ignition by a magneto, lubrication by splashing oil, high-tensile vanadium steel parts, and the famed planetary transmission, which eliminated the gear shift but needed repair at frequent intervals. Introduced near the end of 1908, it sold in two versions at $825 and $850 and in any color the customer wanted so long as that color was black. It continued to be manufactured until 1925. Mass production methods, introduced to the automobile industry by Ford, brought the price to a low point of $295 in 1922. By 1925 a new Model T emerged from the assembly line every 15 seconds.

Henry Ford paid his workers wages high enough to permit them to own their own product starting with the famous $5-a-day minimum in Jan 1914. This, combined with his creation of the assembly line and integration of raw material production into his operation, showed Ford to be the most original manufacturer of his time.

During the 1920s the rise of General Motors and Chrysler seriously eroded the sales of the Model T. Ford closed down the line and, with his engineers, developed the Model A, which was also innovative and even available in four different colors. With safety-glass windows, four-wheel brakes, and hydraulic shocks, the Model A was a great success, but it failed to dominate the market or the imagination the way the Model T did.

1912

American doctor Sidney Russell invents the heating pad, which later grows to become the electric blanket; *See also* 1930 GEN

Corona makes a portable manual typewriter; *See also* 1902 COM

The first automatic telephone exchange in England opens in Epson; it is operated by the Post Office, which has taken over control of telephone communications; *See also* 1892 COM

Siegrist and Fisher develop a subtractive form of color photography with color formers (dyes) embedded in three layers of emulsion; this is a predecessor of the commercial color photography systems called Agfacolour and Kodachrome; *See also* 1908 COM; 1935 COM

Lee De Forest develops the audion amplifier, a three-tube amplifier that increases the strength of an audio signal 120 times; *See also* 1907 ELE

Albert Einstein proposes the theory of interactions of electrons and light that will eventually give rise to the laser; *See also* 1905 ELE; 1957 TOO

Frederick Lowenstein develops the ion controller amplifier; it consists of a De Forest audion tube with a negative voltage applied to the grid (negative bias); *See also* 1907 ELE; 1915 ELE

Victor Kaplan patents his first turbine, a propeller turbine mounted on a vertical shaft; the propeller has movable blades and guide vanes so that its efficiency remains maximum at different workloads; *See also* 1895 ENE; 1919 ENE

1911 cont

German scientist P. Monnartz becomes the first to realize that stainless steels resist corrosion; *See also* 1904 MAT; 1913 MAT

Russian-British-Israeli chemist Chaim Weizmann [b Motol', Russia, Nov 27 1874, d Rehevoth, Israel, Nov 9 1952] discovers how to obtain acetone from bacteria involved in fermenting grain, providing an essential ingredient to Britain for cordite during World War I; *See also* 1889 MAT

About this time the Forbes ship's log is invented, a form of log that protrudes from the bottom of the ship instead of being dragged behind it; *See also* 1878 TRA

US inventor Charles Franklin Kettering [b Loudonville OH, Aug 29 1876, d Loudonville, Nov 25 1958] develops the first practical self-starter for automobiles

The first escalators are introduced at the Earl's Court underground station in London; a man with a wooden leg is hired to ride up and down regularly to demonstrate their safety; *See also* 1887 TRA

1912

The main improvement in the family of azo dyes derived from para red is the introduction by Zitscher and Laska of a fast dye often called naphthol red, a direct descendant of a dye called vaccine red; *See also* 1884 MAT

Polish-American biochemist Casimir Funk [b Warsaw, Feb 23 1884, d New York City, Nov 20 1967] coins the term *vitamin* for a class of substances that Frederick Hopkins had found to be important to health and which had previously been called accessory food factors; *See also* 1901 MED; 1915 MED

The Landis Company introduces the automatic-feed camshaft grinder for the US automobile industry; *See also* 1905 TOO

Nils Gustaf Dalen [b Stenstorp, Sweden, Nov 30 1869, d Stockholm, Dec 9 1937] wins the Nobel Prize for Physics for inventing automatic gas regulators for lighthouses and sea buoys

On Apr 15 the supposedly unsinkable *Titanic* fails on its maiden voyage after striking an iceberg; 1500 people perish; although it is intended to be kept afloat by watertight bulkheads, the rip in its side permits water to flow into enough compartments to sink the ship; *See also* 1985 TRA

Sulzer, a firm in Switzerland, builds the first diesel locomotive; weighing 85 tons, it develops 1200 horsepower; *See also* 1895 TRA

A prototype gyrostabilization system developed by Elmer Ambrose Sperry is installed in the 433-ton torpedo-boat destroyer *Worden*; two gyro wheels weighing 1814 kg (4000 lb) succeed in reducing roll from 30 to about 6 degrees; *See also* 1908 TRA

1913

Henry Ford introduces the first true assembly line, in which cars are carried along a conveyor belt at a speed slow enough for workers to assemble them, but fast enough to reduce the assembly time from 12.5 to 1.5 hours; *See also* 1924 GEN

Light bulbs are improved with better vacuums, based on invention of the Gaede pump for exhausting the air with a rotary cylinder instead of valves or pistons

Irving Langmuir [b Brooklyn NY, Jan 31 1881, d Falmouth MA, Aug 16 1957], while working for General Electric, improves the tungsten lamp by filling the bulb with an inert gas so that atoms of tungsten will evaporate more slowly from the filament; he also develops the coiled tungsten filament; *See also* 1910 GEN; 1934 GEN

Gideon Sundback, hired to improve Whitcomb Judson's C-Curity hook-and-eye fastener, develops a hookless form; the Hookless Fastener Company is founded to take advantage of the new invention, but the company fails because the new fasteners quickly wear out in use; *See also* 1905 GEN; 1914 GEN

The soap pad is introduced

The cascade-tuning radio receiver and the heterodyne radio receiver are introduced; *See also* 1901 COM; 1914 COM

Vacuum triodes are used in repeaters in telephone lines for the amplification of weak signals; the first "audion repeaters" are used on the New York to Baltimore line; *See also* 1903 COM; 1914 COM

Edouard Belin invents the portable facsimile machine (fax), which he calls the Belinograph, but which journalists of the time know as the Belino; capable of using ordinary telephone lines, it quickly replaces the fax machine of Arthur Korn; *See also* 1902 COM; 1914 COM

The term "movie" for motion picture begins to be used; *See also* 1896 COM

British engineer Harry Ralph Ricardo [b London, Jan 26 1885, d Graffham, England, May 18 1974] demonstrates that knock in gasoline engines is caused by the spontaneous detonation of the gas mixture in the cylinder caused by compression; *See also* 1893 ENE; 1921 ENE

Flight and the Wrights

A number of inventors and scientists of the 19th century worked on the problem of heavier-than-air flight with extremely limited success. Some of them, such as Otto Lilienthal and Percy Pilchard, lost their lives in attempts at flight, while others, such as Clément Ader and Samuel Langley, just lost expensive machines. The basic concepts of both the airplane and the helicopter were well known to these would-be aviators, but the technical development of adding an engine to a glider or enlarging a toy helicopter to carry a person was beyond them.

A famous instance concerns Langley's well-publicized attempts, financed by the US government, which ended with crashes in the Potomac River. In 1903 *The New York Times* editorialized over money wasted on Langley because it was clear that people would never be able to fly in heavier-than-air craft. Nine days later, on Dec 17, Orville Wright proved the *Times* wrong. Despite good press coverage at the time, the Wrights' feat was not accepted at first as the beginning of heavier-than-air flight. Two years later the *Scientific American* was labeling the Wright brothers' flights, by then reaching distances of nearly 40 km (14 mi), as a hoax.

Wilbur and Orville Wright, in Dayton OH, had become excited about aviation on reading of Lilienthal's experiments. Just before Lilienthal's fatal glider accident in 1896, Wilbur wrote to the Smithsonian Institution for all available information about flight. Thus began the brothers' systematic study of everything known about heavier-than-air flight.

Eventually the two, who ran a bicycle repair shop for a living, solved all of the problems that bedeviled the earlier experimenters. In Jul 1899 Wilbur invented what became their patented steering system, which, although changed in many details since, is still the basic method by which planes are steered. Previously, all experimenters had tried to steer by shifting the weight of the pilot. Wilbur demonstrated to Orville with a cardboard box how wings could be twisted to change their angle of attack. Details were worked out in wind tunnels and during glider flights.

The Woolworth Building is constructed in New York City; at 232 m (791 ft), it is the tallest office building in the world for the next quarter century, although shorter than the Eiffel Tower; *See also* 1903 FOO; 1930 FOO

The Los Angeles aqueduct, about 346 km (215 mi) long, is completed; it carries water from the Owens River in the Sierra Nevada Mountains; *See also* 230 CE ARC

Carl Gustav Patrik de Laval perfects a vacuum milking machine; *See also* 1890 FOO

German chemist Friedrich Karl Bergius [b Goldschmieden, Poland, Oct 11 1884, d Buenos Aires, Argentina, Mar 30 1949] introduces hydrogenation of coal at high pressure for producing gasoline; the industrial application of the process starts in 1924; *See also* 1911 MAT

Badische Anilin- & Soda-Fabrik (BASF) begins production of ammonium sulfate using the Haber-Bosch process; *See also* 1909 MAT

Henry Brearly develops a stainless steel consisting of an alloy of steel and chrome; *See also* 1911 MAT; 1914 MAT

German surgeon A. Salomen develops mammography for diagnosing breast cancer; *See also* 1911 MED

John Jacob Abel [b Cleveland OH, May 19 1857, d Baltimore MD, May 26 1938] develops the first artificial kidney; *See also* 1943 MED

Emil von Behring introduces a toxin-antitoxin mixture for immunizing children against diphtheria; *See also* 1901 MED; 1923 MED

Bela Schick [b Boglar, Hungary, Jul 16 1877, d New York City, Dec 6 1967] introduces the Schick test for diphtheria; *See also* 1908 MED

Hans Geiger [b Neustadt-an-der-Haardt, Germany, Sep 30 1882, d Potsdam, Germany, Sep 24 1945] and Ernest Rutherford develop the first form of the radiation detector that comes to be known as the Geiger counter; it detects alpha particles (helium nuclei); *See also* 1928 TOO

William David Coolidge invents a hot-cathode X-ray tube; *See also* 1896 MED

American engineer Frederick Kolster [b Geneva, Switzerland, Jan 13 1883] develops a radio-compass system with transmitters on the New Jersey coast; *See also* 1906 COM; 1928 TRA

Igor Sikorsky [b Kiev, Russia, May 25 1885, d Easton CT, Oct 29 1972] builds and flies a multiengine airplane

The Langwies Viaduct carrying the Coire-Arosa Railroad over a deep gorge in Switzerland demonstrates the first successful use of reinforced concrete in a railway bridge; previous bridges of the material were shaken apart by vibrations caused by trains passing over them; *See also* 1904 TRA; 1936 TRA

Flight and the Wrights (Continued)

Langley's main problem was power, however, not steering. After Langley's death, a copy of his airplane fitted out with a more powerful engine flew perfectly well. The Wright brothers solved the power problem as well. Not only did they build a better and lighter engine, they also built a lightweight plane that required less power. Their first plane and its powerful engine together weighed only 340 kg (750 lb).

One of the leading theorists of aviation of the time was Octave Chanute, who had written, in 1894, a detailed account of all previous experiments in *Progress in flying machines*. On May 13 1900, the Wrights entered into a productive correspondence with Chanute, who aided their research but seems not to have made any direct contribution.

Typical of the Wright brothers' systematic attack was their use of US Weather Bureau records to locate Kitty Hawk NC as the best place for glider experiments. Over the next 2 years, during lengthy trips there, they worked at improving their gliders. In 1901 they adjusted the curve of the top of the wings to prevent the kind of accident that killed Lilienthal. In 1902 Orville found a way to improve steering by making the tail movable. Encouraged by their glider successes, the brothers devoted the win-

ter and spring of 1902–1903 to developing the lightweight internal combustion engine.

Back at Kitty Hawk, bad weather and engine problems postponed their first trial until Dec 14. It failed as the engine stalled. Three days later the brothers alternated as pilots on four successful short flights.

The Wright brothers had from the beginning planned to make their fortunes with the airplane, but it took a while. Failing to interest the US government, which had lost $50,000 on Langley, Wilbur Wright took a plane to Europe in 1908, where his flights inspired wonder and admiration in France and Italy. By then others had also learned to fly, among them Louis Blériot, who built his own airplanes and dramatically flew one across the English Channel.

After the Wrights' flights before the crowned heads of Europe and Blériot's headlines, there were no more doubters. Commercial aviation, with the first passengers and freight, began shortly after. World War I produced hundreds of experienced pilots and many advances in aircraft design. Regularly scheduled airlines, taking advantage of the pilot pool, began in Europe in 1919. Orville Wright (Wilbur died young of typhoid fever) finally did make his fortune selling airplanes.

1914

The first modern sewage plant, designed to treat sewage with bacteria, opens in Manchester, England; *See also* 1896 GEN

Henry Ford announces that workers who meet minimum qualifications will be paid at least $5 a day; *See also* 1894 GEN; 1924 GEN

New York socialite Mary Phelps Jacob designs and patents a brassiere made with elastic; the patent is sold to the Warner Brothers clothing manufacturers; Jacob's design was based on an impromptu version of 1913 whipped up from two handkerchiefs; *See also* 1939 GEN

Gideon Sundback develops a much improved version of his hookless fastener, which is first sold as Hookless No. 2 on Oct 28; essentially, Hookless No. 2 is what we know today as the zipper; *See also* 1913 GEN; 1923 GEN

Edwin Howard Armstrong [b New York NY, Dec 18 1890, d New York NY, Feb 1 1954] patents a radio receiver circuit with regeneration (positive feedback) on Oct 6 that he developed while a student at Columbia University; part of the amplified high-frequency signal is fed back to the tuning circuit to enhance selectivity and sensitivity; *See also* 1918 COM

The East and West coasts of the US are linked by a telephone line; signals are amplified by seven repeater stations

Edouard Belin's Belino portable fax machine sends the first remote photo news story (from World War I) over telephone lines; *See also* 1913 COM; 1974 COM

Edward Kleinschmidt [b Bremen, Germany, Sep 9 1875] invents the teletypewriter

The Gulf Refining Company begins distributing free highway maps in Pittsburgh PA; these are quickly nicknamed "gas maps"; highway maps continue to be free at auto service stations until the 1970s, when oil shortages and other economic problems end the practice

William T. Price develops a diesel engine in which a gas mixture is ignited by an electrically heated nichrome filament; this motor requires less compression to operate; *See also* 1893 ENE

Steel for strength

Steel is iron alloyed with a small percentage, often less than 1 percent, of carbon. The first steel was produced by accident in iron making. The superior form of metal was then reproduced by craftspeople who learned to replicate the accident. Such steel was known in the Near East and in India hundreds of years before steel began to be made deliberately in the West. In 1614 a process of forcing carbon into iron was developed; it resulted in layers of steel on the exterior of iron. The resulting cementation or blister steel could be hammered into tools with very hard edges. By the early 18th century René de Réaumur had studied the process carefully and observed the role of carbon. From there it was a short step to melting blister steel and reforming it, producing the first steel that was uniform throughout. After another hundred years, several inventors developed methods for making steel in large quantities directly from iron ore and coke. After that, steel gradually came to be used in all applications requiring great strength or hardness. Steel frames for large buildings and bridges were introduced in the 1880s and have been used ever since.

Shortly before the first large-scale applications of steel, Robert F. Mushet became the first of a large number of metallurgists to improve steel by alloying it with other metals. Mushet steel used tungsten and manganese for extra hardness, important for the cutting edges of lathes and other machine tools. The first corrosion-resistant steel, known as stainless steel, was developed near the beginning of the 20th century, and a variety of stainless steels have been important for many applications since. Chromium alloys, called chrome steel, are both stainless and hard; one use has been in girders exposed to the weather. Nickel steels, developed as early as 1889, were first used for armor plating on battleships. Other alloys commonly include molybdenum, silicon (like carbon, not a metal itself), vanadium, and copper, or combinations thereof. Today's engineers can choose from a large array of specialty steels, each with its own good and bad points.

The firm Krupp in Germany develops a type of stainless steel consisting of steel mixed with 18 percent chrome and 8 percent nickel; *See also* 1913 MAT

Alexis Carrel performs the first successful heart surgery on a dog; *See also* 1893 MED; 1936 MED

Physicist Irving Langmuir invents a mercury-vapor condensation pump that creates a nearly perfect vacuum; *See also* 1910 TOO

The Boston Wire Stitcher Company offers the first desk models of staplers that can be used as modern ones are employed for joining loose sheets of paper; the staples are inserted either loose or in paper wrappings, which makes them difficult to put in and the machines prone to jamming; *See also* 1896 TOO; 1923 TOO

On Aug 14 the Panama Canal, constructed largely under the direction of chief engineer George W. Goethals [b Brooklyn NY, Jun 29 1858, d New York City, Jan 21 1928], is officially opened; it is 82.4 km (51.2 mi) long with a minimum depth of 11.8 m (38.8 ft) and a minimum width of 153 m (500 ft); most of the route is 26.5 m (87 ft) above sea level; *See also* 1904 MED; 1978 TRA

Pilot Tony Jannus runs the St. Petersburg-Tampa Airboat Line in FL for about three months, carrying 1200 passengers during that time across Tampa Bay in a Benoist airboat; this is believed to have been the first passenger airline; it compiles a 100 percent safety rating before going out of business; *See also* 1911 TRA

Elmer Ambrose Sperry invents the autopilot, but it does not go into commercial use until about 20 years later; *See also* 1911 TRA; 1932 TRA

Engineer Robert Hutchings Goddard [b Worcester MA, Oct 5 1882, d Baltimore MD, Aug 10 1945] starts developing experimental rockets; *See also* 1903 TRA; 1919 TRA

Red and green traffic lights are introduced for the first time, in Cleveland OH; *See also* 1918 TRA

Aluminum for lightness

Iron, the main component of steel, is only the second most abundant metal in Earth's crust. Roughly, there are four atoms of aluminum for every three of iron. But aluminum is not available from its ores by simple heating. It was found early in the 19th century by chemical methods and, as early as 1833, produced by electrolysis. Because of the difficulty of producing it, high-priced aluminum was for a time employed in jewelry. The tip of the Washington Monument in the US is aluminum, just as earlier obelisk "little pyramids" were sometimes made from gold. The year after the monument was completed, inventors in the US and France independently discovered a process for making aluminum cheaply, provided electric power could be produced at low cost.

Because aluminum is much less dense than steel (weighs less per unit of volume), it was applied to aircraft. The first rigid-hulled airship, in 1897, used an aluminum frame and sheeting. It leaked, but later airships and soon almost all airplanes employed aluminum in sheathing. The rise of air travel in the early 20th century was one of the main factors in establishing the aluminum industry.

Aluminum has many advantages over other metals. In addition to its low density, it is nearly as conductive of both electricity and heat as copper. Exposed to air, it forms a thin outer coat of aluminum oxide that then prevents further corrosion.

Pure aluminum, however, is weak, bending easily under stress. While this is an advantage in aluminum foil, which has replaced tinfoil for almost all purposes, it is not very helpful in nearly any other use. Early in the 20th century, the first strong alloy of aluminum, called duraluminium or duraluminum, was introduced. Various forms of duraluminium, based on copper, magnesium, and manganese in small percentages, allowed the use of aluminum for structural elements in aircraft and for pistons in airplane engines. Other alloys of aluminum have been developed for special purposes.

Nearly pure aluminum, aside from its use in foil, is most familiar in the form of cans for beverages. Since the end of the 1950s, these have gradually displaced tin-coated steel cans for beer and soft drinks. Increasingly, aluminum cans are being used for other products. Aluminum is also widely employed to reduce weight in household appliances and in cookware, where its combination of corrosion resistance and conductivity make it more popular than copper, which requires polishing with special chemicals to stay shiny and a lining of some other metal to prevent toxic amounts from entering food.

1915

London and Birmingham, England, are connected by underground telephone cables; overhead telephone cables gradually disappear within cities; *See also* 1914 COM

The first transatlantic radiotelephone conversation takes place between Arlington VA and the Eiffel Tower in Paris

The first North American transcontinental telephone call is between Alexander Graham Bell in New York City and Thomas A. Watson in San Francisco; Watson can hear his former employer saying from a distance of 4800 km (3000 mi): "Mister Watson, come here, I want you," repeating his famous first message of Mar 10 1876; *See also* 1914 COM

The radio tube oscillator is introduced; *See also* 1913 COM

American physicist Manson Benedicks discovers that a germanium crystal can convert AC current to DC, a discovery that leads to the development of the chip; *See also* 1906 ELE; 1947 ELE

A.W. Hull patents the dynatron, a vacuum tube in which the increase in the anode voltage results in a decrease of the anode current because of secondary emission of electrons from the anode; *See also* 1907 ELE

Nitrogen: A matter of life and death

We live in a sea of elemental nitrogen, since nearly four out of five air molecules are two nitrogen atoms joined by a chemical bond. Despite this apparent oversupply, the availability of nitrogen in a more usable form has been in one way or another a major problem to humans for the past 10,000 years and perhaps much longer than that. To understand why, it helps to consider nitrogen chemistry, its importance to living creatures, and the normal nitrogen cycle in the environment.

The nitrogen atom has five electrons in its outer shell. This enables it to form compounds with other common elements, notably oxygen and hydrogen, but only with some help. Nitrogen combines with oxygen at the high temperatures in automobile engines, but not at the lower temperatures found in small fires, such as candle flames. Once a combination exists, a relatively small amount of energy can free the oxygen or hydrogen for other uses, releasing the relatively unreactive nitrogen at the same time.

A combination of nitrogen and hydrogen is the "amino" of amino acids, the twenty-odd compounds that form all the proteins in the body. The NH_2 amino radical converts an organic acid of carbon, oxygen, and hydrogen into a unit that easily links via the nitrogen atom with other amino acids. Such polymers are not only the building blocks of life, as we were taught in school, but also the enzymes that cause the processes of life to take place. Furthermore, genes also rely on nitrogen atoms.

Nitrogen is essential for life and it is abundant in air, but there is a catch. Ordinary chemical processes do not remove the nitrogen from the air. With five electrons in its outer shell, the nitrogen atom seeks three more to form a stable molecule. Consequently, two nitrogen atoms forming a molecule arrange to share three of their outer atoms, giving the molecule a triple bond instead of the more common double bond. Each atom treats three of the electrons belonging to the other atom as its own, so each atom can claim eight outer electrons. Such a triple bond is hard to break. Before life existed on Earth, there was only one force strong enough to split the nitrogen molecule—lightning. Freed from each other in a lightning stroke, some of the nitrogen atoms grab an oxygen atom or hydrogen atom (from water vapor) or both. The resulting nitric acid or nitrogen oxides eventually wind up on Earth. Some of the nitric acid reacts with sodium or potassium to form a soluble mineral.

The Corning Glass Works in NY develop Pyrex, a heat-resistant glass consisting of 12 percent boron oxide and 80 percent silicon oxide; *See also* 1884 MAT

Physician Joseph Goldberger [b Girált, (Giraltovce, Slovakia), Jul 16 1874, d Washington DC, Jan 17 1929] establishes that a vitamin deficiency causes pellagra; *See also* 1912 MED

French scientist Paul Langevin (b Paris, Jan 23 1872, d 1946) invents sonar, mainly as a method for ships to detect icebergs; *See also* 1917 TRA

1915

Nitrogen: A matter of life and death (Continued)

Nitrogen compounds from lightning or from mineral deposits were sufficient to get life started nearly 4 billion years ago. Subsequently, some bacteria developed another way to obtain nitrogen from the air. This cannot be easy to do, as no other living creatures have evolved such an ability. Some plants, however, have made suitable living arrangements with the bacteria, providing homes for the bacteria in return for nitrogen rent.

Long before any of this was known by people, farmers had developed processes to put more nitrogen into soil. They learned that growing the same plant in a field each year and removing the plant with its nitrogen to use for food or fodder resulted in poor crops after a few years (unless it harbored nitrogen-fixing bacteria). Practices to sustain nitrogen in the soil developed, notably crop rotation and the use of animal manure as fertilizer. Although animals need nitrogen, their diet nearly always contains far more than is required, so animals regularly excrete nitrogen.

As the human population increased, the need for nitrogen to grow plants grew faster than local supplies. A major source of nitrogen in the 19th and early 20th centuries was guano, highly nitrogenous bird waste from islands off Peru. Lack of rainfall and large nesting populations of seabirds had produced mountains of guano over thousands of years. After exploitation began, the guano was mined faster than it was deposited.

Also during the 19th century, an important new use for nitrogen compounds emerged. Many nitrogen compounds contain large amounts of oxygen bound loosely to the compound. A small amount of heat or even a light tap can provide enough energy to release some of the oxygen, which then combines with other elements, producing more energy along the way. The result is a sudden explosion. Such a reaction is at the heart of gunpowder, where potassium nitrate gives the mixture its kick. In 1846 an even more explosive combination of nitrogen and cellulose was found by accident; this discovery set off a wave of development that led to nitroglycerin, dynamite, TNT, plastic explosives, and other blasting compounds. As a result, nitrogen was coveted by generals and farmers alike.

Getting nitrogen from the air was the obvious solution, but the only way known to do this during the 19th century was by imitating lightning. Although generation of electricity in large amounts began at the end of the century, efforts to produce nitrogen this way were not very effective.

Shortly before World War I, two patriotic Germans solved the problem of fixing nitrogen, with Fritz Haber working out the basic chemistry and Karl Bosch developing a manufacturing process. Bosch's first big plant was finished shortly after the start of the war. Before the Haber-Bosch process, the best source of nitrogen for explosives had been potassium nitrate, found in the driest desert in the world in Chile—a long way from Germany and easily defended. If it had not been for the Haber-Bosch process, Germany would have run out of ammunition in 1916 and have had to surrender. Instead, the war went on until Germany was defeated by superior force. (Haber had also developed poison gases that almost turned the tide in Germany's favor; a few years after the war, Haber had to leave his beloved Germany because he was a Jew.)

Since World War I the Haber-Bosch process has contributed both to fertilizer and explosives production. Even as fertilizer, nitrogen compounds have proven to be a mixed blessing. Runoff from heavily fertilized farmlands has polluted lakes and streams.

The British introduce nonrigid airships known as blimps (thought to be named for their designation as "type B, limp") to use as scout vehicles during World War I; *See also* 1900 TRA; 1937 TRA

The British army develops the tank largely as a way to overcome German barbed wire emplacements; *See also* 1874 FOO

The first all-metal airplane, the Junkers J-1, is built in Germany; it is also the first plane to have cantilever wings that are directly attached to the fuselage instead of supported by braces and wires

Fokker aircraft become the first airplanes to be equipped with a machine gun that can fire between the blades of a moving propeller

Detroit blacksmith August Fruehauf invents the tractor trailer, a truck with the cab and engine separate from the main body of the truck

In Germany, the Mittelland Canal is started; *See also* 1914 TRA; 1938 TRA

1916

1917

A 50 kW electronic alternator designed by Ernst F.W. Alexanderson [b Uppsala, Sweden, Jan 25 1878] starts operating in New Brunswick NJ; *See also* 1907 COM; 1918 COM

1918

The Kaiser Wilhelm Geschutz, a German cannon, achieves a long-distance record for conventional gunpowder-powered artillery by shelling Paris from 122 km (76 mi) away; *See also* 1966 TOO

Edwin H. Armstrong develops the superheterodyne radio receiver; the incoming signal is mixed with the signal of an internal oscillator that is tuned simultaneously with the receiving circuit; the signal is transformed to a lower fixed frequency (the intermediate frequency), amplified, and demodulated; the principle behind this receiver is the basis for all radio receivers now in use; *See also* 1914 COM

A 200 kW alternator starts operating at Station NFF, the naval station at New Brunswick NJ; it is the most powerful radio transmitter of its time; *See also* 1917 COM

J. Abraham and E. Bloch build a calculating machine based on the binary numbers 0 and 1; 0 corresponds to an open switch and 1 to a closed switch; *See also* 1867 GEN; 1937 ELE

FOOD &
SHELTER

MATERIALS

MEDICAL
TECHNOLOGY

TOOLS &
DEVICES

TRANSPORTATION

1916

Artificial detergents are developed in Germany in order to preserve fats for the manufacture of soap; *See also* 1957 MAT

Kotaro Honda [b Aichijen Prefecture, Japan, Feb 1870, d Tokyo, Feb 12 1954] discovers that adding cobalt to tungsten steel produces increased magnetic strength, leading to the development of powerful magnetic alloys such as alnico

American engineer John Fisher develops the first modern washing machine; *See also* 1910 TOO

Windshield wipers for automobiles are introduced in the US

1917

Clarence *"Bob"* Birdseye [b Brooklyn NY, Dec 9 1886, d New York City, Oct 7 1956] begins to develop freezing as a way of preserving foods after experiencing success with naturally frozen foods in Labrador; *See also* 1906 FOO; 1923 FOO

Henry Ford introduces the Fordson tractor—lightweight, cheap, and reliable; *See also* 1901 FOO; 1919 FOO

The first commercial product made from Bakelite, the first plastic, is introduced—the gearshift knob for a Rolls-Royce automobile; *See also* 1909 MAT

Physicians in the US learn that children develop lead poisoning by eating paint chips that have peeled off of interior walls, especially in poorer neighborhoods, where paint is more likely to be allowed to peel; *See also* 1890 MED; 1955 MED

Ralph Parker [b Malden MA, Feb 23 1888, d Sep 4 1949] develops a vaccine against Rocky Mountain spotted fever; *See also* 1897 MED

Max Mason [b Madison WI, Oct 26 1877] invents the submarine detector; *See also* 1915 TRA

The Quebec Railway Bridge over the St. Lawrence River in Canada, which took 17 years to build, partly because of major accidents in 1907 and 1916, opens; at 335 m (1100 ft), it is the longest steel cantilever bridge ever built

1918

Willis Polk is the first architect to use the curtain wall construction technique, pioneered by Joseph Paxton in 1849, in a large urban structure, the Halladie Building in San Francisco CA; the technique is later made famous by Mies van der Rohe in office buildings in New York City in the 1950s; *See also* 1909 FOO; 1958 FOO

Kelvinator launches a mechanical refrigerator for home use that is the first successful entry into the field; *See also* 1881 FOO

Tooth implants, driven directly into patients' jaws instead of being surgically implanted, are developed; *See also* 1889 MED

The electric beater for mixing foods is introduced; *See also* 1919 FOO

Traffic lights incorporating a yellow warning light are introduced in Great Britain; *See also* 1914 TRA

Bell invents a high-speed hydrofoil boat; *See also* 1900 TRA

1919

Paper cups manufactured by a successor to the Public Cup Vendor Company are renamed Dixie Cups

Women are allowed to vote in the US

The Treaty of Versailles is signed, formally ending World War I

The British Patent Act of 1919 reinterprets provisions of previous acts that require patentees to grant licenses when a patent is not "being worked" in the UK and that revoke patents that are worked only outside the UK after a few years; *See also* 1907 GEN

Shortwave radio is developed; *See also* 1921 COM

The Radio Corporation of America (RCA) is founded; *See also* 1921 COM

Charles Francis Jenkins [b Dayton OH, Aug 22 1867, d Jun 6 1934] applies for a patent on a television system using spinning prismatic rings; *See also* 1911 COM; 1922 COM

William H. Eccles [b Ulverston, England, Aug 23 1875] and F.W. Jordan publish a paper on flip-flop circuits; first used in electronic counters, flip-flop circuits become important components in computers around 1940; *See also* 1918 ELE; 1937 ELE

Hugo Koch invents the Enigma, an encoding machine that provides for 22 million combinations; it is used in World War II by Germany; *See also* 1940 ELE

A turbine designed by Victor Kaplan is installed in an Austrian textile mill, producing 25.8 horsepower; *See also* 1912 ENE; 1920 ENE

1920

Alfred P. Sloan [b New Haven CT, May 23 1875, d New York City, Feb 17 1966] of General Motors introduces the concept of a modern business "corporation," delegating operating decisions to divisions, but centralizing financial control and planning

The first regular licensed radio broadcast, by station KDKA, operated by the Westinghouse Electric and Manufacturing Company in Pittsburgh PA is heard; KDKA is also the first station to transmit presidential election results; *See also* 1906 COM; 1948 COM

Lee De Forest develops a system to record sound synchronized with pictures on a separate track on photographic film used for cinema; he employs a photoelectric cell for sound reproduction; this system is used a few years later in the first talking motion pictures; *See also* 1926 COM

Nasavischwilly proposes the use of magnetic tape for the recording of sound; *See also* 1927 COM

Associated Electrical Industries is the first to reheat steam for reuse in turbines on a commercial scale; the reheating prevents loss of the latent heat in steam after it has done some work by passing through the turbine; nearly all electrical generators that use turbines in later years adopt this concept; *See also* 1895 ENE

German aerodynamic engineer Albert Betz proves that the maximum amount of energy from the wind that can be extracted by a windmill is 59.3 percent; *See also* 1929 ENERGY

Several installed Kaplan turbines show damaged blades; Victor Kaplan himself discovers that the pitting on these blades is caused by cavitation—the formation and collapse of low-pressure bubbles in the liquid; *See also* 1919 ENE

1919

Universal Company, introduces the electric blender in the US; this version is primarily intended for commercial, not home, use; *See also* 1918 TOO; 1937 FOO

The NE State Legislature institutes tests that tractors must pass before they can be sold in the state; of the first 103 tractor models tested, only 39 pass; *See also* 1917 FOO; 1924 FOO

A.A. Griffith [b England, 1893, d 1963] conducts experiments on the fracture of ordinary glass to find the surface tension of a solid; *See also* 1920 MAT

Louise Pearce [b Winchester MA, Mar 5 1885, d Aug 10 1959] discovers a compound, which becomes known as tryparsamide, that cures sleeping sickness

British physicist and engineer Robert Alexander Watson-Watt [b Brechin, Scotland, Apr 13 1892], patents his "radiolocator," a device for locating ships or planes by means of radio waves; because he uses shortwave radio waves instead of microwaves, the precision is low; *See also* 1924 TOO; 1931 TOO

The *Aquitania* becomes the first merchant ship to navigate with a gyrocompass; *See also* 1911 TRA

Lockheed develops the hydraulic brake system for automobiles; *See also* 1902 TRA

The British airship R 34 is the first airship to cross the Atlantic Ocean; *See also* 1902 TRA; 1927 TRA

A method of reaching extreme altitudes by Robert Hutchings Goddard suggests sending a small vehicle to the Moon using rockets; adverse press comments make Goddard disinclined to expose himself to further ridicule; *See also* 1914 TRA; 1924 TRA

1920

The Remington company recognizes that the use of chlorate in the primer for bullets is the chief cause of rusting in gun barrels, and begins to produce bullets that are chlorate-free, ending the necessity for constant cleaning of small arms

The last blast furnace in England still using charcoal to smelt iron, at Backbarrow near Windermere, is modernized to use coke; *See also* 1735 MAT

A.A. Griffith publishes the basis for a theoretical treatment of the strength of solids and fracture mechanics; *See also* 1919 MAT

Johnson & Johnson introduces the Band-Aid, a sterile-packaged individual bandage using tape to hold an absorbent pad in place over a minor cut

Marshal B. Lloyd invents a power loom for weaving flat sheets of wicker, providing the first fiber "fabric" that can be created by weaving it apart from a framework; the new form of wicker quickly becomes popular in US middle-class furniture until replaced about 10 years later with furniture based on metal instead of wood and wicker; *See also* 1879 MAT

Retired US Army officer John T. Thompson [b Newport KY, Dec 31 1860, d Great Neck NY, Jun 21 1940] patents his submachine gun, later famous as the tommy gun; *See also* 1884 TOO

1921

Playwright Karel Câpek coins the term *robot* to describe the mechanical people in his play *RUR*; *See also* 1962 TOO

RCA starts operating "Radio Central" on a site in Port Jefferson, Long Island NY; the transmitter is of an arc type, but because better vacuum-tube transmitters are becoming available, the project is not completed; *See also* 1919 COM

The American Radio League establishes contact via shortwave radio with Paul Godley in Scotland, demonstrating for the first time that shortwave radio is well suited for long-distance communication; *See also* 1919 COM

Albert W. Hull [b Southington CT, Apr 19 1880, d 1966] invents the first form of magnetron, an electron tube placed in a magnetic field that produces microwaves; at that time normal vacuum tubes could not oscillate at microwave frequencies; *See also* 1932 COM; 1937 ELE

Thomas Midgley, Jr. [b Beaver Falls PA, May 18 1889, d Worthington OH, Nov 2 1944] discovers that tetraethyl lead prevents knock in gasoline engines; this makes gasoline with the antiknock additive suitable for use in high-compression engines; *See also* 1913 ENE; 1978 MAT

Robots, fantasy and fact

The idea of a mechanical humanoid figure that performs humanlike tasks has fascinated people since antiquity. Philon of Byzantium, who lived during the 3rd century BC, described such automatons. In the Far East inventors built mechanical puppets that played music and moved like humans. From the Renaissance on, the builders of humanoid figures (and of animals) reached a high level of mechanical skill, incorporating clock-work and pneumatic techniques. Well-known examples are Vaucanson's flute player that could play a genuine flute and humanoid figures that could write and draw built by Pierre Droz and his son in the 1770s.

The term *robot* was coined in 1921 by the playwright Karel Câpek in his play *R.U.R.* (the title stands for Rossum's Universal Robots, the name of the manufacturer of imitation people). Câpek chose to call his mechanical humanoids robots from the Polish word for menial worker or drudge. In the play, the robots ultimately destroy humankind, a gloomy view shared by many who believe that technology will lead one day to destruction.

There are no effective mechanical humans yet, but machines that can replace human workers at some tasks are increasingly common. Such machines have taken over the title robot. Several hundred thousands of these 20th-century robots paint, weld, assemble cars, mount electronic parts on printed circuits, and perform a multitude of operations in manufacturing. Such industrial robots have been in use for over a quarter of a century: Unimation built the first one in 1962; General Motors installed robots in its assembly lines the same year.

Robots also serve for exploration and work in hazardous environments. They repair defects inside nuclear reactors, explore the deep-sea bottom, or perform experiments on the surface of a distant planet.

Modern robot technology has several origins. It partially derives from the numerically controlled machine tool, first developed in the late 1940s. The manipulators used for handling radioactive materials, called waldos, are also predecessors of industrial robots. Like robots, such handlers have a name that is derived from a science fiction tale. Waldos are remotely controlled by human operators.

1922

George Squier [b Dryden MI, Mar 21 1865, d Mar 24 1934] patents a method of transmitting music over wires and forms the company that will become the Muzak Corporation in 1934; Muzak becomes the main purveyor of "piped-in" music to stores, elevators, and offices

FM (frequency-modulated) radio is described in terms of the mathematics of waves by American mathematician John R. Carson; *See also* 1933 COM

Philo T. Farnsworth describes an electronic television system; *See also* 1919 COM; 1925 COM

A Boston chemist obtains gelatin from the ears of female pigs, spins it into thread, and uses the thread to weave two purses, one of which is now in the US Smithsonian Institution's National Museum of American History (establishing that if one is clever enough it *is* possible to make a silk purse from a sow's ear)

Cultured pearls are introduced

Coco Chanel selects the fifth of a number of bottles of perfume developed by Ernest Beaux, which soon becomes famous as Chanel No. 5; *See also* 1889 MAT

Robots, fantasy and fact (Continued)

More recently, artificial intelligence (AI) techniques have been used for programming robot control systems. Cybernetics, the study of feedback mechanisms in living organisms and in machines, a subject developed by the mathematician Norbert Wiener, has also inspired robot builders.

The development of robots has progressed slowly and in definite steps. Early robots were able to learn certain movements, for example, those required for spray painting an object. An operator, holding the manipulator arm, would perform the motions needed to paint a given object. The robot would store the motions its arm had gone through in its computer memory. Whenever a new object for painting was presented, the robot would repeat the motions exactly. However, a second object had to be carefully held in the same position as the first, or most of the paint would miss.

Frederick Grant Banting [b Alliston, Ontario, Canada, Nov 14 1891, knighted 1934, d Musgrave Harbour, Canada, Feb 21 1941], Charles Best [b West Pembroke ME, Feb 27, 1899, d Toronto, Ontario, Mar 31 1978], John McLeod, and James Collip extract insulin from the human pancreas and start experiments on dogs in an effort to develop a treatment for diabetes; *See also* 1922 MED

Alexander Fleming [b Lochfield, Scotland, Aug 6 1881, knighted 1944, d London, Mar 11 1955] discovers the antibacterial substance lysozyme in saliva, mucus, and tears; *See also* 1928 MED

C.O. Nylen and G. Holmgren introduce microsurgery using the operating microscope they invent

Canadian-American medical student John Augustus Larson [b Shelbourne, Nova Scotia, Dec 11 1892, d Sep 21 1965] invents the polygraph (lie detector)

1921

Robots, fantasy and fact (Continued)

"Intelligent" robots are those that can react to different environmental situations; they are equipped with sensors and even computer vision. The Hopkins Beast, an experimental robot developed at Johns Hopkins University during the 1960s, moved along corridors until it found a wall socket to charge its batteries. When charged, it would travel farther until it found a new electric outlet. An intelligent robot painter would recognize that an object was in the wrong position and correct the position, or that the object presented was not the one to be painted, and reject it.

Today the most important technologies under investigation for the development of robots are artificial vision and other sensory systems. Equipped with such systems and with sensory feed-back, robots could manipulate, for example, machine parts placed randomly on a conveyor belt. An example of an intelligent robot is WABOT–2, developed at Waseda University in Japan during the 1980s. The experimental robot sight-reads music and plays it on an organ while operating the foot pedals.

Just as the use of computers has branched out from number crunching to a wide variety of applications, so it is probable that the use of robots will grow from manufacturing tasks to a host of activities, such as picking fruit, cleaning windows, and waiting on tables. Other possible applications envisioned for robots are more significant, such as robot prison guards or robot surgeons. Whether such developments would enhance the quality of life for humans remains as questionable today as it was in 1921, when Câpek raised his own doubts.

1922

Frederick Banting and Charles Best's *Internal secretions of the pancreas* is published; *See also* 1921 MED

A model 39-B airplane built by Aeromarine Plan & Motor Company makes the first ever landing on a US aircraft carrier, the USS *Langley*; *See also* 1910 TRA

1923

The vibraphone is invented in the US; it is a form of xylophone that uses electrically powered resonators to affect tone; *See also* 1931 GEN

The big breakthrough for Gideon Sundback's Hookless No. 2 slide fastener comes when the device is used by B.F. Goodrich on its line of rubber galoshes; a B.F. Goodrich salesman calls Hookless No. 2 the zipper, because he can close it and open it quickly (zip it); *See also* 1914 GEN

Vladimir Zworykin [b Mourom, Russia, Jul 30 1889, d Jul 29 1982] files a patent for an entirely electronic television camera tube, the iconoscope; *See also* 1922 COM; 1925 COM

George Eastman introduces 16-mm negative film for amateur use; within a year, Bell & Howell and the Victor Animatograph Company bring 16-mm cameras and projectors onto the market; *See also* 1911 COM

Seismic methods are used in Mexico for the first time to search for petroleum deposits; *See also* 1900 ENE

Elmer Sperry

Among inventors Elmer Sperry stands out by his entrepreneurship combined with a strong sense of independence. Throughout his career he declined many job offers from industry and preferred to work alone, founding his own manufacturing companies.

Sperry's interest in invention was roused when he was 15 years old and visited in 1876 the Centennial Exhibition at Philadelphia, which celebrated technology and industry. In 1880 he made his first invention, an arc light with electric generator that he installed in the local Baptist church on Christmas Eve. He then founded his first business, the Sperry Electric Light, Motor, and Car Brake Company, selling his arc lighting system throughout the Midwest.

Finding it difficult to compete with the large electric firms, Sperry then turned to full-time invention. Most of his inventions—he obtained 400 patents throughout his lifetime—centered on improvements of existing systems. He used a standard approach to each problem: First he studied the existing patent literature on the system he sought to improve and looked for the weakest point of that system. Often he found that the controls of a particular device were not as good as customers would have liked. Then Sperry set out to improve those controls, often trying several alternative solutions. In this way he invented an automatic carbon feed for arc lights, an automatic output control for electric generators, a motorman's control for electric streetcars, a speed control for electric automobiles, and an automatic focusing device for searchlights.

Working with controls, Sperry began to specialize in using the effect of rotating wheels, which tend to maintain their orientation in space without regard to external forces. Such wheels are most familiar as tops and especially the scientific tops called toy gyroscopes.

In 1908 Sperry patented a system for stabilizing ships with large gyroscopes. He then developed a similar stabilizer for aircraft. One gyroscope registered an airplane's roll and another one the pitch. These gyroscopes acted on the controls of aircraft via a pneumatic system, stabilizing flight. For these systems Sperry developed sophisticated feedback systems, thus anticipating technology based on cybernetics three decades later. He is best known for his invention of the gyrocompass, a device now belonging to the standard equipment of every commercial and military aircraft.

1924

The Engines Branch of Morris Motors Ltd in the UK introduces the first continuous-flow assembly line designed to handle large and complex metal parts; *See also* 1913 GEN

Zenith produces its first portable radio for consumer use; it weighs 6.6 kg (14.6 lb), costs $230, and fits into a suitcase; *See also* 1962 COM

Still photographs are transmitted by radio from New York City to London; *See also* 1925 COM

The German company Leitz introduces the Leica, the first 24- x 36-mm camera; *See also* 1923 COM

The spiral-bound notebook is produced

FOOD & SHELTER	MATERIALS	MEDICAL TECHNOLOGY	TOOLS & DEVICES	TRANSPORTATION
"Bob" Birdseye launches his first commercial venture in frozen foods in New York City; the Birdseye Seafood Company, goes bankrupt but leads the way to further and more successful operations and eventual merger into the newly formed General Foods in 1929; *See also* 1917 FOO; 1939 FOO	John B. Tytus develops continuous hot-strip rolling of steel	Albert Calmette [b Nice, France, Jul 12 1863, d Paris, Oct 29 1933] and Camille Guérin [b Poitiers, France, Dec 22 1872, d Paris, Jun 9 1961] develop the tuberculosis vaccine BCG (Bacillus Calmette-Guérin); *See also* 1908 MED French bacteriologist Gaston Ramon introduces a more effective vaccine against diphtheria; *See also* 1913 MED; 1930 MED	The Boston Wire Stitcher Company offers a simplified desk model stapler that is the first to attain widespread popularity in offices; soon after, staples begin to be glued together in long rows, making them easy to insert and preventing the machines from frequent jamming; *See also* 1914 TOO	Spanish inventor Juan de la Cierva [b Murcia, Spain, Sep 21 1895, d in a crash at Croydon Aerodrome, near London, Dec 9 1936] develops the basic idea of the autogiro: Instead of a wing, it has a freely rotating rotor that provides the lifting force; *See also* 1908 TRA The Wilhelmina Canal for barge traffic, mostly carrying coal from Limburg to the west, is completed in the Netherlands; *See also* 1915 TRA; 1927 TRA The Royaulcourt Tunnel on the Nord Canal is completed; *See also* 1927 TRA

FOOD & SHELTER	MATERIALS	MEDICAL TECHNOLOGY	TOOLS & DEVICES	TRANSPORTATION
International Harvester introduces the Farmall tractor, the first tractor with closely spaced front wheels that allow it to be used in cultivating row crops and in making sharp turns at the ends of rows; *See also* 1919 FOO; 1955 FOO	Kimberley Clark introduces the first version of Kleenex, known as "Celluwipes" Urea-formaldehyde plastics (resins) are invented; *See also* 1909 MAT; 1926 MAT	Acetylene is used as an anesthetic; *See also* 1846 MED Willem Einthoven of the Netherlands wins the Nobel Prize in physiology or medicine for his invention of the electrocardiograph; *See also* 1903 MED	Edward Victor Appleton [b Yorkshire, England, Sep 6 1892, d Edinburgh, Scotland, Apr 21 1965] is the first to determine the distance to a remote object by radio ranging, a precursor of radar; he locates the Heaviside layer of the ionosphere at 100 km (60 mi) above Earth's surface; *See also* 1904 TOO; 1931 TOO The self-winding watch is patented; *See also* 1927 TOO	Henry S. Hele-Shaw [b Essex, England, Jul 29 1854, d Jan 30 1941] and T.E. Beacham develop the variable-pitch propeller for airplanes *The rocket into interplanetary space* by Hermann Oberth [b Hermanstadt, Transylvania (in Romania), Jun 25 1894] gives the first truly scientific account of space-research techniques; it is the first book to contain the notion of escape velocity (the speed necessary to escape the pull of gravity caused by a planet or other astronomical body); *See also* 1919 TRA; 1926 TRA

1925 The Bell System founds the Bell Telephone Laboratories, employing 3000 researchers; *See also* 1911 GEN

Incandescent lamps with inside-frosted bulbs are made

Television advances: John Logie Baird [b Helensburgh, Scotland, Aug 13 1888, d Bexhill, England, Jun 14 1946] produces a television picture of a recognizable human face, with gray scales; Charles Jenkins transmits moving television silhouettes over radio waves; Vladimir Zworykin files for a patent on a color television system; the patent is granted in 1928; *See also* 1924 COM; 1926 COM

Astronomer Henri Jacques Chrétien [b Paris, Feb 1 1870, d Washington DC, Feb 6 1956] invents the anamorphic objective lens; it compresses images laterally and is used from 1952 on for CinemaScopic films; *See also* 1911 COM; 1952 COM

AT&T introduces commercial wirephoto service; *See also* 1902 COM; 1926 COM

1926 The pop-up toaster is introduced in the US

Hugo Gernsback [d Apr 19 1967] founds *Amazing Stories*, the first magazine of science fiction

The movie *The Jazz Singer*, starring Al Jolson, introduces the era of talking motion pictures; *See also* 1920 COM

John Logie Baird produces television images of moving objects using the Nipkow disk, and succeeds in transmitting pictures over telephone lines between London and Glasgow, Scotland; the images are silhouettes; *See also* 1884 COM; 1928 COM

RCA opens the first commercial radiophoto transmission service across the Atlantic; *See also* 1925 COM

John Logie Baird

When John Logie Baird experimented in England with television, it was still a time when inventors and scientists could perform wonders with string and sealing wax. The famous advances of 1932 at the Cavendish Laboratory at Cambridge University, England—discovery of the neutron and splitting of the atom—were done with instruments largely built from surplus telephone hardware.

Baird began experimenting with television in 1923 after failing health caused him to give up an engineering job with an electrical company. After briefly selling patented socks and soap and an attempt to manufacture jam in Trinidad, he moved to Hastings, England. There he started experimenting with a television system based on the Nipkow disk. Such a disk contains a spiral of holes and, when spinning, scans an entire image.

Baird's first apparatus was makeshift: He cut the cardboard Nipkow disk out of a hatbox and mounted the motor driving the disk on a tea chest. The projection lamp resided in a biscuit tin and the proper focusing of light was achieved by bicycle lamp lenses. Knitting needles and string held everything in place. A photoelectric cell placed behind the Nipkow disk generated a current that quickly varied in intensity with light from the object scanned. The current was sent to a neon lamp. The image was reconstructed by observing the lamp through matching rotating disk.

In 1925 Baird moved his apparatus to London, where Gordon Selfridge hired him to demonstrate television in Selfridge's department store. A breakthrough came when he showed his system to members of the Royal Society. The event was covered in the press and led to financial backing. Baird started experimenting with the transmission of pictures over telephone lines, sending images from London to Glasgow, improving on an American record of 320 km (200 mi).

Baird demonstrated a color television system in 1928, and increased the number of scanning lines of his cameras from 30 to 240, improving image definition. He explored infrared and stereo television and built the first recording device for storing images on phonograph records, Phonovision. He also started a company that manufactured television receivers, of which he sold a thousand. The BBC, at first reluctant, allowed Baird to start a regular television service in 1936, but after three months dropped him in favor of a system based on electronic scanning developed by Marconi EMI.

1925

A wheel-type can opener that rides around the rim of the can on a serrated wheel is patented; it is similar to a common type of modern hand can opener, and a predecessor of the electric version in its mode of operation; *See also* 1870 FOO

Harry George Ferguson [b Dromore, Ireland, Nov 4 1884, d Oct 25 1960] patents a control system that prevents tractors from turning over when a plow or other tool being pulled hits an obstruction *See also* 1924 FOO

Cemented carbides are introduced for use in cutting tools; these are metal carbides that are ground into small particles, mixed with cement, and then sintered; in particular, tungsten carbide permits much higher cutting speeds than previously possible; *See also* 1900 MAT; 1926 MAT

The first chromium dyes, especially suitable for dyeing wool, are developed; these are greatly improved in the 1940s by slightly modifying their chemistry

The world's first public road with entrances and exits rather than intersections, the Bronx River Parkway in New York City and Westchester County, opens; since 1913 a total of 454 parcels of land had been acquired and 370 buildings torn down or moved to make room for the 25-km (15.5-mi) park and roadway planted with 30,000 trees and 140,000 shrubs; some trees and shrubs are used to separate northbound and southbound lanes, another innovative practice at the time; *See also* 1940 TRA

1926

Tungsten-carbide cutting tools, based on the concept of cemented carbides, are introduced in Germany under the name *Widia*; *See also* 1925 MAT; 1927 MAT

Hermann Staudinger [b Worms, Germany, Mar 23 1881, d Freiburg-im-Bresgau, Germany, Sep 8 1965] begins work that leads to the realization that plastics are all formed from long chains of small groups of atoms called monomers; in plastics, the monomers are joined together, so the material is said to be a polymer, or polymerized; *See also* 1924 MAT; 1930 MAT

Theodor Svedberg [b Valbo, Sweden, Aug 30 1884, d Stockholm, Sweden, Feb 25 1971] wins the Nobel Prize in chemistry for his development of and work with the ultracentrifuge

Robert Hutchings Goddard launches the first liquid-fuel propelled rocket, which reaches a height of 56 m (184 ft) and a speed of 97 km (60 mi) per hour; *See also* 1924 TRA; 1927 TRA

Jean A. Grégoire builds in France the first front-wheel driven car, the Tracta; *See also* 1934 TRA

1927

J.A. Neill improves magnetic sound recording and creates the tape recorder by replacing the steel wire—used in the sound recorder invented by Valdemar Poulsen in 1893—by a diamagnetic tape covered by a metallic layer; *See also* 1893 COM; 1931 COM

Boris Rtcheouloff applies for a patent for a video tape recorder; *See also* 1939 COM

John Logie Baird creates the first videodisk system; *See also* 1979 COM

Herbert Hoover, the American Secretary of Commerce, and Walter Sherman Gifford, president of AT&T, hold a conversation via the first picturephone known using television cameras and receivers equipped with Nipkow disks; the device allows transmission of pictures as people speak to each other over the telephone; *See also* 1884 COM; 1936 COM

AMI Corporation introduces the first jukebox with electrical amplification and a choice of records (8 choices available); *See also* 1889 COM

The Schulz Player Piano Company introduces the Auto-typist, a typewriter that can produce multiple copies of the same document by using the basic mechanism of the player piano, with the letters to be typed punched into a paper roll; this is not the first typewriter to use a punched paper roll in this way, nor is it the last, but it is the version that makes the idea popular; *See also* 1872 COM; 1974 COM

Ralph V.L. Hartley [b Spruce NV, Nov 30 1888] introduces the concept of information as a measure for the quantity of data in a message; *See also* 1938 COM

The pentode, a vacuum tube with five electrodes, is introduced; *See also* 1907 ELE

Vannevar Bush [b Everett MA, Mar 11 1890, d Belmont MA, Jun 28 1974] and coworkers develop at MIT in MA an electromechanical analog computer, the Differential Analyzer, which can solve differential equations; *See also* 1876 TOO; 1936 ELE

Georges-Jean-Marie Darrieus [b Toulon, France, Sep 24 1888] patents in France a windmill with a vertical axle; a working windmill is built 2 years later with a rotor about 20 m (66 ft) high; it delivers 10 kw of electricity; *See also* 1883 ENE; 1929 ENE

Industrial research

The first industrial laboratories appeared in Germany in the 1870s and 1880s. The synthetic dye industry was quickly growing and becoming very competitive, forcing manufacturers to ensure the continuous development of new dyes to keep abreast of the competition.

In the US, the first industrial labs were created to solve problems arising from the application of electricity in technical products. In 1876 Thomas Alva Edison established a laboratory in Menlo Park (Edison NJ) with the aim of producing new inventions. At one point 80 scientists worked there, among them for a brief period, Nikola Tesla. Most of the inventions are credited to Edison, although clearly everyone helped. This was the birthplace of the electric light, the phonograph, and the first motion pictures.

General Electric was in 1901 the first corporation to set up an industrial laboratory. Physical chemist Irving Langmuir, who joined the laboratory in 1909, first did research to extend the life of light bulbs, but soon diversified in several directions, including fundamental or basic research with no immediate practical goal. Doing such basic research at an industrial laboratory later became a major activity. Corporations were proud of their contributions to basic science, but also recognized that research into fundamentals sometimes led to the development of entirely new and unexpected products.

1927

Buckminster Fuller [b Milton MA, Jul 12 1895, d Los Angeles CA, Jul 1 1983] moves to a Chicago slum and develops the Dymaxion house, a transportable dwelling suspended from a single mast that costs about the same as an automobile; *See also* 1935 TRA

Krupp in Germany and General Motors start using electric furnaces for the production of tungsten carbide mixed with cobalt; *See also* 1907 MAT

The "octane number," a value associated with gasoline expressing its resistance against knock, is introduced; *See also* 1921 ENE; 1930 MAT

The artificial rubber Buna is invented

About this time Richard Drew of what is now the 3-M Corporation invents the first version of the cellophane tape that is familiar under its brand name as Scotch tape

Antonio Moniz obtains X-ray pictures of the arteries of the brain by injecting a contrast medium into the blood vessels; *See also* 1897 MED

Philip Drinker and Louis Shaw [b Sep 25 1886, d Aug 27 1940] develop the iron lung, a device for mechanical artificial respiration

Julius Wagner von Jauregg [b Wels, Austria, Mar 7 1857, d Vienna, Sep 27 1940] wins the Nobel Prize in physiology or medicine for the treatment of some forms of paralysis using malaria inoculation to induce the fever

Rolex introduces the first waterproof watch; it is called the Oyster; *See also* 1924 TOO

Charles A. Lindbergh [b Detroit MI, Feb 4 1902, d Kipahulu HI, Aug 26 1974] makes the first nonstop solo flight across the Atlantic Ocean, in 33.5 hours; *See also* 1919 TRA; 1932 TRA

Using ideas about rockets and space from a 1924 book by Hermann Oberth, *Verein für Raumschiffart* (The Society for Space Travel) is founded in Germany; among its early members are Wernher von Braun [b Wirsitz, Germany (in Poland), Mar 23 1912, d Alexandria VA, Jun 16 1977], who will develop the first rockets to travel in space, and Willy Ley [b Berlin, Germany, Oct 2 1906, d New York City, Jun 24 1969], who will write books that make rocket concepts easily understood by the average US citizen; *See also* 1926 TRA; 1932 TRA

The Maas-Waal Canal for barge traffic is completed in the Netherlands; *See also* 1923 TRA

The Holland Tunnel opens, linking NY and NJ; *See also* 1882 TRA

Industrial research (Continued)

Several other large companies established their own industrial laboratories during the first years of the 20th century. Bell Laboratories became one of the most famous. In a restricted sense, the American Bell Company was doing research in the 1880s to improve telephone connections. This effort had led to the development of mechanical and vacuum tube repeaters for amplifying telephone signals. The number of engineers and scientists doing research at Bell increased steadily; by 1914 their number was 550 and by 1924, 3000. That year the company founded Bell Telephone Laboratories. The number of employees increased until in the early 1990s the staff counted about 30,000 people.

Scientists and engineers at Bell Labs were responsible for an impressive series of inventions and discoveries. Karl Guthe Jansky discovered radio waves coming from the center of the galaxy in 1932, while studying interference in shortwave telephony receivers. Arno Penzias and Robert Wilson, in a similar study of interference, discovered the cosmic background radiation in 1964. Claude Shannon developed information theory into an important mathematical tool at Bell Labs.

In addition to such basic research, Bell Labs produced a number of practical inventions, ranging from early computers to recent fiber-optic technology. The most far-reaching of these at Bell Labs was the invention of the transistor by William Shockley, John Bardeen, and Walter Brattain in 1947.

The breakup of the AT&T monopoly by US courts led to a reshaping of Bell Labs. The part that went to the new regional telephone companies, a lab called Bellcore, has only one scientist doing basic research. The remainder of Bell Labs is much scaled down.

Research facilities modeled on Bell Labs, such as those at IBM or Microsoft, have been a major employer of scientists outside the universities. More recently, the shoe has been on the other foot. Several successful university scientists have founded industries that are heavily science-based. Some companies, especially biotechnology firms using recombinant DNA techniques, must perform several years of pure research before coming up with their first product. In these cases it is difficult to tell where the industry ends and the research laboratory begins, although in some ways these new companies are not that different from Thomas Edison's pioneering laboratory/business at Menlo Park.

1928

Joseph Schick invents the modern form of electric razor; *See also* 1908 GEN

Walter Diemer invents bubble gum

Public concern about the dangers of radioactivity leads to the formation of the International Commission on Radiological Protection, an independent, nongovernmental body of experts

Hugo Gernsback's radio station WRNY in New York City, after encouraging listeners to build their own television sets using Nipkow disks, begins on Aug 21 to broadcast television shows; the first day's programming features a fitness show, concerts, a cooking demonstration, and a lecture by Gernsback; Gernsback goes broke in 1929 and the whole operation is abandoned; *See also* 1884 COM; 1936 COM

General Electric begins television broadcasts; *See also* 1928 COM

Agfa, a German producer of photographic materials, invents a film for instantaneous development; the idea is extended by Edwin H. Land some 20 years later as the basis for his Polaroid camera; *See also* 1947 COM

Wallace J. Eckert and Leslie John Comrie independently begin using a punched-card system similar to Hollerith's to automate astronomical calculations; *See also* 1911 ELE

1929

Raymond Loewy [b Paris, Nov 5 1893, d Monte Carlo, Monaco, Jul 14 1986] redesigns the Gestetner office duplicating machine in one of the first modern examples of industrial design for esthetics as well as ergonomic efficiency; *See also* 1934 GEN

Hermann Oberth proposes orbiting giant mirrors to direct sunlight toward the nighttime portion of Earth for lighting the surface

Harry Williams adds a device to prevent tilting to pinball machines

NBC's first television station operates with 60 scanning lines at 20 frames per second; *See also* 1928 COM; 1931 COM

Bell Labs develops a color television system with 50 scanning lines at 17.7 frames per second; *See also* 1940 COM

The British Broadcasting Corporation (BBC) starts experimental television broadcasts using John Logie Baird's system; twelve images per second are transmitted at a wavelength of 216 m; *See also* 1926 COM; 1936 COM

Gunter Krawinkel introduces in Germany the first video telephone booth; *See also* 1927 COM; 1936 COM

German engineer Felix Wankel [b Luhran, Germany, Aug 13 1902] patents a rotary engine; however, it is not until the 1950s that the Wankel engine becomes practical

A.A. Griffith conducts experiments with a gas turbine for airplane propulsion; *See also* 1908 ENE; 1930 ENE

Georges-Jean-Marie Darrieus erects a 19-m (63-ft) two-bladed windmill in France; Meanwhile, Finnish engineer S.J. Savonius patents a vertical-axle wind turbine, called an S-rotor; *See also* 1927 ENE; 1931 ENE

Georges Claude develops the first power plant to use the difference in temperature between the upper and lower layers of the ocean to generate electricity

1928

Securit, a type of safety glass that is heat treated, is patented; when shattered it holds together instead of scattering into pieces; *See also* 1905 MAT

Eugéne Houdry [b Domont, France, 1892, d Jul 18 1962] develops a method of cracking crude oil using a silica-alumina catalyst; he obtains gasoline with octane numbers between 91 and 93; *See also* 1911 MAT

Edwin Herbert Land [b Bridgeport CT, May 7 1909, d Cambridge MA, Mar 1 1991] develops a transparent substance that polarizes light; this can be used as a camera filter or in various other ways

Alexander Fleming discovers penicillin in molds; its clinical use in therapy starts only in the 1940s, when Howard Florey and Ernst Chain further develop it and it is learned how to manufacture it in quantity; *See also* 1921 MED

Greek-American George Papanicolau [b Coumi, Greece, May 13 1883, d Feb 19 1962] develops the Pap test for diagnosing uterine cancers; *See also* 1913 MED

Dutch radiologist Ziedes de Plantes develops a method using a moving X-ray source that can keep a single plane of a patient's body in focus; this is a predecessor of the CT scan; *See also* 1896 MED; 1967 MED

American inventors Joseph W. Horton [b Ipswich MA, Dec 18 1899] and Warren Alvin Marrison [b Invary, Canada, May 21 1896] develop the first quartz crystal clock; because of its piezoelectric effect, a quartz crystal controls the frequency of an electronic oscillator precisely; *See also* 1944 TOO

The radiation counter devised by Hans Geiger is substantially improved in collaboration with German physicist S. Müller; subsequently, careful physicists refer to the device as a Geiger-Müller counter; *See also* 1913 TOO

The radio beacon is introduced for navigation; *See also* 1913 TRA; 1942 TRA

The Bach Air Yacht with three engines is built; it is an early airliner with room for two crew members and ten passengers

1929

Construction is started on the Empire State Building in New York City; it is completed in 1931; *See also* 1931 FOO

Boron carbide is discovered; it displaces carborundum (silicon carbide) as the hardest artificial substance; *See also* 1891 MAT; 1955 MAT

E.A. Murphy and W.H. Chapman of the Dunlop Rubber Company develop a form of foam rubber using latex

Psychiatrist Hans Berger [b Neuses, Germany, May 21 1873, d Jena, Germany, Jun 1 1941] develops the electroencephalogram (EEG)

Manfred J. Sakel [b Nadvorna, Austria, Jun 6 1900, d Dec 2 1947] introduces insulin shock for the treatment of schizophrenia; *See also* 1937 MED

Christiaan Eijkman of the Netherlands and Sir Frederick G. Hopkins of England win the Nobel Prize in physiology or medicine for their work with vitamins; *See also* 1897 MED

General Motors introduces the synchronized gear box, which allows the changing of gears without the use of the accelerator in conjunction with the clutch; *See also* 1939 TRA

Robert Goddard launches the first instrumented rocket; it carries a barometer, a thermometer, and a small camera; *See also* 1927 TRA; 1945 TRA

Hermann Oberth's *Wege zur Raumschiffarht* (Way to space travel) describes the concept of the multistage rocket; *See also* 1927 TRA

Umberto Nobile [b Italy, 1885, d Rome, Jul 30 1978] crosses the North Pole in an airship; *See also* 1919 TRA

The Grand'Mère Bridge in Quebec, Canada, is the first to use the prestressed twisted-strand cables developed by US civil engineer David Steinman [b 1886, d 1960]; *See also* 1869 ARC

1930

The electric blanket, a larger form of the heating pad, is introduced; *See also* 1912 GEN

Photoflash light bulbs are introduced for use in photography

A tape recorder using magnetized plastic tape is developed in Germany by the I.G. Farben corporation

Audiences at the London Coliseum watch television; *See also* 1929 COM

Clay tablets unearthed at the site of the Mesopotamian city of Uruk and inscribed with some of the earliest form of writing known show no evidence of having derived from pictographs, suggesting that writing may have had some different origin; *See also* 1738 COM

The photovoltaic effect of light on selenium is rediscovered by B. Lange; one of several photoelectric effects (interactions of light and electricity), the photovoltaic effect is a current produced between electrodes when light shines on one of the electrodes; *See also* 1876 MAT; 1931 ELE

The first heat pump, a kind of reversed refrigerator, is installed in the home of T.G.N. Haldane

The Brown Bovery Company builds a 2000-horsepower explosion gas turbine based on the design by Hans Holzwarth; however, the company gives up construction of this type of turbine 3 years later; *See also* 1908 ENE

Engineer Frank Whittle [b Coventry, England, Jun 1 1907] patents the jet engine, eleven years before an aircraft flies under jet power; *See also* 1680–89 TRA; 1937 ENE

1931

Adolph Rickenbacker, Barth, and Beauchamp invent the electric acoustic guitar by attaching a microphone to the instrument and connecting it to an amplifier and loudspeaker; *See also* 1923 GEN; 1935 GEN

The first patent on stereophonic recording and reproduction of sound is issued; *See also* 1877 COM; 1958 COM

The Columbia Broadcasting System (CBS) starts television broadcasting; *See also* 1929 COM; 1932 COM

Lee De Forest applies for a patent for video recording for film; *See also* 1920 COM

The photovoltaic effect of light on selenium is rediscovered for the second time in 2 years, this time by L. Bergmann; *See also* 1930 ELE; 1937 COM

Truman S. Gray [b Spencer IN, May 3 1906, d Cambridge MA, Nov 7 1992] invents the photo-electric integraph, a calculating machine that solves complex mathematical problems by turning them into rays of light; *See also* 1990 ELE

A 30-m (100-ft) two-bladed windmill is erected at Yalta, USSR (Russia); French engineer G.J.M. Darrieus patents a vertical-axle windmill of a type sometimes called an "eggbeater"; the Darrieus rotor wind engine is reinvented in the 1970s in Canada; *See also* 1929 ENE; 1941 ENE

The oil-filled electric cable, developed by L. Emanueli of Italy, is first used for carrying high-voltage electricity; *See also* 1891 ENE

1930

The Chrysler Building in New York City, at 319 m (1046 ft), briefly replaces both the Eiffel Tower and the Woolworth Building as the world's tallest structure; however, it is soon overshadowed by the nearby Empire State Building; *See also* 1913 FOO; 1931 FOO

The Postum Company begins marketing frozen foods for the first time; *See also* 1923 FOO; 1939 FOO

Sliced bread is introduced

About this time lead begins to be added to commercially sold gasoline to prevent knocking; *See also* 1921 ENE; 1978 MAT

Several plastics are introduced: I.G. Farben in Germany develops polystyrene; Waldo L. Semon [b Demopolis AL, Sep 10 1898] of the B.F. Goodrich Company invents polyvinyl chloride; also about this time melamine plastics (resins) are invented; *See also* 1926 MAT; 1933 MAT

Continuous flow glass production is introduced; *See also* 1943 MAT

Ernest H. Volwiler [b Hamilton OH 1893, d Lake Forest IL, Oct 3 1992] and Donalee L. Tabern [b US 1900, d 1974] formulate Nembutal (pentobarbital sodium), a barbiturate used for inducing hypnotic sleep; *See also* 1863 MED; 1936 MED

Hans Zinsser [b New York City, Nov 1878, d Sep 4 1940] develops an immunization against typhus; *See also* 1923 MED; 1937 MED

Otis Barton designs a hollow steel ball suspended by a cable for undersea exploration, which he calls a bathysphere; *See also* 1905 TRA; 1934 TRA

Engineer Cedric Bernard Dicksee [b England, 1888, d 1981] develops a diesel engine suitable for road vehicles; it is an 8.1-L, six-cylinder engine that is installed in trucks; *See also* 1915 TRA

1931

The Empire State Building in New York City is completed; with a height of 381 m (1250 ft), it is the tallest building until 1972; *See also* 1930 FOO; 1972 FOO

Du Pont introduces Freon, the first commercial chlorofluorocarbon; Freon is the trade name for the highly inert gas dichlorodifluoromethane, but similar gases developed later and also used in refrigeration or as spray propellants are also called Freons; *See also* 1974 MAT

Karl Bosch and Friedrich Bergius of Germany win the Nobel Prize in chemistry for their invention of high-pressure methods of chemical production; *See also* 1909 MAT

Ernest Goodpasture [b Montgomery County TN, Oct 17 1886, d Nashville TN, Sep 20 1960] grows viruses in eggs, making the production of such vaccines as the one for polio possible for viral diseases; *See also* 1907 MED

Experiments by Karl Guthe Jansky [b Norman OK, Oct 22 1905, d Red Bank NJ, Feb 14 1950] show radio waves from space, paving the way for radio telescopes; *See also* 1937 TOO

British Post Office engineers observe that shortwave radio receivers are disturbed when aircraft fly within 6 km (4 mi), but fail to make the necessary mental connection that would convert this idea into radar; *See also* 1924 TOO; 1935 TOO

Othmar Hermann Ammann [b Schaffhausen, Switzerland, Mar 26 1879, d New York City, Sep 22 1965] designs and supervises construction of the 1067-m (3500-ft) George Washington Bridge connecting NY and NJ over the Hudson River; its span is double that of the previous record holder for length; *See also* 1883 TRA; 1964 TRA

The Bayonne Bridge over Kill van Kull between NY and NJ in the New York Harbor opens with the longest span of any arch bridge, 503 m (1662 ft)

1932

Radio Corporation of America (RCA) demonstrates a television receiver with a cathode-ray picture tube, making experimental broadcasts from the Empire State Building in New York City; *See also* 1923 COM; 1936 COM

Scientists working for Electrical and Musical Industries, Ltd. (EMI) patent an electronic television camera, the basis of the system eventually adopted by the British Broadcasting Corporation (BBC); *See also* 1923 COM; 1936 COM

The first car radio is installed in a Studebaker by the German firm Blaupunkt; *See also* 1924 COM

Guglielmo Marconi discovers that he can detect radio waves of very high frequency, known as microwaves; the first application of such waves is in radar, about 10 years later; *See also* 1935 TOO

Wernher von Braun

Wernher von Braun began his remarkable career early. As a young student he assisted rocket pioneer Hermann Oberth in his experiments. When he obtained a doctorate in 1934, von Braun's dissertation topic was combustion in rocket engines. Two years earlier, he had started work on rockets for the German military. In 1938 he became the technical director of the German military rocket center at Peenemünde, located on a small island on the Baltic coast. There he led the development of a workable liquid-fuel rocket engine that powered the V–2 rocket.

After two failures, the first V–2 rocket made a successful test flight in Oct 1942. Its engine supplied a thrust of 25,000 kg (55,000 lb) and quickly accelerated the rocket to a height of 96 km (60 mi) at supersonic speed. It continued as a *ballistic* missile, falling freely as inertia carried it toward a distant target. The V–2 was equipped with a warhead and became fully operational in Sep 1944. It is believed that the Germans fired a total of 4320 V–2 rockets, of which about 1200 were aimed at London and 2100 were aimed at the Belgian cities of Antwerp, Brussels, and Liège, which had been retaken by Allied forces. The Germans also developed prototypes of three- and four-stage rockets and toyed with the idea of using them for the construction of a space station.

After the German defeat, von Braun and about 120 rocket specialists surrendered to the Americans. They were taken to the US and settled in the desert at White Sands NM. There experiments with captured German V–2 rockets continued, often carrying instruments for the exploration of the upper atmosphere and celestial objects, until the supply of V–2s was exhausted in 1951.

Von Braun was assigned to the development of American liquid-fuel rockets and in 1952 became technical director of the US Army's ballistic missile program. He led the development of the Redstone, the first US ballistic missile. Von Braun later led the design team for the four-stage Jupiter rocket, which placed the first American artificial satellite, Explorer 1, in orbit after repeated failures of the Vanguard rocket developed by the US Navy to launch a satellite.

From 1960 on, von Braun led the team developing the huge Saturn rocket that placed the first astronauts on the Moon in 1969. By then he had become a familiar figure to most Americans and was closely identified with the US space program.

1933

High-intensity mercury-vapor lamps are introduced; *See also* 1901 GEN

Unemployed because of the Depression, Germantown PA vacationer to Atlantic City Charles Darrow invents the popular board game Monopoly

Walther Meissner and R. Ochsenfeld discover that a magnetic field is expelled from the interior of a superconductor (the Meissner effect or perfect diamagnetism)

Edwin Howard Armstrong patents wide-band frequency modulation, commonly called FM radio; probably invented as early as 1931, FM radio was not widely accepted until the late 1950s; *See also* 1906 COM; 1941 COM

Franklin V. Hunt [b Manchester NH, Sep 3 1883], John Pierce [b Des Moines IA, Mar 27 1910] and J.A. Lewis introduce lateral engraving of sound records, which improves sound quality; *See also* 1931 COM; 1962 COM

1932

FOOD & SHELTER

The Zuider Zee in the Netherlands is enclosed

Tractors with rubber tires are introduced; *See also* 1908 FOO

MATERIALS

The shape memory effect is discovered in an alloy of gold and cadmium; when the alloy is brought out of its original shape by bending, it returns to its original shape when heated; *See also* 1987 MAT

MEDICAL TECHNOLOGY

German chemist Gerhard Domagk, [b Lagow (in Poland), Oct 30 1895] discovers the first sulfa drug, Prontosil; he finds it kills streptococci and is very effective against blood poisoning; *See also* 1910 MED; 1935 MED

Armand Quick [b Theresa WI, Jul 18 1894, d Jan 26 1978] introduces the Quick test to measure the clotting ability of blood

TOOLS & DEVICES

C.E. Wynn-Williams develops the thyratron, an electronic tube used for counting electric pulses

TRANSPORTATION

The German General Staff hires astronomy student Wernher von Braun to undertake the development of rocket missiles; *See also* 1927 TRA; 1934 TRA

Amelia Earhart makes the first solo transatlantic flight by a woman, completing the trip on May 22; *See also* 1927 TRA; 1933 TRA

Auguste Piccard [b Basel, Switzerland, Jan 28 1884, d Carmel CA, Mar 11 1970] becomes the first human to enter the stratosphere; his balloon climbs to 16,201 m (53,153 ft); *See also* 1783 TRA; 1948 TRA

An automatic pilot for aircraft is introduced for civilian use; *See also* 1914 TRA; 1947 TRA

The century-old Welland Canal in Canada, the bypass for Niagara Falls, is enlarged for the second time (the first was in 1887) and reopens on Aug 6; *See also* 1829 TRA

1933

FOOD & SHELTER

On Sep 8 work starts on the Grand Coulee Dam across the Columbia River; the dam becomes the largest concrete structure ever built; it is 168 m (550 ft) high and 1272 m (4173 ft) long; its designer is John Lucian Savage, chief engineer of the US Bureau of Reclamation; *See also* 1936 FOO

MATERIALS

William Francis Giauque [b Niagara Falls, Ontario, May 12 1895, d Oakland CA, Mar 28 1982] and coworkers use a magnetic field to induce cooling in liquid helium to 0.004°C (0.007°F) above absolute zero; *See also* 1908 MAT

Imperial Chemical Industries of Great Britain, utilizing basic research developed by Dutch chemist Antonius M.J.F. Michels [b Amsterdam, Holland, Dec 31 1889] over a period of more than a dozen years, produces the plastic polyethylene (polyethene); *See also* 1930 MAT; 1939 MAT

MEDICAL TECHNOLOGY

Grantley Dick-Read [b Beccles, England, Jan 26 1890, d Jun 1959] writes *Natural childbirth*, advocating exercises and procedures for childbirth without drugs

TOOLS & DEVICES

Ernst Ruska [b Heidelberg, Germany, Dec 25 1906, d May 30 1988] builds the first electron microscope that is more powerful than a conventional light microscope; he obtains a magnification of 12,000 power; *See also* 1840 TOO; 1935 TOO

TRANSPORTATION

Russian aeronautics engineer and designer of manned space probes Sergei Pavlovich Korolev (b Zhitomir, Ukraine, Dec 30 1906, d 1966) builds the first Soviet liquid-fuel rocket; *See also* 1926 TRA

American aviator Wiley Post is the first to make a solo round-the-world flight; it takes him 7 days, 18 hours, and 49 minutes; *See also* 1932 TRA

1934

The coiled tungsten filament is itself coiled; the new coiled-coil tungsten filament produces a light that is 20 percent more efficient; *See also* 1913 GEN; 1937 GEN

Fluorescent lamps are under development by major US lighting corporations as a low-voltage alternative to gas-discharge lighting; *See also* 1904 GEN; 1938 GEN

Raymond Loewy redesigns the refrigerator for Sears, Roebuck & Company, using aluminum shelves; the new appliance is a hit when exhibited at the 1937 Paris International Exhibit and influences all subsequent kitchen appliance design; *See also* 1929 GEN

Alan Mathison Turing

Alan Mathison Turing was a young British mathematician who formulated his ideas about computers in order to solve a purely philosophical problem, the Entscheidungsproblem that mathematician David Hilbert stated in 1928. This problem was to determine whether every statement in mathematical logic could be proven either true or false. If every mathematical proposition could be proven by an entirely mechanical process, for example, by an automatic machine, the problem would be positively solved. Turing devised such an imaginary machine. In 1935 (published in 1936 with corrections added in 1937) Turing was able to characterize exactly which kinds of propositions could be proved. These turned out to be virtually all problems of mathematical interest, but not all problems whatsoever. Thus, Turing's work demonstrated that the solution to the Entscheidungsproblem was negative.

Turing's theoretical device is known as a Universal Turing Machine. As presently characterized, it consists of an endless paper tape that runs under a read/write head. The paper tape is divided into frames that can be empty or that can contain a symbol, such as a letter or a number. The machine performs tasks by moving the tape back and forth while the head reads, writes, or deletes the symbol in each frame, according to a limited set of instructions. Despite the very few operations described, a Turing Machine can calculate the solution to any problem that can be calculated by a human using a much greater range of tools.

Between 1935 and 1938 Turing was in the US, obtaining his doctorate at Princeton University. When Turing returned to England, he found himself involved with real computers during World War II. He became one of the scientists and scholars working in secrecy at the British government Code and Cipher School of the Department of Communications, housed unobtrusively in a few wooden barracks at Bletchley Park, north of London. It was called a "school" as part of the disguise of its true purpose. The main task of the Code and Cipher School was deciphering coded German messages, especially those created with Enigma machines. The Enigma was a typewriterlike enciphering machine that employed four rotors to create random permutations of letters. The machine was very efficient at encoding messages, and at the beginning of the war the British needed a few weeks for deciphering each message. Thus, the actions described in a message had usually been carried out before the messages were decoded. Turing designed an electromechanical deciphering machine, code-named "Bombe," which greatly reduced the decoding time. Turing was also part of the team that developed the Colossus computers. The Colossus machines were a series of electronic computers that incorporated as many as 1500 vacuum tubes. Use of the Colossus reduced the time to decipher a German message to a few minutes, cutting substantially British losses at sea.

After the war Turing worked on the design of the Automatic Computing Engine, or ACE. However, only a scaled-down version of it was built in 1950.

While working with computers, Turing became strongly interested in machine intelligence. In 1950 he wrote an article on artificial intelligence entitled "Computing machinery and intelligence." To show that computers might be able to express intelligence, he devised a test consisting of sample dialogues between a computer and a human interrogator. Called the Turing test, these dialogues suggested that a computer might be so programmed that in principle the interrogator could not determine whether he was dealing with a machine or another human. This article discussed several other facets of artificial intelligence, a scientific discipline that was only recognized as such 6 years later.

Turing did not live to see the development of modern computers. He was prosecuted for his homosexuality in a period a few years before the British government decriminalized homosexual acts. The government forced Turing to endure a hormonal treatment intended to change his sexual orientation. Instead, Turing committed suicide in 1954.

1934

Freeze-dried coffee is manufactured in Switzerland; *See also* 1906 FOO; 1940 FOO

Cambridge engineer Sir Geoffrey I. Taylor [b London, Mar 7 1886, d Jun 27 1975] explains that the ductility of metals comes from dislocations called line defects in the crystal structure of the metals; *See also* 1941 MAT

Quinacrine (also known as mepacrine, Atabrine, or Atebrin) is introduced as an effective treatment for malaria; *See also* 1891 MED; 1981 MED

George R. Minot, William P. Murphy [b Stoughton WI, Feb 6 1892], and George H. Whipple [b Ashland NH, Aug 28 1878, d Rochester NY, Feb 1 1976] win the Nobel Prize in physiology or medicine for the discovery and development of liver treatment for anemia

Glass for the 5-m (200-in.) reflecting lens for the Hale telescope at Mount Palomar Observatory in CA is poured on Dec 2; a test and backup disc had been poured starting on Mar 25, a process that takes 15 days just to fill the mold, followed by many days of controlled cooling; the Hale becomes the most effective telescope on Earth for more than 50 years; *See also* 1936 TOO; 1948 TOO

Wernher von Braun and his staff in Germany successfully launch their first rockets powered by alcohol and liquid oxygen, the A2 series; *See also* 1932 TRA; 1937 TRA

The first streamlined car to be offered to the public, the Chrysler Airflow, designed by Carl Breer, is introduced; despite its rounded profile and fenders, it is not a commercial success; *See also* 1935 TRA

Citroën introduces a car with front-wheel drive, the 7A, also known as the Traction; it has superior road-holding properties; *See also* 1926 TRA

The steamship *Queen Mary* is launched on Sep 26 and begins trials; it is the fastest ocean liner of its time and one of the largest; her transatlantic record will be 3 days, 20 hours, 42 minutes; *See also* 1938 TRA

Charles *William* Beebe [b Brooklyn NY, Jul 29 1877, d near Arima, Trinidad, Jun 4 1962] and Otis Barton set a depth record by diving to 1001 m (3038 ft) below the ocean's surface in a tethered sphere called a bathysphere; *See also* 1930 TRA; 1948 TRA

Vannevar Bush

Vannevar Bush was an American engineer best known for his work on mechanical computers. He became interested in mechanical devices at an early age. After a brief stint as an engineer at the General Electric Company, Bush obtained a PhD in engineering from MIT. His research at MIT into the transmission of electric power required the numerical solution of differential equations. Solving differential equations is often an arduous task.

Utilizing his interest in mechanical devices, Bush set out to build a machine that would solve differential equations. His machine was of the analog type. Analog devices are like phonograph records or other recording devices that directly convert changes in one medium (such as mechanical grooves in records or patterns of magnetism) into changes in another medium, such as electrical currents or vibrations of air molecules. They are contrasted with digital devices, which use numbers, in one form or another, to record discrete changes.

Around 1925 Bush developed the Product Integraph, a mechanical analog calculator that could perform certain steps needed for the solution of differential equations concerning simple rates of change. The basic methods used in such machines had been originally developed by applied mathematicians such as James Maxwell and Lord Kelvin. However, because of insufficiently precise construction, the earlier machines were not very accurate.

Many problems require differential equations that go beyond simple rates of change. For the solution of higher order differential equations, Bush devised in 1930 the Differential Analyzer. It was a versatile, almost entirely mechanical, analog computer that performed calculations for the design of electrical networks. Subsequent versions incorporated electric circuits that replaced some of the mechanical components, making the device easier and faster to operate. Bush's analog computers were widely used during World War II, but were subsequently replaced by the much faster digital electronic computers.

Bush himself started investigating digital computers in 1937 and is said to have proposed the construction of a program-controlled electronic computer. His research contributed to the Rapid Arithmetic Machine Project at MIT, which contained electronic circuits, such as registers and counters, that later became widely used in digital computers.

1935

B.A. Adams and E.L. Holmes synthesize a number of resins from coal that can be used to remove salts of all kinds from water; the synthetic materials are known as ion-exchange resins

In Oklahoma City, newspaper editor Carl Magee proposes the parking meter, which appear in town this year; H.C. North, a clergyman, gets the first ticket for not putting in his coin, but manages to talk his way out of the fine, claiming he had just gone to get change

Gibson Guitar Company begins to sell electric acoustic guitars; *See also* 1931 GEN; 1946 GEN

The beer can is introduced in NJ

Kodak introduces the Kodachrome process of color photography; *See also* 1912 COM; 1936 COM

International Business Machines (later IBM) introduces its version of the electric typewriter, which becomes a leader in offices; *See also* 1902 COM; 1961 COM

German leader Adolf Hitler uses taped magnetic recordings to broadcast his speeches; because of the absence of the characteristic periodic noise of recordings on disks, these speeches sound live over the radio; *See also* 1931 COM

1936

The British Broadcasting Corporation (BBC) starts regular public daily television broadcasts in Great Britain, the first regular electronic public television service anywhere (Gernsback's 1928 broadcasts used mechanical scanning); *See also* 1932 COM; 1937 COM

The Olympic Games in Berlin are televised

RCA installs its first television studios in the Empire State Building in New York City; *See also* 1932 COM

The Reichspost in Germany develops a picturephone service between Berlin, Leipzig, Nuremberg, and Hamburg; pictures are scanned at 180 lines and transmitted via special coaxial cables; *See also* 1927 COM; 1992 COM

Frans Michel Penning invents a means for confining electrical currents within radio tubes; this is later modified for containing electrons and named the Penning trap; *See also* 1973 ELE

Konrad Zuse [b 1910] begins work on a primitive form of digital computer using parts from the German equivalent to an Erector Set; in 1939 he succeeds in obtaining funding from the German government to develop further machines; *See also* 1927 ELE; 1938 COM

1935

The first combine that can be operated by a single person is introduced in the US; *See also* 1838 FOO; 1938 FOO

Jesse W. Beams [b Belle Plain KS, Dec 25 1898] obtains the first separation of isotopes by centrifuging; *See also* 1944 MAT

Wallace Hume Carothers [b Burlington IA, Apr 27 1896, d Philadelphia PA, Apr 29 1937] patents nylon, which he invented in 1934; however, its roots at Du Pont de Nemours and Company, where Carothers worked, go back to a program started in 1927; *See also* 1938 MAT

John Gibbon [b Philadelphia PA, Sep 29 1903] and his wife develop the first prototype of the heart-lung machine

Edward C. Kendall isolates cortisone, a substance present in the cortex of the adrenal gland

Gerhard Domagk uses the first sulfa drug, Prontosil, on his youngest daughter to prevent her death from a streptococcal infection, in the first use on a human being; its success in this and other instances make Prontosil famous worldwide as the first "wonder drug"; *See also* 1932 MED

British scientists led by Robert Alexander Watson-Watt conduct experiments leading to the first radar; they detect faint signals from passing aircraft by placing a van loaded with shortwave receivers about 16 km (10 mi) from the powerful shortwave transmitting station at Daventry, England; by Jun they have a cathode-ray apparatus on which planes can be seen; by Sep they can measure the height of an aircraft flying 24 km (15 mi) away; *See also* 1931 TOO; 1937 TOO

The first commercial electron microscope becomes available in England; *See also* 1933 TOO; 1938 TOO

Lockheed introduces the Douglas DC-3, the prototype of the modern passenger plane; powered by two engines of 1200 horsepower it reaches 300 km (186 mi) per hour and can carry up to 30 passengers in a pressurized cabin; more than 13,000 planes are built, of which a large part are for military use during World War II

Buckminster Fuller introduces his three-wheeled Dymaxion automobile at the Chicago World's Fair; it is a streamlined car designed to maximize gain from minimum input of energy; *See also* 1934 TRA

1936

Boulder Dam, later christened Hoover Dam for a while, is completed on the Colorado River, creating Lake Mead, the world's largest reservoir; *See also* 1933 FOO

Alexis Carrel, working with Charles A. Lindbergh, develops a form of artificial heart that is used during cardiac surgery; *See also* 1914 MED

Ernest H. Volwiler and Donalee L. Tabern formulate sodium pentothal, used for inducing hypnotic sleep; it is also known as "truth serum" because of the state of suggestability it induces; *See also* 1930 MED

Daniele Bovet [b Neuchâtel, Switzerland, Mar 23 1907], Leonard Colebro, and coworkers discover that the wonder drug Prontosil breaks down in the body and that the part that kills streptococci is a known chemical, sulfanilamide; they show that sulfanilamide is as effective as Prontosil; *See also* 1932 MED; 1937 MED

The 500-cm (200-in.) telescope designed by George Ellery Hale is delivered to its site at Mount Palomar in southern CA in Apr, and grinding and polishing operations on it begin, but World War II causes its final installation to be delayed; *See also* 1934 TOO; 1948 TOO

British scientists raise the range of their early radar equipment to the point that they can detect aircraft flying 120 km (75 mi) away; *See also* 1935 TOO; 1937 TOO

German engineer Heinrich K.H. Focke develops the first practical helicopter, the Fa-61, which successfully flies on Jun 26; by changing the angle of the rotor axis, the Fa-61 can fly forward or backward; *See also* 1908 TRA

The first prototype of the Spitfire, designed by Reginald Joseph Mitchell [b Talke, near Stoke-on-Trent, England, May 20 1895, d June 11 1937] is flown by Captain J. Summers; more than 19,000 Spitfires are built, of which the most advanced reach speeds of 740 km (460 mi) per hour

A bridge completed at Aue, Germany, is the first to use prestressed concrete; *See also* 1904 TRA

1936 cont

The Agracolour process of color photography is developed, based on color dyes first introduced by Siegrist and Fisher; *See also* 1912 COM; 1939 COM

Penguin Books introduces the first paperback books; *See also* 1440–49 COM

Magic bullets (part 1)

At the end of the 19th century, medical researchers had identified several microorganisms responsible for diseases such as syphilis, anthrax, tuberculosis, and cholera. To observe these bacteria in the microscope, one had to stain them (see "Color and chemistry," p 194). German bacteriologist Paul Ehrlich began to search for stains that might kill bacteria. In 1907 he discovered that Trypan red, a dye that stains trypanosomes, the organisms that cause African sleeping sickness, also kills them. Ehrlich then started tests with arsenic compounds. His student Sahachiro Hata discovered in 1909 that one of these compounds, later known as Salvarsan, killed the bacterium that causes syphilis. Ehrlich called these compounds "magic bullets" because they killed harmful pathogens without affecting the patient.

1937

The coiled tungsten filament for incandescent lamps becomes commercially available; *See also* 1934 GEN

American law student Chester Carlson [b Seattle WA, Feb 8 1906, d New York City, Sep 19 1968] invents xerography, the first method of copying to be based on properties of selenium; his first efforts are extremely crude, however, and many years of development lie ahead before he has a commercial product; *See also* 1959 COM

Alec Reeves in England describes a transmission system in which analogue sound is transformed into electrical pulses by a sampling process; the receiver transforms these pulses back into an analogue sound signal; *See also* 1979 COM

The British Post Office issues the first television transmission standards; *See also* 1936 COM

Alan Mathison Turing [b London, Jun 23 1912, d Wilmslow, England, Jan 7 1954] writes *On computable numbers* which describes the "Turing machine," an imaginary machine developed in 1935 that can solve all problems that are computable; in fact, the definition of computable is that the problem can be solved by a Turing machine

John V. Atanasoff [b Hamilton NY, Oct 4 1903] starts work on the first electronic computer, a machine for solving systems of linear equations; the first operational prototype is completed in Oct 1939; an operational version known as the ABC, which fails frequently because of problems with the punched-card input, is working by 1942; *See also* 1936 ELE; 1943 ELE

Mathematician Georges Stibitz [b York PA, Apr 30 1904], while working for Bell Telephone Laboratories, develops the first binary circuit, a combination of batteries, lights, and wires based on Boolean algebra that can add two binary numbers; he calls the device "model K" or "kitchen adder" because he built it in his kitchen; this circuit becomes instrumental in the development of subsequent electromechanical computers at Bell Labs; *See also* 1867 GEN; 1938 ELE

Brothers Russel H. [b Washington DC, Apr 24 1898, d Jul 28 1959] and Sigurd Fergus Varian develop the klystron, a vacuum tube capable of generating microwaves and which will be used in radar transmitters; *See also* 1921 ELE; 1939 ELE

Frank Whittle and A.A. Griffith build the first working jet engine in England; the same year in Germany, von Ohain and M. Müller independently develop a similar engine; *See also* 1930 ENE; 1941 TRA

Magic bullets (part 1) (continued)

Unfortunately, very few new magic bullets for the treatment of other infections were found during the next years. In 1932 German biochemist Gerhard Domagk discovered that a dye containing the sulfonamide radical, called Prontosil Rubrum, was effective against streptococcal infections. Strangely enough, Prontosil did not kill streptococci in the test tube, but did so when injected into a mouse. Domagk had not tested Prontosil on humans, but when his daughter contracted streptococcal blood poisoning by pricking herself with a needle, and was near death, Domagk injected her with Prontosil Rubrum and saved her life.

Chemists discovered that Prontosil changed in the body and produced a fragment called sulfanilamide, the active agent against certain bacteria. They then started looking for other compounds that would chemically resemble sulfanilamide, and would kill a wider range of bacteria. Several such sulfa drugs were found. They were effective against a series of diseases that formerly were treatable only with difficulty.

1936 cont

Ferdinand Porsche [b Bohemia, (the Czech Republic), 1875, d 1951] introduces the Volkswagen ("people's car"), known as the Beetle because of its shape; it becomes the car with the largest production run in history, overtaking that of the Model T Ford at 15,000,033 units in 1972; *See also* 1941 TRA

1937

The City of Greater New York begins construction of a major system of reservoirs on the Delaware River and its branches; water from these reservoirs will be taken by aqueduct to the Croton Reservoir system, from which it will supply New York City; *See also* 1913 FOO

The Miracle Mixer, soon renamed the Waring Blendor (with an "o") after bandleader Fred Waring—Frederick J. Ossius, inventor of the mixer, was the brother-in-law of Waring's publicity director—is introduced at the National Restaurant Show in Chicago IL; the Blendor is sold primarily to bars to make frozen daiquiris and not extensively promoted for home use until 1947; *See also* 1919 FOO

Norman de Bruyne develops the composite plastic Gordon-Aerolite, which consists of high-grade flax fiber bonded together with phenolic resin; *See also* 1930 MAT

Carl von Linde and M. Fränkl start an experimental steel plant using Bessemer convertors supplied with pure oxygen instead of air for burning the carbon in iron

Pharmocologist Daniele Bovet, working at the Pasteur Institute in France, develops the first antihistamine

Italian doctors Ugo Cerletti [b Conegliano, Italy, Sep 12 1877, d Jul 27 1963] and Lucio Bini [b Rome, Sep 18 1908] develop the first form of electroconvulsive therapy (ECT), often known as shock treatment, for treating schizophrenia; *See also* 1929 MED

Microbiologist Max Theiler [b Pretoria, South Africa, Jan 30 1899, d New Haven CT, Aug 11 1972] introduces a vaccine against yellow fever; *See also* 1930 MED; 1942 MED

Based on the model of the successful drug sulfanilamide, scientists create sulfapyridine, the second sulfa drug; *See also* 1936 MED; 1939 MED

Grote Reber [b Wheaton IL, Dec 22 1911] builds in his backyard the first intentional radio-telescope, a 9.4-m (31-ft) diameter dish, with which he begins to receive signals from space in 1938; *See also* 1931 TOO; 1946 TOO

The British Government begins construction of a chain of twenty radar "watching" stations along the east coast of Great Britain; *See also* 1935 TOO

The zeppelin *Hindenburg* burns in a hydrogen fire while trying to land in Lakehurst NJ; the crash, which kills 36 of the 97 aboard, is captured on film, and a live radio broadcast of the disaster is recorded; the sight and sound is so horrifying that the crash effectively ends rigid airship development, although the US continues to make non-rigid airships, called blimps, for coastal patrol; *See also* 1900 TRA; 1915 TRA

The first rocket tests are performed at the research station at Peenemünde, Germany; one of the leaders of the team is Wernher von Braun; *See also* 1934 TRA; 1938 TRA

The 1280-m (4200-ft) Golden Gate Bridge in San Francisco CA is completed; it is designed by engineer Joseph Strauss [b Cincinnati OH, Jan 9 1870, d Los Angeles CA, May 16 1938] with consultation by Othmar Ammann; *See also* 1931 TRA

1938

General Electric Company and the Westinghouse Electric Corporation introduce the first fluorescent lamps; *See also* 1934 GEN

Hungarian Lazlo Biró patents the first ballpoint pen that is commercially successful; *See also* 1895 COM

George Harold Brown develops the vestigial sideband filter for use in television transmitters, doubling the horizontal resolution of television pictures at any given bandwidth

US engineer Thomas Ross develops the first machine that can learn from experience; it is a mechanical mouse that runs on toy train tracks and learns by trial and error to find its way through mazes; *See also* 1960 ELE

A symbolic analysis of relay and switching circuits by Claude Elwood Shannon [b Gaylord MI, Apr 30 1916] is a founding document of the mathematical theory of information; *See also* 1927 COM; 1949 COM

German scientist Konrad Zuse completes the Z_1; it is the first working computer to use a binary code; *See also* 1937 ELE; 1940 ELE

1939

American Julius S. Kahn proposes injecting an inert gas in a disposable container; upon release, it would become an aerosol (a fine spray of vaporized materials); *See also* 1941 GEN

Warner Brothers clothing company in Bridgeport CT introduces cup sizing for brassieres; *See also* 1914 GEN

The hand-held electric slicing knife is introduced in the US

Two new television camera tubes are invented: Albert Rose and Harley Iams develop at the RCA Laboratories the *orthicon* in which the mosaic plate on which the image is projected is scanned by an electron beam that resupplies electrons that have been emitted by photoemission and that are recuperated by a collector; Vladimir K. Zworykin, Harley Iams, and Morton develop the *image iconoscope* which has a semitransparent photocathode on which is projected the image to be televised; *See also* 1923 COM; 1946 COM

The first commercial system of color photography in which a negative is produced appears; it is a success with skilled amateur photographers and cinematographers; *See also* 1936 COM

John Turon Randall [b England, Mar 23 1905] and Henry Albert Boot [b Birmingham, England, Jul 29 1914] develop the first practical magnetron, a vacuum tube that produces radio waves of greater power and higher frequency than possible with other means; their work builds on various other forms of magnetron developed in the 1920s and 1930s; secretly brought to the US, the Randall-Boot magnetron becomes the central component of radar systems; *See also* 1937 TOO

Georges Stibitz and Samuel B. Williams build the Complex Number Computer (also called the Bell Telephone Lab Computer Model 1 or BTL Model 1), consisting of more than 400 relays; they connect it to three telex units, thus introducing the principle of operating a computer via a terminal; *See also* 1937 ELE

1938

In the US, the self-propelled combine is introduced; *See also* 1838 FOO

The Nestlé Corporation, after 8 years of development, produces the first satisfactory "instant" coffee, or dried coffee powder that can be reconstituted as coffee by adding hot water; *See also* 1901 FOO

Katherine Burr Blodgett [b Schenectady NY, Jan 10 1898] discovers a method for covering glass with extremely thin films that eliminate reflection of light; applications include lenses and scientific instruments

Nylon, invented in 1934 by Wallace Hume Carothers of Du Pont, goes on sale in the US in the form of toothbrush bristles; by 1939 it is available in limited amounts as yarn for stockings; *See also* 1935 MAT; 1940 MAT

Archer J.P. Martin [b London, Mar 1 1910] and Richard L.M. Synge [b Liverpool, England, Oct 28 1914] begin the work that leads to paper chromatography, using a solvent in a column of silica to separate amino acids; *See also* 1906 MAT; 1944 MAT

English surgeon Philip Wiles develops the first total artificial hip replacement, using stainless steel; *See also* 1905 MED

The radio altimeter is introduced; it reports a plane's altitude on the basis of radio signals from a known source; *See also* 1904 TOO

Manfred von Ardenne gives a demonstration of the scanning electron microscope; *See also* 1935 TOO

Germany's liquid-fueled rocket experiments, under the direction of Wernher von Braun, succeed in producing a rocket that can travel 18 km (11 mi); *See also* 1934 TRA; 1942 TRA

The steamship *Queen Elizabeth*, the largest luxury liner ever built, is launched on Sep 26; because of World War II, she is fitted out to be a troop carrier at first and does not make her first civilian voyage until Oct 1946; *See also* 1934 TRA

In Germany, the Mittelland Canal is completed, giving Berlin water connections to the North Sea, Basle, the Oder, and the Baltic; *See also* 1915 TRA

1939

The first precooked frozen foods are marketed under the Birds Eye label; *See also* 1923 FOO

Paul Müller [b Olten, Switzerland, Jan 12 1899, d Basel, Switzerland, Oct 13 1965] discovers that DDT is a powerful insecticide; *See also* 1874 FOO; 1972 FOO

Teijiro Yabuta [b Shiga Prefecture, Japan, Dec 16 1888] and Hayashi extract gibberellin A from a soil fungus and find that it is a potent growth hormone for rice; *See also* 1952 FOO

Powder metallurgy is introduced

On Sep 1 the first full-scale manufacture of polyethylene (polyethene) starts in Great Britain; *See also* 1933 MAT

René Jules Dubos [b Saint-Brice, France, Feb 2 1901, d Feb 20 1982] searches for and finds two compounds produced by a soil bacterium that kill other bacteria—antibiotics; these are the first antibiotics deliberately sought for this property; *See also* 1877 MED; 1940 MED

Scientists create sulfathiazole, the third sulfa drug; *See also* 1937 MED; 1941 MED

Gerhard Domagk of Germany wins the Nobel Prize in physiology or medicine for discovery of the first sulfa drug, Prontosil; *See also* 1932 MED

German engineer Pabst von Ohain's jet engine becomes the first such engine actually to fly an airplane, the Heinkel He-178; the experimental plane reaches 500 km (360 mi) per hour; *See also* 1937 ENE

Pan American institutes the first regular commercial flights across the Atlantic Ocean; *See also* 1927 TRA

Igor Sikorsky constructs the first helicopter designed for mass production; it flies for the first time on Sep 14; *See also* 1923 TRA; 1944 TRA

E.L. Thomson introduces automobiles with an automatic clutch; *See also* 1929 TOO

1940

The first US color television broadcast takes place using a system developed by Peter Carl Goldmark for the Columbia Broadcasting System (CBS); *See also* 1929 COM; 1953 COM

Georges Stibitz and Samuel B. Williams demonstrate the remote operation of a computer in New York City from Dartmouth College; they use a telex linked to a Bell Telephone Lab Computer Model 1, also called *Complex Calculator*; *See also* 1975 COM

Bletchley Park in the English countryside houses a concentration of intellect to face the challenge of the German encoding machine Enigma; among the scientists working at the center are Alan Mathison Turing, Wynn-Williams, and T.H. Flowers; they develop a series of computers dedicated to decoding messages from German forces; *See also* 1919 ELE; 1943 ELE

Konrad Zuse completes the Z_2; it is a binary-coded computer similar in design to the Z_1, but using telephone relays instead of mechanical logical circuits; *See also* 1936 ELE; 1941 ELE

1941

Americans L.D. Goodhue and W.N. Sullivan introduce the first aerosol containers, following a suggestion of Julius S. Kahn; the containers produce vapors containing insecticides and are soon christened "bug bombs"; *See also* 1939 GEN

The first television broadcast license for a system using 525 scanning lines is issued by the Federal Communications Commission (FCC) in the US; the same year the FCC grants broadcasters permission to use commercials for financing programs; the National Broadcasting Company (NBC) and the Columbia Broadcasting System (CBS) start transmitting programs and commercials in New York City; *See also* 1936 COM; 1948 COM

On Oct 10 the Radio Club of Columbia University, under the direction of FM's inventor Edwin Armstrong, opens the first regularly scheduled FM station, WKCR; *See also* 1933 COM

Konrad Zuse's Z_3 computer, containing 2600 relays, is the first to use a punched tape for data entry; it also makes use of an error-detecting code; it is the first working universal computer controlled by a program; *See also* 1940 ELE; 1943 ELE

Beauchamp E. Smith, Palmer C. Putnam, and coworkers, in a series of experiments between 1937 and 1946, find that their windmill on Grandpa's Knob near Rutland VT with two blades and a tip-to-tip diameter of 53.3 m (175 ft), erected this year, is the most efficient; it delivers 1250 kw of electricity to the grid until it loses a blade in 1945 and is shut down; *See also* 1931 ENE

1940

Chemists working for Imperial Chemical Industries in England discover that plant hormones can be used as selective herbicides; *See also* 1896 FOO; 1945 FOO

Freeze-drying, developed earlier for medicines, is used for food preservation for the first time in the US, although previously used for coffee in Switzerland; *See also* 1934 FOO; 1941 FOO

In May, mass-produced nylon stockings become available; *See also* 1938 MAT

In Great Britain, John Rex Whinfield [b Feb 16 1901] and J.T. Dickson invent the artificial fiber terylene, better known in the US as Dacron; the fiber is more resistant to heat than nylon and more resistant to wear than rayon (although not so wear-proof as nylon), and holds color better than either; *See also* 1935 MAT; 1944 MAT

Howard W. Florey [b Adelaide, Australia, Sep 24 1898, knighted 1944, d Oxford, England, Feb 21 1968] and Ernst Boris Chain [b Berlin, Germany, Jun 19 1906, d Ireland, Aug 11 1979] at Oxford University obtain penicillin in a purified form and show that it can be used as an antibiotic; *See also* 1928 MED; 1945 MED

Herbert M. Evans [b Modesto CA, Sep 23 1882] uses radioactive iodine to prove that iodine is used by the thyroid gland

In Los Angeles CA the Arroyo Seco Parkway opens; it is destined to become the first part of the Los Angeles freeway system; *See also* 1925 TRA

At noon on Oct 1 the Pennsylvania Turnpike, the first modern highway in the US, opens; local feed-and-tallow dealer Homer D. Romberger takes the first ticket; by Oct 6 the highway has its first traffic jam; *See also* 1925 TRA

The Tacoma (WA) Narrows Bridge collapses as a result of wind stress, resulting in serious rethinking of future bridge design

1941

Freeze-dried orange juice is supplied to the American Army; *See also* 1940 FOO

Norton and Loring use X-ray diffraction methods to show that the relation between stress and extension known as Hooke's law is caused by openings in the crystal structure of a solid, just as Hooke contended in 1679; *See also* 1670–79 MAT

Canadian-American surgeon Charles Branton Huggins [b Halifax, Nova Scotia, Sep 22 1901] shows that administration of female sex hormones can be used to control prostate cancer; *See also* 1903 MED

Microbiologist Selman Abraham Waksman [b Priluki, Russia, Jul 22 1888, d Hyannis MA, Aug 16 1973] coins the term *antibiotic* to describe substances that kill bacteria without injuring other forms of life; *See also* 1940 MED; 1943 MED

Scientists create sulfadiazine, the fourth sulfa drug; *See also* 1939 MED

In May, the *Gloster E28* or *Meteor* becomes the first British airplane powered by a jet engine, an outgrowth of the engine design first patented by Frank Whittle in 1930; *See also* 1937 ENE; 1944 TRA

In Jul the Willys-Overland Company produces the first Jeep, based on designs made in 1940 and selected by the US Army in Jun 1941 over other designs; *See also* 1936 TRA; 1945 TRA

The Caproni-Campini CC2, a jet plane built by Secondo Campini flies from Milan to Rome with a velocity of 500 km (310 mi) per hour

1942

Computers: From analog to digital

Information can be presented in a continuous fashion, like the points on a line, or in terms of individual and discrete elements, as in the dots used to make up a photograph in a newspaper. Continuous information is referred to in engineering as analog, because early devices that used continuous information did so by employing one physical system as an "analogy" for another. For example, changes in electrical voltage might be used as an analogy for changes in speed, allowing a voltmeter to be used as a speedometer. Discrete information in engineering is represented by whole numbers (integers), which in turn are shown in a numeration system by combinations of distinct signs called digits. Thus, the alternative to an analog system is a digital system, in which information is represented by discrete elements. Increasingly, engineers have turned to digital systems, partly because they are less noisy than analog systems, partly because the mathematical theory of digital information, developed by applied mathematicians such as Ralph Hartley and Claude Shannon, includes error correction techniques, and mostly because all digital information can be represented easily by switches of one type or another.

Except for the analog computers used in some special applications, all electronic computers today are digital. They use some form of on-and-off, or flip-flop, switches to represent numbers and characters and to perform calculations. The flips and flops are interpreted as zeros (0's) and ones (1's), handled as what mathematicians know as the binary numeration system, originated by Gottfried Leibniz in the 17th century. For example, 0's can be represented by a negative voltage and 1's by a positive voltage.

1943

Georges Stibitz, a mathematician at Bell Labs, was the first to use binary numbers in a calculating device, although his device was electromechanical, and not electronic. In 1937 Stibitz built at his home a very simple adding device that could sum up two binary numbers. Subsequently, he built simple binary multipliers and dividers, and together with Sam Williams of Bell Labs he developed the complex calculator (also known as the Bell Telephone Lab Computer Model 1, or BTL Model 1). The calculator, completed at the end of 1939, contained 400 telephone relays, and a teletype machine was used for data input and output. Two departments at Bell Labs that had to perform intensive calculations were connected to the BTL Model 1 via two remote teletype machines that functioned as computer terminals. In 1940 Stibitz linked the computer via telegraph lines to a terminal at Dartmouth College in New Hampshire, where mathematicians at a congress could enter calculations and receive a result a minute later.

William B. Shockley starts his research at Bell Labs on semiconductors that results in the development of the transistor in 1947; on Dec 29 1939 Shockley records in his laboratory notebook that it should be possible to replace vacuum tubes with semiconductors; *See also* 1947 ELE

Physicist John V. Atanasoff and his student Clifford Berry build an electronic calculator called ABC (Atanasoff Berry Computer); this computer uses vacuum tubes, logic circuits, and a memory and is designed for the solution of systems of linear equations; *See also* 1940 ELE; 1945 ELE

The Colossus, a computer with 1500 vacuum tubes, designed by T.H. Flowers and M.H.A. Newman in a team headed by Alan Turing, is completed at Bletchley Park; it starts work on deciphering German coded messages; it is the first all-electronic calculating device; *See also* 1940 ELE; 1949 ELE

On Dec 2 the message "The Italian navigator has entered the New World" is used to signal start-up of the first nuclear reactor, a controlled chain reaction in a pile of uranium and graphite at the University of Chicago; it is designed by Enrico Fermi [b Rome, Italy, Sep 29 1901, d Chicago IL, Nov 28 1954]; *See also* 1946 ENE

1942

After the yellow fever vaccine used to inoculate thousands of US Army personnel is shown sometimes to cause hepatitis B, a new vaccine is developed that does not require the use of human blood serum; *See also* 1937 MED

Dorothy I. Fennel discovers a powerful new penicillin species, *Penicillium fennelliae*; *See also* 1940 MED; 1947 MED

On Jan 6 William R. Hewlett [b US, 1913] patents the variable frequency oscillation generator, more familiarly known as the audio oscillator, a device for generating high-quality audio frequencies; the audio oscillator had actually been developed several years earlier and eight of them had been purchased in 1939 from the newly formed Hewlett-Packard Company by Walt Disney for use in the film *Fantasia*

A loran (long-range navigation) system begins operation for the first time along the Atlantic seaboard of North America; *See also* 1928 TRA

In Mar work starts on the Alcan Highway, now known as the Alaska Highway, from Dawson Creek, British Columbia, to Fairbanks AK; the 2450-km (1523-mi) highway is built by about 18,000 workers in eight months, twelve days, crossing a hundred rivers and five mountain ranges along the way

Wernher von Braun and his staff in Germany successfully launch for the first time the prototype of the V-2 rocket, the AS-4, on Oct 3; the 12-ton rocket travels 200 km (125 mi) in 296 seconds, reaching a speed of 5300 km (3300 mi) per hour and a height of 97 km (60 mi); *See also* 1938 TRA; 1944 TRA

1943

Chemists working for Imperial Chemical Industries in England find that benzene hexachloride has four forms, only one of which is a potent insecticide; *See also* 1939 FOO

Continuous casting of steel is developed by German engineer S. Junghans; *See also* 1930 MAT

Wilhelm Kolff [b Leiden, the Netherlands, Feb 14 1911] develops the first kidney dialysis machine; it cleanses the blood outside the body and is used for patients with nonfunctional kidneys; *See also* 1913 MED

Selman A. Waksman discovers the antibiotic streptomycin, produced by a mold that grows in soil; previous antibiotics worked only against Gram-positive bacteria, but streptomycin is effective against Gram-negative bacteria; *See also* 1941 MED; 1944 MED

Jacques-Yves Cousteau [b St. André-de-Cubzac, France, Jun 11 1910] and Emile Gagnan invent the Aqualung, commonly known as scuba (for self-contained underwater breathing apparatus) gear; it consists of a steel bottle of compressed air and a pressure-control valve that supplies air at normal pressure to the diver

1943
cont

John W. Mauchly [b Cincinnati OH, Aug 30 1907, d Jan 8 1980] publishes his report "The use of a high-speed vacuum tube device for calculating," arguing that computers using vacuum tubes would be substantially faster than existing ones using electromechanical relays; in Apr, Mauchly, John Presper Eckert, and J.G. Brainerd follow this with the design of the ENIAC computer; *See also* 1942 ELE; 1946 ELE

Russian physicist Vladimir I. Veksler [b Ukraine, Mar 4 1907, d Moscow, Russia, Sep 22 1966] develops phase stability, the basic idea for the synchrotron—independently invented in 1945 by US physicist Edwin McMillan; this idea enables construction of particle accelerators that move particles at the same speed; *See also* 1946 ELE

1944

J. Presper Eckert [b Philadelphia PA, Apr 9 1919] and John Mauchly develop the mercury delay line store, an early form of computer memory that stores data as acoustic pulses running down a tube filled with mercury; in Jan Eckert proposes construction of the Magnetic Calculating Machine, in which numbers and instructions would be stored on spinning disks or drums covered with a magnetizable material and read by coils placed close to its surface; *See also* 1956 ELE

Herman Goldstine [b Chicago IL, Sep 13 1913], meeting John von Neumann [b Budapest, Hungary, Dec 3 1903, d Washington DC, Feb 8 1957] accidentally, tells him about plans to build the ENIAC; von Neumann becomes interested and involved in the design of early computer architecture; *See also* 1943 ELE; 1945 ELE

The electromechanical computer Mark I (officially the Automatic Sequence Controlled Calculator), devised by Howard H. Aiken [b Hoboken NJ, Mar 9 1900, d Mar 14 1973] and a team of engineers from IBM, is completed at Harvard; Mark I uses paper tape for programming, but breaks down frequently

Computers: From telephone relays to vacuum tubes

In the early 1940s in the US, Bell Labs developed several calculators based on telephone relays (the earliest, BTL Model 1, was complete at the end of 1939). The BTL Model V, the version of 1946, contained 9000 relays and weighed 10 tons. Its performance, considered very good at the time, was inferior to that of a present-day pocket calculator from a stationery store: A multiplication of two seven-digit numbers took 1 second and their division 2.2 seconds.

In Germany, independently, Konrad Zuse, after having built a mechanical computer called Z_1 in the sitting room of his parents, also used telephone relays in a computer called Z_2. A friend, Helmuth Schreyer, had proposed the use of vacuum tubes for switching elements, but because of the war conditions, vacuum tubes were difficult to obtain and Zuse used relays instead. The Z_3, also based on electromechanical relays, was completed in 1941; it became the first universal computer controlled by a program.

Although many engineers at that time preferred electric relays because they were more reliable than vacuum tubes, it was the vacuum tube that emerged victorious from the decade 1940 to 1950, mainly because of the tremendous speed at which logic circuits equipped with vacuum tubes could perform calculations.

Theoretical physicist John V. Atanasoff and his assistant Clifford Berry had already completed a prototype of an electronic computer in 1942 with vacuum tubes that in many aspects reflected the design of modern computers. In England, during the years 1940 to 1944, mathematician Alan Turing headed a top-secret team that developed a series of vacuum-tube computers called Colossus. These were dedicated to the task of deciphering German coded messages transmitted from Enigma coding and decoding machines, thought to produce an unbreakable code. The Colossus prevailed, however, demonstrating the power of vacuum tubes and also aiding the Allied side in winning World War II. Shortly after the war, all plans for electromechanical computers were abandoned, first for vacuum tubes and somewhat later for transistors.

The first yarn is made using the artificial fiber terylene, better known in the US as Dacron; *See also* 1940 MAT

The work of Archer Martin [b London, Mar 1 1910] and Richard Synge [b Liverpool, England, Oct 28 1914] culminates in a classic paper by Consden, Gordon, and Martin that spells out the details of how to accomplish paper chromatography; *See also* 1938 MAT

Chemist Lars Onsager [b Oslo, Norway, Nov 27 1903, d Coral Gables FL, Oct 5 1976] works out the details of the gaseous diffusion method of separating uranium-235 from uranium-238 about this time; *See also* 1935 MAT

Alfred Blalock [b Culloden GA, Apr 5 1899, d Baltimore MD, Sep 15 1964] and Helen Brooke Taussig [b Cambridge MA, May 24 1898, d May 20 1986] perform the first "blue baby" operation, correcting blood supply to the lungs of a female infant

Benjamin Minge Duggar [b Gallion AL, Sep 1 1872, d New Haven CT, Sep 10, 1956] and coworkers discover the antibiotic Aureomycin, the first of the tetracyclines, as a result of checking many soil samples for antibacterial action; *See also* 1939 MED

The Greenwich Royal Observatory installs its first quartz-crystal clock, providing ten times the accuracy of the previous pendulum system; *See also* 1928 TOO

Early in the year, the German armed forces begin to use the V-1 flying bomb, propelled by a jet engine and controlled by an autopilot, against Great Britain

In Sep the first V-2 rockets explode in London; they bring with them a ton of explosives that travel at supersonic speeds, so that the explosions occur completely without warning, before the rockets can be heard; *See also* 1942 TRA

The Germans introduce a rocket-powered airplane, the Me 163B-1 Komet, into World War II, but its habit of exploding spontaneously makes it a poor weapon; *See also* 1942 TRA

Igor Sikorsky builds the first modern helicopter, the VS 36 A; it has an enclosed cockpit for the pilot and adjustable pitch for the rotor blades; *See also* 1939 TRA

1945

GENERAL

The US explodes the first nuclear weapon, known as the atomic bomb, at 5:30 a.m. local time on Jul 16 at Alamogordo air base in NM; on Aug 6 a somewhat different device is dropped on Hiroshima, Japan, followed by another nuclear weapon used against Nagasaki, Japan, on Aug 9

COMMUNICATIONS

Author Arthur C. Clarke [b Somerset, England, Dec 16 1917] proposes a geosynchronous satellite; such a satellite would hover over the same spot on Earth because it revolves at the same speed as Earth's rotation; *See also* 1965 COM

The Eidophor is the first practical light-tube (valve) video projector

ELECTRONICS & COMPUTERS

Grace Murray Hopper [b New York City, Dec 9 1906, d Jan 3 1992] coins the term "bug" to indicate a fault interfering with the running of a computer program

John von Neumann begins computer research at the Institute for Advanced Studies in Princeton NJ and writes "First draft of report on the EDVAC" (EDVAC means Electronic Discrete Variable Computer), outlining the "von Neumann architecture" for computers; *See also* 1944 ELE; 1948 ELE

ENERGY

Rolls-Royce in England develops a jet engine with an afterburner; extra fuel is injected into a special chamber at the rear of the engine, delivering additional thrust; because of high fuel consumption afterburners are mainly used on military aircraft; *See also* 1941 TRA

1946

GENERAL

The Japanese Union of Scientists and Engineers (JUSE) is founded; its aim is to revive Japanese industry using statistical quality control

On Jul 5 the bikini bathing suit, a brief halter and separate bottom, appears in France; the style gradually spreads to beaches around the world, reaching the US in the 1960s; it is named after a US test site for atomic weapons in the South Pacific

The Fender Guitar Company introduces the modern electric guitar with a solid body and various built-in controls; *See also* 1935 GEN

COMMUNICATIONS

Maurice V. Wilkes [b Dudley, England, Jun 26 1913] develops an early version of Assembler, a mnemotechnic language that simplifies considerably the programming of computers

The first mobile telephones are introduced

The first zoom lens introduced by Zoomar; *See also* 1899 COM

Albert Rose, Paul K. Weimer and Harold B. Law develop the *image orthicon*, an improved version of the orthicon that becomes widely used by television companies because of its excellent sensitivity to low light levels; *See also* 1939 COM

ELECTRONICS & COMPUTERS

John Blewett [b Toronto, Canada, Apr 12 1910], his attention called to a short letter in *Physical Review* by Russian physicists D. Iwanenko and I. Pomeranchuk, learns that large particle accelerators theoretically release electromagnetic radiation caused by the acceleration of electrons or other particles traveling in curved paths; Blewett demonstrates that this radiation, now known as synchrotron radiation, actually exists; *See also* 1943 ELE

On Feb 14 John William Mauchly and John Presper Eckert demonstrate ENIAC, the first all-purpose, all-electronic computer, to scientists and industrialists; ENIAC, developed secretly during World War II, does not use binary numerals, but has its vacuum tubes arranged to display decimal numerals; it draws so much electricity that it causes the lights in a nearby town to dim each time it is used; its early press releases claimed such feats as multiplying 360 ten-digit numbers or extracting a square root "in a single second"; *See also* 1945 ELE; 1949 ELE

ENERGY

On Dec 24 the first Soviet nuclear reactor goes into operation under the direction of Igor Vasilevich Kurchatov, [b Sim, Russia, Jan 12, 1903]; *See also* 1942 ENE

Herbicide 2, 4-D, the first modern plant poison, is introduced; *See also* 1940 FOO

Fluoridation of a water supply to prevent dental decay is introduced in the US; *See also* 1874 MED

Sir Alexander Fleming, Sir Howard W. Florey, and Ernst Boris Chain of England win the Nobel Prize in physiology or medicine for the discovery of penicillin and research into its value as a weapon against infectious disease; *See also* 1928 MED; 1940 MED

Willys-Overland begins producing Jeeps for civilian as well as military use; *See also* 1941 TRA; 1946 TRA

The White Sands proving ground for US rocket research is established in NM; *See also* 1929 TRA

1945

The Gaggia company in Italy introduces an espresso machine that uses steam under pressure to produce strong coffee of low caffeine content; *See also* 1908 FOO

Alfred Charles *Bernard* Lovell [b Oldland Common, England, Aug 31 1913, knighted 1961] starts construction of a radio-telescope with a fixed reflecting disk 66 m (218 ft) in diameter; *See also* 1937 TOO; 1957 TOO

The innovative Tucker Torpedo automobile, with its engine in the rear and a third front headlight, is introduced by Preston Tucker [b Capac MI, Sep 21 1903, d Jan 7 1957]; by 1947 the company has produced only 51 cars, and collapses; *See also* 1945 TRA

1946

Antibiotics

Since antiquity people have tried to cure diseases by eating, drinking, or applying substances, most often plant extracts we now label herbal medicines. These were sometimes effective, but more often not. An odd assortment of other agents were tried as well, including spiderwebs, animal organs, and molds or fungi. Like the herbal medicines, these were sometimes effective.

In the first part of the 20th century, systematic research revealed inorganic chemicals that were more effective and reliable than the natural remedies of old. The most powerful of these was a group of related medicines called sulfa drugs.

The strongest agents against microbes, however, had always existed in nature, just as traditional healers had believed. Louis Pasteur and Jules François Joubert observed in 1877 that anthrax bacilli were killed by bacteria from the air. In 1928 Alexander Fleming discovered by accident that on a microbial culture plate that had been contaminated by a mold, the mold spots formed bacteria-free zones around them.

He identified the mold as a very common one, belonging to the *Penicillium* group, and tested it on several bacteria. Fleming found that it inhibited the growth or even killed certain bacteria while it had no effect on other ones. Fleming could not obtain the active agent in the mold, which he called penicillin, in sufficient quantities, and his discovery remained largely ignored until 1939, when Howard Walter Florey and Ernst Chain succeeded in isolating and purifying penicillin. The development of penicillin as an antibiotic was a major breakthrough, and penicillin quickly supplanted many sulfa drugs then in use. It proved effective against several infections and saved many lives at the close of World War II.

Subsequently, several other antibiotics were prepared from molds, many of them attacking microorganisms resistant to penicillin. For example, streptomycin, developed by Selman Abraham Waksman in 1947, cures tuberculosis, which resists penicillin. Modern antibiotics are often chemically modified.

OVERVIEW

The Electronic Age

For most of the quarter century immediately following the close of World War II, there was a technological threat that was unlike any other known in the past. Terrible weapons had been developed that could destroy the world, perhaps completely. The fission and fusion bombs developed in the 1940s and 1950s were so much more destructive than earlier weapons that it was hard to relate them to ordinary explosives. Furthermore, the new bombs not only blew things up and started massive fires, they melted or vaporized materials that were thought completely stable; seeing a photograph of the shadow of a human permanently in place on a concrete wall after the bombing of Japan with fission devices reminded viewers that this was a brand-new force. On top of everything else, the area bombed was altered by becoming radioactive. It was already known that radioactivity could cause people to sicken and die. Even scarier, it could affect future generations by causing mutations of germ cells.

As if the threat of the new bombs were not enough, the long-range rockets developed during the war were soon improved to where the rockets could drop bombs almost without warning thousands of miles away from where they were launched.

For the first time since the first millennium CE, a great many people began to believe that the end of life on Earth was imminent. The means were clearly identified as the result of science and technology, even though the political situation at the time, known as the cold war, was the major occasion for concern.

During the first years after the war, a speedy economic recovery from the Great Depression in the US and the ravages of war in Europe took place. Industry started growing rapidly, and many new types of industry, such as plastics and semiconductors, appeared.

By 1945 vacuum-tube technology had reached a high degree of perfection, making the swift development of electronics and communications possible. Early computers, automatic systems, television, scientific instruments, all relied initially on vacuum tubes. The full impact of the discovery of the transistor in 1947 was felt about 10 years later, when semiconductor devices became cheap to manufacture. The transistor and, later, integrated circuits quickly displaced the vacuum tube in most of its applications.

War spurs invention

Many inventions, such as the jet engine, the rocket, and nuclear energy, were the direct result of technological developments during World War II. The war also speeded up the creation of new materials, such as artificial rubber, plastics, and synthetic fuels. In addition, there was extraordinary progress in the electronics field. Many of the technologies developed during World War II for defense needs, such as radar and computers, entered civilian life. They made possible the huge growth of communications and civilian aviation, the automatization of industry, and space exploration.

The jet engine and the liquid-fuel rocket engine became fully operational during World War II, powering the first large rockets (the German V–2) and fighter planes, both in England and Germany. These inventions led to the huge development of civil aviation and space exploration during the 1960s. More than ever, new developments were the result of the convergence of several new technologies. The development of new, heat-resistant materials, used in jet engines, and the availability of microelectronics contributed to the extraordinary growth of civil aviation. Without computers, space travel also would have been impossible.

Changes in society

Since the Industrial Revolution, technology has exerted an important influence on the way people live. While the

Industrial Revolution initiated a general movement of the population toward cities, the availability of the automobile had the reverse effect: People started leaving cities and living in the forever sprawling suburbs. Traveling by car, especially in the US, became very cheap, and also contributed to the decrease in rail passenger services.

Technology also influenced and even rearranged the traditional divisions between professions and the work force. The introduction of automation in manufacturing allowed many manufacturing processes to be done by less skilled workers. Also, the new, more complicated technology associated with automated manufacturing required more know-how. Technical know-how became the domain of an increasingly powerful but small group of people. During the 1960s and 1970s, automation also reached the office, with the same results. For example, when the first computers appeared in industry, banking, and administration, they were completely mystifying to the average worker, and the few computer "specialists" gained considerable earning power. Computers started doing many office tasks that required special skills.

The main consequence of increasing automation was that the middle group of skilled workers lost ground. The work force split into two groups: the specialized worker who had the knowledge to deal with sophisticated and automated machinery, and the unskilled worker still required for many menial tasks, such as feeding pieces to be transformed to automatic machines.

Environmental problems

Environmental problems caused by industry have existed since the Middle Ages and became extensive during the 19th century. Instead of localized pollution problems, industrialized centers, such as England's Midlands, experienced regional pollution. Even then, however, the fact that industrial pollution formed a threat to human lives became fully understood only after World War II. Earlier in the 20th century, burning coal for heating and the generation of electricity was the main cause of atmospheric pollution. Sometimes this had catastrophic consequences, such as the notorious London smog—although the danger of smog was not recognized until a particularly bad episode in 1952 took 4000 lives. With the advent of oil and natural gas, coal heating almost completely disappeared in the US and Europe (in London it was outlawed).

The switch to oil and gas, however, introduced new pollutants into the atmosphere. Again the consequences were sometimes severe, such as the photochemical smog in Los Angeles, mainly caused by the exhaust fumes from cars. At the same time, concerns arose about the natural reserves of energy sources. It became clear that oil and gas reserves, and also those of coal, were in danger of becoming depleted in the near future. As a response to these concerns, a group of scientists and economists founded in 1968 the Club of Rome. The Club of Rome sponsored a team at the Massachusetts Institute of Technology (MIT) to study the future development of several aspects of human activity, such as population growth, food production, industrialization, nonrenewable resources, and pollution. The report, *The Limits of Growth,* published in 1972, was based on computer models; it predicted catastrophic developments, such as a shortage of food and other resources if the then existing trends continued. Fortunately, many such predictions have proven to be too pessimistic, but the report set in motion several ecological movements and resulted in positive change. For example, many rivers and the air in some cities are less polluted today than they were in the 1960s and 1970s.

Technology and world power

One consequence of World War II was that governments understood that the scientific and technological development of a country is a crucial factor in its defense capabilities. In the US, government became the major funding agency of scientific and technical research and development, and in the Soviet Union (Russia), the military became the largest employer of engineers and scientists. For example, during US fiscal year 1963, the federal government financed about 60 percent of the research and development by industry; most of it was research that was expected to benefit the military. The close relationship between the military and industry became known as the military-industrial complex, a term coined by US President Dwight Eisenhower. Eisenhower's farewell speech of Jan 17 1961 warned the nation against the dangers of the power of industry supplying the military with weapon systems. Despite this warning, both the American and the Soviet military-industrial complexes became the driving forces of the arms race that marked the cold war era.

Many scientists who during World War II had worked on nuclear bombs started or joined movements against the nuclear arms race. The Union of Atomic Scientists, for example, was founded in 1946 by a group of scientists, several of whom had worked on the Manhattan Project. Their aim was to warn governments and the public against the dangers of a nuclear arms race.

The scientists' warnings had little effect at the time. However, a new factor attenuated the arms race somewhat by diverting attention to a different race, the race in space. Competition between powers, especially the US and the Soviet Union, for successes in space seemed for a time to be what American philosopher William James had termed "the moral equivalent to war."

Major advances

Communications. The postwar years saw enormous growth in communications technology. Television spread around the world in less than a decade, and radio broadcasting became a suitable medium for the new high-fidelity recordings as a result of the introduction of frequency-modulation (FM) radio. Microwave technology, initially developed for radar, became used for the transmission of telephone conversations over land. The telephone systems of cities were linked by microwave transponders, easily recognizable by their dish- or horn-shaped microwave antennas, which allowed a strong increase in message capacity. Coaxial cables also increased the capacity of telephone lines drastically. Telephone traffic between Europe and the US initially depended on short wave radio. The first transatlantic telephone cable in 1956, followed by several other cables in the 1960s, made a telephone call between London or Paris and New York less of an exotic event, although true ease in international calling did not come until satellite transmission began.

Communications satellites, introduced during the 1960s, expanded enormously telephone communication links between and even within continents. The first communications satellite, Echo I, was a huge metallized balloon that circled Earth and that could relay radio communications by bounced radio waves so the waves could travel over long distances. Subsequently, active communications satellites were placed in geostationary orbits; their operation could be compared to the microwave transponders used on Earth to link cities.

Electronics and computers. Until the mid 1950s all electronic devices functioned with vacuum tubes, and early computers were sometimes equipped with several thousands of them. Failures were frequent because of the limited lifetime of the heating filaments in vacuum tubes. Vacuum tubes were also power hungry, and the impact of electrons on the anode produced excessive heat.

The invention of the transistor in 1947 by John Bardeen, Walter Brattain, and William Shockley revolutionized electronics. The transistor became an ideal substitute for the vacuum tube for most purposes because it was more reliable, required much less power, and produced little heat. Electronic devices based on transistors could be built in compact sizes, yet with greater complexity of design. One of the first civilian applications of transistors was in hearing aids, making miniaturization practical for the first time.

The first computers built with transistors appeared around 1954. By 1959 practically all computer manufacturers used transistors throughout, making computers much smaller and more reliable. During the 1960s the transistor displaced vacuum tubes in television sets, radios, stereographic photographs, and most scientific instruments.

In the early years of their development transistors could not replace all functions performed by tubes—for example, in the 1950s no transistors were available that could operate at very high frequencies. But by the end of the 1960s, the capabilities of transistors generally surpassed those of vacuum tubes, even in high-frequency applications, such as high-fidelity amplifiers. The transistor, because of its small size, is also more suitable for incorporation into printed circuits than vacuum tubes are. A printed circuit uses printing technology to deposit a wiring pattern in conducting material on a board, thus making it possible to reproduce wiring by mass production, just as printing with movable type moved book manufacture into mass production. Printed circuits with transistors soon were used in all electronic devices, not only because they could be mass produced, but also because they were sturdy and the wiring errors of manual production were eliminated.

A development as important as the introduction of

transistors was that of the integrated circuit, a kind of printed circuit that goes beyond boards. Developed independently by Jack Kilby at Texas Instruments and Robert Noyce at Fairchild Semiconductor, the technique consists of embedding electronic elements and connections on a small wafer of silicon. In this way several transistors can be placed on such a chip, increasing miniaturization and reliability further by decreasing the number of components to be placed on a printed circuit board. Using photolithographic techniques, yet another borrowing from printing technology, such chips could be produced cheaply in quantity. Soon they replaced complete circuits, such as logical gates for computers, or complete electronic amplifiers. The greatest breakthrough in chip technology came during the early 1970s with the introduction of the first computer memory chips and microprocessor chips, heralding the development of the microcomputer a few years later.

Although the progress in electronic components, especially in transistors and integrated circuits, affected all areas of electronics, the development of computers and digital electronics can be termed a technical revolution. Computers progressed from cumbersome electronic renderings of electromechanical machines to highly flexible and user-friendly universal machines. Their application spread from number crunching into all areas of human activity, such as administration, banking, defense systems, civil aviation, space exploration, and petroleum exploration. Software, which was nonexistent when the first computers were built, developed quickly, and from mid-1950 on several high-level computer languages became available. The 1960s and 1970s can be termed the era of the mainframe, large computers with much hardware, typically housed in a "computer center" and operated by computer specialists. Further developments in the 1970s and 1980s, especially in integrated-circuit technology, brought the computer out of the computer center and made it a versatile tool accessible to everyone.

Energy. Coal had been the main energy source in the industrialized world since the Industrial Revolution. This situation changed during World War II. Although coal remained the single largest energy source until about 1970, its place has now been taken over by petroleum. Natural gas and, in eastern Europe, lignite are two other fossil fuels that today form a substantial part of the energy supply. Because of discoveries of natural gas reserves, coal gas was phased out.

Petroleum (oil) became widely used because it was easy to handle and a cheap source of power; it displaced coal in most countries. Petroleum also was the starting material for a large number of new chemical products, of which plastics became the most important.

The chemical industry became the fastest growing industry during the postwar years.

During the 1950s oil became in many countries the main source of energy for domestic heating. Oil is not only easier to manipulate and to transport than coal, but oil became much more available because of several new techniques developed for exploration on land and at sea, and for obtaining it at offshore locations. Offshore drilling for oil had already started before World War II, and the first offshore oil was obtained in 1947. During the 1960s and 1970s, a substantial part of oil was obtained from floating drilling platforms at sea (almost 15 percent of all the oil supply in 1970). However, the political crisis in 1973 in the Middle East caused important increases in energy prices, and several countries became convinced that they had to follow an energy policy that allowed them to be self-sufficient. Nuclear energy was viewed by several countries as a desirable alternative, and nations such as France embarked on extensive development of nuclear power.

The atomic bomb had convincingly proved that enormous amounts of energy can be liberated by nuclear reactions. To use this energy for peaceful purposes, the energy had to be released in a controlled way. This can be done by slowing down nuclear chain reactions so that they occur in a sustained but constant rate. In a nuclear chain reaction, neutrons produced by fission reactions cause new fission reactions to occur; by absorbing these neutrons with a substance called a moderator, it is possible to fine-tune the fission reaction so that the chain reactions take place at a constant, but slow, rate.

The first nuclear reactors developed in the 1940s and 1950s used uranium as fuel and graphite as moderator; they were cooled by carbon dioxide gas. These reactors were mainly built in Europe, especially France and England. In addition to using the reactors to generate power, both nations also used them as a source of plutonium for the development of nuclear weapons. The US developed pressurized water reactors (PWR). Water, maintained under pressure to prevent it from boiling, serves simultaneously as moderator and cooling fluid. The heated water is led through separate steam generators (heat exchangers) that produce the steam for driving the turbines. The reactors do not generate significant amounts of plutonium. Other types sometimes used are boiling water reactors in which steam is generated in the reactor itself and reactors based on the properties of deuterium oxide (heavy water).

Food and shelter. Although the farming community had long resisted large-scale production techniques, postwar agriculture was mainly marked by increased mechanization. Tractors came into general use, and livestock farming became more automated. The use of improved seed and of herbicides and pesticides also made farming much more efficient. A consequence of this development is that the farming population started decreasing strongly in Europe and the US, a trend that is still continuing today.

Concern about herbicides and pesticides in farming, including the possible consequences to health as well as loss of effectiveness on repeated use, led to the development of organic farming. Thousands of farmers began to switch to agriculture based on the use of organic compost, crop rotation, shallow plowing, and the avoidance of chemical fertilizers, pesticides, and herbicides, a trend that became much more prominent after 1973. Of course, much of this trend was simply a return to agricultural practices of earlier in the 20th century.

Around the developed world, modern transport has meant that fruits and vegetables produced in suitable climates became available year-round. For example, northern hemisphere nations obtain seasonal produce from the southern hemisphere, and vice versa. Even within large nations, such as the US, most food was no longer produced locally, but was shipped all over the country from a few locations. The increased use of commercial greenhouses has led, however, to the cultivation of a much wider variety of vegetables in the northern countries of Europe, such as tomatoes.

One of the main developments in architecture during this period was the modern "glass" office building—actually a steel and reinforced concrete building with a curtain wall or a supported wall of glass. Often the glass was tinted to make the building look blue or golden. In New York City the Lever House and the Seagram Building, only a few blocks apart on Park Avenue, started the trend. Soon almost all office construction in the heart of major cities around the world used similar designs, creating a uniformity that was known as the International Style. Such buildings were often raised by ingenious construction techniques featuring giant cranes that rose with the building.

At the same time, a more modest revolution was taking place at the level of personal housing. Ever since Roman times, people had increasingly moved into large multiple dwellings (apartment houses) in cities. After World War II the first reversal of such urban migration took place as builders established inexpensive, often prefabricated, housing developments outside of cities. Now people could afford to leave the giant multiple dwellings for a home of their own, and large numbers took advantage of the opportunity. The suburbs came into being. Soon the toll on the inner city began to be felt, especially as shopping centers in the suburbs began to satisfy the basic needs of the people who had moved from the city.

Materials. Petroleum has not only played an important role in the supply of energy, but it has also become the most important raw material for an extraordinary number of new products, including plastics, solvents, resins, and detergents. During the 1960s, the chemical industry grew three times as fast as the total growth of industries. The chemical industry has a large influence on daily life. While before World War II, most objects in use had a natural origin—wood, glass, and rubber are examples—after the war, most manufactured objects were made of plastics and other synthetic materials. Cotton and wool were displaced by polyesters and nylon, and the chemical industry became one of the major suppliers of the textile industry.

Medical technology. The introduction of penicillin and other effective antibiotics had a big influence on the practice of medicine. Many diseases that had been difficult to treat, such as tuberculosis and syphilis, became controllable. Researchers also developed several wide-spectrum antibiotics that could fight many different species of bacteria. A better understanding of immunology led to the development of new vaccines, especially vaccines against viral diseases such as influenza, measles, and polio. It also led to the better understanding of immunological rejection mechanisms, and the development of drugs that prevent organ rejection. These developments made organ transplants possible. The most spectacular achievement was the first transplant of a human heart by Christiaan Barnard in 1967. Surgeons also started to perform liver and lung transplants around that time. Advances in endocrinology led to the development of the contraceptive pill.

Developments in electronics also contributed to medicine. The pacemaker, introduced in 1960, allowed regulation of the heartbeat by synchronizing it with electrical pulses. The combination of computer technology with X-ray diagnosis led to the development of the computerized tomography scanner (CT scanner) in 1972.

Tools and devices. The automation of manufacturing practices and machine tools was also facilitated by the development of chips and computers. By the 1950s milling machines, typesetting apparatus, and other devices were being controlled by paper tape or by punch cards, but in the 1960s these controls were gradually replaced by electronic controls. The first industrial robots also appeared in the 1960s, although they did not become effective until after 1973, and were always more popular in Japan than in the Western nations. Also by the mid 1960s, the logical precursor of the machine tool was invented. Known as CAD, computer-assisted design allows a tool designer to picture the part and make adjustments visually that otherwise would require hours of calculation. In some systems, the computer can then control the machine tool to produce the finished part.

By far the most influential new device of the period was the laser (light amplification by stimulated emission of radiation), an implement for producing coherent light of great intensity. Based on a concept first put forward by Einstein nearly 40 years before, the first breakthrough was not with light, but with microwaves. The maser, which uses the same principles as the laser, was most useful as a detector for faint microwaves, but a number of people quickly recognized that the maser could be adapted to produce electromagnetic waves as well as detect them. Such production was most available at low wavelengths of light and in the infrared. Within a half-dozen years of the maser, people were working on its light analog; by 1960 the first working lasers were developed, although ultimately the patent credit went to an earlier inventor who actually had little impact on the field at its inception.

In the beginning, the laser was a wonderful toy for which the actual uses were uncertain. Soon it found applications in cutting and welding at all levels from eye surgery to machining parts. Farmers even used lasers to make sure fields were level. For the laser, an important transition came when lasers based on carbon dioxide were introduced in 1970. These were the first really powerful lasers. Like many developments during this period, however, the greatest applications were to come in the years after 1973.

Other devices of the period included those that allowed a number of new ways of observation, ranging from the traditional—the Hale reflecting telescope on Mount Palomar, inaugurated right after World War II—to the unexpected, observations of atoms with the field ion microscope, something most scientists had thought would never be possible. Several new devices greatly improved observation of cells in biology, ranging from ways to see the insides of living cells to the scanning electron microscope that produces dramatic, three-dimensional-appearing views of surfaces.

Transportation. Both military and civilian air transportation underwent a profound change during World War II. Airliners inherited the powerful piston engines developed for heavy bombers, resulting in a series of four-engined planes that could cross the Atlantic Ocean in less than 20 hours. Jet engines, developed during the war, were installed on military planes in the late 1940s and civilian planes in the 1950s, reducing the Atlantic crossing to about 7 hours. In most industrialized countries airlines underwent a similar development to that of railways during the 1840s and 1850s. Seating capacity reached 500 in large commercial craft termed jumbo jets; airport facilities also were expanded. Flying became much cheaper and airlines became serious competition for railways, especially in the US, and to a lesser degree in Europe.

Because jet engines can develop much more thrust than can be achieved with propellers, weight became less of a problem and large jet planes became increasingly used for freight, such as perishable foods.

Transatlantic passenger ships were victims of the development of civilian airlines. During the 1950s and early 1960s ocean liners still could compete with airlines, but they practically disappeared during the 1970s. Several famous large passenger ships, such as the *France,* were converted into cruise ships, starting them on new, successful careers.

The jet engine was not the only new source of transportation power. The use of nuclear energy to power submarines was soon recognized as a great advantage because it would allow submarines to operate for extended periods without the need for refueling or resurfacing. The first nuclear submarine, the *Nautilus,* was commissioned in 1954. Nuclear submarines equipped with nuclear missiles became during subsequent years an important element in the arms race since they were mobile launching platforms whose location could not be detected by an enemy.

A few civilian ships powered by nuclear reactors were built during this period. The first was the ice breaker *Lenin,* built by the Soviet Union; it started operating in the Arctic seas in 1959. The *Savannah,* launched by the US in 1962, the *Otto Hahn,* launched by Germany in 1969, and the *Mutsu,* launched by Japan in 1967, all proved to be not economically viable. The construction of such ships has since been abandoned.

Work toward a nuclear airplane started in the US in 1948, but was abandoned in 1961 after millions of dollars had been spent on the project. The biggest obstacle was that no airplane could be built to carry a nuclear shield that would protect the pilots against radiation from the reactor. Ideas for a nuclear rocket engine were also abandoned because of the nuclear pollution of the atmosphere such an engine would produce.

During the postwar years, both the Soviet Union and the US continued developing rockets that were based on the German V–2 rocket design. The aim of rocket development was initially military—strategic medium-range missiles appeared in the Soviet Union around 1955 and in the US around 1958. The deployment of intercontinental missiles followed in 1959. The launch of the first artificial satellite, Sputnik, in 1957 by the Soviet Union started "the race in space" between the US and the Soviet Union. Progress was spectacular: The first astronauts circled the Earth in the early 1960s, and man landed on the Moon in 1969, less than 12 years after the launch of the first artificial satellites. The first inhabited space station, *Salyut 1,* was launched in 1971 by the Soviet Union. *Skylab,* a manned orbital laboratory, was launched in 1973 by the US.

1947

GENERAL

The US Atomic Energy Commission is founded on Jan 1; its main responsibility is the development of nuclear arms; *See also* 1945 GEN; 1952 GEN

COMMUNICATIONS

Physicist Dennis Gabor [b Budapest, Hungary, Jun 5 1900, d London, Feb 9 1979] develops the basic concept of holography, although the technique does not become truly practical until after the invention of the laser

Edwin H. Land announces his invention of a camera and film system that develops pictures inside the camera in about a minute; the system is based on an invention made in 1928 by the German company Agfa, which never commercializes the idea; *See also* 1928 COM; 1963 COM

ELECTRONICS & COMPUTERS

Howard Aiken completes the Mark II computer at Harvard; its design is almost entirely electromechanical, and it cannot compete with the electronic computers in existence; *See also* 1944 ELE; 1948 ELE

On Dec 16 John Bardeen [b Madison WI, May 23 1908, d Jan 30 1991] and Walter Houser Brattain [b Amoy, China, Feb 10 1902, d 1987], working at Bell Labs, perform the experiment that results in the first recognition of the transistor effect; they report this discovery to the management at Bell Labs on Dec 23; the transistor is announced to the general public in 1948; *See also* 1942 ENE; 1948 ELE

ENERGY

The two first nuclear reactors built by Canada in Chalk River, the ZEEP and the NRX, become operational; *See also* 1946 ENE; 1952 ENE

The first peacetime nuclear reactor in the US starts construction at Brookhaven NY on Aug 11; *See also* 1942 ENE; 1948 ENE

Ralph Miller invents a version of the internal combustion engine in which the cycle of piston movements has a shorter compression than expansion stage; the Miller-cycle engine provides higher power for its size and fuel consumption, but its complexity limits its use to a few boats and motors for electrical generators; *See also* 1877 ENE; 1993 ENE

Andrei Sakharov [b Moscow, (Russia), May 21 1921, d Moscow, Dec 14 1989] and F.C. Frank propose the use of negative muons to produce fusion reactions in a mixture of deuterium and hydrogen; this possibility is rediscovered in 1957 by Luis Alvarez; *See also* 1958 ENE

1948

GENERAL

A Communist coup in Czechoslovakia (the Czech Republic and Slovakia) and a Soviet blockade of Berlin (Germany) mark the active start of the cold war between the US and the USSR, although its roots date back to the years immediately following World War II; *See also* 1949 GEN

Five days of smog in Donora PA kill 20 and leave 14,000 injured; *See also* 1952 GEN

Adidas in Germany develops sneakers, made from war-surplus canvas and fuel-tank rubber; *See also* 1909 GEN; 1962 GEN

COMMUNICATIONS

The magnetic drum for data storage in computers is introduced; it consists of a rapidly spinning drum covered with a magnetic film on which the data is encoded as tiny magnetic domains; *See also* 1944 COM; 1951 COM

Physicist Peter Mark Goldmark [b Budapest, Hungary, Dec 2 1906, d Westchester County NY, Dec 7 1977] develops the first long-playing record in the US

The Gerber standard for television—625 scanning lines at 25 frames per second—is adopted in Europe; *See also* 1941 COM

ELECTRONICS & COMPUTERS

Physicist William Bradford Shockley [b London, Feb 13 1910, d Stanford CA, Aug 12 1989], American physicist Walter Houser Brattain, and American physicist John Bardeen announce the discovery of the transistor, a tiny device that works like a vacuum tube, but uses less power; *See also* 1947 ELE; 1951 ELE

Manchester University's Mark I prototype begins operating; designed by Tom Kilburn, it is a stored-program electronic computer; the first to use von Neumann architecture, it stores data in Williams tubes (a type of cathode-ray tube); *See also* 1945 ELE; 1951 ELE

ENERGY

The first experimental nuclear reactor in France, the Zoe or EL-1, built at Fontenay-aux-Roses, goes into operation; *See also* 1947 ENE; 1950 ENE

A solar-heated house, designed by architect Eleanor Raymond [b Cambridge MA, Mar 23 1887] and engineer Maria Telkes is built in Dover MA; employing flat-plate collectors, it is ultimately deemed a disappointment because of lack of heat storage on cloudy days; *See also* 1949 ENE

Chloramphenicol, a powerful antibiotic, is discovered; its use is now restricted because of dangerous side effects; *See also* 1944 MED

1947

The first airplane to reach supersonic speed in the US, the Bell X-1, an experimental rocket plane, is flown by Charles E. (Chuck) Yeager [b Myra WV, Feb 13 1923]; *See also* 1944 TRA; 1967 TRA

The Hale telescope at Mt Palomar

As an individual object, the greatest technological achievement of the first half of the 20th century was just a big piece of glass, the mirror for the Hale telescope at Mt Palomar in California. Although the Soviet Union tried to top it during the cold war, their slightly larger Zelenchuksaya telescope of 1976 came nowhere near the Hale reflector's cold perfection.

Astrophysicist George Ellery Hale is mainly remembered for his ability to put together financing and technical teams needed to produce large and effective telescopes. His first success, in 1897, was the world's largest refractor, the Yerkes telescope of the University of Chicago, with a 101.6-cm (50-in.) lens. Hale's next triumphs were a 152.4-cm (60-in.) reflector followed by a 254-cm (100-in.) reflector, both for Mt Wilson Observatory in California. Each mirror in turn was the world's largest. In 1928 Hale proposed his greatest project in an article in *Harper's Magazine,* calling for a reflector of "200 inches or, better still, to twenty-five feet." But 25 feet (762 cm, or 300 in.) appeared beyond the technological limits of the time, so Hale and his backers settled for 200 inches, only twice the diameter, four times the area, and eight times the size of the then largest reflector.

The main problem in building the telescope was casting the glass mirror, although many other technical obstacles had to be surmounted as well. Several test castings in increasingly large sizes resulted in flawed pieces of Pyrex. The second full-sized mirror, however, proved to be flawless. It was transported from Corning NY to Pasadena CA by rail at speeds of no more than 40 km (25 mi) per hour. It was ground and polished over the next decade, with time out for World War II. Final tests in 1948 revealed a tiny deviation in one section of the mirror, about equal to one wavelength of green light. Additional polishing resolved this last detectable problem.

A C-54 transport plane from the US Air Force crosses the Atlantic Ocean relying entirely on an automatic pilot

The first truly streamlined automobile, the Studebaker, as designed by Raymond Loewy, is introduced in the US; although not a great success, it influences all subsequent automobile design; *See also* 1934 TRA; 1948 TRA

Hyman George Rickover [b Makov, Russia (Poland), Jan 27 1900, d Arlington VA, Jul 8 1986] convinces the US Navy to begin building nuclear submarines; *See also* 1942 ENE; 1954 TRA

The tubeless tire is introduced by Goodyear in the US; *See also* 1888 TRA; 1953 TRA

New York City opens the Fresh Kills Landfill on Staten Island; by the 1990s it is the largest artificial structure on Earth, with a volume greater than the previous leader, the Great Wall of China, but built entirely of garbage

Sir Charles Frank postulates that a screw dislocation will cause crystals to grow more easily and rapidly; *See also* 1934 MAT; 1952 MAT

John Franklin Enders [b West Hartford CT, Feb 10 1897, d 1985], Thomas Huckle Weller [b Ann Arbor MI, Jun 15 1915], and Frederick C. Robbins [b Auburn AL, Aug 25 1916] learn how to grow mumps viruses in chick tissue using penicillin to prevent bacterial contamination; the same technique works for the polio virus; *See also* 1931 MED; 1952 MED

The atomic clock is introduced; *See also* 1944 TOO; 1969 TOO

The 500-cm (200-in.) telescope designed by George Ellery Hale, often referred to by the name of its site on Mount Palomar in southern CA, but later renamed the Hale telescope, becomes operational; J.A. Anderson gets the first look using the primary mirror in Jan, but the official dedication is on Jun 3; the Hale telescope is the largest of its time and remains the most effective for the next 40 years; *See also* 1936 TOO; 1990 TOO

1948

Auguste Piccard builds his first bathyscaphe for underwater exploration; *See also* 1930 TRA; 1960 TRA

The US Atomic Energy Commission starts work on the design of a nuclear plane; *See also* 1942 ENE; 1956 TRA

General Motors' Cadillac automobile introduces tail fins on its models in the US, starting a trend toward using such fins as design elements that lasts throughout most of the 1950s; *See also* 1947 TRA

1948 cont

After a walk in the woods with his dog, Swiss engineer George deMestral steals an idea from the cockleburs in his socks and the dog's coat and invents the fastener Velcro; *See also* 1914 GEN; 1955 GEN

The first cable television systems in the US appear

Ampex develops the first US audio recorder using magnetic tape; *See also* 1930 COM

IBM introduces the Selective Sequence Electronic Calculator (SSEC); mainly electromechanical, it has electronic circuits for performing calculations and data storage; *See also* 1911 ELE; 1952 ELE

First-generation computers

As was the case in England in the development of the Colossus computer, defense needs were the driving force behind the development of the first large electronic computer built in the US. The Electronic Numerical Integrator and Computer (ENIAC), although completed at the end of 1945, when World War II was over, was initially designed for the calculation of trajectories of projectiles. Developed by the University of Pennsylvania's Moore School of Electrical Engineering for the Ballistic Research Laboratory in Aberdeen MD, it was a general-purpose decimal machine containing 18,000 vacuum tubes. Its design was related to the Differential Analyzer built by Vannevar Bush, except that the mechanical components, such as counters and adders, were replaced by electronic ones. The logic of the machine was integrated into its hardware, and by changing the setup of the hardware it could be adjusted to perform different tasks. Such "programming" consisted of plugging cables in plug boards and setting hundreds of switches.

Because ENIAC used vacuum tubes instead of relays, it was about a thousand times faster in performing calculations than contemporary electromechanical machines. For example, ENIAC could calculate the trajectory of a shell in 20 seconds, which is faster than the 30 seconds or so that a real shell takes to reach its target. Many of the vacuum tube circuits in ENIAC were derived from those already in use in nuclear and cosmic ray particle counters. The computer had a very small programmable memory consisting of flip-flop circuits that could store twenty words along with several permanently wired memory circuits called function tables, which could be modified for specific calculations.

Even while ENIAC was being completed at the Moore School, its designers, John Mauchly, Presper Eckert, and Herman Goldstine, were aware of its limitations, and started work with the mathematician John von Neumann on an entirely new design of computer. The basic concept, now known as von Neumann architecture, separates logic functions entirely from hardware; that is, most instructions for the execution of calculations are not permanent, called hardwired or read-only memory (ROM), but are stored in a temporary memory called random-access memory (RAM). Instructions can be placed anywhere in memory and even modified by the computer itself when needed. Besides a random-access read/write memory, a von Neumann computer contains a central processor and uses binary numbers and Boolean algebra for processing and storing data.

Unlike ENIAC, which in some ways was a parallel processor (working on different aspects of a task at the same time, or in parallel), computers with von Neumann architecture process data serially; that is, one instruction comes after another and is executed only when the preceding one is completed. Von Neumann architecture was implemented in virtually all subsequent computers for the next quarter century. Such vacuum tube computers based on the von Neumann architecture are now known as the first generation of computers. The second generation was born with the incorporation of the transistor in computer circuitry.

In 1944 von Neumann published his ideas, and those of his colleagues working on ENIAC in the *First draft of a report on the EDVAC*. Von Neumann's ideas would have been first implemented on a planned computer to be called EDVAC (Electronic Discrete Variable Computer). Because of patenting problems, its completion became delayed, however, and other computers became the first to use von Neumann architecture.

First-generation computers (Continued)

The EDVAC design called for a memory that would be able to store at least 1024 32-bit words ("words" in the binary numeration system consisting of 32 zeros (0's) and ones (1's), with each binary digit known as a bit). The construction of such a memory was at that time somewhat of a technological hurdle.

One type of available memory derived from the mercury delay line, a device developed for early radar to measure the time between the emission and the reception of the reflection of a radar signal. The delay line consisted of a metal tube filled with mercury. At each end of the tube a transducer was mounted, one that operated as a tiny speaker for emitting sound pulses and another that served as a microphone. A series of sound pulses could be permanently stored in such a tube by amplifying the signals picked up at one end and feeding them back at the other; the signals would travel around as in a merry-go-round. Such a delay line could store up to a thousand bits as pulses and constituted the memory of several of the early first-generation machines.

Another type of memory was derived from television technology. Information was stored on the inside of a special type of cathode-ray tube as tiny, electrically charged spots, and then read by another electron beam.

Both of these types of memory were unwieldy and not entirely reliable. They were soon replaced by ferrite-core memories that not only equipped the later first-generation computers, but also were used by the second-generation computers during the 1950s and 1960s. Ferrite-core memories used magnetic domains for recording information, a technology that, in different forms, is still the most common today. Ferrite-core ROM was also used in first-generation computers, but that was replaced by semiconductor technology in the second generation.

A group at Merck & Company headed by Karl August Folkers [b Decatur IL, Sep 1 1906) discovers that certain bacteria require vitamin B_{12} for growth, enabling the group to use the bacteria as indicators so they can isolate the pure vitamin; *See also* 1912 MED

Elizabeth Hazen and Rachel Brown discover the antibiotic Nystatin, the first safe fungicide; it has a wide field of application, including treatment of athlete's foot and the restoration of books or paintings attacked by fungus; *See also* 1935 MED

Philip Showalter Hench [b Pittsburgh PA, Feb 28 1896, d Ocho Rios, Jamaica, Mar 30 1965] discovers that cortisone can be used to treat rheumatoid arthritis

Physiologist Walter Rudolf Hess [b Frauenfeld, Switzerland, Mar 17 1881, d Locarno, Switzerland, Aug 12 1973] describes in his book *Das Zwischenhirn* (The diencephalon) his technique of using small electrodes to stimulate specific regions of the brain; Hess used this techique to identify various regions in the brains of dogs and cats; *See also* 1929 MED

American optician Kevin Touhy develops the modern corneal contact lens when he accidentally breaks off the corneal part while making a lens of the older type, which also covers the whites of the eyes; *See also* 1887 MED; 1958 MED

The first transfer machines for the manufacture of engine blocks start operation in Detroit MI; they perform 550 tooling operations in 15 minutes, accelerating the production of automobile engines considerably; *See also* 1912 TOO; 1954 TOO

Paul Kirkpatrick [b near Wessington SD, 1894, d Palo Alto CA, Dec 26 1992] and Albert Baez develop the X-ray reflection microscope for examination of living cells; this microscope continues in use in medicine and in astronomy, where it is used to take X-ray photographs of galaxies; *See also* 1938 TOO; 1957 TOO

During the airlift to Berlin in the winter of 1948-1949, about a hundred DC-3 aircraft transport about 150,000 tons of goods on 22,000 flights; *See also* 1935 TRA

1949

China falls to Communist forces and the People's Republic of China is formed; *See also* 1948 GEN; 1950 GEN

The Soviet Union (Russia) in Aug explodes its first atomic bomb, using virtually all the plutonium it has on hand and following US plans for the bomb smuggled to the Soviet Union by such spies as Klaus Fuchs; *See also* 1945 GEN; 1951 GEN

After a lab technician installs the accelerometers backward on Major John Paul Stapp's rocket sled, Captain Edward Aloysius Murphy pronounces: "If there's more than one way to do a job and one of those ways will end in disaster, then someone will do it that way": Major (later Colonel) Stapp later abridges this to the more common form he termed Murphy's law: "If anything can go wrong, it will."

The first modern photocomposition system, the Lumitype 200, is introduced by Photon in the US; commercialized in 1954, it can set between 30,000 to 50,000 characters per hour; *See also* 1894 COM

John William Mauchly develops the Short Code, which allows computers to recognize mathematical codes consisting of two numbers; it is considered to be the first high-level programming language; *See also* 1946 ELE; 1954 COM

Claude Shannon publishes *The mathematical theory of communication*, based on his dissertation from 1938; he argues that information is a measurable quantity and establishes the basic rules governing all kinds of communication, including electronic forms; *See also* 1938 COM

John William Mauchly and John Presper Eckert build BINAC (the Binary Automatic Computer); it is the first electronic stored-program computer in the US, storing data on magnetic tape; it goes into operation in Aug; *See also* 1946 ELE; 1951 ELE

ILLIAC I, a computer using von Neumann architecture, as outlined in unpublished works by John von Neumann of 1945 and 1946, is built at the University of Illinois in Champaign-Urbana; ORDVAC, also using von Neumann architecture, is built by the US Army at its Aberdeen Proving Ground; *See also* 1946 ELE; 1952 ELE

EDSAC (Electronic Delay Storage Automatic Calculator) goes into operation at Cambridge University; one of the first stored-program computers to operate, it contains only 3000 vacuum tubes but is six times faster than previous machines; data are stored in mercury delay lines; *See also* 1946 ELE

An experimental solar oven designed by Felix Trombe is built at Odeillo in the French Pyrenees; using a paraboloid mirror to concentrate sunlight, it produces an output of 50 kw; *See also* 1948 ENE; 1954 ENE

1950

On Jun 25 North Korean forces invade South Korea, crossing boundaries set up by the Allies at the end of World War II; UN forces under the command of US General Douglas MacArthur repel the invasion; later in the year, China intervenes on the side of North Korea and pushes back the UN troops; *See also* 1949 GEN; 1953 GEN

The Japanese Union of Scientists and Engineers (JUSE) publishes *Elementary principles of quality control* to encourage competiveness in Japanese industry; *See also* 1946 GEN; 1956 GEN

Diner's Club introduces the first charge card, a prototype of the credit card (a charge card does not come with a line of credit, but must be paid off in full monthly); it is devised by Ralph Scheider

Paul K. Weimer, Stanley V. Forge, and Robert R. Goodrich develop the vidicon, a television camera tube; the principle of the tube is similar to that of the orthicon tube; *See also* 1946 COM

A form of Kodachrome film for color photography is introduced; chemical masks in the emulsion for the negatives improve the quality of prints; *See also* 1935 COM

Alan Turing proposes in an article, "Can a Machine Think," in the journal *Mind*, a test to determine whether a computer has real intelligence; in the Turing test, as it comes to be known, a computer in one room that can communicate with humans in another room must be able to convince the humans that it is intelligent; *See also* 1937 COM

John William Mauchly and John Presper Eckert found the first company setting out to commercialize computers; *See also* 1949 ELE; 1951 ELE

Start-up of the SAGE (the Semi Automatic Ground Environment) system by the US Air Force begins; SAGE will collect data from radar stations and other sources that then are processed in the Whirlwind computer system in real time; *See also* 1951 ELE

The first nuclear reactor in England goes into operation at Windscale; *See also* 1948 ENE; 1953 ENE

George A. Stephen [b 1922, d Kildeer IL Feb 11 1993], frustrated by a brick backyard "fireplace" grill that cooked unevenly and smoked too much, invents an all-metal enclosed grill; the company he works for, Weber Brothers, a metal-parts maker, manufactures the new grill, and soon the Weber grill and imitations of it come to dominate outdoor cooking in US suburbs

1949

In the US, General Mills and Pillsbury both begin marketing prepared cake mixes; *See also* 1930 FOO

Pilot plants producing yarn from the artificial fiber Dacron, better known in Great Britain, where it was invented, as terylene, begin operation in Wilton, England; *See also* 1944 MAT; 1955 MAT

X rays from a synchrotron are used for the first time in medical diagnosis and treatment; *See also* 1946 ELE

The Comet, the first jet airliner, designed by Geoffrey de Havilland [b Surrey, England, Jul 27 1882, d London, May 21 1965], makes its first flight on Jul 27; Comets enter service in May 1952, but are withdrawn 2 years later after two crashes caused by material fatigue; they are returned to service in 1958, becoming the first passenger jets to cross the Atlantic Ocean; *See also* 1941 TRA; 1968 TRA

The prototype of the Aerocar Model I, designed by M.B. Taylor, is completed in Oct; six "flying automobiles" built by Aerocar Inc. travel over 320,000 km (200,000 mi) on roads and log over 5000 hours in the air; *See also* 1968 TRA

1950

Embryos are transplanted in cattle for the first time; *See also* 1901 FOO; 1983 MED

The artificial sweetener cyclamate is introduced; *See also* 1879 FOO; 1965 FOO

Archaeologist Vere *Gordon* Childe [b Sydney, Australia, Apr 14 1892, d Mt Victoria, Australia, Oct 19 1957] renames the Agricultural Revolution the Neolithic Revolution to emphasize that the changes of that period (roughly 10,000 BP) involved more than just farming

Robert Wallace Wilkins [b Chattanooga TN, Dec 4, 1906) introduces the treatment of high blood pressure with reserpine, following the practice of using the drug in the form of snakeroot in India

The first automobile to use a gas turbine for power is introduced by Rolls-Royce in an experimental version; it attains speeds of 240 km (150 mi) per hour; Rover also introduces the gas-turbine Jet I in England; *See also* 1929 ENE; 1963 TRA

1951

The Soviet Union (Russia) explodes the first atomic (nuclear fission) bomb of its own design; a previously tested atomic weapon had been based on stolen US plans; *See also* 1949 GEN; 1953 GEN

Marion Donovan develops a throwaway diaper made of a shower curtain and absorbent padding; the idea is rejected by industry and Donovan creates her own company for the manufacture of "Boaters"

A video tape recorder (VTR) that uses magnetic tape is developed by Armour Research and demonstrated to Ampex, which starts its own research program; *See also* 1939 COM; 1956 COM

The military supercomputer Atlas is equipped with magnetic drums with a capacity of one megabyte; *See also* 1948 COM

Transcontinental television is inaugurated in the US

The first book on computer programming is published

Grace Murray Hopper develops the first compiler, called A0, which translates the codes used by programmers into binary machine code; *See also* 1952 COM

Fred Waller [b Brooklyn NY, Mar 10 1886, d May 18 1954] introduces Cinerama, the projection of films on a wide, curved screen by three projectors, which gives an effect of three-dimensions; this is the first of three types of motion picture systems with enhanced dimensionality introduced about this time; all are mainly intended to give movies a competitive advantage over television; *See also* 1900 COM; 1952 COM

John William Mauchly and John Presper Eckert build UNIVAC I, the first electronic computer to be commercially available and the first to store data on magnetic tape; it incorporates 100 mercury delay lines and 5000 vacuum tubes; in Mar the first UNIVAC I is installed at the US Census Bureau; *See also* 1950 ELE; 1954 ELE

Ferranti introduces the Mark I, the first commercial computer based on the Mark I computer developed by Tom Kilburn at Manchester University; *See also* 1948 ELE

The Whirlwind is the first computer capable of real-time processing of data; it is used in the SAGE system for processing data from radar stations; *See also* 1950 ELE

Buffer memory, which temporarily stores data from or to slow peripherals, thus freeing the central processor for other tasks, is introduced by Remington

William Shockley, Stanley Morgan, Morgan Sparks, and Gordon Teal [b Dallas TX, Jan 10 1907] develop the *p-n* junction transistor using crystal growth techniques; Western Electric starts the commercial production of transistors; *See also* 1948 ELE; 1954 ELE

1952

The US explodes its first thermonuclear device, the prototype of a fusion weapon that will be known as the hydrogen bomb; a fission bomb is used to start hydrogen fuel fusing into helium in a 2-story, 50-ton device that soon leads to bombs that can be carried on airplanes and in guided rocket missiles; *See also* 1945 GEN; 1953 GEN

Great Britain explodes its first nuclear-fission bomb; *See also* 1949 GEN; 1957 GEN

The CBS television network uses a UNIVAC computer to predict the results of the US presidential election; UNIVAC's first prediction of a landslide is right on the mark but not believed by its operators; they quickly reprogram it so that it incorrectly predicts a close contest; *See also* 1951 ELE

UHF (ultra-high frequency) television broadcasting is authorized in the US; *See also* 1941 COM

IBM introduces the 701, also called the Defense Calculator; it is a 36-bit computer of the von Neumann type, developed by Bob Overton Evans and coworkers, and equipped with an electrostatic memory of 4096 words; a total of nineteen IBM 701 computers are built; *See also* 1948 ELE; 1953 ELE

John von Neumann's IAS computer is completed at the Institute of Advanced Studies in Princeton NJ; it is based on von Neumann architecture and has parallel-processing capabilities; *See also* 1949 ELE

Westinghouse Electric Corporation builds the first breeder reactor at the US Atomic Energy Commission's laboratories in Arco ID, known as Experimental Breeder Reactor 1, or EBR-1; a breeder reactor produces more plutonium than the uranium it burns for fuel; contemporary accounts say it is like getting 2 pounds of coal for each one burned and proclaim the new reactor the forerunner of an era of cheap energy; *See also* 1950 ENE; 1955 ENE

1951

Fred Joyner of Eastman Kodak, measuring the speed of light through plastics for an aircraft-cockpit project, finds that the 910th compound he measures sticks the lenses of the refractometer together so tightly that they cannot be unstuck; when a panicked Joyner takes the problem caused by the cyanoacrylate compound to his supervisor, Harry Coover [b Newark DE, Mar 6 1919], Coover remembers he had to give up his own World War II experiments with cyanoacrylates because the chemicals stuck to everything they touched; Coover thinks he and Joyner may have found a new kind of glue; indeed, they have discovered the first superglue, although it takes 7 years more before the superglue is introduced commercially

US surgeon John H. Gibbon, Jr. develops the heart-lung machine; *See also* 1935 MED; 1953 MED

Antabuse, a drug that prevents alcoholics from drinking, is introduced

Reuben Kahn [b Kovno, Lithuania, Jul 26 1887] introduces a "universal reaction" blood test for detecting several disorders at an early stage; *See also* 1932 MED

Chrysler introduces power steering for automobiles

Great Britain introduces the "zebra" street crossing, an important contribution to pedestrian safety; when there is a pedestrian on the striped crossing, cars going both ways must halt; *See also* 1918 TRA

1952

Artificial ice cream in which vegetable oils replace butter fat is introduced in the US; *See also* 1961 FOO

P.W. Brian and other scientists in Europe and the US find that among the gibberellins in soil fungi, the active growth hormone is one that is named gibberellic acid; *See also* 1939 FOO

William *Conyers* Herring [b Scotia NY, Nov 15 1914] and John K. Galt [b Portland OR, Sep 1 1920] demonstrate that thin "whisker" crystals of tin are much stronger than is tin in bulk; *See also* 1948 MAT

Pilkington introduces the "floating glass" method for the continuous production of glass sheet; the glass floats on liquid tin on which it is stretched before being cooled down

A polio epidemic in the US strikes, affecting 47,665 persons; Jonas Edward Salk [b New York City, Oct 28 1914] develops a killed-virus vaccine against polio; it is used for mass inoculations starting in 1954 and successfully prevents the disease, although it is later superseded by a live-virus vaccine; *See also* 1948 MED; 1957 MED

The Massachusetts Institute of Technology (MIT) develops the first numerically controlled tooling machine; it is a standard three-geared milling machine with a separate numerical control unit; the positions of the tools are controlled by data stored on tape; *See also* 1818 TOO; 1954 TOO

Jacques-Yves Cousteau and Fernand Benoit direct the first investigation of an ancient wreck using Scuba apparatus; *See also* 1943 TRA; 1985 TRA

The steamship *United States* breaks the transatlantic record set by the *Queen Mary* by more than 10 hours, with a new record of 3 days, 10 hours, 40 minutes; *See also* 1934 TRA

1952 cont

A long-lasting temperature inversion traps coal smoke along with fog in London, England, forming a smog that takes 4000 lives, mostly of the elderly and infirm; *See also* 1948 GEN; 1953 GEN

On May 1 the children's toy Mr. Potato Head, a collection of plastic eyes, ears, mouths, and so forth that can be stuck into a potato to produce a face, becomes the first children's toy ever advertised on television; in the next 40 years about 50 million are sold, most including a plastic potato to avoid having to attach parts with sharp spikes that could injure children

Truly three-dimensional-appearing motion pictures are shown in the US in a system called Natural Vision, although viewers must wear special polarizing glasses; the first film is *Bwana Devil*; *See also* 1893 COM

CinemaScope wide-screen movies are introduced commercially, the last of the partial 3D systems of 1951-52 and the only one to have, along with its imitators, a lasting effect; images are compressed laterally by an anamorphic objective during filming and decompressed during projection onto a wide screen; *See also* 1925 COM; 1951 COM

The first hearing aids equipped with junction transistors appear on the market; Bell, because of its commitment to helping the deaf since the time of Alexander Bell, waives all patent royalties on the application of the transistor in hearing aids; *See also* 1951 ELE; 1964 ELE

The first accident at a nuclear reactor occurs at Chalk River in Canada, where a technician's error causes a hydrogen explosion in the nuclear core; there are no casualties; *See also* 1947 ENE; 1955 ENE

The transistor

Traditionally, the inventor was regarded as distinct from the scientist. Samuel Morse, Alexander Bell, and Thomas Edison were inventors but not trained scientists; often their unflagging efforts compensated for lack of formal scientific training. In the 20th century, invention became more the domain of the scientist. For example, John Fleming and Lee De Forest, the inventors of, respectively, the vacuum diode and triode tubes, were scientists first and inventors second.

The transistor can be viewed, as can the laser, as an invention of physicists. William B. Shockley, who joined Bell Labs in 1936 and landed in the vacuum tube department, had a background in solid-state physics. He succeeded in convincing management to allow him to move to the semiconductor laboratory, where the solid state was studied. But his stint with vacuum tubes seems to have steered him to the idea of a semiconductor amplifying device. As early as 1939 he wrote in his lab notebook, "It has today occurred to me that an amplifier using semiconductors rather than vacuum tubes is in principle possible."

In 1946 Shockley, Walter Houser Brattain, an experimental physicist who also had started out in Bell Labs working on vacuum tubes, and John Bardeen, a theorist, joined forces to develop a semiconductor amplifying device. They started experimenting with p- and n-type germanium and silicon semiconductors. P-type semiconductors have an excess of positive charge carriers or "holes" (an electron vacancy that can travel through the crystal structure that acts like a positive charge). N-types have an excess of negative charge carriers, which are electrons.

Bardeen and Brattain put together the first transistor in Dec 1947. It was a point-contact transistor consisting of a single germanium crystal with a p- and an n-zone. Two wires made contact with the crystal near the junction between the two zones like the "whiskers" of a crystal-set radio. Although a very primitive transistor, it was capable of amplifying a signal from a microphone.

A few months later Shockley devised the junction transistor. It was a true solid-state device in that it did not need the whiskers of the point-contact transistor. The junction transistor consisted of a germanium crystal in which a layer of p-type was sandwiched between two layers of n-type.

AT&T, owner of Bell Labs, licensed the transistor very cheaply to other manufacturers, and, in the spirit of Alexander Bell's commitment to helping the deaf, waived patent rights completely for the use of transistors in hearing aids, which became the first application of the new technology. Manufacturers initially had considerable difficulties in producing transistors because of the high demands of purity of the semiconductor crystals. Soon transistors became reliable and cheap, and they supplanted vacuum tubes in most applications during the late 1950s and the 1960s.

Jonas Salk and Albert Bruce Sabin

Poliomyelitis, commonly called polio, was unknown before 1840. It became a feared disease that singled out children and often caused paralysis or death. Franklin D. Roosevelt was paralyzed from the waist down by the disease in 1921, making polio well known in the US. In the late 1940s and early 1950s a major epidemic of the disease struck the US.

Jonas Salk, a medical doctor, had worked on the influenza vaccine during the 1940s. As the head of the Virus Research Laboratory at the University of Pittsburgh School of Medicine, he began the study of the virus causing polio. He confirmed the existence of three types of polio viruses and started growing them in cultures of monkey kidney tissue. He prepared vaccines by killing these viruses with formaldehyde. The killed viruses did not cause the disease, but did stimulate the production of antibodies that protect the human organism against the disease.

In 1952 Salk tested the vaccine on children who had already experienced polio and thus were immune to the virus. The vaccine increased the amount of antibodies in these children, elevating their immunity. The first trial on children who never had polio took place in 1954; it showed the vaccine to be safe and effective.

Throughout the 1950s the incidence of polio was reduced drastically worldwide. Albert Bruce Sabin, a medical doctor who also had studied the polio virus, prepared a vaccine containing a live strain of a virus. This virus also stimulates the production of antibodies but is too weak to cause the disease itself. The Sabin vaccine, which (unlike the Salk vacine) can be administered orally, was tested in large-scale trials in 1957 in the US and other countries. It was adopted in the Soviet Union and Europe in 1959, and was approved in the US in 1961. It has two advantages over the Salk vaccine—it can be administered orally and it provides some immunity to persons who come in contact with the one who takes the vaccine.

Virginia Apgar [b Westfield NJ, Jun 7 1909, d Aug 7 1974] introduces the Apgar score which predicts the health of a newborn baby by measuring pulse, respiration, muscle tone, color, and reflexes; its use soon becomes universal in the US as well as in much of the rest of the world; *See also* 1944 MED

British doctor Douglas Bevis develops amniocentesis, a method of examining the genetic heritage of a fetus while it is still in the womb; *See also* 1958 MED

Robert Wallace Wilkins discovers that reserpine is a tranquilizer, the first one found; he had been using it to treat high blood pressure; *See also* 1950 MED

Per-Ingvar Branemark observes that a titanium microscope he is using to study bone tissue has bonded to the bone; following up on this discovery, he learns by 1965 how to make dental implants that bond to the jawbone; *See also* 1918 MED; 1967 MED

The world's first sex-change operation is performed on George Jorgenson, who becomes known to the world as Christine

The CF-100 airplane built by Avro Aircraft of Canada becomes the first straight-wing combat aircraft to exceed the speed of sound (Mach 1), which it does in a steep dive on Dec 18; *See also* 1947 TRA

1952 cont

1953

GENERAL

An armistice in the Korean War is signed on Jul 27; this ends the fighting, although it does not lead to a peace treaty; *See also* 1950 GEN; 1954 GEN

The Soviet Union (Russia) explodes on Oct 12 its first thermonuclear, or fusion, weapon, also known as a hydrogen bomb; unlike an earlier US thermonuclear device, the Soviet version is already configured as a weapon; *See also* 1952 GEN; 1954 GEN

The European Patent Organization is formed; members can apply for a European patent in any of the languages of the member states, including English; *See also* 1883 GEN; 1970 GEN

Smog in New York City is believed responsible for 200 deaths; *See also* 1952 GEN; 1972 GEN

COMMUNICATIONS

Commercial color television begins in the US with a few broadcasts toward the end of the year on the Columbia Broadcasting System (CBS); most major programs on the National Broadcasting Company (NBC) network are in color; Radio Corporation of America, owner of NBC, had developed the compatible (so called because broadcasts could also be received by existing black-and-white only sets) color television system used in conjunction with the US National Television System Committee; CBS, which had developed a separate, incompatible system, abandons its technology for the compatible one; *See also* 1929 COM

Jay Forrester [b Anselmo NE, Jul 14 1918] develops the ferrite core memory for computers; *See also* 1953 ELE

ELECTRONICS & COMPUTERS

The IBM 650 computer, also called the Magnetic Drum Calculator, is introduced; it is the first computer to be manufactured in large numbers; it is derived from the IBM 701 and has a memory of 1000 10-byte words; 1500 units are sold before it is taken out of production in 1969; *See also* 1952 ELE; 1956 ELE

Kenneth Olsen [b US, 1926] builds the first computer equipped with a ferrite-core memory, a storage device developed by Jay Forrester earlier in the year; the computer is called Memory Test and performs tasks for the SAGE military operation; *See also* 1951 ELE; 1953 COM

ENERGY

Great Britain makes plans to build its first nuclear power plant for production of electric power on an industrial scale; the carbon-dioxide-gas cooled reactor at Calder Hall will also produce plutonium for use in nuclear weapons; *See also* 1950 ENE; 1956 ENE

Nuclear energy stored in graphite, a phenomenon first detected but not recognized in 1949 by N.J. Pattenden, begins to be employed as an adjunct power source at the Windscale (now Sellafield) nuclear power plant in England; *See also* 1950 ENE; 1957 ENE

1954

GENERAL

Communists take over in Hanoi, Indochina (Vietnam), as the French give up their rule; *See also* 1953 GEN; 1959 GEN

The first true hydrogen bomb is exploded by the US on Bikini Atoll in the South Pacific; the power of the blast is expected to be 7 megatons, but proves to be 15; *See also* 1953 GEN; 1957 GEN

Nuclear physicist Robert Oppenheimer is labeled a security risk by the US and denied access to nuclear information

The United Kingdom Atomic Energy Authority is founded; its aim is the development of civilian applications of nuclear power; *See also* 1947 GEN; 1966 GEN

COMMUNICATIONS

John Backus [b Philadelphia PA, Dec 3 1924] publishes a preliminary report, *Specifications for the IBM mathematical FORmula TRANslating system—FORTRAN*, which marks the beginning of the development of true programming languages; *See also* 1949 COM

Earl Masterson develops the Uniprinter for use with computers; called a line printer, it is capable of executing 600 lines per minute; *See also* 1884 COM; 1970 COM

ELECTRONICS & COMPUTERS

Gordon Teal at Texas Instruments introduces the silicon transistor, which is much cheaper to manufacture than previously used germanium-based versions; *See also* 1951 ELE; 1961 ELE

The first transistor radio, the Regency, appears on the market; it is not only smaller than any tube radio, but at $49.95, cheaper as well; *See also* 1951 ELE

The UNIVAC 1103A is the first commercial computer equipped with a ferrite-core memory; it is 50 times faster than the UNIVAC I of 1951; *See also* 1953 ELE

ENERGY

Bell Telephone scientists D.M. Chapin, Calvin Souther Fuller [b Chicago IL, May 25 1902], and G.L Pearson, following an idea put forward this year by Paul Rappaport, develop the silicon photovoltaic cell, which can produce electric power from sunlight; *See also* 1839 GEN; 1981 ENE

The Soviet Union completes the first small nuclear reactor that is intended primarily to produce electric power; *See also* 1952 ENE; 1957 ENE

The US Atomic Energy Act allows private companies to build nuclear reactors and to maintain stocks of nuclear fuel; *See also* 1947 GEN; 1957 ENE

The Association for Applied Solar Energy is founded in the US; the association publishes the periodical *The Sun at Work*; *See also* 1949 ENE; 1977 ENE

1953

FOOD & SHELTER

Percy L. Spencer of Raytheon, having observed that a microwave device in a Raytheon laboratory melts a candy bar in his pocket, and having experimented almost immediately with microwaved popcorn, patents the "high-frequency dielectric heating apparatus" that we know as the microwave oven; in its original form it is a giant, costly device aimed at restaurant use; *See also* 1937 ELE; 1955 FOO

MATERIALS

For the first time, sales of detergents in the US are higher than sales of soaps; *See also* 1916 MAT; 1957 MAT

British chemists I.D. Rattee and W.E. Stephen rework the chemistry of fiber-reactive dyes from basic principles, finding simple ways to produce the dyes for the first time; *See also* 1894 MAT; 1956 MAT

MEDICAL TECHNOLOGY

The first kidney tranplant is performed in Paris; the graft fails after 21 days because of rejection; *See also* 1963 MED

John H. Gibbon, Jr. uses his heart-lung machine to keep Cecelia Bavolek alive while operating successfully on her heart; it is the first use of the machine on a human being; *See also* 1951 MED; 1967 MED

Frederick Sanger (b Rendcombe, England, Aug 13 1918) becomes the first to determine the molecule-by-molecule structure of a protein, insulin; *See also* 1921 MED

TOOLS & DEVICES

Physicist Charles Hard Townes [b Greenville SC, Jul 28 1915] develops the maser (short for microwave amplification by stimulated emission of radiation), the precursor of the laser; in the device microwaves are amplified by stimulating emission at the same wavelength in ammonia gas molecules that have been excited (that is, boosted to a higher energy level); *See also* 1912 ELE; 1958 TOO

TRANSPORTATION

Michelin of France and Pirelli of Italy introduce radial-ply tires; *See also* 1947 TRA

1954

FOOD & SHELTER

TV dinners are introduced in the US; *See also* 1939 FOO

Frédéric, Jean, and Henri Lescure reinvent the pressure cooker, originally invented by Denis Papin in the 17th century; *See also* 1680–89 FOO

MATERIALS

Manufacturers in the US introduce bags of prepared asphalt so that home owners can build their own blacktop driveways

MEDICAL TECHNOLOGY

Chlorpromazine (Thorazine) is introduced for the treatment of mental disorders; *See also* 1952 MED

TOOLS & DEVICES

Various manufacturers build highly automated factories; Ford builds a 40-worker plant for engine blocks that has the same output as an older factory needing 117 workers; Raytheon's new radio plant replaces 200 workers with 2, who between them assemble 1000 radios a day; and a machine for forming cups from aluminum strips replaces 55 workers with 1

TRANSPORTATION

The first atomic-powered submarine, the USS *Nautilus*, built by Admiral Hyman G. Rickover, is commissioned; sea trials start Jan 17 1955; *See also* 1947 TRA; 1957 ENE

The first vertical-take-off plane is developed in Great Britain, where it is known as the "flying bedstead"; *See also* 1939 TRA

Malcolm MacLean starts using containers for transporting goods between New York City and Houston; these containers can be loaded onto trucks and ships; container transport spreads quickly throughout the world by 1965 and now displaces most other forms of transport of goods; *See also* 1915 TRA; 1964 TRA

1955

GENERAL

The first international conference of scientists on the dangers of nuclear armaments is held in Pugwash, Nova Scotia; *See also* 1954 GEN; 1956 GEN

Velcro is patented; *See also* 1948 GEN

The first conference in Geneva, Switzerland, on the peaceful uses of nuclear power takes place; *See also* 1954 GEN; 1958 ENE

COMMUNICATIONS

Scientist Narinder S. Kapany [b Moga, India, Mar 1927] introduces the optical fiber; he discovers that a glass fiber surrounded by a cladding can conduct light over great distances with little loss of intensity; *See also* 1966 COM

The Semi-Automatic Business Related Environment (SABRE), introduced by IBM, connecting 1200 teletypewriters, is the first large network linked to a database; it is used by American Airlines for passenger reservations; *See also* 1940 COM; 1958 COM

Jack Gilmore completes the TX.O, a man-machine interface using a cathode-ray tube for display, an optical pen, a flexowriter, and function keys; it is the precursor of the modern video terminal; *See also* 1960 ELE

ELECTRONICS & COMPUTERS

Bell Telephone builds the first computer that uses transistors instead of electron tubes; *See also* 1954 ELE; 1956 ELE

ENERGY

Electricity generated by nuclear power is used for the first time in the US on Jul 17 in Arco ID—from an experimental reactor at the US National Reactor Testing Station—and on Jul 18 in Schenectady NY—from a prototype nuclear submarine reactor in West Milton NY; *See also* 1952 ENE; 1957 ENE

The core of the nuclear reactor EBR-1 is destroyed on Nov 29 when an operator accidentally fails to drop the cooling rods in response to reactor overheating; *See also* 1952 ENE; 1957 ENE

Lewis Strauss [b Charleston WV, Jan 31 1896, d Jan 21 1974], chairman of the US Atomic Energy Commission, predicts that electricity produced by nuclear power plants will soon be "too cheap to meter"; *See also* 1952 ENE; 1992 ENE

1956

GENERAL

The Japanese Union of Scientists and Engineers (JUSE) starts a morning radio program on quality control to encourage competiveness in Japanese industry; *See also* 1950 GEN; 1962 GEN

Norman Cousins, editor of the *Saturday Review* founds SANE (National Committee for a Sane Nuclear Policy) in the US; *See also* 1955 GEN; 1963 GEN

Bette Nesmith (mother of Michael Nesmith, actor on "The Monkees" television program), after using white paint for years to cover up her bad typing, starts a cottage business, supplying "Mistake Out" to secretaries at the company where she works; her business slowly develops into the large company known as Liquid Paper Inc.

COMMUNICATIONS

The first transatlantic telephone cable, linking Scotland with Newfoundland, is put into operation on Sep 25; before this cable telephone conversations between the US and Europe were possible only by shortwave radio telephone; *See also* 1927 COM; 1988 COM

John Backus and a team at IBM complete FORTRAN I, the first full-fledged computer programming language; previously, computer programs had to be installed in machine language; *See also* 1954 COM; 1959 COM

Stanislaw Ulam and P. Stein program a computer to play a form of chess on a 6 x 6 board; the program, called MANIAC I, becomes the first computer program to beat a human in a game; *See also* 1958 COM

Ampex brings its first video tape recorder to market; *See also* 1951 COM; 1979 COM

ELECTRONICS & COMPUTERS

Univac introduces the first commercially available computer of the second generation; it uses transistors instead of vacuum tubes, making operation less costly and more reliable; *See also* 1955 ELE; 1959 ELE

IBM introduces the Model 305 Business Computer; it can store and access 20 megabytes on four disk-storage elements; *See also* 1953 ELE; 1960 ELE

IBM introduces the hard disk for the storage of data; the system called RAMAC (random access method of accounting and control) makes use of indexes for locating information on disk; *See also* 1948 COM; 1970 COM

ENERGY

Christopher Hinton [b Tisbury, England, May 12 1901] opens Calder Hall in England on Oct 17; it is the first large-scale nuclear power plant designed for peaceful purposes, producing 4.2 megawatts of electricity; *See also* 1953 ENE; 1957 ENE

1955

Tractors outnumber horses in the US for the first time in history; *See also* 1924 FOO

Tappan introduces the first microwave oven designed for home use; although aimed at the home market, it is still a large and costly luxury item; *See also* 1953 FOO; 1967 FOO

R.C. Brian, R.F. Homer, J. Stubbs, and R.L. Jones in Great Britain find that a substance prepared by fellow chemist R.J. Fielden is a powerful herbicide; the herbicide, christened diquat, can kill established weeds without touching seedlings that have not yet emerged; *See also* 1945 FOO; 1967 FOO

The first artificial diamonds for industrial use are produced in the US by a team headed by Percy Bridgman along with staff members of the General Electric Company; the diamonds are produced from carbon by pressure and heat and are immediately put on sale; *See also* 1929 MAT; 1958 MAT

W.F. Lorch uses renewable cartridges of ion-exchange resins in a device that purifies and softens water; *See also* 1892 MED

Rilsan, an artificial fiber, is invented; *See also* 1949 MAT

US paint manufacturers limit lead in paint to 1 percent or less; *See also* 1917 MED; 1971 MED

On Jan 11 Lloyd H. Conover [b US, 1923] patents tetracycline, the first antibiotic made by chemically modifying a naturally produced drug; it soon becomes one of the most useful antibiotics for treatment of a number of infectious diseases; *See also* 1947 MED

The field ion microscope, developed by American physicist Erwin Wilhelm Mueller, becomes the first instrument that can picture individual atoms; it accomplishes this by releasing ions from the tip of a needle cooled nearly to absolute zero; the ions strike a fluorescent screen, producing an image of the needle tip; *See also* 1903 TOO; 1978 TOO

Christopher Cockerell patents the hovercraft, also known as an air-cushion vehicle or a ground-effect machine; it can travel over water or flat solid surfaces on a cushion of air that it generates; the first experimental hovercraft is tried out in 1959; *See also* 1877 TRA; 1968 TRA

1956

Fiber-reactive dyes are introduced commercially for the first time; developed primarily for artificial fibers and for cotton, these are easy to apply at low cost and quite fast; *See also* 1953 MAT

Birth control pills are used in a large-scale test conducted by John Rock and Gregory Pincus [b Woodbine NJ, Apr 9 1903, d Boston, Aug 22 1967] in Puerto Rico; *See also* 2000 BC MED; 1961 MED

The kidney dialysis machine, developed by Wilhelm Kolff in 1943, comes into use in the US; *See also* 1943 MED; 1964 MED

Werner Forssmann [b Berlin, Germany, Aug 29 1904, d Schopfheim, W Germany, Jun 1 1979], Dickinson Richards [b Orange, NJ, Oct 30, 1895, d 1973] and André F. Cournand [b Paris, France, Sep 4, 1895, d 1988] win the Nobel Prize in physiology or medicine for their use of the catheter for study of the heart and circulatory system

The Lip company in France produces the first commercial watch to run on electric batteries; *See also* 1924 TOO

On Jan 31 the Air Force ground-tests its first nuclear-powered jet engine—intended for a bomber; it produces mildly radioactive exhaust as well as too much direct radiation for unshielded personnel to remain near it; *See also* 1948 TRA; 1961 TRA

1957

The Common Market is established in Europe

The US Atomic Energy Commission launches Project Plowshare, officially the Division of Peaceful Nuclear Explosives; various plans are developed for using nuclear explosions to create harbors, improve mines, and so forth; because of the dangers of nuclear contamination, very few of these plans are carried out; *See also* 1954 GEN; 1967 GEN

Great Britain explodes its first hydrogen bomb on Christmas Island; *See also* 1954 GEN; 1960 GEN

Smith-Corona makes a portable electric typewriter, but at 8.3 kg (18.3 lb) it weighs more than its manual predecessors; *See also* 1912 COM

John McCarthy founds the Artificial Intelligence Department at MIT; *See also* 1956 COM; 1959 COM

H.A Simon, Allen Newell and C. Shaw develop the program GPS (General Problem Solver); it is derived from Logic Theorist and is an early form of artificial intelligence program; *See also* 1956 COM; 1959 COM

Gordon Moore, Robert Noyce, and others who left Shockley Semiconductor Laboratory when Shockley refused to allow them to pursue research into silicon transistors, found Fairchild Semiconductor; *See also* 1954 ELE; 1958 ELE

Plans for a 1623 calculator built by Wilhelm Schickard (Shickardt) are discovered in correspondence between Schickard and Johannes Kepler, for whom a copy of the calculator was to have been built; Schickard's is the first mechanical calculator known to have been built, preceding a more famous adding machine built by Blaise Pascal by almost 20 years; *See also* 1640–49 TOO

Nuclear wastes stored in the Soviet Union in a remote mountain region of the Urals explode; radioactive contamination affects a large region, causing several villages to be evacuated permanently; *See also* 1955 ENE; 1986 ENE

On Oct 7 nuclear energy stored in graphite, while undergoing "controlled" release at the Windscale (now Sellafield) nuclear power plant in England, gets out of hand and destroys the core of the reactor, releasing radioactivity into the environment; the British government attempts to suppress news of the accident; *See also* 1953 ENE; 1986 ENE

French engineers improve the safety of manufacture for fuel rods for nuclear power plants by developing a way to weld titanium and tungsten in a vacuum with electron beams; *See also* 1901 TOO

The US Army Power Package Reactor I begins producing 2000 kw of electricity at Fort Belvoir VA; *See also* 1952 ENE

In Oct the General Electric Boiling Water Reactor, the first power reactor to be supported entirely with private money, starts producing 5000 kw of electricity for the Pacific Gas and Electric grid; *See also* 1954 ENE

In Dec the Shippingport Atomic Power Station, designed and constructed in part by US Admiral Hyman G. Rickover, opens in PA; it is the first commercial nuclear power plant in the US; *See also* 1954 ENERGY; 1989 ENERGY

Grace Murray Hopper

Grace Murray Hopper was teaching mathematics at Vassar at the start of World War II. Like many patriotic Americans, she wanted to join the armed services. Although her weight was too low and the military preferred to employ mathematicians as civilians, she convinced the US Navy in 1943 to let her join.

After completing Midshipman School she joined Howard H. Aiken at the Bureau of Ordnance Computation Project at Harvard. Aiken had built the Mark I computer, a huge electromechanical calculator. During the war, Hopper designed programs for that computer so that it performed complex calculations for the military. In 1945 she and the team at Harvard started developing Mark II. It was in this computer that a moth, in the words of Hopper, "had been beaten to death" by a relay, stopping the computer. The moth ended up Scotch-taped in their lab logbook, with the note "First actual case of bug being found." Even then the word *bug* denoted something that caused the computer to fail or to produce incorrect calculations.

After having worked on the Mark III computer, Hopper joined the Eckert-Mauchly Computer Corporation, which was building the UNIVAC computer, the first large computer sold commercially. During the 1940s and the early 1950s, computers were programmed in machine code, which contains the actual step-by-step instructions the computer follows. Such programming was a difficult and tedious task, requiring a good knowledge of how the hardware of a computer works. Machine-code programming had been Hopper's first job when she started working with computers during World War II.

Hopper created the first compiler for UNIVAC. A compiler is a program that translates instructions written in a simpler form into machine code. Hopper's compiler turned programs written in simple English into machine code. Saying that she was going to communicate with a computer in plain English produced general disbelief. Managers had accepted that computers understand numbers; but computers that understand English seemed inconceivable. Although it made the work of programmers much easier, convincing them to use the compiler was not an easy task. They had gotten used to writing machine code. Hopper developed a distaste for the phrase "but we've always done it that way."

Hopper rejoined the Navy in 1967, and retired in 1986 as a rear admiral, the Navy's oldest officer on active duty.

West German workers manufacturing the herbicide 2, 4, 5-T develop a skin disease later named chloracne; their cases lead to the first recognition that dioxin frequently contaminates such herbicides; *See also* 1988 MED

John Bardeen, Leon N. Cooper [b New York City, Feb 28 1930], and John R. Schrieffer [b Oak Park IL, May 31 1931] formulate the BCS theory that explains superconductivity in metals; *See also* 1911 MAT; 1986 MAT

A link between exposure to asbestos and the development of lung cancer, suspected since 1935, is definitely established

Wisk, the first liquid laundry detergent, is introduced in the US; *See also* 1916 MAT

Alick Isaacs [b Jul 17 1921] and Jean Lindenmann discover interferons, natural substances produced by the body that fight viruses; *See also* 1980 MED

Microbiologist Albert Bruce Sabin [b Bialystok, Poland, Aug 26 1906, d Washington DC, Mar 3 1993] develops a polio vaccine based on live, weakened viruses; *See also* 1952 MED

The high-speed dental drill is introduced, making it possible to work on teeth painlessly without anesthesia for simple fillings; it is driven by a tiny turbine powered by pressurized air

Allan Cormack [b Johannesburg, S Africa, Feb 23 1924], begins development of the CT (computed tomography) or CAT (computerized axial tomography) scan at the University of Capetown in South Africa; *See also* 1960 MED

Columbia University physics graduate Gordon Gould has the idea on Nov 11 that will translate into the laser; Gould does not, however, apply for a patent until 1959, and by then others have also begun to work on lasers; Gould's patent claims are not accepted until after 1986; *See also* 1953 TOO; 1958 TOO

Astronomer Bernard Lovell and engineer H.C. Husband supervise construction of a steerable 76-m (250 ft) radio telescope at Jodrell Bank, England, which comes into operation this year; *See also* 1946 TOO; 1993 TOO

Vladimir Zworykin patents an instrument for observing living cells, using ultraviolet light and a color television screen; *See also* 1948 TOO

The first artificial satellite, Sputnik 1, is launched by the Soviet Union on Oct 4; it is about 58 cm (23 in.) in diameter and weighs about 84 kg (184 lb); Sputnik 2 is launched on Nov 3, carrying the live dog Laika into space (but not bringing it back); *See also* 1680–89 TRA; 1959 TRA

Satellites into space

In 1957 science watchers were excited by the first major program of worldwide scientific cooperation, the International Geophysical Year (IGY), a two-year study of Earth and its environment in space. In the West, articles about the program sometimes mentioned that as part of the effort, the US would launch an artificial satellite. Few noted that the Soviet Union also planned such a launch, possibly because they doubted that this achievement was within Soviet technological capabilities.

The physics involved in putting a satellite into orbit about Earth had been known since Isaac Newton's time. Indeed, Newton was the first to suggest that an artificial satellite was possible. All that was required to launch a satellite was a rocket of sufficient power.

Both the US and the Soviet Union had taken German rocket engineers captive near the end of World War II. These engineers had since been employed in each country, along with other scientists, in developing large rockets that could travel great distances.

US scientists and the public around the world were astonished when the Soviet Union became the first to succeed in putting a satellite in orbit. People all over the world who had previously shown little interest in the IGY satellite program became very excited by Sputnik 1 (sometimes translated as "fellow traveler 1"). When US launch efforts failed at first, many developed a new respect for Soviet technology. Ultimately, the US turned to its German rocket scientists for success, as the Soviet Union had also done.

1958

Stereo records are introduced commercially; in the system that becomes the standard, each sound channel is engraved on one side of a single groove; earlier systems sometimes used two separate grooves; Duke Ellington experimented with stereo recording much earlier, but the results were originally released as monaural recordings; *See also* 1931 COM; 1961 COM

Bell introduces the Modem dataphone, which allows the use of telephone lines for transmitting binary data; *See also* 1955 COM; 1966 COM

Alex Bernstein and Michael Roberts develop a chess program that runs on an IBM 704 and plays like a fair amateur; *See also* 1956 COM

Jack Kilby [b Jefferson City MO, Nov 8 1923] of Texas Instruments and Robert Noyce [b Denmark IA, 1927] of Fairchild Semiconductor Corporation separately invent the integrated circuit, a device that carries several electronic components on one silicon chip; integrated circuits become in a few years the main components of computers; *See also* 1951 ELE; 1970 ELE

RCA launches the BIZMAC computer; it can access 200 magnetic tapes via a network of smaller computers; it is used for the storing and processing of large data bases; *See also* 1956 ELE; 1970 ELE

The first large-scale nuclear power station in the Soviet Union, built in Troitsk, Siberia, becomes operational; *See also* 1954 ENE; 1986 ENE

Previously secret work toward fusion power is revealed at the second international conference on peaceful uses of the atom; *See also* 1947 ENE

1959

A revolutionary government led by Fidel Castro takes over Cuba; although at first believed to be friendly to the US, Castro soon reveals himself to be Communist and allies Cuba with the Soviet Union; *See also* 1954 GEN; 1960 GEN

A. Leo Oppenheim [b Vienna, Austria, Jun 7 1904, d Jul 21 1974] analyzes a clay envelope found at Nuzi, north of Babylon (in Iraq), in 1920; it is inscribed with a list of various kinds of sheep, 49 in all; inside the envelope had been 49 counters or tokens (lost by Oppenheim's time); Oppenheim uses this and other evidence to propose that the tokens had been an accounting system for sheep, just as some peasants today keep track of their sheep with small stones; *See also* 3500 BC COM

The first commercial Xerox copier is introduced; *See also* 1937 COM

Grace Murray Hopper and Charles Phillips invent COBOL, the Common Business Oriented Language, a computer language designed for business uses that becomes the most widely employed language for that purpose; *See also* 1956 COM; 1960 COM

John McCarthy completes the first version of LISP (List Processing), a programming language especially designed for applications in artificial intelligence; *See also* 1956 COM; 1965 COM

Jean Hoerni [b Geneva, Switzerland, Sep 26 1924] produces at Fairchild Semiconductors a "flat," or planar, transistor in which the junctions are insulated with silicon; Robert Noyce uses this technique to produce an integrated circuit; this becomes the main method used for integrated circuits; *See also* 1958 ELE; 1961 ELE

A number of manufacturers announce transistorized computers; they include the IBM 7090, IBM 1401, IBM 1620, NCR (National Cash Register) 304, and the RCA 501; *See also* 1956 ELE; 1960 ELE

The first reactor to produce electricity for commercial use in France, the G-2 in Marcoule, begins operation with a capacity of 25 megawatts; *See also* 1948 ENE

Francis Bacon builds a fuel cell in which hydrogen and oxygen react in a mixture of potassium hydroxide in water; *See also* 1839 ENE; 1965 ENE

1958

Mies van der Rohe [b Ludwig Mies van der Rohe, Aachen, Germany, Mar 27 1886, d Chicago IL, Aug 17 1969] designs the Seagram Building in New York City with bronze walls and tinted glass curtain wall construction; it becomes the prototype of office building construction in the 1960s and 1970s; *See also* 1918 FOO

A method for producing diamond from methane, later called chemical vapor deposition, is patented in the US, but the diamond produced by the method is mixed with graphite, making it of limited use; *See also* 1955 MAT; 1977 MAT

Kaiser Aluminum introduces the first aluminum cans; previous cans had been made from steel plated with tin; *See also* 1887 MAT; 1959 MAT

Ian Donald of Scotland is the first to use ultrasound to examine unborn children; *See also* 1952 MED

Clarence Walton Lillehei [b Minneapolis MN, Oct 23 1918] introduces the external pacemaker for controlling heart action; *See also* 1959 MED

Bifocal contact lenses are introduced; *See also* 1948 MED; 1965 MED

American physicist Arthur L. Schawlow shows that the principle of the maser can be applied to produce and amplify visible light in a device called a laser (short for light amplification by stimulated emission of radiation); the first working laser is built in 1960; *See also* 1953 TOO; 1960 TOO

The US launches its first satellite and first of the Explorer series on Jan 31; when radiation counters aboard US satellites Explorer 1 and 3 mysteriously fail, James Van Allen suspects the cause is high levels of radiation; he designs a better radiation counter for Explorer 4, launched Jul 26, which results in discovery of the Van Allen radiation belt that surrounds Earth; *See also* 1957 TRA; 1962 TRA

The US Navy nuclear submarine *Nautilus* travels under the Arctic ice cap; *See also* 1929 TRA; 1959 TRA

1959

The ship *Methane Pioneer* is the first to demonstrate that liquid methane can safely be transported across oceans, carrying 32,000 barrels from Louisiana to England

Coors Beer is first marketed in 7-oz aluminum cans, the first aluminum beer cans; *See also* 1958 MAT; 1963 MAT

Ake Senning implants a pacemaker (heart stimulator); *See also* 1958 MED

Norman J. Holter [b Helena MT, Feb 1 1914] introduces a portable electrocardiograph for continuous recording of an electrocardiogram to study heart patients; *See also* 1903 MED; 1991 MED

On Dec 29 Richard P. Feynman [b New York City, May 11 1918, d Los Angeles CA, Feb 15 1988] at the annual meeting of the American Physical Society gives a talk entitled "There's plenty of room at the bottom"; he proposed etching or evaporating materials to make tiny transistors and connections, using light or electrons to create transistors, developing structures built one atom at a time, and building circuits with as few as seven atoms; although many listeners did not expect these devices to come to pass, virtually all Feynman's ideas are put to practical use over the next 30 or so years; *See also* 1987 TOO

The Soviet space probe Lunik 2 becomes the first human-produced object to reach the Moon when it crashlands on Sep 14; *See also* 1957 TRA; 1961 TRA

The St. Lawrence Seaway opens on Jun 26, connecting the St. Lawrence River and the Great Lakes for oceangoing ships; some of it is an enlargement of the first St. Lawrence Seaway, built by the Canadians in 1803, which was only 4.25 m (14 ft) deep; the new seaway is 8.23 m (27 ft) deep; *See also* 1848 TRA; 1992 TRA

Lenin, the first nuclear-powered icebreaker built by the Soviet Union, is commissioned; *See also* 1958 TRA; 1963 TRA

British industrialist Hugh Kremer sets up a prize of £50,000 for successful human-powered flight over a prescribed course 1 mi long in the shape of a figure 8; *See also* 1977 TRA

1960

The halogen lamp is introduced; an incandescent lamp bulb is filled with a halogen gas that regenerates the filament, allowing it to glow at a higher temperature, producing a brighter light; *See also* 1913 GEN

France explodes its first nuclear-fission bomb in the Sahara; *See also* 1957 GEN; 1968 GEN

An American U-2 spy plane flown by Francis Gary Powers is shot down over the Soviet Union, increasing tensions between the US and the USSR; *See also* 1959 GEN; 1963 GEN

Max V. Mathews [b Columbus NE, Nov 13 1926] of Bell Labs writes the first music synthesizer program, MUSIC; *See also* 1969 GEN

Echo, the first passive communications satellite, is launched on Aug 12 as a result of efforts by John Robinson Pierce (b Des Moines IA, Mar 27, 1910), who believes in the future of communications via satellite; *See also* 1957 TRA; 1962 COM

European and American computer scientists meeting in Paris agree on a set of standards for the programming language ALGOL (ALGOrithmic Language), based on plans put forward at a 1958 conference in Zurich; although interest in ALGOL is great in Europe, little interest comes from the US, where COBOL is preferred; *See also* 1956 COM

Paul Baran of the Rand Corporation develops the principle of packet switching, a way of interchanging data between computers that uses discrete bundles of information for each interchange; *See also* 1964 COM

The Royal McBee Corporation develops the first typewriter that punches a paper tape automatically while the original document is being typed, eliminating the need for separate tape punching; *See also* 1927 COM

Remington Rand, builder of UNIVAC, completes the Livermore Advanced Research Computer, or LARC, for Lawrence Livermore Laboratories; with 60,000 transistors, it is the first large scientific computer to use transistors

IBM launches Project 360, developed by Gene Amdahl [b Flandreau SD, Nov 16 1922]; the 360 stands for a guarantee of compatibility between computer peripherals over 360 degrees; *See also* 1956 ELE; 1964 ELE

In Nov Digital Equipment Corporation led by Kenneth Olsen introduces the PDP-1, a computer with a maximum memory of 26,000 bytes, making it state of the art; it is the first commercial computer with keyboard input and a monitor to show what the user has entered; because of its size and configuration, it is the predecessor of the minicomputer; *See also* 1959 ELE

Frank Rosenblatt completes the Perceptron computer at Cornell University; it is the first computer that can learn new skills by trial and error, using a type of neural network that simulates human thought processes; *See also* 1938 COM; 1991 ELE

The world's first boiling-water reactor is built in the US; the reactor heats water under pressure to produce steam to power a generator; *See also* 1963 ENE

The beginning of programming

The first large computers, such as the ENIAC, incorporated logic that was part of the circuitry; therefore, they could only be "programmed" to a small extent by changing wiring. These were followed by programmable computers with von Neumann architecture. Instructions were independent of the circuitry. Early programs were entirely written in machine language: Each instruction gave the central processing unit (CPU) a specific task to do, typically one such as: Move a word from one specified location to another; or change a bit in a specified location from a 0 to a 1 unless 1 already. Such programming can be very tight and efficient, but the writing becomes tedious for longer programs and also requires a thorough knowledge of computer hardware.

The introduction of interpreters was a major breakthrough. Interpreters translate a program written in an easier programming language into machine language. Particular locations in the CPU do not have to be specified, and commands are more closely connected to operations. A big disadvantage of such an interpreter is that it translates the program line by line into machine language, slowing down execution.

A compiler offers a better method for translating programming language into machine language. A compiler translates a whole program into machine language first. When the compiled program is run on the computer, execution time is shorter, since there is no translation step going on. Grace Hopper developed one of the first such compilers during the early 1950s.

John Charnley implants a two-part joint replacement made of plastic and cobalt-chrome; *See also* 1938 MED

William H. Oldendorf [b Schenectady NY, Mar 27 1925, d Los Angeles CA, Dec 14 1992] begins work about this time in what comes to be known as CT or CAT scan; working independently of researchers in South Africa and Great Britain at the University of California in Los Angeles, Oldendorf writes medical papers and obtains patents in the field, but fails to attract enough interest for development; *See also* 1957 MED

About this time some women enlarge or reshape their breasts with a semiliquid plastic in a bag that is implanted in the skin; the plastic material is silicone, familiar to many by its trade name as the toy Silly Putty; about 2 million women use the implants over the next 30 years; *See also* 1992 MED

Theodore Harold Maiman [b Los Angeles CA, Jul 11 1927] develops the first laser in May, using a ruby cylinder with polished end surfaces that act as mirrors; it is the first light source that can supply coherent light; *See also* 1958 TOO; 1965 TOO

In Oct Jane Goodall [b London, Apr 3 1934] observes tool-making by a chimpanzee, calling the practice to the attention of science; she sees a chimp named David Graybeard take a blade of grass and shape it so that it can be poked into a termite mound for the purpose of removing termites; David proceeds to eat the termites so obtained

The *Trieste*, the second bathyscaphe built by Auguste Piccard, is piloted by naval Lieutenant Don Walsh and Piccard's son Jacques to a new depth record in the ocean—10,920 m (35,800 ft); *See also* 1948 TRA; 1985 TRA

The beginning of programming (Continued)

During the 1950s several experimental and mostly mathematically oriented languages appeared. The first one that gathered wide success was FORTRAN (FORmula TRANslator). Developed in 1956 it underwent several changes and is still used today in technical and scientific applications.

Several other languages with different orientations appeared. One of them, COBOL, was developed in 1959 by Grace Hopper for business applications. BASIC, a programming language developed by the mathematicians John Kemeny and Thomas Kurtz at Dartmouth University in 1965, was first used as an educational tool, but became in the late 1970s the most popular language for the personal computer.

BASIC, and to a lesser degree FORTRAN, was often criticized because of its use of loops and branches. This produced the inconvenience that when a program grew to any length, it became more and more entangled in these loops and branches, a phenomenon called spaghetti. A minor change anywhere in the program could have uncontrollable repercussions in the whole program, requiring tedious searches and rewriting, a process called debugging.

Languages that allowed structured programming were a solution to the spaghetti problem. ALGOL, developed by an international committee, and Pascal, written by Niklaus Wirth, are such languages. These languages require that the coding be organized in logical groups, making reading and amending programs much easier.

In these languages, the software itself and the data were still entirely separated. In a more recent approach, data and logical procedures are grouped into discrete units called objects. The first of these object-oriented languages was Simula I, developed in 1966. Its successor, Simula, introduced the concepts of "object" and "class," in which classes are objects that have similar properties. Groups of objects, or classes, can be related by inheriting characteristics from one group to its successors. SMALLTALK, the successor of Simula, introduced communication between objects; a later version of SMALLTALK incorporated the concept of inheritance. These concepts, which became well known with the advent of object-oriented languages, had been invented early in the 20th century by mathematicians who could not have foreseen their eventual application to computers.

Object orientation has been termed a revolution in software development. Because objects can be used independently in software development, they can also be reused as building blocks, allowing much faster development of software systems. Several new object-oriented languages have been written; among the better-known are C++, Eiffel, Object Pascal, and Clascal.

1961

GENERAL

Investigations in Scandinavia and the US Adirondack Mountains confirm that acidity is increasing in small lakes, killing some species; the cause of the increase is believed to be acid precipitation resulting from air pollution, which becomes known as acid rain; *See also* 1964 GEN

COMMUNICATIONS

John McCarthy introduces the concept of time-sharing for the MAC, or Multiple Access Computer, which allows several operators to use a computer at the same time; *See also* 1955 COM; 1962 COM

IBM introduces the Selectric typewriter, in which characters are printed on paper by a rotating ball while the carriage remains fixed; *See also* 1935 COM

In the US, FM radio stations begin to broadcast in stereophonic sound; *See also* 1958 COM; 1986 COM

ELECTRONICS & COMPUTERS

IBM completes the 7030 computer, containing 169,100 transistors, for Los Alamos Laboratories; nicknamed "Stretch," is it supposed to be a hundred times faster than IBM's then current 704 mainframe, but it falls short and is only about 30 times as fast; *See also* 1960 ELE

ENERGY

Recognizing that peak electric loads differ in France and Great Britain, both by time of day and day of the week, engineers plan and build the Cross-Channel Cable for transferring extra electric power from one side of the English Channel to the other; the cable opens in Dec and is capable of transferring up to 160,000 kw of power at 200,000 volts; *See also* 1963 ENE

1962

GENERAL

The New Balance company introduces the first running shoes to have ripple soles and a rubber wedge for shock absorption between the sole and the upper; *See also* 1948 GEN; 1967 GEN

The Japanese Union of Scientists and Engineers (JUSE) produces a magazine on quality control and encourages the establishment of quality control associations throughout Japan; in 1970 there are 20,000 associations, in 1980 their number increases to 100,000; *See also* 1956 GEN

COMMUNICATIONS

Philips introduces the Compact Cassette, also known as the audiocassette, for recording sound on magnetic tape; *See also* 1930 COM; 1991 COM

Telstar, the first active communications satellite, is launched on Jul 10; it relays the first transatlantic television pictures; each transmission lasts only 20 minutes because the ground stations in England and Maine can "see" the satellite simultaneously only for that time; *See also* 1960 COM; 1964 COM

The Aviation Supply Office in Philadelphia PA introduces time-sharing for inventory control on its computer; *See also* 1961 COM

ELECTRONICS & COMPUTERS

Brian D. Josephson discovers that a supercurrent will flow between two superconductors in close proximity separated by a thin oxide barrier (or any short distance); the current flow, or DC Josephson effect, has since been exploited in various devices, especially the SQUID, a very sensitive magnetometer; *See also* 1983 ELE

D.N. Nasledov reports the first observation of laser activity by a semiconductor, the emission of infrared light from a gallium arsenide diode; *See also* 1970 ELE

Paul K. Weimer of RCA invents the thin-film transistor; *See also* 1951 ELE; 1974 ELE

ENERGY

A 104-kiloton underground nuclear explosion in NE creates Sedan Crater in a Project Plowshare experiment in the use of fission for earth moving; the crater is 390 m (1280 ft) in diameter and 98 m (320 ft) deep; *See also* 1957 GEN; 1967 GEN

In Great Britain the world's first advanced gas-cooled reactor, fueled by enriched uranium, is built; *See also* 1956 ENE

The nondairy coffee creamer is introduced; *See also* 1952 FOO; 1987 FOO

Jack Lippes introduces an inert plastic IUD (intrauterine device) for birth control; *See also* 1956 MED

Alan Shepard, Jr. accomplishes the first US spaceflight, the Mercury 3 suborbital mission of the *Freedom 7* space capsule, a flight of 5 minutes on May 5; *See also* 1958 TRA; 1962 TRA

1961

The space race

From the beginning of the Space Age in Oct 1957, the Soviet Union and the US took two different approaches to solving technological problems. The Soviet way was called brute force, because it relied on powerful rockets. Early US rockets, except for close imitations of German V–2 rockets, were unreliable and not so powerful. The Soviet approach was the clear winner at first. The Soviets had the first satellite and, in 1959, the first probe to approach the Moon and go into orbit about the Sun. Three US attempts in 1958 to get to the Moon failed, but US scientists came close about two months after the Russians. Shortly, a Soviet rocket actually hit the Moon, and another went into orbit about it.

The space race was closely watched all over Earth. In the early 1960s, the immediate goal was to put a human into orbit. With their more powerful rockets, the Russians won that round as well. It was nearly a year after cosmonaut Yuri Gagarin made the original one-orbit flight that astronaut (later senator) John Glenn duplicated the feat. Before that, on Mar 25 1961, US President John F. Kennedy tried to shift the focus of the race. He made it a national goal to put humans on the Moon by the end of the decade.

It was clear that human spaceflight was all that mattered in the race. When the US became the first nation to succeed in having a probe reach another planet (after unsuccessful tries by both contenders), no one took that to mean that the US had leaped in front in the race.

But, as a skilled boxer wears down a puncher, the US steadily piled up points with the judges. When in 1969 the first astronauts set foot on the Moon, Kennedy's strategy proved correct. By landing humans on the Moon, the US had won in the eyes of the world. The Soviet Union did not even try to duplicate the feat, but instead worked toward the development of a permanent space station, a goal that remains controversial to this day in the US.

Yuri Gagarin [b Gzhatsk, USSR, Mar 9 1934, d near Moscow, Mar 27 1968] of the Soviet Union makes the first orbital flight in a spacecraft in Vostok 1 on Apr 12, making a single orbit in a flight of 1 hour 48 minutes; *See also* 1959 TRA; 1963 TRA

On May 14 the US tests the engine for a nuclear-powered, robot-guided ramjet, designed to operate in a similar fashion to later, nonnuclear cruise missiles; *See also* 1948 TRA

US President John F. Kennedy cancels the Air Force nuclear bomber project because he thinks cheaper and less problematic long-range missiles can accomplish the same goals; *See also* 1956 TRA; 1964 GEN

Silent spring by Rachel Carson [b Springdale PA, May 27 1907, d Silver Spring MD, Apr 14 1964] alarms environmentalists and makes the general public aware of the danger of introducing pesticides into the environment; *See also* 1961 GEN

W. Buehler discovers that Nitinol, an alloy of titanium and nickel, "remembers" the shape possessed when hot; after cooling into a different shape, it springs back when heated again; NASA uses this material for the construction of an antenna of an artificial satellite, which unfolds when heated with an electric current; *See also* 1932 MAT; 1987 MAT

Lasers are used in eye surgery for the first time; *See also* 1960 TOO; 1985 MED

Unimation in the US markets the world's first industrial robot; originally patented by Georg C. Devol in 1961; General Motors installs these robots on its assembly lines; *See also* 1954 TOO; 1964 TOO

On Feb 20 John Glenn, Jr. [b Cambridge OH, Jul 18 1921] completes the first orbital flight by an American, the Mercury 6 mission of the *Friendship 7*, in three orbits; *See also* 1961 TRA; 1965 TRA

1962

The US space probe Mariner 2 is launched Aug 27; it becomes the first object made by humans to voyage to another planet, Venus; *See also* 1959 TRA; 1976 TRA

1963

A direct telephone link, called the hot line, is established between the White House in the US and the Kremlin in the Soviet Union; *See also* 1960 GEN

The US, the Soviet Union, and Great Britain agree to stop atmospheric tests of nuclear weapons; *See also* 1956 GEN

"Instant" color film, or self-developing color photography, is introduced by the Polaroid Land Company after a program of research headed by Elkan R. Blout [b New York City, Jul 2 1919]; Howard G. Rogers [b Houghton MI, Jun 21 1915] is credited with developing the basic system; *See also* 1947 COM

D. Gregg of the Stanford Research Institute and 3M develops a videodisk that stores images for several minutes; *See also* 1927 COM; 1989 COM

Lotfi Zadeh [b Baku, Russia, Feb 4 1921] starts his work on fuzzy logic at the University of California at Berkeley; the first application is 25 years later, when in Japan the Sendai underground train system (subway) is equipped with computers based on fuzzy logic; *See also* 1986 TRA

NASA launches the IMP satellite; it is the first satellite to contain integrated circuits; *See also* 1958 ELE; 1964 ELE

Semiconductor diodes that use electron tunneling go on sale only 6 years after the tunneling effect is discovered by Leo Esaki [b Osaka, Japan, Mar 12 1925]

In Canada the first CANDU (Canadian deuterium uranium) reactors begin operating; they use uranium in a pressure tube surrounded by heavy water (deuterium oxide); *See also* 1947 ENE; 1971 ENE

The world's largest pumped storage station is built at Blaenau Ffestiniog in Snowdonia national park in Wales; by pumping water to the top of a mountain when there is electricity to spare, the water's ability to use gravitational force to do work can be stored until needed to power generators that rectify changes in power output on the electric grid or supply power during peak need periods

1964

An undeclared war between the US and North Vietnam, fought in North and South Vietnam, accelerates as the US accuses North Vietnam of attacking a US destroyer and retaliates with air raids in North Vietnam; *See also* 1975 GEN

China explodes its first nuclear fission bomb aboveground; *See also* 1960 GEN; 1967 GEN

The US Congress passes the Wilderness Act, setting up the National Wilderness Preservation System; *See also* 1961 GEN; 1965 GEN

The US military cancels its plans for a nuclear-powered, low-flying missile because its radioactivity makes it too dangerous; the military spent $260 million on the project; *See also* 1961 TRA

The American Standard Association adopts ASCII (the American Standard Code for Information Interchange), which, using seven bits, allows the transmission of 128 different characters; the ASCII standard is accepted in 1966 by the International Standard Organization

First plans are drawn for the Advanced Research Project Agency (ARPA) of the US Defense Department to establish Arpanet, a computer network that can connect different types of computers and that uses packet switching for communication; the network becomes operational in 1969; *See also* 1960 COM; 1973 COM

The International Telecommunications Satellite Organization (Intelsat) is set up by the US and eleven other countries to develop a global commercial telecommunications satellite system; *See also* 1962 COM; 1965 COM

Control Data Corporation introduces the CDC 6600, designed by Seymour Cray [b Chippewa Falls WI, 1925]; the 6600 is often considered to be the first supercomputer to become a commercial success; it has a speed of 9 megaflops (that is, 9 million multiplications per second); *See also* 1968 ELE

The IBM 360 computer, designed by Gene Amdahl, is introduced; the computer is compatible with a wide range of peripherals and becomes a commercial success; *See also* 1960 ELE

Zenith develops a hearing aid equipped with an integrated circuit; the chip is an integrated amplifier that is also used in NASA's IMP satellite; the hearing aid is the first commercial product incorporating such a device; *See also* 1952 ELE

1963

FOOD & SHELTER

Ermal Fraze of the Dayton Reliable Tool and Manufacturing Company, Dayton OH, obtains the patent on a method of attaching a ring to a scored portion of the top of a beverage can so that the resulting tab can be pulled away; although not the first easy-opening can, this is the first such beverage can; it is quickly adopted by major canners of beer and soda; *See also* 1987 FOO

MATERIALS

Major brands of beer are marketed in 12-oz aluminum cans; *See also* 1959 MAT

MEDICAL TECHNOLOGY

James Daniel Hardy performs the first lung transplant; *See also* 1953 MED; 1967 MED

Thomas Starzl [b Le Mans IA, Mar 11 1926] and Francis Moore [b Evanston IL, Aug 17 1913] perform the first liver transplant; *See also* 1953 MED; 1967 MED

A man receives the kidney of a chimpanzee in an operation at Tulane Medical School in New Orleans LA; he survives for nine months; *See also* 1905 MED; 1992 MED

The first vaccine against measles is introduced; it uses a killed virus and is not reliably effective; *See also* 1937 MED; 1967 MED

TOOLS & DEVICES

Friction welding is invented; *See also* 1898 MAT

TRANSPORTATION

The *Savannah*, the first nuclear merchant ship, is officially launched, following sea trials that began in Mar 1962; it travels over 800,000 km (500,000 mi), but, because of high running costs, the ship is decommissioned in 1971; *See also* 1954 TRA; 1967 TRA

The first Rolls-Royce automobile with a diesel engine is produced; it is the first passenger car to use a diesel; *See also* 1950 TRA

Valentina Tereshkova, the first woman cosmonaut, is launched on 48-orbit Soviet space mission Vostok 6 on Jun 16, a dual launch with Vostok 5; *See also* 1961 TRA; 1964 TRA

1964

MATERIALS

Permanent press clothing is introduced, based on polyester fibers such as Dacron; *See also* 1949 MAT

MEDICAL TECHNOLOGY

Home kidney dialysis is introduced in Great Britain and the US; *See also* 1956 MED

Baruch S. Blumberg [b New York City, Jul 28 1925] discovers the Australian antigen, the key to the development of a vaccine for hepatitis B; *See also* 1980 MED

TOOLS & DEVICES

The first industrial robot in Europe is installed in a Swedish foundry for lifting castings from molds and placing them on a cooling bed; *See also* 1962 TOO

IBM develops the first CAD (computer aided design) system, the system 2250; *See also* 1987 TOO

Verner E. Suomi [b Eveleth MN, Dec 6 1915] invents the spin-scan weather satellite, the main form of weather satellite used from 1966 through the end of the 20th century; it spins a hundred times a second for stability, scanning a bit more of Earth's surface during each revolution, making a complete image in about 20 minutes; *See also* 1966 TOO

TRANSPORTATION

A high-speed rail link is opened between Tokyo and Osaka in Japan; the "bullet" trains run with an average speed of 160 km (100 mi) per hour; *See also* 1846 TRA; 1981 TRA

The 1289-m (4620-ft) Verrazano Bridge, the longest suspension bridge in the world at that time, opens to traffic in New York City; it is designed by Othmar Ammann; *See also* 1937 TRA; 1981 TRA

Containerships are introduced, simplifying international trade; *See also* 1954 TRA; 1968 TRA

Soviet cosmonauts Vladimir Komarov, Konstantin Feotistov, and Boris Yegorov, the first multiperson crew in space, are launched on the sixteen-orbit Voskhod 1 mission on Oct 12; *See also* 1961 TRA; 1967 TRA

1965

The US Congress passes the Highway Beautification Act, restricting advertising visible to drivers on US highways *See also* 1964 GEN; 1966 GEN

John Kemeny and Thomas Kurtz [b Oak Park IL, Feb 22 1928] develop BASIC (Beginners All-purpose Symbolic Instruction Code), a computer language for novices; it becomes the main programming language used by owners of personal computers, although most commercial programs for personal computers are written in more sophisticated languages; *See also* 1960 COM; 1971 COM

Joseph Weizenbaum at MIT develops ELIZA, a computer program for the study of natural language communication between man and machine; it can carry on a conversation, questioning the user about his or her psychological problems; *See also* 1963 COM; 1968 COM

Early Bird (Intelsat 1) is the first commercial telecommunications satellite placed on a geostationary orbit by Intelsat; it can relay 240 telephone conversations simultaneously; *See also* 1964 COM

The Soviet Union launches its first domestic communications satellite, Molniya; *See also* 1964 COM

The first portable consumer video recorder, including camera, is introduced by Sony; *See also* 1963 COM; 1979 COM

Burroughs, implementing a design from a team led by Dan Slotnick [b New York City, Nov 12 1931, d Oct 25 1985], develops the ILLIAC IV (Illinois Automatic Computer), the first computer to use a parallel, non-von Neumann design (later known as massively parallel architecture); it has 64 identical scalar computers operating in parallel; later massively parallel computers use microprocessors instead of separate computers; *See also* 1983 ELE

Digital Equipment Corporation (DEC) introduces the PDP-8 (Programmed Data Processor), the first minicomputer; a minicomputer, compared with the mainframe, is simple to operate and low cost ($18,000); it has 4K of ferrite-core memory; the introduction of minicomputers augurs the beginning of widespread computerization in business and education

Designers produce chips with a component density of 1000 per square cm (middle scale integration or MSI); *See also* 1964 ELE; 1973 ELE

The most successful application of fuel cells begins; they are used for on-board power in the US piloted space program, beginning with the Gemini missions; fuel cells are later used in the Apollo series of Moon missions and in the space shuttles; *See also* 1959 ENE

Geologists discover oil and natural gas under the North Sea; *See also* 1923 ENE; 1969 ENE

1966

The US Congress passes the Rare and Endangered Species Act, setting up rules by which the Fish and Wildlife Service can list and protect species in danger of extinction; *See also* 1965 GEN; 1967 GEN

The European Nuclear Energy Agency is founded in Mol, Belgium; *See also* 1954 GEN

Charles Kao and Georges Hockham of Standard Telecommunications Laboratories in England demonstrate that fibers of very pure glass can be used for the transmission of light carrying data over long distances, replacing traditional copper wire and electric currents for this purpose; *See also* 1955 COM; 1977 COM

John Van Green introduces the Carterphone, a device for the transmission of digital data over telephone lines; it consists of a portable modem and acoustic coupler; *See also* 1958 COM; 1969 COM

The world's largest electricity plant using tidal power, the Rance Tidal Works, is installed near the mouth of the Rance River on France's Channel coast (near the isle of Jersey); a dam 750 m (2,500 ft) long contains 24 turbines and generators that produce a total of 240 megawatts of electric power

1965

The artificial sweetener aspartame, eventually marketed as Nutrasweet in the US, is invented at Searle Laboratory in the US; *See also* 1950 FOO; 1981 FOO

Astroturf, a sort of artificial grass made mainly from nylon, is used to cover the football playing field at the Astrodome in Houston TX; in subsequent years most new and some existing football and baseball fields will use similar artificial turf as a playing surface; *See also* 1935 MAT

Royal Crown (RC) Cola becomes the first soft drink to use aluminum cans; *See also* 1963 MAT; 1967 MAT

Morris E. Davis reports that estrogen therapy prevents atherosclerosis and osteoporosis in postmenopausal women; *See also* 1941 MED

Soft contact lenses are invented; *See also* 1958 MED; 1988 MED

On Jul 13 Benjamin A. Rubin [b New York City, Sep 27 1917] patents the bifurcated vaccination needle, useful mainly for delivering smallpox vaccine in small doses with little difficulty; it becomes one of the main tools used in the worldwide eradication of the disease

Joseph Giordmaine and Robert Miller develop the continuously tunable laser; *See also* 1960 TOO; 1970 TOO

The Mont Blanc Tunnel, 13 km (8 mi) long, linking Chamonix, France, with Courmayeur, Italy, opens for automobile traffic; *See also* 1882 TRA; 1976 TRA

Ford Motors invents the sodium-sulfur electric battery for use in automobiles; although it is lighter and more powerful than conventional lead-acid batteries, it is also much more expensive to manufacture; *See also* 1900 ENE; 1991 ENE

Virgil I. Grissom and John W. Young, the first American multi-person space crew, are launched on the three-orbit Gemini 3 mission on Mar 23; *See also* 1962 TRA; 1969 TRA

1966

Harry M. Meyer, Jr. [b Palestine TX, Nov 25 1928] and Paul D. Parkman [b Auburn NY, May 29 1932] develop a live-virus vaccine for rubella (German measles); *See also* 1963 MED; 1967 MED

The first spin-scan weather satellite is lofted by the US; *See also* 1964 TOO

An all-time height record for a gunpowder-powered missile is set on Nov 19 by the US Defense Department's High Altitude Research Program (HARP); the HARP gun fires a 94-kg (185-lb) projectile to an altitude of 180 km (111.8 mi); *See also* 1918 GEN

Fuel injection for automobile engines is developed in Great Britain; *See also* 1893 ENE

1967

Syukuru Manabe [b Japan, Sep 21 1931] and R.T. Wetherald warn that human activities that increase the amount of carbon dioxide in the air, such as the burning of fossil fuels, are causing a greenhouse effect that will raise global temperatures; *See also* 1966 GEN; 1968 GEN

The General Conference of Weights and Measures in Sèvres, France, redefines the second, which had been based on the rotation of Earth, in terms of the transition level between the two lowest levels of cesium as measured by an atomic clock; specifically, the second is the amount of time corresponding to 1,192,631,770 cycles of the electromagnetic wave tuned to the cesium atoms' transitions; *See also* 1889 GEN; 1983 GEN

China explodes a thermonuclear, or hydrogen, bomb; *See also* 1964 GEN

Project Gasbuggy, part of Project Plowshare of the US Atomic Energy Commission, explodes a nuclear charge underground near Farmington NM in hopes of stimulating production of natural gas; the attempt and another one 2 years later fail; *See also* 1957 GEN

William J. Bowerman [b 1909], track coach of the University of Oregon, develops the first track shoes with nylon uppers; with some friends, he forms Nike to manufacture and sell the new shoes; *See also* 1962 GEN

Overseas direct dialing from New York to London and Paris begins on Mar 1 with 80 New York customers undergoing a three-month trial; in Jun 1966 three calls are dialed from Philadelphia in the first demonstration of the new technique; *See also* 1971 COM

Hell Digiset and RCA Videocom introduce photocomposition machines that use cathode-ray tubes; *See also* 1949 COM

Ray Milton Dolby [b Portland OR, Jan 18 1933] develops a method to eliminate background sound in recordings; "Dolby sound" also improves fidelity in other ways, and it becomes a standard feature on many recorded music players as well as in motion pictures; *See also* 1958 COM

Olof Soderblom introduces the concept of a computer network in which computers are connected to a closed loop rather than a centralized system; this concept will be taken over by IBM for the design of its networked computer systems; *See also* 1955 COM

Looking into people (part 2)

While X rays and radioactive tracers revolutionized diagnosis, both were far from perfect. Images were often difficult to see, interpretation was complicated, and both technologies employed damaging radiation.

Although the comic book hero Superman could focus his X-ray vision on details, medical X rays originally involved no discrimination whatsoever. Radiation passed through the body and collected on a flat piece of film or fluoroscope screen. Images from different planes within the body were merged into a single plane.

As early as 1917, however, mathematicians had shown that images from several different angles could be combined to give three-dimensional views. An early effort to accomplish something like this was made by Dutch radiologist Ziedes de Plantes. He used a moving X-ray source to obtain a single "slice" of the target. The mathematical method for building an image uses a large number of such slices to reconstruct the target object.

After World War II, the invention of powerful computers made mathematical reconstruction feasible. Several attempts to accomplish this were made in England, South Africa, and the US. Today this three-dimensional method, originally called computer-assisted tomography (CAT), but now known as computed tomography (CT), has become a common diagnostic tool.

The CT scan produces better images, but still uses damaging X rays. Another approach solved that problem and achieved more as well. Chemists had found a way to make specified atoms or parts of molecules produce characteristic radio waves on demand. Weak radio waves and strong magnetic fields, used to induce the radio waves, are not thought to be harmful.

Powerful electromagnets cause the nuclei of atoms to line up, somewhat the way that iron filings can be aligned by an ordinary horseshoe magnet. When the field is removed, the nuclei return to their normal disarray, in the process emitting radio waves at frequencies that vary depending on the particular nucleus and its state in a molecule.

In the early 1970s several inventors, notably Raymond Damadian, created machinery that could use this effect on whole human bodies. The result, now known as magnetic resonance imaging (MRI), reveals soft tissue that would not be otherwise observable. Medical doctors renamed the technique, known as nuclear magnetic resonance to chemists, out of fear that patients would think it had something to do with nuclear fission or fusion. Actually, MRI provides the safest way to look inside the body.

The US Department of Agriculture starts a test project in irradiating wheat and other foods to kill insects; *See also* 1955 FOO; 1968 FOO

The National Academy of Sciences reports in Jun that the practice of adding antibiotics to animal food, while producing greater yields, may leave traces in meat and may increase drug resistance in bacteria; *See also* 1941 MED

Using an improved electron tube introduced in Japan, Raytheon introduces the first small, affordable microwave oven; *See also* 1955 FOO

Coca-Cola and Pepsi-Cola bottlers begin to use aluminum cans; *See also* 1965 MAT

Surgeon Christiaan Neething Barnard [b Beaufort, South Africa, Nov 8 1922] performs the first partially successful human heart transplant on Dec 3; the recipient of the new heart is Louis Washkansky, who lives for eighteen days; *See also* 1963 MED; 1968 MED

Cleveland OH surgeon Rene Favaloro develops the coronary bypass operation; *See also* 1953 MED

Godfrey N. Hounsfield [b England, 1919] uses medical X rays to reconstruct a three-dimensional image in the essential development of the CT (CAT) scan, a method previously discovered by Allan Cormack and William H. Oldendorf; *See also* 1960 MED; 1971 MED

Clomiphene is introduced to increase fertility; it also results in an increase in multiple births

A 20-year study of fluoridation in Evanston IL shows that dental cavities have been reduced by 58 percent as a result of adding fluorides to the water supply; *See also* 1945 MED

A live-virus vaccine against measles that is more effective than the previous killed-virus version becomes available; *See also* 1963 MED

Bonding using composite resins is developed as an alternative to filling teeth with metal alloys; *See also* 1952 MED

Albert M. Cohen [b Boston MA, Jun 15 1918] and coworkers report that LSD (lysergic acid diethylamide) produces breaks in chromosomes

The tanker *Torrey Canyon* runs aground near the Cornish coast in Great Britain, in one of the worst ecological disasters; it spills over 119,000 tons of oil into the sea; *See also* 1872 TRA; 1968 TRA

The *Mutsu*, a nuclear-powered ship, is launched in Japan; however, it remains immobilized by Japanese fishermen for about 2 years because of their fear of nuclear contamination; in fact, nuclear leaks do occur when the ship starts operating; *See also* 1963 TRA; 1969 TRA

The Silver Bridge over the Ohio River collapses in Dec, killing 46; *See also* 1940 TRA

A US rocket plane reaches the speed of 7232 km (4520 mi) per hour; *See also* 1947 TRA

The eighteen-orbit Soviet Soyuz 1 mission starts off on Apr 23; Vladimir M. Komarov is killed when his first parachute fails; this is the first fatality of the space program; *See also* 1964 TRA; 1971 TRA

1968

GENERAL

The US Congress passes the Wild and Scenic Rivers Act, protecting parts of some rivers from commercial exploitation; *See also* 1967 GEN; 1970 GEN

France explodes a thermonuclear, or hydrogen, bomb; *See also* 1960 GEN

COMMUNICATIONS

Douglas Engelbart [b Portland OR, Jan 30 1925] demonstrates during an autumn conference on computers in San Francisco the computer mouse, a device that moves the cursor (pointer) on a computer monitor and issues simple "execute" commands; the mouse is later adopted by Apple for the Lisa and Macintosh computers and then spreads throughout the industry; *See also* 1983 COM

Author Arthur C. Clarke writes *2001: A space odyssey*; the science fiction novel features the computer Hal, which understands the human voice and can converse with its operators; *See also* 1971 COM

Edward Feigenbaum [b Weekawken NJ, Jan 20 1936] at Stanford and geneticist Joshua Lederberg develop DENDRAL, an expert system for the identification of chemical substances based on the results of spectrometric analysis; *See also* 1959 COM; 1970 COM

ELECTRONICS & COMPUTERS

Control Data Corporation launches the CDC 7600 supercomputer, designed by Seymour Cray, the most powerful computer until the appearance of the Cray supercomputer in 1976; many regard the 7600 as the second commercially successful supercomputer, following the 6600, also designed by Cray; the 7600 has a speed of about 40 megaflops; *See also* 1964 ELE; 1970 ELE

Burroughs introduces the B2500 and B3500, the first computers incorporating integrated circuits; *See also* 1959 ELE

Gordon Moore and Robert Noyce, both leaving Fairchild Semiconductor, found Intel Corporation with the intention of manufacturing integrated circuits; *See also* 1957 ELE; 1970 ELE

ENERGY

The Soviet Union opens an 800-kw tidal power station near Murmansk; *See also* 1966 ENE

1969

GENERAL

Jazz pianist Paul Bley is the first person to perform on a music synthesizer in front of a live audience; previous uses of synthesizers were only on record

MUSIC V, developed by Max Mathews, allows a computer to play musical scores that are stored as digital data on tape; *See also* 1960 GEN

COMMUNICATIONS

IBM Corporation improves its magnetic automatic typewriter by substituting a card instead of a tape; the new version is known as the Mag Card Selectric; *See also* 1960 COM; 1972 COM

The RS-232-C standard for data exchange between computers and peripherals, such as modems, is introduced for the serial transmission of data; *See also* 1964 COM; 1972 COM

ELECTRONICS & COMPUTERS

Some banks in the US begin to install individual automated teller machines (ATMs), but they are not linked with other machines; *See also* 1977 ELE

"Bubble memory" devices are created for use in computers; unlike conventional memory devices, bubble memory continues to remember even when the computer is turned off

ENERGY

Geologists discover oil in Alaska; *See also* 1965 ENE; 1973 ENE

1968

The Nuclear Materials Equipment Corporation begins sterilizing bacon and potatoes with radiation from radioactive cobalt-60 as a means of preserving them; *See also* 1898 FOO

James L. Goddard, commissioner of the US Food and Drug Administration, refuses in Apr to permit canned ham that has been radioactively sterilized to be used for human consumption by the US Army; *See also* 1967 FOO

English researchers report in Apr that oral contraceptives can cause blood clots in susceptible women; *See also* 1956 MED

Christiaan N. Barnard performs a second human heart transplant; this time the patient, Philip Blaiberg, lives 74 days with his new heart; *See also* 1967 MED; 1969 MED

French-American physiologist Roger Guillemin [b Dijon, France, Jan 11 1924] and Polish-American biochemist Andrew Victor Schally [b Vilna, Lithuania, Nov 30 1926] discover in this year and the next a simple substance made by the brain that affects the hormones produced by the pituitary gland

J. *Desmond* G. Clark [b London, Apr 10 1916] proposes that stone tool industries be grouped into five modes that are descriptive rather than based on specific sites; the system names are Mode I (simple flakes and cores), Mode II (flakes are produced by direct percussion), Mode III (wide use of prepared cores), Mode IV (blades and burins are dominant), and Mode V (microliths)

The first supertankers for carrying petroleum are put into service; *See also* 1964 TRA

The first supersonic airliner, the Soviet Tupolev TU-144, is demonstrated on Dec 31; *See also* 1947 TRA; 1976 TRA

Regular hovercraft service starts across the English Channel; *See also* 1955 TRA

The luxury ocean liner *Queen Elizabeth II* is launched; *See also* 1938 TRA

Aerotechnik Entwicklung und Apparatebau of Germany builds the WGM2 one-person, cheap, easy-to-fly helicopter; with its open seat, flying it is rather like riding a bicycle; *See also* 1939 TRA

1969

The first home yogurt maker is marketed

British physiologist Robert Edwards and surgeon Patrick Steptoe perform the first successful fertilization *in vitro* of a human ovum; the first "test-tube baby" resulting from this technique is born in England in 1978; *See also* 1978 MED

In Texas, Denton Cooley [b Houston TX, Aug 22 1920] and Domingo Liotta replace the diseased heart of Haskell Karp with the first artificial heart to be used in a human being; Karp lives for nearly three days; *See also* 1968 MED; 1982 MED

The scanning electron microscope reaches practical use after 15 years of development; *See also* 1938 TOO

An atomic clock built by the US Naval Research Laboratory based on the natural vibration of ammonia molecules achieves a precision of 1 second in 1.7 million years; *See also* 1948 TOO

The US Apollo 11 mission, with Neil A. Armstrong, Michael Collins, and Edwin E. Aldrin, Jr. aboard, is launched on Jul 16; the first lunar landing is achieved on Jul 20, as well as limited inspection, photography, evaluation, and sampling of lunar soil; *See also* 1965 TRA; 1970 TRA

The 25,000-ton German nuclear-powered ship *Otto Hahn* is launched; *See also* 1967 TRA

1970 The US environmental movement makes itself felt on Apr 22 when about 20 million Americans participate in ceremonies such as teach-ins and rallies to celebrate the first Earth Day; *See also* 1968 GEN; 1972 GEN

The Patent Cooperation Treaty permits patents to be filed in English at such designated receiving offices as the US Patent and Trademark Office; the patentee, by paying appropriate fees, can then obtain patent protection in other countries signatory to the convention (provided there are not prior patents); *See also* 1953 GEN

The Pentagon starts developing the Global Positioning System consisting of 21 satellites for the military; by receiving the radio signals from three or more satellites, a person can determine a position on Earth within 23 m (75 ft)

Intel introduces a memory chip that can store 1024 bits of data, replacing the voluminous ferrite core memories used by computers; the chip has $9 million in sales during the first year it is on the market; *See also* 1953 COM

The floppy disk is introduced for storing data used by computers; *See also* 1956 ELE; 1978 COM

The daisy wheel printer, an impact printer that uses a metal wheel with spokes that are tipped with letters to print with an inked ribbon, is introduced for use with computers; *See also* 1961 COM; 1971 COM

Kenneth Thomson and Dennis Ritchie at Bell Labs develop Unix, an operating system for both small- and medium-sized computers that becomes the standard for multitasking and multiuser systems; *See also* 1972 COM

Xerox establishes at Stanford University in CA the Palo Alto Research Center (PARC), a facility for noncommercial computer research; work at PARC leads to the development of Ethernet and the concept of "icons"; *See also* 1973 COM

Edward Shortlife at Stanford develops MYCIN, a medical expert system based on 500 "if-then" rules; *See also* 1968 COM

Terry Winograd develops SHRDLU, a dialogue system that allows a computer to converse about a tabletop world that contains building blocks; it permits simple manipulation of blocks and a description by the computer of results of certain manipulations; *See also* 1968 COM

Control Data Corporation introduces the STAR 100 computer, which has a vectorial architecture—a non-von Neumann design in which information is processed as vectors instead of as numbers; this allows a faster speed when a problem can be expressed in vector form; *See also* 1971 ELE

Charles A. Burrus [b Shelby NC, Jul 16 1927] develops the light-emitting diode (LED); *See also* 1962 ELE; 1987 ELE

RCA introduces the metal-oxide semiconductor (MOS) technology for the fabrication of integrated circuits, making them cheaper to produce and allowing greater miniaturization; *See also* 1958 ELE; 1976 ELE

Carbon-dioxide lasers are introduced for industrial cutting and welding; *See also* 1963 TOO; 1978 TOO

Yi Hoh Pao develops a system to cut concrete using a water jet with a speed of 800 m (2600 ft) per second; *See also* 1907 TOO

The first of the jumbo jets, the Boeing 747, with a two-story cabin capable of carrying more than 400 passengers, goes into service across the Atlantic on Jan 21; *See also* 1949 TRA; 1972 TRA

The US Apollo 13 mission starts off on Apr 11 with James A. Lovell, Jr., Fred W. Haise, Jr., and John L. Swigart, Jr. in its crew; its third lunar landing attempt is aborted due to loss of pressure in liquid oxygen in the service module and fuel cell failure; *See also* 1969 TRA; 1973 TRA

The first Chinese and Japanese artificial satellites are launched; *See also* 1958 TRA

The integrated circuit, or chip

By the end of the 1950s, electronics manufacturers were faced with circuits of increasing complexity. Computers, for example, contained tens of thousands of transistors. Assembly often required hundreds of thousands of interconnections, all of which had to be soldered by hand.

A transistor is essentially a replacement for a vacuum tube. Electronic devices, such as computers and radios, use several different types of component. Some are very fast switches. Others act as electronic gates, allowing certain messages to get through while rejecting others, or restricting flow of current to a single direction. In a computer, such gates become logic circuits that, for example, combine two statements into one using *and*, *or*, or *if-then*. Another component amplifies a signal. After the invention of the transistor, various configurations of transistors were created to perform these functions, components still referred to by such traditional names as resistor, capacitor, diode, and so forth. In each case the method was to add specific impurities to different regions of a semiconductor chip, a process called doping.

The idea of placing several components on a single chip came to two people who worked independently. Jack Kilby, an electronics engineer, in 1958 joined a team at Texas Instruments that studied ways to reduce the size of computer circuits. Kilby created a semiconductor chip that carried an oscillator made of several components, such as switches, resistors, capacitors, and diodes, all made from doped semiconductors. Although these components were on a single chip, Kilby had not developed a suitable way to connect them, and had to create interconnections in the traditional way. He demonstrated his oscillator-on-a-chip to executives from Texas Instruments on Sep 12 1958 and filed for a patent a few months later.

Robert Noyce, a physicist, was the head of research at the recently founded Fairchild Semiconductor. That company had developed the "planar" technology for the manufacture of transistors. In this method, large numbers of transistors were created on a single wafer, which subsequently was cut up to yield the single transistors. Noyce realized that this technology would also be suitable for creating an entire circuit on such a wafer. But unlike Kilby, he found a way to make connections between the components on the wafer by laying down conducting tracks. Noyce filed a patent for his integrated circuit one year after Kilby did, and was granted it in Apr 1961. Kilby's patent application was rejected because he had not solved the problem of interconnecting the components on the chip.

A legal battle between Kilby and Noyce ensued, which ultimately was won by Noyce by decision of the Court of Customs and Patents Appeals in Nov 1969. Fairchild Semiconductor and Texas Instruments, however, had already agreed to share the licensing of integrated circuits in 1966. Noyce and Kilby, regarded by the technical community as coinventors, a view both have accepted, jointly received the National Medal of Science for their invention.

1971

Direct telephone dialing, as opposed to operator-assisted calling, begins between parts of the US and Europe on a regular basis; *See also* 1967 COM

US Centron introduces dot matrix printers; these become widely applied for both text and graphics; *See also* 1970 COM; 1976 COM

Niklaus Wirth develops Pascal (named for Blaise Pascal, who invented the first calculator), a popular language used on home computers; *See also* 1965 COM; 1972 COM

Raj Reddy [b Katoor, India, Jun 13 1937], at Carnegie Mellon University, develops Hearsay, a software program for speech recognition by computer; his project, SUR (Speech Understanding Research), is funded by DARPA, the research arm of the US armed services; *See also* 1968 COM; 1986 COM

Gary Boone and Michael Cochran develop the TMS 1000 microprocessor, the first microprocessor to be patented; the 1000 combines input circuits, memory, a central processor, and output circuits all on one chip; the chip is widely applied in pocket calculators; *See also* 1972 ELE

Ted Hoff, S. Mazor, and F. Fagin develop the Intel 4004 microprocessor; it contains 2300 transistors on a 7 mm by 7 mm silicon chip and can process 4 bits at a cycle rate of 60,000 per second; *See also* 1968 ELE; 1972 ELE

Texas Instruments introduces the first pocket calculator, the Pocketronic; it can add, subtract, multiply, and divide only; it weighs more than 1 kg (about 2.5 lb) and costs around $150; *See also* 1889 TOO; 1974 ELE

John Cocke and coworkers at IBM recognize that a computer can function with just the most commonly used instructions, omitting the less commonly used altogether; this thinking produces by the 1980s the RISC (Reduced Instruction Set Computer) processor; *See also* 1975 ELE

Seymour Cray leaves Control Data Corporation to found his own computer-manufacturing company, Cray Research, a company specializing in vectorial supercomputers; *See also* 1970 ELE; 1989 ELE

In the Soviet Union the first semiindustrial scale magnetohydrodynamic (MHD) power generator is announced in Apr; such a generator produces electricity by passing a conducting fluid through a magnetic field; the Soviet generator is designed to produce 25 megawatts and to use natural gas; a German MHD generator goes into operation in Apr also, but produces only 1 megawatt

In Canada, the first nuclear power station that is cooled by ordinary water goes into service; *See also* 1963 ENE

Scientists and defense

The outcome of World War II was influenced by certain technologies developed by the Allied forces. Radar produced by powerful tubes called magnetrons gave the British an advantage in the air over the Germans. The US obtained the magnetron from the British and equipped its warships with radar, gaining an advantage over the powerful Japanese navy. The British used electronic computers to decipher German messages. The US finished the war with the atomic bomb. Convinced by such wartime successes, governments of large industrialized nations started funding military research and development (R&D) on a very large scale, initiating what subsequently became known as the arms race.

The Manhattan Project, which developed the first US atomic bomb, had demonstrated that theoretical scientists could play an important role in defense. Therefore, the military started funding large numbers of basic research projects at universities. The US Office of Naval Research funded research projects in biology and oceanography. The US Defense Advanced Research Projects Agency (DARPA) was for a long time virtually the only agency funding artificial intelligence research.

The Strategic Defense Initiative (SDI—also known as star wars), introduced in 1983, kindled a debate among US scientists on whether universities should be involved in military research projects at all. Many scientists' objections were based on the principle of not wanting to see science become a tool of destruction. Others decried the secrecy rules imposed on scientists at universities that participated in SDI projects.

The arms race and SDI recruited large numbers of scientists and engineers. By 1984, for example, 14 percent of the total of 4 million scientists and engineers in the US worked in military R&D. The end of the cold war in the late 1980s reduced levels of weapons research both in the East and West. In the US, several weapons laboratories turned to work on civilian research projects.

Raymond V. Damadian [b New York City, Mar 16 1936] applies for a patent for magnetic resonance imaging (MRI, also known as NMR) to detect tumors; *See also* 1973 MED; 1977 MED

The US government passes the Lead Paint Poisoning Prevention Act, which requires that interior paint applied before 1955 be stripped from buildings; the act is never fully implemented, however; *See also* 1955 MED; 1980 MED

The US Food and Drug Administration asks doctors to stop prescribing diethylstilbestrol (DES) to control bleeding in expectant mothers because of evidence that the drug predisposes their daughters to cancers of the reproductive tract

It is shown in the US that electric currents can speed the healing of fractures; *See also* 1987 MED

In England, the first completely sterile hospital units are introduced to protect patients at special risk from infection

The diamond-bladed scalpel is introduced in Great Britain

Chemists at the Research Triangle Institute in NC isolate the active ingredient of a powerful anticancer drug made from the bark of Pacific yew trees; they name this ingredient taxol; the anticancer activity of extracts from the bark had been known since the 1960s; *See also* 1992 MED

1971

Soviet cosmonauts Vladimir A. Shatalov, Alexei S. Yeiseyev and Nikolai N. Rukavishnikov begin the Soyuz 10 mission on Apr 23; the craft docks with *Salyut 1*, the first space station; *See also* 1967 TRA; 1986 TRA

Soviet cosmonauts Georgi Dobrovolsky, Viktor I. Patsayev, and Vladislav N. Volkov begin the Soyuz 11 mission on Jun 6; all three are killed during reentry; *See also* 1967 TRA; 1986 TRA

1972

The Club of Rome publishes *The limits of growth,* also known as the Meadows report, announcing that Earth will face environmental disasters if pollution and depletion of Earth's resources continue at current levels; *See also* 1970 GEN; 1973 GEN

In Jun the UN Conference on Human Environment meets in Stockholm, Sweden, to examine global environmental issues; *See also* 1970 GEN

The US Congress passes a Clean Air Act, allocating $95 million to local, state, and national air pollution control efforts

Dennis Ritchie and Kenneth Thompson at ITT Bell Labs develop C; later versions of C become highly successful and are widely used for writing software packages; *See also* 1971 COM

Alain Colmeraurer develops PROLOG, a computer language developed for applications in artificial intelligence; *See also* 1959 COM

Threshold Technologies introduces the first speech-recognition system, the VIP 100; the system can recognize a limited number of words when each word is pronounced separately; *See also* 1968 COM; 1973 COM

Alan Kay introduces SMALLTALK, a computer language especially adapted for graphics, including windows and icons; it is also one of the first object-oriented languages; *See also* 1970 COM; 1973 COM

Telenet Communications Corporation establishes a computer network with terminals all over the world; TYMNET also establishes a communications network; *See also* 1969 COM

Word processing is introduced: Lexitron puts the first word processing system on the market; VYDEC Corporation introduces a word processing system that is immediately more successful; Wang Laboratories offers a word processing system that uses the IBM Selectric typewriter as a printer with text stored on magnetic tape; the Wang version continues to be successful until replaced by microcomputer-based systems; *See also* 1964 COM; 1978 COM

Philips Corporation of the Netherlands introduces a disk-laser recording system called Laservision; *See also* 1965 COM; 1982 COM

Intel develops the first 8-bit microprocessor chip, the 8008; it is used in the Mark-8 "personal minicomputer," but because of a weak design it is replaced by the 8080 chip the following year; *See also* 1971 ELE; 1973 ELE

Odyssey developed by Magnavox is the first video game; *See also* 1972 ELE

Nolan Bushnell invents a video game with a liquid crystal screen; the toy, called Pong, sells so well that Bushnell founds Atari, which later starts producing home and personal computers

In Germany, an experimental power station uses coal that is converted to gas before being burned to produce electric power; *See also* 1985 ENE

1972

The World Trade Center in New York City surpasses the Empire State Building, also in New York City, in height; *See also* 1931 FOO; 1973 FOO

In the US, the use of DDT is restricted to protect the environment, especially eagles, hawks, and other predatory birds whose eggshells are dangerously thinned, lowering the birds' reproductive rate; *See also* 1939 FOO

A coworker of Hideki Shirakawa accidentally prepares a form of polyacetylene that has metallic properties, such as electric conductivity; *See also* 1977 MAT

The first experimental computerized axial tomography (CAT scan, later known as CT scan) imager for medical purposes is introduced in Wimbledon, England; it detects a brain tumor in a living patient on Oct 4; *See also* 1971 MED; 1979 MED

The first Earth-resources satellite, Landsat I, is launched

A European consortium, originally the Groupement d'Intérêt Economique, flies the first A300B Airbus, a commercial jet airplane that soon becomes one of the leading passenger carriers around the world; *See also* 1970 TRA; 1986 TRA

The last section of the current London Bridge across the Thames, a three-span structure of prestressed concrete, is opened; *See also* 1831 ARC

The air-cargo version of the Boeing 747 (jumbo jet) starts service on the Frankfurt-New York line; *See also* 1970 TRA

The coming of future shock

Technological change in the period starting after World War II began coming too fast for humans to adjust to it. Older people in the late 1940s had seen the coming of electric lights and telephones when they were children, airplanes when they were teenagers, and radios when they were in their twenties and thirties. While this was faster change than experienced in previous periods of history, there was still time for people to adjust to the new. Furthermore, not a lot of change occurred at the level of daily living in the Great Depression of the 1930s, and, although war in the 1940s brought many people into greater awareness of new technology, the home front was not much affected until after the war was over.

Starting in 1946, however, aspirin and mustard plasters suddenly were replaced with antibiotics, radio with television, commercial travel by train and bus with easy-to-afford commercial flight, the old icebox with the refrigerator and home freezer, and the file cabinet and papers with computers and punch cards. After a couple of decades of this kind of change, the pop sociologist Alvin Toffler coined the term "future shock" to describe people's reaction to rapid-fire technological change.

While economic activity was traditionally expressed in terms of the manufacture of goods and food, a new component, that of the service industry, started to play an important role. A new type of society, called by American sociologist Daniel Bell the Post-Industrial Society, started taking shape. Technological change during this period came so fast that many people could not follow or conceptualize it. One result was that the gap between science and technology, and the lack of understanding of it by the public, increased.

Although technology (and science) is viewed by many as part of our culture, its development is only understood by a small part of the population. The author C.P. Snow referred to this problem as that of the "Two Cultures," the literary and the scientific, with very little understanding between the two.

Satellites systems began to produce global television, allowing viewers all over Earth to watch the same program at once and further affecting the way they perceived the world. This capacity was a factor in making the Vietnam War a political issue in the US in ways that previous wars had not been. Now people away from the front lines had daily access to pictures of death and destruction. Pleasanter scenes, such as major sporting events, also could be shared around the world. This trend continues to accelerate with the gradual increase in the capacity of cable television to handle more information than was available to earlier generations.

OVERVIEW

The Information Age

A combination of new developments in electronics, information storage, methods for manipulating genetic information, communication and display techniques, Earth satellite capabilities, and trends in society led to a revolution that began in the early 1970s and appears to be continuing today. The common thread binding the different pieces of this revolution is information, a word that encompasses but surpasses data. Its modern sense derives from the scientific work of Ralph V.L. Hartley in the 1920s and Norbert Wiener and Claude E. Shannon in the 1940s, but as often is the case with scientific concepts that cross over into public discourse (entropy and chaos are two recent examples), the idea of information has become somewhat of a catchall for much more than was intended by the scientists. Nevertheless, it is in this broader sense that the period since 1972 can be dubbed an Information Age.

The effect of a transition from goods to information can be felt in all enterprises. For example, the importance of traditional manufacturing (metalworking and construction) has sharply declined relative to the newly emerged communications and information technology industries. Information and information processing are new "products" in which the trade has increased enormously in the last two decades.

Information and society

Information processing now occupies an important position in our culture. Computer technology even fascinates many people for its own sake. The value society attaches to information has increased not only because information has become a commodity, but because information technology and computers are considered the cutting edge of technological progress. Software and computers form the magic of the modern world. More than any other technology, the computer's possibilities seem unlimited. It is the new Eldorado for the inventive

spirit. The mantle of the inventor entrepreneurs, such as Thomas Edison, Alexander Bell, and Henry Ford, is worn today by the information entrepreneurs—Steve Jobs of Apple and NeXT, Seymour Cray of supercomputer fame, or William Gates of Microsoft.

Information and information technology have also colored our view of the world, forming the new metaphor of our reality. The existence of metaphors for the understanding of the world is not a new phenomenon; during Galileo's and Newton's times, the world was viewed as clockwork and people as complicated mechanical machines. Today, many picture the human mind as a type of complicated computer. Parallel processing and neural networks are considered emulations of the way the nervous system operates, and many researchers in artificial intelligence (AI) believe that AI will ultimately lead to understanding of the human mind.

The acceptance of this new metaphor has widespread consequences. There is an inherent belief that even the most difficult problems will find a solution in information gathering and processing and in the power of software; examples are the use of expert systems in medicine and the way that modern warfare is based on weapons in which information and information processing are the most important aspects. Naturally, new terms have made their appearance, the most striking of which is probably "knowledge engineering."

The economic organization of society is not the only structure to undergo change in the transition to the Information Age. Such concepts as "democracy" and "hierarchy" are also transformed. Democracy moves away from the old system of political bosses and power structure based on money and remote decision making. In the US, President Clinton had not been in office a hundred days before it was possible to use electronic forums to find out every speech, document, and movement of the presidency, and even to leave "Dear Bill" messages in the White House computer system. Hierarchy in business similarly moves away from the present system of management through levels of supervisors and forepersons. The recent adoption of computer networks by businesses embodies this trend: The structure of a company becomes reflected in the structure of its electronic communications system. Hierarchical structures are replaced by passwords and by restricted mailing lists for electronic communications (E-mail) systems in large offices.

Computers take over

In the early 1970s there was a growing interest in what computers could do combined with an expectation of great things just around the corner. But the great things would not come. Computer-assisted instruction (CAI) based on time-share computers and programmed learning were exciting, but CAI failed to make any real inroads in education. Word-processing machines that seemed to be able to solve many of problems of publishing, especially in editing complex manuscripts that previously had been typed over and over, could not cope. Early word processors could not do columns or any formats other than straight text; to do even pure text required sending every keystroke through the one or two persons in the office who had spent weeks learning how to work the machines; and early word processors were broken more often than they were up and running. A computer could determine the best path for a complex manufacturing process or best price for a product, but you needed to hire the person who developed the program to run it. Even with the right person and a powerful computer on hand, everyone had to line up for computer time and time from the one person who knew how to make it work.

Wonderful progress was around the corner and many people seemed to know what it would be, but no one seemed to know how to make it happen.

In 1975 there were two major events in the world of computers. The first supercomputer, the Cray-1, was announced, and the MITS Altair 8800 home computer kit was put on sale. With a tiny memory of 256 bytes in the Altair versus a superfast speed of 100 million operations a second in the Cray, it should have been easy to tell which event heralded the future. As we all now know, the supercomputer, while still important, did not change the world the way the personal computer, heralded by the lowly Altair, did.

Information goes beyond computers and television

Human beings had learned to manipulate heredity in various ways at least by the start of the Neolithic, the time we term the Agricultural Revolution. Classic examples include wheat bred to retain its seeds on the stalk and animals bred to convenient sizes and dispositions. But until 1973, most of the methods used were decidedly "low tech," although artificial insemination and the use of frozen embryos, as well as the use of chemicals to induce chromosomal changes in plants, were already practiced extensively in industrialized parts of the world.

The year 1973 marked the engineering birth of biological manipulation of information with new technological means, which we now call genetic engineering. Twenty years earlier, the work of James Watson, Francis Crick, Rosalind Franklin, and Maurice Wilkins had established that information in the form of a three-letter code is the basic stuff of life. This code, physically present in the chemical structure of deoxyribonucleic acid (DNA), directs growth and development of all living creatures (except for a few viruses that use a slightly different variation). Starting in 1973 biologists learned ways to edit the messages in DNA.

About the same time biologists also began to learn that our immune system is also an information processing system, one that normally fights disease by transmitting information from cell to cell. Although it has yet to reveal all its mysteries, the immune system has become one of the major battlefields of medicine. Suppressing it was essential in transplanting organs or fighting autoimmune diseases; enhancing it selectively led to new vaccines or ways to fight cancer; and copying it made possible monoclonal antibodies, one of the great successes in gaining information. Monoclonal antibodies are lines of very specific chemicals that react only with target substances (usually proteins), thus providing ways to identify the substances even in tiny amounts or to carry specific chemicals to the places where such substances exist. Just as medicine began to learn how the immune system worked, a new threat to the system, acquired immunodeficiency syndrome (AIDS), arose. AIDS is a syndrome in which parts of the immune system are destroyed by an infectious virus, leaving the victim to die of one disease or another.

Even machines came to be replaced by information. In the traditional methods of making a part by casting and machining, the basic mold was made by hand; a series of generations of molds and parts led inexorably toward the assembly line. In the 1980s this pattern began to change with the development of stereolithography. The basic mold was created as pure information using computer-assisted design. The next generation was formed from molten plastic by laser beams directly controlled by the information in the computer; this is the stereolithography stage. The plastic part could be used directly or, if one modified the technique somewhat, could be made into a paper mold into which metal could be sprayed one layer at a time (this prevents the paper from catching fire). The result is one copy in steel or zinc of the information put into the computer. An advantage of using pure information for a first generation comes when small errors appear in the finished prototypes—they can be eliminated by a process like copyediting and the part can be "reprinted," just as a book manuscript is corrected. Another advantage is that the finished part can be sent anywhere in the form of pure information on a computer diskette or, even faster, over the telephone. While not quite at the "Beam me up, Scotty" level of the Star Trek shows, for inanimate machine prototypes this is nearly the same thing.

The environment in the Information Age

The early part of this period was a high-water mark for the environmental movement in Western countries. The first Earth Day was celebrated on Apr 22 1970. The 1970s and early 1980s gave birth to laws intended to clean up the environment, protect endangered species, and restore wilderness regions. As the 1980s pro-gressed, however, the dichotomy between economic progress and environmental concerns became more prominent, especially in the US, where conservative presidents from 1980 to 1992 tended to put business interests ahead of all other concerns.

The nuclear accidents at Three Mile Island in PA and Chernobyl in the Soviet Union cast a pall over the use of nuclear reactors for power. No new nuclear reactors for production of electric power have been commissioned in the US since Three Mile Island in 1979, and a few have been put out of service as a result of age. Nuclear reactors are recognized as good in the sense that they do not use fossil fuels or contribute to air pollution when they are working properly. On the other hand, they are feared because of the possibility of accidents—it is not known yet how many will die early as a result of Chernobyl's meltdown—and they produce waste that is almost impossible to store safely. Consequently, there has been a serious effort to search for a better way to produce energy. Nuclear fusion has long been held out as that way, but little progress has been made in developing it in the past 20 years. Environmentalists favor solar power or wind power, but industry has not moved very far in this direction either. It remains to be seen how this power dilemma will be resolved.

Critics of the Information Age

Many fear that some aspects of a society based on information technology will form a threat rather than inspire social progress. For example, the use of large personal databases by government, insurance companies, and even marketing organizations can be viewed as infringement on personal privacy.

Others believe that the superinflation of information forms a different threat. For example, excessive use of computers in teaching young children may impoverish the development of intellectual capabilities as a result of the specific and narrow fashion in which a computer stores and delivers information. For these critics the computer replaces "knowledge" with mere data: Ideas contain data, but data contain no ideas.

Criticism of basing society on information sometimes revolves around what some see as intrinsic limits to information processing. Star Wars—the US Strategic Defense Initiative (SDI) —is often used as an example. In 1983 US President Ronald Reagan announced the start of SDI, a satellite-based high-technology plan to provide an impenetrable barrier to enemy intercontinental ballistic missiles. Although several versions were designed and a few tests were conducted, it soon became apparent to most scientists that technical problems probably would keep the planned shield from

Many inventions change the details of the way we work or live, but a few change our lives in dramatic ways that are unexpected. It is usually the case that even the expected changes do not happen in the way people have predicted. If you read the descriptions of the future made right after World War II, you find that experts expected the helicopter to replace the automobile for short trips, and the same experts failed to anticipate the impact of jet aircraft on long-distance population mobility, managing to predict both trends incorrectly.

One of the unexpected changes of the Information Age has been time shifting. This takes two main forms. One of them is that it is no longer necessary for an individual to plan a specific time for an event. Instead, the event occurs when the person is ready for it. The other form is that events for which there had in the past been a lag, during which something else could happen, now take place instantaneously.

The second form of time shifting, the loss of lag, is less a new development than it is the final (or near final) culmination of a trend that goes back to the domestication of horses and that has continued at an exponentially accelerating pace since. The steps from exchange of information over a period of months to a period of days to one of hours to instantly have come faster than we have had time to think about in recent years. In the 1980s business changed suddenly from use of mail services that took days or even weeks to the use of overnight mail. In the 1990s, everything shifted into the present through faxes, modems, electronic mail (all traveling on telephone lines), and increased use of the older instantaneous system, the telephone for voice communication. It got more difficult to kite a check or to pretend that you were about to meet a deadline, because data could be and was expected to be transmitted instantly.

At the same time, data storage permitted time shifting of the first type to occur—moving an event from a certain time to a more convenient one. Although faxes can transmit almost instantaneously, they also can be sent to a machine in the dead of night to be read whenever the receiver feels like it. Electronic mail is likewise both instantaneous and shiftable. Unlike old-fashioned voice communication over a telephone, there is never a busy signal or nobody home with electronic mail; the message is delivered instantly, but can be picked up at the receiver's convenience. Even today's business telephones encourage time shifting by using powerful answering machines called voice mail.

Not all time shifting is business related, either. The advent of moderately easy to use video recording has allowed families to time shift their entertainment. People do not have to schedule a trip to the motion pictures; they pick up a video and watch it when they are ready. If they want to watch a program scheduled for an inconvenient time, they can tape it for replay later. Similarly, bankers' hours no longer matter to the customer, who can deposit checks, pay bills, transfer money, or obtain cash at any time of the day or night from an automated teller machine (ATM).

One factor that makes time shifting possible has been the acceptance of standards. Facsimile transmission is an old technology that was introduced about the same time as radio and motion pictures, but for a long time it was used only by newspapers or other specialists in transmitting photographs. In the 1970s and 1980s international standards were set for fax transmission, which, along with the less expensive devices made possible by microprocessors, ensured that a person with no special training or experience could successfully send or receive a fax on the first try. This should be contrasted with the use of the modem for communication directly between computers. There are so many different standards available that it often takes a day or two to get two computers to agree on a set of protocols before the modems can be used. Sometimes necessary standards are set by official bodies, as with the fax, and other times they are set by the public, who, for example, choose the VHS standard for videotape over several other incompatible systems.

being effective. In particular, difficulties in programming the complex software for the project seemed unsolvable. Reliable information for computers to use in launching nuclear weapons could not be produced fast enough. Furthermore, every experience with software in the past has shown that unanticipated problems almost always become apparent during actual operation. Since actual operation of the SDI system would entail a nuclear attack, there was no way to test the software programming. Although Republican administrations continued to work on the project, as soon as there was a president from the other party, SDI was scaled back and quietly dropped.

Major advances

Communications. The Information Age saw sweeping revolutions in communications, many, but not all, derived from computers. As their name suggests, the first computers were thought of primarily as very powerful calculators. Although early large computers were soon handling masses of data beyond simple calculation, the true possibilities of the tool in such fields as design and education were just beginning to become evident about the same time that the first personal computers were marketed.

When people began to play with their own comput-

Many mechanical devices are becoming entirely replaced by electronic ones in recent times. A good example is the typewriter: A classical mechanical typewriter consists of a large number of mechanical parts. A modern, electronic typewriter has very few mechanical parts: Everything is achieved with a few tiny motors and logic circuitry. The electronic typewriter also demonstrates the decrease of the one-to-one relationship between a specific technology and its application. With minor changes, the electronic typewriter can be changed into a computer printer. Some typewriters incorporate such a dual function in their design. Because an electronic typewriter contains a microprocessor, its circuitry is also very close to that of a simple computer.

The kind of transition from a mechanical-based operation to one based on or controlled by electronics is often called high technology (abbreviated as high tech). This is a relative term. Look at the process of beating egg whites to make a meringue, for example. At the lowest end of low technology we find beating with a stick or, slightly higher, a fork. A specialized tool, the whisk, improves the process. Slightly higher technology, appropriate to the Industrial Revolution, is a geared beater propelled manually. The Electric Age brought the same beater propelled by an electric motor, perhaps with a number of speeds available. At the highest end of high technology, the stiffness of the egg whites would be measured with an electronic device and a feedback mechanism would adjust the speed along a continuum. Although only the last

beater would be labeled high tech, the relative nature of the term is also clear.

High technology is not necessarily the most appropriate way to accomplish an end. In the example of the high-tech eggbeater, critics could point out that the final product is no better than that achieved by a good cook with a whisk—perhaps not even as good—and little better than could be achieved over a longer period of time with a stick picked up from the ground. Such a high-technology eggbeater may be an example of overkill, especially in poor countries where electricity is often in short supply. Similarly, reformers often look for low-technology solutions to problems, such as simple solar heaters for cooking in countries that are short of fuel and electricity, or passive solar architecture to preserve the environment in developed regions.

Even low technology has its critics, however. Global warming, caused by the low-tech approach of burning lignite, coal, oil, and natural gas to produce energy of all sorts, is feared by many scientists. The production of excessive carbon dioxide, the intermediary in the projected global warming, could be curtailed by high-tech solar panels that convert sunlight directly into electricity; the electricity could be made into a portable fuel by using it to disassociate water for its hydrogen. Although such electricity and hydrogen production has been demonstrated, it is far from being commercially useful. If made to succeed, however, it would be a high-tech solution to a low-tech problem.

ers, they soon found new ways to use the devices. Before 1973 there were no computer word processors or spreadsheets at all. By 1980 the personal computer had a third major tool, the relational database—which, because of its generality, was more useful than the specific databases developed earlier for multiple-user computers. Similarly, the graphic programs for personal computers that followed word processing, spreadsheets, and databases brought a fourth new general tool into the mid-1980s. All four of these tools relieved the average computer user from programming chores. Although learning how to use the more complex tools most effectively might be as daunting as programming, the same tools could also be used to solve problems with little or no additional training.

One way computer use was simplified was by the introduction of pictures as control methods to replace words. The idea of using icons instead of complicated menus or commands goes back to 1973, although it did not become popularly used until the mid-1980s, and at that time only on the Macintosh computer family. Icons are small pictures displayed on the computer screen to

suggest commands; for example, a file can be erased by using a mouse to drag the file folder over to a picture of a wastebasket and dumping it in. In the early 1990s, icons became the communication device of choice for most personal computers. Throughout this period, with increasing likelihood of usefulness, voice recognition and production by computers (and other electronic devices) was the promised way to make operation of electronic devices even easier. Also, the keyboard—still something of an innovation as a data interface in the 1960s, despite the success of office typewriters and related instruments—was gradually wearing out its welcome. In addition to the mouse (for pointing and dragging), a number of manufacturers brought out input devices that accepted handwriting.

Not all changes in communications were directly linked to computers, although all had microprocessors embedded somewhere. The way light was used altered the world as well. Retail merchandising changed with the laser scanner, and new uses for the Universal Product Code, that pattern of lines read by a scanner, emerged throughout the period. Displays of all types were improved, a major revolution that went almost unnoticed. Light was used for storage in laser-based compact disks (CDs) and optical disks. The CD, invented for recording music, soon was adapted for recording information of all types. Laser printers replaced not only the impact printers used with home computers, but many of the typesetting machines used for producing books and magazines. Laser printers combined with innovative software made everyone potentially a publisher, although the real advantages of desktop publishing were primarily commercial in terms of speed and cost.

Related to several of the trends in communications was the continued shift from analog to digital or, perhaps more accurately if not colloquially, from continuous data to discrete data. At first only computers were digitalized. Today, CDs and related tape-recording methods use digital sampling to approximate continuous sound very effectively and astronomers and space probes use digital techniques to approximate continuity in images. Combining these ideas, it has become apparent that one of the next breakthroughs will be digitized high-definition television. An advantage of the digital methods is that the spaces between the digits can be used to convey other information, rather as the television comedian Kevin Nealon does in the role of "Mr. Subliminal."

Electronics and computers. For most people, the Information Age is intimately tied up with the sudden rise of the personal computer. Although it may not have become the home appliance envisioned by many at the start, the personal computer, just a gleam in the eye in 1973, has become essential to office operations of all types in the advanced industrial nations, especially in the US and Canada. Each year new versions appear, faster and with more capabilities than those of the year before. The new personal computers are smaller and lighter each year, and more and more workers find that the laptop or notebook computer that they carry with them from place to place is most useful for their purposes. The small computers have become so advanced in design that many people prefer to use them instead of similarly equipped desktop computers even when working in the same place each day.

Since 1975 the most powerful computers have been so much faster than the kind employed by ordinary businesses that they are called supercomputers. The supercomputers' great speed is essential in such tasks as modeling complex change or designing complicated parts. Even as the supercomputers were becoming faster and more complex, an entirely different idea surfaced. Instead of putting every scrap of information through a single microprocessor at a faster and faster pace, one could use a number of microprocessors working on different parts of the same problem to gain speed. Computers built this way, using what is called massively parallel architecture, have already outsped some the fastest supercomputers for some kinds of problems. The future, however, probably does not belong to either type alone, as some problems resolutely resist breaking into the bits and pieces used by massively parallel computers.

Another route to computing speed has been to utilize the speed of light, faster than the speed of an electric current in a wire. Light also has other advantages. Electrons are fermions, which means that two of them cannot be in the same "place" at one time. Light particles are bosons, happy to share the same place with each other. As a result, it is much easier to design light-based equipment in which everything happens more or less at once. Although the technical problems seem to be formidable, there has been considerable progress toward the all-optical computer in the 1980s and 1990s. Meanwhile, fiber optics can be used with electronic equipment to improve data transmission in mixed systems.

It is easy to forget that modern electronic devices go considerably beyond the computer. Everything from ATMs and automobiles to vacuum cleaners and washing machines is built today with microprocessors that act as small, special-purpose computers inside more traditional machines. The subtle effects of this transition from the mechanical to the electronic are not always perceived at the gross level until one of the devices breaks down and a chip has to be replaced. For the most part, however, the embedded microprocessors simply make old functions happen better and more smoothly. In some devices, popular mostly in Japan so far, the microprocessors use a system known as fuzzy logic to make many decisions that otherwise would have to be made by the operator of, say, a subway car or a washing machine.

Energy. Use patterns of energy in the developed world changed dramatically starting in 1973, the year of the first crisis caused by the Organization of Petroleum Exporting Countries (OPEC). The OPEC oil embargo steeply raised prices of petroleum and products based on petroleum. A second OPEC action in 1978 led to even higher prices that have remained at their new level since. Although the initial shock of the higher prices and petroleum scarcity caused governments to attempt to find alternative fuels and to promote fuel conservation, the novelty wore off and new oil finds in Alaska and the North Sea contributed to a return to the status quo of using high-priced oil recklessly.

A second major change, starting in the 1980s, was the turn away from nuclear energy as a power source in the US and some other countries, although in France, especially, nuclear power plants continued to replace coal-fired generators. Although safety concerns were a factor, residual conservation reduced the need for new electricity-generating plants at the same time that the economics of plant construction and maintenance shifted in favor of coal-fired plants. One factor that saw the demise of what had once been called electric power "too cheap to meter" was the cost of dealing with far-reaching safety regulations and of handling waste fuel. In Eastern Europe, where these costs were minimized by fewer regulations and fewer scruples, nuclear plants continued to be built. The Chernobyl disaster, followed by the collapse of communism—not directly connected, of course—revealed a network of poorly maintained plants that are viewed in the West as time bombs set to go off in the near future.

Although a number of innovative alternatives to nuclear power and fossil fuel were pursued, including everything from improved old-fashioned windmills to high-tech means of tapping the difference in temperature between the surface and the depths of the ocean, the only alternative that has continued to grow in use, efficiency, and theory is solar power. Each improvement in solar cells, which convert sunlight to electricity directly, is widely heralded, but actual applications of the new technology have proven too costly for widespread use so far, except in watches, calculators, and similar low-energy devices that work well with existing solar-cell technology.

Food and shelter. Food for health was on people's minds in Western societies. People were afraid at one time or another to eat sugar, fat, meat, cholesterol, raw fish, salt, solid margarine, even apples (that had been sprayed with the chemical Alar). In a counterpoint to the feared foods, people were encouraged to consume such foods as oat bran, olive oil, fish, broccoli, orange vegetables, and even red wine and fois gras in large quantities. The primary reason for the new mandates, which seemed to change on a weekly basis, was better health. Although the guidelines were confusing, health statistics suggested that for those trying to follow them, some overall benefits were occurring.

A whole different set of prejudices revolved around what the food was put in before it was eaten. The fast-food chains that had come to be a major influence on diet were a common target of packaging reform, with the McDonald's chain switching its packaging for burgers several times in hopes of satisfying both environmentalists and hungry customers. The debate between those favoring foam plastic cups and those favoring paper cups was never resolved.

Change tends to take place slowly in architecture, even when everything else seems to have been speeded up.

Since 1973 the main changes have been stylistic—away from the purely functional and unadorned—and size-related. Early in the Information Age the world got a new tallest building in the Sears Tower in Chicago IL and a new tallest structure in the CN Tower of Toronto, Canada. Even the world's largest church, which had been St. Peter's in Rome since 1626, was exceeded in size.

The characteristic new shelters of the Information Age, in North America at least, were indoor malls. These too reached sizes that far surpassed any shopping areas previously built, especially in northern climates where spending a day out of the winter cold was an attraction in itself.

Materials. Two totally unexpected types of material have dominated technology news since the mid-1980s. One is a class of ceramics that become superconducting at temperatures much higher than those known earlier. The other is a class of carbon molecules and compounds that is typified by buckminsterfullerene, a carbon molecule of 60 atoms in the shape of a soccer ball.

Superconductors are, as their name suggests, materials that conduct electricity with little resistance—theoretically with no resistance at all. They also have other unusual electromagnetic properties. The first superconductors to be found were metals that became superconducting near absolute zero for reasons that are now well understood. Ceramic superconductors exist at temperatures as much as 100°C (180°F) higher. Although such materials have great possibility as the base of a new technology, this potential has failed to be realized because of difficulties in fabrication and other problems.

The carbon molecules are called fullerenes when they are pure carbon. Because of the way carbon atoms bond to each other, fullerenes are various shells surrounding empty space. In the laboratory fullerenes and related compounds have done such wonderful feats as superconducting and forming diamonds. They have yet to reach a commercial product.

Medical technology. A list of all the individual medical advances during this period would be quite long. Along with electronics, medicine has shown the greatest advances in technology during the period since World War II.

Increased access to information about a patient's inner state was the key development in medical technology in this period. New in the early 1970s, CT (computed tomography) and MRI (magnetic resonance imaging) scans became, especially in the US, routine noninvasive procedures for all sorts of medical problems. Each uses complex mathematical techniques to reconstruct clear views of internal organs, with the CT scan relying on a version of conventional X rays, and the MRI scan using a newer technology in which magnetic forces cause individual atoms to broadcast radio waves that can be detected outside the body. To some degree the new technologies are complementary, since CT scans are better for looking at bones, while MRIs can image soft tissues that are transparent to X rays.

Gradually, genetic engineering has gone from the laboratory into the clinic. Although still not an everyday occurrence, except for some substances produced by genetic techniques, more and more chronic or congenital diseases were in experimental treatment by genetic-based methods during the early 1990s.

While diagnostic programs based on expert knowledge of disease and artificial intelligence techniques have been in use for some time, the gradual replacement of actuality by information about actuality, or virtual reality, has only recently come to advanced medical technology. The entire human anatomy is now available, every muscle, bone, and organ, in a CD-ROM version, so that a medical student can manipulate "Information Man" instead of an actual cadaver. Since robot surgery also became a reality in the 1990s, it is possible to picture these two linked together, with a real doctor performing a procedure on Information Man that the computer then uses to direct a robot to perform the same procedure on a real human.

Tools and devices. More of the devices that affect everyday life came to be designed or manufactured in Japan, even when the basic concepts or science behind the ideas originated in the US or Western Europe. This lag affected everything from innovations in automobile engine design to consumer electronics.

The most exciting tool of the period was the laser, which continued to be a source of new ways to do old things and to accomplish some goals no one had dreamed of previously. Lasers very quickly moved from science museum demonstrations and light shows at rock concerts into homes, factories, and hospitals. Often the only way that a user would recognize that a laser was involved was if the word was used in the name, as in a laser printer.

Another major development that has had unexpected applications is the scanning tunneling microscope and its various spin-offs. Not only has this tool and the closely related atomic force microscope revealed details about surfaces, molecules, and atoms that were once thought to be forever beyond imaging, it has also led to tools that manipulate individual atoms or molecules. The full force of this technique is yet to come, but technology appears to be nearing a point predicted by Richard Feynman in the late 1950s, a point at which useful devices can be built from a number of molecules

that could be counted on the fingers of two hands. This trend has been dubbed nanotechnology, since it includes parts thousands of times smaller than those of the microtechnology that powered the US space effort and led to the notebook computer.

Transportation. Transportation in this period revolved around the old bridal tradition of "something old, something new, something borrowed, something blue."

An old idea brought back from the brink of extinction was the passenger train, speeded up by conventional means in Japan and France, with new computerized technology for better cornering available from Sweden. In Germany and Japan various experimental trains are suspended from the track and propelled by magnets. Trains appear to be more efficient than airplanes for moving people and goods intermediate distances. In the US and England, however, the extensive network of and investment in old track seems to be holding back development in this area.

Perhaps the oldest idea of all is human-powered transportation. Human-powered flight became a reality in the Information Age, although it was not widely used. Human-powered bicycles, however, remained the most widely used mechanized transport system worldwide.

The reusable space shuttle of the US is something new, but it is a troubled means of transportation. It almost seems as if the more experience people have with the shuttles, the less reliable they become. By the early 1990s there was a movement to find better ways to get into and out of near-Earth space.

Environmental concerns caused a number of automobile manufacturers to borrow and update almost every idea ever tried for replacing the gasoline engine in motor cars with some other means of propulsion—except perhaps the steam engine. Although none of the experimental vehicles had reached the general public by the early 1990s, plans were underway for electric cars and internal combustion engines powered with methane or other non-gasoline hydrocarbons before the end of the decade. Electric automobiles powered by fuel cells instead of batteries, a brand new technology, may take a little longer to reach the roads. Although the fuel cell is well over a hundred years old, its only use in transportation so far has been to power devices aboard US spacecraft in the 1970s. Experimental solar-powered electric cars have also been tested, notably in competitive races in Australia and the US.

The blue and deep black under the seas is being explored by a new generation of submersible devices, some with human pilots and some with robots. These devices are revealing not only new facts about life and geology at the bottom of the sea, but also are photographing or retrieving parts of old shipwrecks.

1973

GENERAL

Stanley N. Cohen and Herbert W. Boyer perform the first experiments in genetic engineering, demonstrating that deoxyribonucleic acid (DNA) molecules can be cut with chemicals called restriction enzymes; selected pieces of the DNA are then inserted into the bacterium *Escherichia coli*, where they perform as new genes; *See also* 1974 GEN

The first Endangered and Threatened Species List is issued by the US Fish and Wildlife Bureau; *See also* 1966 GEN; 1977 GEN

Daniel Bell publishes *The coming of post-industrial society*, describing the changes that society is undergoing because of the introduction of advanced technology at all levels of human activity

COMMUNICATIONS

Alan Kay develops the first "office computer"; it is based on SMALLTALK software, which features the use of a mouse, icons, and graphics; *See also* 1972 COM; 1974 ELE

Thomas B. Martins and R.B. Cox develop a system that allows a computer to understand spoken commands; *See also* 1968 COM; 1975 COM

The British Post Office starts up Prestel, a Viewdata service for the public; it supplies data from a large number of organizations and businesses via television screens; *See also* 1980 COM

Scientific-Atlanta introduces the first mobile Earth station for communications satellites; *See also* 1979 COM

ELECTRONICS & COMPUTERS

The ENIAC patent is invalidated, thereby crediting John Vincent Atanasoff as the originator of the modern computer; *See also* 1946 ELE

Ten thousand components are placed on a chip of 1 sq cm in area using a technique termed large-scale integration (LSI); *See also* 1965 ELE; 1974 ELE

Hans Dehmelt [b Germany, 1922] and assistants Philip Ekstron and David Wineland succeed in capturing a single electron and holding it in place, in a device called a Penning trap, for as long as ten months at a time; *See also* 1936 ELE; 1980 COM

ENERGY

After a year of steeply raising prices for reduced supplies of petroleum, the Organization of Petroleum Exporting Countries (OPEC) uses war between Israel, Egypt, and Syria as an excuse to embargo oil shipments to any countries giving assistance to Israel, especially the US and the Netherlands; total petroleum production is also cut; in the US, gasoline prices jump and there are long lines at service stations; although much of the energy crisis is viewed as both temporary and psychological, this is the beginning of the end of inexpensive energy around the world; *See also* 1859 ENE; 1979 ENE

1974

GENERAL

Genetic engineering, in which the genes of one organism are inserted in another, is viewed with alarm by a committee of 139 scientists from the US National Academy of Sciences led by Paul Berg [b New York City, Jun 30 1926]; in Jul the committee calls for a halt to specified research until the implications are better understood; *See also* 1973 GEN; 1976 GEN

Noticing a need for bookmarks that do not slip out but are easily removable, 3-M engineer Art Fry develops what will become the familiar Post-It Notes, slips of paper with glue just strong enough but not too strong

COMMUNICATIONS

The IBM Corporation's Memory typewriter can store up to 50 pages of type on a magnetic medium and play it back automatically at 150 words per minute; *See also* 1969 COM; 1978 COM

Electronic scanners that use a laser to read bar codes on merchandise for price and to report on inventory are introduced in a few US supermarkets; in 1980 there is a general introduction of the technique; *See also* 1960 TOO

Panasonic demonstrates high-definition television with 1125 scanning lines; *See also* 1948 COM; 1989 COM

The UN sets the first international fax standard (the Group 1); it allows facsimile messages to be transmitted at about one page in 6 minutes; *See also* 1913 COM; 1980 COM

The first championship for computers playing chess is held in Stockholm, Sweden; *See also* 1958 COM; 1977 GEN

ELECTRONICS & COMPUTERS

Hewlett Packard introduces the programmable pocket calculator; *See also* 1971 ELE

Jonathan Titus describes in *Radio Electronics* how to build the Mark-8 "personal minicomputer"; it uses the Intel 8008 microprocessor and is the precursor of the Altair 8800; *See also* 1972 ELE; 1975 ELE

David Ahl develops a microcomputer consisting of a video display, keyboard, and central processing unit; Ahl's employer, DEC, shows no interest; *See also* 1958 ELE; 1975 ELE

A computer memory chip (D–RAM, for dynamic random access memory) with 4 kilobits (4096 bits) of memory becomes commercially available; it will be used in the first personal computers; *See also* 1970 ELE; 1976 ELE

T. Peter Brody and coworkers at Westinghouse show that thin-film transistors can be used effectively to control liquid-crystal displays; *See also* 1962 ELE; 1975 COM

ENERGY

Brazil starts up its $5 billion gasohol program to replace gasoline with alcohol in cars; by 1980 there are 750,000 automobiles running on gasohol in Brazil; *See also* 1860 ENE

Windows that change opacity as a result of an electric current (electrochromic windows) are proposed as a way of saving energy; *See also* 1958 FOO; 1991 FOO

1973

On May 3 the Sears Tower in Chicago IL becomes the tallest structure in the world, surpassing the World Trade Center in New York City; the Sears Tower reaches a height of 442 m (1450 ft); *See also* 1972 FOO; 1974 FOO

The World Trade Center in New York City is completed; the two towers are 417 m (1368 ft) and 415 m (1362 ft) high; they are the second tallest buildings after the Sears Tower in Chicago; *See also* 1931 FOO

The push-through tab on soft drink and beer cans is introduced; *See also* 1963 FOO; 1975 FOO

John Malland at Aberdeen University in Scotland introduces nuclear magnetic resonance scanning (NMR or, more recently, MRI for magnetic resonance imager) for medical diagnosis; *See also* 1977 MED

Alfred Brickel builds the world's largest kaleidoscope, 312 cm (123 in.) long and 190 cm (75 in) in diameter, weighing 200 kg (500 lb); *See also* 1816 TOO

On Mar 6 the US launches Pioneer 11, the first space probe to reach the vicinity of Saturn; *See also* 1962 TRA; 1976 TRA

On May 25 the first US space station, *Skylab 2*, is launched with Charles Conrad, Jr., Joseph P. Kerwin, and Paul J. Weitz in its crew; the Skylab Orbital Assembly is established in Earth orbit and medical and other experiments are conducted; *See also* 1971 TRA; 1986 TRA

On Nov 3 the US launches Mariner 10, the first space probe to observe two planets, Venus and Mercury, and the only probe ever to observe Mercury; *See also* 1962 TRA; 1976 TRA

1974

The CN Tower in Toronto, Canada, becomes the tallest structure in the world, surpassing the Sears Tower in Chicago IL; however, the Sears Tower continues to be the world's tallest building, since the CN Tower is essentially for radio and television transmission; the tower reaches a height of 550 m (1805 ft); *See also* 1973 FOO

Frank *Sherwood* Rowland [b Delaware OH, Jun 28 1927] and Mario Molina warn that chlorofluorocarbons (also called Freons), commonly used as spray propellants and in refrigeration, may be destroying the ozone layer in the atmosphere; the ozone layer protects against excesssive ultraviolet radiation; such radiation can cause skin cancer; *See also* 1931 MAT; 1976 MAT

G. *Samuel* Hurst [b Pineville KY, Oct 13 1927] and coworkers apply for a US patent on their method for using a laser to count small numbers of atoms or even individual atoms in a sample; *See also* 1960 TOO; 1975 MAT

Track shoes in which rubber midsoles are replaced with ethylene vinyl acetate (EVA) foam are introduced; *See also* 1967 GEN

1975

The undeclared war between the US and North Vietnam finally concludes as North Vietnamese forces take Saigon (Ho Chi Minh City) and US forces evacuate; *See also* 1964 GEN

Xerox develops Ethernet, an access protocol for local area networks that have a ring structure; packets of data carry an address allowing the system to direct it to its destination; *See also* 1960 COM

Michael Shrayer develops the Electric Pencil on the Altair computer; it is the first word processing software package for a personal computer; *See also* 1972 COM; 1978 COM

E.M.I.-Threshold Technology develops a computer that can understand spoken commands and reply with a synthetic voice; it is based on a system developed by Thomas B. Martins and R.B. Cox; its Threshold 600 voice recognition unit can recognize up to 100 words just 300 milliseconds after hearing one, but the processing time increases to a second when the machine's full 512-word vocabulary is allowed; *See also* 1973 COM; 1978 COM

IBM introduces the laser printer, which creates an output comparable to that obtained by photocomposition; *See also* 1971 COM; 1976 COM

The first liquid-crystal displays for pocket calculators and digital clocks are marketed in Great Britain; *See also* 1979 COM; 1985 COM

The first pictures from the surface of Venus are received, from the Russian probes Venera 9 and Venera 10; *See also* 1962 COM; 1990 TOO

Edward Roberts introduces the first personal computer, the Altair 8800 in kit form, in the US; it has 256 bytes of memory; *See also* 1974 ELE; 1977 ELE

IBM designs a microcomputer code named 5100; it is not distributed because of the belief that there will be no market for microcomputers; *See also* 1974 ELE; 1981 ELE

John Cocke and coworkers at IBM begin the 801 project, the development of a minicomputer that employs what later comes to be known as RISC architecture; *See also* 1971 ELE; 1988 ELE

Floating Point Systems creates the 120B computer equipped with an array processor; it has great impact on scientific calculating

Engineers from several electronics firms found Zilog Corporation, with the aim of manufacturing microprocessor chips; *See also* 1976 ELE

The McDonald's fast food chain replaces its cardboard, foil, paper wrapping for a Big Mac hamburger with a one-piece polystyrene "clamshell"; *See also* 1988 FOO

A patent issued to Omar Brown but assigned to Ermal Fraze for a "can with an inseparable tear strip" covers the ancestor of the present pull-tab used on beverage cans; *See also* 1973 FOO; 1980 FOO

Ruth Siems of General Foods develops Stove Top Stuffing as an alternative starch for the American diet

Samuel Hurst and coworkers demonstrate that they can identify a single cesium atom in a sample that otherwise consists of 10^{19} argon atoms; *See also* 1974 MAT

Nathaniel C. Wyeth becomes the first senior engineering fellow at the Du Pont Corporation, largely because of his success about this time in developing the plastic soda bottle, made from polyethylene terephthalate (PET); this innovation has since largely come to replace glass bottles for many purposes; *See also* 1888 MAT; 1939 MAT

César Milstein and Georges J.F. Köhler announce in Great Britain their discovery of how to produce a single line of identical antibodies, termed monoclonal antibodies; because they all represent the same lineage or clone of antibodies, each binds with the same chosen protein or antigen; in years to come, monoclonal antibodies find a variety of uses in identifying specific molecules; *See also* 1984 MED

A method of cooling particles called "optical molasses" is proposed; the particle would be placed in the intersection of three laser beams, each designed to resist any motion an atom makes in the direction of a beam; *See also* 1933 MAT; 1985 TOO

Charge-coupled devices (CCD's), invented in the 1960s at Bell Labs, are installed in astronomical telescopes for the first time by NASA's Jet Propulsion Laboratory and the University of Arizona in the US; *See also* 1948 TOO

Roy Clampitt and coworkers in the UK develop the first liquid-metal ion source, which soon is used to produce effective high current focused ion beams; *See also* 1978 TOO

The Soviet Union launches Venera 9 on Jun 8; it returns the first photographs from the surface of Venus; *See also* 1973 TRA

The first cooperative US-Soviet space mission, the Apollo Soyuz Test Project, is launched on Jul 15; a three-man Apollo spacecraft docks with the two-man Soviet *Soyuz 19*; *See also* 1973 TRA

The laser

Albert Einstein in 1917 explained the principle of the laser. He calculated that when an excited molecule is hit by an electromagnetic particle (photon), the molecule will fall to a lower energy level and emit an identical photon moving in the same direction. The net result is two photons, where one existed before, amplifying the signal. Charles Townes used the principle of stimulated emission to amplify microwaves in 1953, inventing the maser (an acronym for Microwave Amplification by Stimulated Emission of Radiation).

Arthur Schawlow proposed using this method for amplification of light in 1958. The first laser (Light Amplification by Stimulated Emission of Radiation) was built by Theodore Maiman two years later.

The laser produces absolutely monochromatic light, that is, light of a single frequency. A second important property of lasers is that the light waves all travel in exactly the same direction and are in step. Light with these properties is called coherent light. One important property of a coherent light beam is that it does not spread when traveling through space. Therefore, a laser can carry a lot of energy into a small spot, heating the target spot by discharging the energy.

Lasers have found a wide range of technical applications. The focusing powers allow use as cutting tools in industry, even cutting through steel. Extremely short pulses of laser light transmit information via fiber-optic cables. Physicians use lasers in surgery, including eye operations and clearing of clogged arteries.

Lasers have also become ubiquitous in our daily life. Every compact disc player incorporates a laser, and a laser beam scans the bar code on packaged goods. Lasers are also used to produce holograms—three dimensional images such as those found on credit cards.

Military engineers have dreamed of lasers as weapons. Because the energy remains bundled in a narrow beam, a powerful laser could destroy an enemy warhead in space. Success in this specific endeavor is lacking, but use of lasers for guiding bullets or rockets is common.

1976

Expert committees in the US and Great Britain introduce guidelines for genetic engineering research, imposing safe laboratory practices; many other countries subsequently introduce similar guidelines; *See also* 1974 GEN; 1978 MED

Genentech, the first commercial company devoted to development of products through genetic engineering, is established in south San Francisco CA; *See also* 1974 GEN; 1980 GEN

The US Congress passes the Toxic Substances Control Act, intended to keep dangerous chemicals out of the environment; *See also* 1957 MAT; 1984 MAT

JVC introduces the VHS (video home system) format for videotape; *See also* 1989 COM

Gay Kildall develops CP/M (Control Program for Microcomputers), an operating system widely used in 8-bit personal computers; *See also* 1977 COM

IBM develops the ink-jet printer; *See also* 1975 COM

Ryszard Michalski [b Kalusz, Poland, May 7 1937], at the University of Illinois, develops AQII, an expert system for the diagnosis of diseases of soybean crops; *See also* 1968 COM

Chuck Peddle develops the 6502 microprocessor, which employs MOS technology and will be used in the Apple II computer; *See also* 1977 ELE

Zilog develops the 8-bit Z80 microprocessor, which can address 64 kilobytes of memory and has a clock speed of 2.5 MHz (increased to 8 MHz in 1983); it becomes widely applied in 8-bit computers; *See also* 1975 ELE; 1981 ELE

A computer chip with 16 kilobits (16,384 bits) of memory becomes commercially available; it will later be used in first IBM personal computer; *See also* 1974 ELE; 1981 ELE

Seymour Cray of Cray Research completes the Cray–1, the first supercomputer with a vectorial architecture; in order to shorten connections between components, he designs the computer in cylindrical form; *See also* 1971 ELE; 1982 ELE

1977

The US signs the Convention of International Trade in Endangered Species (CITES); the CITES organization identifies endangered species and prohibits commerce in them or products based on those species; *See also* 1973 GEN; 1986 GEN

Ann Moore patents the Snugli, a baby carrier that allows the parent to carry a baby while walking or working; *See also* 1872 GEN

Joe Condon and Ken Thomson complete at Bell Labs a dedicated computer for playing chess; the computer contains 325 chips and evaluates 5000 chess positions per second; *See also* 1974 COM

AT&T, various Bell subsidiaries, Western Electric, and ITT begin to experiment with the use of fiber-optic cables in GA, IL, and England; *See also* 1955 COM; 1978 COM

Tandy Corporation and Commodore Business Machines introduce personal computers that have built-in monitors; other available microcomputers rely on separate television sets; both the Tandy and Commodore computers use tape cassettes for data and program storage; *See also* 1960 ELE

Paul Allen and William Gates found Microsoft, which becomes in the next few years the most important producer of software for microcomputers; *See also* 1981 COM

Steven P. Jobs [b 1955] and Stephen Wozniak [b Sunnyvale CA, 1950] introduce the Apple II, the first personal computer available in assembled form; it remains the best-selling computer until the introduction of the IBM personal computer; *See also* 1975 ELE; 1981 ELE

Xerox introduces the Star 8010, an office computer based on the Alto developed a few years earlier by Alan Kay; *See also* 1972 COM

The first linked automatic teller machines (ATMs) are introduced by a Denver-based credit card processor; *See also* 1969 ELE

The first solar-powered water heater is installed; *See also* 1954 ENE; 1978 ENE

Peter E. Glaser proposes the construction of a solar collector with an area of 100 km² (40 sq mi) placed in a geostationary orbit so that it beams the collected energy to Earth in the form of radio waves; such a device could supply about 10 gigawatts of energy; *See also* 1954 ENE; 1981 ENE

Construction of the largest commercial fast breeder reactor, the Superphénix, with an output of 1200 megawatts, is started in France; *See also* 1952 ENE; 1985 ENE

1976

The US General Services Administration opens an environmental demonstration building in Saginaw MI in Oct; in addition to a solar collector in the roof for heating and cooling and double-glazed windows with overhanging roofs, the building features large masses of soil, called Earth berms, piled against walls to reduce heat loss; *See also* 1979 FOO

The stadium and track built for the Olympic Games in Montreal, Canada, in Jul are made from concrete pieces that are glued together and then stressed by applying tension after the pieces have been set in place; *See also* 1977 FOO

The US National Academy of Sciences reports that chlorofluorocarbons, or Freons, used in various spray cans can deplete the ozone layer in the atmosphere, resulting in increased ultraviolet radiation at the surface level of Earth; *See also* 1974 MAT; 1978 MAT

Stephanie L. Kwolek [b New Kensington PA, Jul 31 1923] develops Kevlar, an aramid plastic fiber as strong as steel, used in making tires, bulletproof vests, boat shells, and components for the aerospace industry; *See also* 1930 MAT

The very hard metal Coromant is introduced; it allows rapid tooling of metal parts; *See also* 1883 MAT; 1898 MAT

Erwin Neher and Bert Sakmann develop the patch-clamping technique to study the traffic of ions across cell membranes; by tightly sealing a very thin glass pipette against a cell membrane, one can isolate a small patch of it and study the ion channels it contains

The US Food and Drug Administration (FDA) bans the use of chloroform in drugs and cosmetics because it is suspected of being a carcinogenic substance; *See also* 1831 MED

The American physicist John M.J. Madey builds a free-electron laser (FEL); the photons are amplified in a beam of free electrons generated by an accelerator that passes through an alternating magnetic field; each time the electrons are deflected by the field they emit photons, which in turn stimulate the emission of new photons by the electrons; *See also* 1970 TOO; 1980 TOO

In Apr workers complete the safety bore for what will be the world's longest vehicular tunnel, through the St. Gotthard pass in the Swiss Alps, south of Zurich; the main tunnel is 16.4 km (10.2 mi) long; *See also* 1965 TRA; 1990 TRA

The French-English Concorde becomes on May 24 the first supersonic airliner to operate a regularly scheduled passenger service; *See also* 1968 TRA

US space probes Viking 1 and 2 soft-land on Mars and begin sending back direct pictures and other information from the surface of the planet; on Jun 19 Viking 1 becomes the first spacecraft ever to soft-land on a planet other than Earth; *See also* 1973 TRA; 1977 TRA

1977

The Georges Pompidou National Center for Art and Culture opens in Paris on Jan 31; the building is constructed of concrete slabs enclosed in glass and supported by a network of pipes; *See also* 1987 FOO

In Jun the US Air Force starts building the world's largest structure made entirely from wood that is glued together, a ramp and platform 38 m (125 ft) high intended for use in testing the effects of electromagnetic pulses (EMP) caused by nuclear explosions; metal parts would be affected by EMP and interfere with the tests; *See also* 1976 FOO

Hideki Shirakawa, Alan G. McDiarmid [b Masterton, New Zealand, Apr 14 1927], and Alan J. Heeger [b Sioux City IA, Jan 22 1936] prepare a highly conductive form of polyacetylene by doping it with iodine; such a plastic could replace metals such as copper in electrical applications; *See also* 1972 MAT; 1987 MAT

Researchers in the Soviet Union find a way to make diamond from methane without the graphite impurities; *See also* 1958 MAT; 1981 MAT

The last recorded case of smallpox found in the wild is in Somalia; however, the virus is retained for research purposes in several laboratories; *See also* 1796 MED

Andreas R. Gruentzig [b Dresden, Germany, Jun 25 1939, d Oct 27 1985] invents balloon angioplasty, a method for unclogging diseased arteries; *See also* 1967 MED; 1985 MED

Raymond V. Damadian builds his first medical magnetic resonance imager (MRI) based on nuclear magnetic resonance spectroscopy techniques; *See also* 1973 MED

Paul MacCready [b New Haven CT, Sep 29 1925] flies the human-powered airplane the *Gossamer Condor* and wins the 50,000-pound Kremer Prize; *See also* 1959 TRA; 1979 TRA

Space probes Voyager 1 and 2 are launched on a journey to Jupiter and the outer planets; *See also* 1976 TRA; 1985 TRA

The world's largest steel-arched bridge, 924 m (3030 ft) long with a main span of 518 m (1700 ft), crosses the New River Gorge in WV

1978

Apple brings out the first disk drive for use with personal computers; *See also* 1970 COM

Olivetti of Italy and Casio of Japan introduce electronic typewriters equipped with memory for storing text; *See also* 1974 COM

John Barnaby develops Wordstar for the CP/M operating system; marketed by MicroPro and adapted for DOS, it becomes the most widely used word processing package in the early 1980s; *See also* 1975 COM; 1979 COM

Texas Instruments introduces the TMC 0280 speech synthesizer chip; together with the TMC 0350 read-only memory chip of 131,072 bits and the THC 0270, the system can pronounce 165 words; it forms the heart of the Speak & Spell learning aid for children; *See also* 1975 COM; 1980 COM

The Nippon Electricity Company (NEC) introduces the Voice Data Input Terminal; it can recognize 120 words spoken in groups of up to five words; words are compared to reference patterns stored in memory; the processing time per word is 300 milliseconds; *See also* 1975 COM; 1980 COM

The first telephone link via optical fiber in Europe is established between Martlesham and Ipswich in England; *See also* 1977 COM; 1985 COM

DEC introduces a 32-bit computer with virtual address extension (VAX); it can run larger programs than its semiconductor memory can hold; VAX machines become an industry standard for scientific and technical applications, using an operating system called VMS; *See also* 1968 ELE

Intel introduces its first 16-bit processor, the 8086; IBM uses a slightly simplified version, the 8088, for the central processor unit in its first marketed personal computer, the PC; *See also* 1972 ELE; 1981 ELE

To celebrate renewable resources, 25 million people around the world celebrate Sun Day; *See also* 1970 GEN; 1981 ENE

Composites

Traditionally, scientists have classified materials in three classes: metals, plastics, and ceramics. Metals are tough, used mainly for their strength and stiffness; plastics are light and cheap to manufacture; and ceramics are hard, good electrical insulators, and resistant to heat and to corrosion. In recent years, scientists have created materials that do not fit the traditional classification and that have entirely new properties. High-temperature superconductors are striking examples: They are ceramics that are insulators at room temperature but become superconducting at low temperatures.

Composite materials are combinations of materials from different classes that have properties different from or better than either of their parents. The idea is not new. When the Israelites in Egypt made bricks from clay and straw, the product was a composite.

The composite of glass fibers and resin, popular for construction of boats and automobile bodies, is commonly called fiberglass. This material combines the low weight and corrosion resistance of plastic with the tensile strength of glass fibers. Similar composites of carbon fiber and resin have the toughness of steel but are four times lighter; these composites are now widely used in aircraft construction and such specialized applications as fishing rods.

Although not usually thought of as composites, the essential properties of semiconductor devices result from inserting small amounts of one kind of atom into bulk materials of a different atom, a process called doping. The modern electronics industry is based almost entirely on silicon doped with such atoms as arsenic or boron.

In Feb the Illinois Regional Library for the Blind and Physically Handicapped and the Community Library in Chicago, a building designed by Stanley Tigerman, opens; it is laid out for ease in memorization—helpful to the blind—and for travel in wheelchairs

The world's largest commercial trade area, the Dallas Market Center in TX, is created by adding cantilevered precast concrete pads and lighter curtain walls to extend the building to cover 67 hectares (165 acres); *See also* 1918 FOO

Chlorofluorocarbons (Freons) are banned as spray propellants in the US on grounds that they damage the ozone layer in the atmosphere; the ozone layer screens out most of the harmful ultraviolet radiation from the Sun; *See also* 1976 MAT; 1990 MAT

The US government institutes the first steps toward banning lead from gasoline, mainly to protect the platinum catalysts in catalytic converters in autombile exhausts, not to protect the environment; *See also* 1921 ENE; 1984 MED

L. E. Lyons, Hugh McDiarmid, and Neville Mott [b Leeds, England, Sep 30 1905] experiment with organic semiconductors, consisting of carbon compounds doped with oxygen; *See also* 1970 ELE; 1988 MAT

The first human baby conceived outside the body—Louise Joy Brown, called the test-tube baby—is born to Lesley Brown in Great Britain; in a technique developed by Robert G. Edwards and Patrick C. Steptoe, the ovule is extracted from the mother to be, inseminated in a petri dish, and then reimplanted; *See also* 1969 MED

David Botstein [b Zurich, Switzerland, Sep 8 1942], Ronald W. Davis [b Marda IL, Jul 17 1941], and Mark H. Skolnick propose at a conference in Apr that DNA sequencing be used to develop gene markers for various genetic diseases; *See also* 1974 GEN; 1986 MED

About this time, the diamond-studded saw for cutting stone is introduced at marble quarries, replacing helicoidal-wire cutters; *See also* 1895 TOO

Robert L. Seliger and coworkers at Hughes Research Laboratory develop the first scanning ion microscope with a liquid-metal ion source; *See also* 1959 TOO; 1975 TOO

The US government agrees to turn the Panama Canal and the Canal Zone over to Panama on Dec 31 1999 under treaties that go into force in 1979; *See also* 1914 TRA

In Apr the world's longest floating bridge, 1850 m (6074 ft), opens across the Demerara River in Guyana

---

The post-industrial society

During the Industrial Revolution of the 18th and 19th centuries, the economic base changed from agriculture and food production to manufacturing. Manufacture of goods became the main source of wealth, making engineers and inventors key contributors.

A similar change begin around the middle of the 20th century. In many economically advanced countries, services started to surpass manufacturing as the main source of income. This change gave rise to a new form of society, which in the early 1960s American sociologist Daniel Bell named the post-industrial society. In such a society wealth is generated by services instead of production of goods.

Bell identified five characteristics of a post-industrial society: 1) The majority of the labor force is engaged in the supply of services and not in manufacturing and food production. 2) Society requires an increased number of professionals and people with technical qualifications, especially in science and engineering. 3) Theoretical knowledge of science and technology becomes the basis for economic power and social policy, apparent in the development of high technology. 4) Technology and modeling become important in planning for the future. For example, technology assessment becomes important in studying the influence of new energy sources on the environment. 5) Intellectual technology takes over intellectual functions such as decision making. The widespread proliferation of computers and information networks are examples of the new form of intellectual technology described by Bell.

These developments decreased the importance of the factory worker and the clerical worker, whose tasks were taken over by computers. The advent of small, powerful computers in the 1970s began to make the US into the society Bell had described.

1979

Philips and Sony bring the videodisk to market; images and sound are recorded digitally as tiny pits in the surface of the disk; *See also* 1982 COM; 1989 COM

The first commercial network of cellular telephones is set up in Tokyo, Japan; later in the year Bell Labs tests a cellular system on 2000 users in Chicago IL; *See also* 1946 COM; 1983 COM

The International Maritime Satellite Organization (Inmarsat), providing communication and navigation services via satellite to ships, is founded; *See also* 1973 COM

Daniel Bricklin [b 1951], a computer scientist and accountant, and programmer Robert Frankston develop VisiCalc (Visible Calculator), the first spreadsheet program for the microcomputer; VisiCalc enables personal computer users to develop business applications without learning a programming language; by 1985 800,000 copies are sold; *See also* 1978 COM; 1980 COM

The High Order Language Working Group (HOLWG), established by the Pentagon, studies the possibility of a standard language that would replace the thousand or so languages used by the US Defense Department; Jean Ichbiah and coworkers develop ADA (named for Lady Ada Lovelace, the first programmer) for this purpose

P.G. LeComber and coworkers from the University of Dundee show that amorphous silicon can be used to build thin-film transistors that control liquid-crystal displays; *See also* 1975 COM; 1985 COM

The first digital telephone exchange in Great Britain, at Glenkindie, near Aberdeen, becomes operational; the last electromechanical exchange in Great Britain closes in 1990; *See also* 1912 COM

Control Data introduces its Cyber 203 supercomputer; *See also* 1976 ELE

Motorola introduces the 68000 microprocessor chip; it has a 24-bit capacity for reading memory and can address 16 megabytes of memory; it will be the basis of the Macintosh computer developed by Apple; *See also* 1984 ELE

Steven Hofstein invents the field-effect transistor, using metal oxide technology (MOSFET); *See also* 1970 ELE

The nuclear reactor at Unit 2 of Three Mile Island in PA loses its water buffer and undergoes a partial meltdown through operator error on Mar 28; a small amount of radioactive material escapes the containment dome, but the reactor itself is irreparably damaged; no one is injured; *See also* 1957 ENE; 1986 ENE

Sharp price rises in petroleum by OPEC (the Organization of Petroleum Exporting Countries) and a revolution in oil-producing Iran (followed by an embargo starting Nov 12 as a result of Iran's seizure of the US embassy in Teheran) result in gasoline prices in the US rising above $1 a gallon and in scattered shortages; *See also* 1973 ENE; 1985 ENE

Perpetual motion (part 3): "It keeps on going. . ."

The law of conservation of energy offers the strongest argument against the idea of perpetual motion machines. German mathematician Gottfried Leibniz, in his *Essay on dynamics* (1692), first stated clearly that energy could not be created out of nothing. In 1847, before the Academy of Sciences in Berlin, German physicist Hermann von Helmholtz presented his famous law of the conservation of energy: The quantity of energy in a system remains constant; only the form of energy can change. Thus, you can change thermal energy into mechanical energy, but you cannot obtain a net gain in energy.

Nevertheless, physicists have not succeeded in convincing everybody that perpetual motion is impossible. As recently as 1979, American inventor Joseph Newman filed a patent for an energy machine, a motor that produces more energy than the electrical energy it consumes.

Sometimes other devices are put forward as akin to perpetual motion. The type of nuclear reactor called a breeder is said to produce more fuel than it consumes, providing useful energy at the same time. Ordinary nuclear reactors are quite wasteful, losing much energy in unproductive nuclear reactions. A breeder reactor collects the energy that would be otherwise lost and converts it into fuel. This is not really a perpetual source of energy, just a more efficient one.

Work begins on the Rogun Dam in Tadzhikstan, Russia, near the borders of Russia, Afghanistan, and China; at 335 m (1098 ft), it is the highest dam in the world; *See also* 1936 FOO; 1983 FOO

The 21-story Chevron Plaza Building is started in New Orleans LA; it is designed from precast reinforced concrete with a main objective of saving energy in its heating, ventilating, and air-conditioning systems; *See also* 1976 FOO; 1987 FOO

Work begins on the 44-story apartment tower over New York City's Museum of Modern Art; it becomes one of the world's tallest buildings constructed from cast-in-place concrete

Human insulin is synthesized by genetic engineering methods; *See also* 1953 MED; 1980 MED

In Mar Herbert Needleman releases an influential study that shows that low levels of lead in children's blood and teeth correlate with lower IQ test scores, suggesting the low-level lead poisoning can reduce intelligence; *See also* 1978 MAT; 1980 MED

Allan MacLeod Cormack of South Africa and the US and Godfrey N. Hounsfield of England win the Nobel Prize in physiology or medicine for their invention of computed axial tomography; *See also* 1972 MED

The first of the European Space Agency's Ariane rockets successfully lifts off on Dec 24 from the Guiana Space Center in Kourou, French Guiana; the rocket is designed to launch satellites into orbit for various customers, including the space agency and commercial users; *See also* 1985 TRA

Paul MacCready's human-powered airplane the *Gossamer Albatross* flies across the English Channel; *See also* 1977 TRA; 1988 TRA

Nuclear power

After World War II, governments instituted a search for peaceful uses for nuclear energy. Comic books on "the atom" informed children that the atomic energy in a steamship ticket was enough to circle the globe hundreds of times, raising the expectations of a generation. Scientists did design nuclear reactors that some expected would become sources of unlimited power. According to Lewis Strauss, the chairman of the Atomic Energy Commission (AEC) in 1954, electricity from nuclear power would be "too cheap to meter."

The initial use of nuclear reactors was not peaceful, however. The first energy-producing reactors powered nuclear submarines; the first was launched by the US in Jan 1954.

The Soviet Union completed in Jun 1954 a small reactor that delivered the first peaceful application of nuclear energy, but not in practical amounts. In the US, experimental and submarine reactors were connected to the electric power grid in Jun 1955. In England a 50,000-kilowatt reactor, known as Calder Hall, started operation in Oct 1956.

Since then, nuclear reactors have been built in many countries. In some, such as Belgium and France, nuclear power is now the main source of energy. This development made these countries independent from the use of imported oil, supplies of which can be disturbed by political crises.

Enthusiasts have heralded nuclear energy as ecologically sound because it does not produce carbon dioxide that causes global warming. However, the partial meltdown of a nuclear reactor at the Three Mile Island power plant near Harrisburg PA in 1979 and the reactor explosion in Chernobyl (Ukraine) in 1986 showed that nuclear energy carries with it the potential for disaster. A second problem is cost; despite predictions, nuclear energy is more expensive to generate than energy from fossil fuel. A third major problem is disposal of radioactive waste fuel.

During the 1970s, engineers began to design nuclear reactors to incorporate passive safety systems. Such systems use cooling methods that rely on gravity alone and do not require pumps that can break down. In one design, reactor vessels are submerged in tanks of boron-laced water, which floods the reactor immediately in the case of a problem, stopping the nuclear reactions.

1980 Martin Cline and coworkers transfer genes from one mouse to another and succeed in having the genes function in the new organism; *See also* 1976 GEN; 1981 GEN

France launches Minitel, an experimental telephone inquiry system consisting of terminals placed in homes; the system will be established throughout France and offer a large number of on-line services; *See also* 1973 COM

Votrax develops the SC–01 single-chip voice synthesizer with unlimited vocabulary; a separate chip stores phonemes as six-bit words, and a special circuit selects and connects the phonemes, creating words; *See also* 1978 COM; 1986 COM

IBM achieves 91 percent accuracy with a voice recognition system using an IBM System/370 Model 168 computer; the system has a 1000-word vocabulary, can process the words at a normal speaking pace, and can display the spoken words on a screen; *See also* 1978 COM; 1982 COM

Wayne Ratliff develops dBase II, a database software package that includes a programming language derived from Vulcan I, a database program developed for mainframes at the Jet Propulsion Laboratories in the late 1960s; the package and later versions of it become the principal electronic filing system for personal computers; *See also* 1979 COM; 1981 COM

Hans Dehmelt and scientists at the University of Heidelberg in Germany succeed in photographing a single barium atom by itself; they name the atom Astrid; *See also* 1988 MAT

The current international fax standard (Group 3) is set; it allows facsimile messages to be transmitted at about one page per minute or faster; *See also* 1974 COM; 1990 COM

IBM introduces the 5120 microcomputer, a business-oriented computer; it is not commercially successful; *See also* 1975 ELE

The US Supreme Court rules that a microbe developed by General Electric for oil cleanup can be patented; *See also* 1790 GEN; 1988 MED

Alternative energy sources

Many scientists believe that Earth's atmosphere is warming as a result of carbon dioxide produced by fossil-fuel power stations and cars. The major alternative to fossil fuel, nuclear power, also has drawbacks. Researchers and engineers have been looking at alternative energy sources that do not tax the environment or deplete natural fuel reserves.

Waterpower, since the introduction of efficient turbines during the last century, is still the most important alternative energy source. In some countries, hydraulic power contributes a sizable part of the electricity supply: in Canada about 65 percent and in Italy 25 percent.

Wind power, also one of the oldest energy sources, is undergoing a strong revival since the 1980s. In the Netherlands the sleek rotating blades of modern wind turbines are now as much a mark of the landscape as the traditional windmills. The largest wind turbines can generate more than 2 megawatts of power. More commonly, smaller wind turbines producing about 50 to 100 kilowatts each are grouped in so-called wind farms. Wind supplies a much smaller fraction of electrical energy than hydroelectric power. Even in Denmark, where nuclear power is outlawed, wind power supplies only 2.5 percent of all electrical energy.

Solar energy also supplies a small fraction of the world's total energy demand. There are two main ways to collect solar energy.

The most common solar method since the 1970s has been to let the Sun heat water in tubes mounted in special panels on roofs. During the winter, water from these solar collectors heats the building and supplies hot water. In some versions, daytime heat is stored in a large, insulated tank of water from which it is recovered at night.

The other solar energy method, the use of photovoltaic cells, converts solar energy directly into electricity. During the late 1980s and early 1990s, scientists developed solar cells that convert more than 35 percent of sunlight falling on them into electricity. Solar cells are still too expensive for large solar power stations, but they are now used to supply energy to devices at remote sites.

1980

The City of David Archaeological Project rediscovers Warren's Shaft, a vertical well that formed part of the waterworks for Jerusalem from before David's time; cleaned out, the shaft provides access to the entire waterworks; *See also* 1867 ARC

About this time the present pull-tab used on beverage cans is introduced, based on a version of the pull-tab first produced by Coors Beer; *See also* 1975 FOO

Dornier Medical Systems of Munich, W Germany, develops the lithotripter, a machine that uses sound waves to break up kidney stones while the stones are still in the kidney

The successful production of human interferon in bacteria is announced by Charles Weissmann of the University of Geneva; *See also* 1957 MED; 1988 MED

The pharmaceutical firm Eli Lilly starts the production and testing of human insulin, using genetically altered bacteria; the insulin can be used by diabetics that are allergic to insulin obtained from animals; *See also* 1979 MED; 1982 MED

Scientists at the New York Blood Center develop a successful experimental vaccine against hepatitis B; *See also* 1981 MED

The US government bans paints containing lead; *See also* 1979 MED; 1984 MED

Hughes Aircraft Corporation introduces a cutting machine for textiles using a laser beam; *See also* 1970 TOO

A miniature Nd:YAG rod laser pumped by a diode laser is developed; *See also* 1976 TOO; 1984 TOO

The space shuttle

During the 1960s, the US National Aeronautics and Space Administration (NASA) launched large numbers of space vehicles. It seemed like a good idea to replace expendable, expensive rockets with a reusable space booster. After a study of several designs, NASA in 1972 decided to construct a Space Transportation System (STS) based on the space shuttle. The shuttle takes off vertically, orbits Earth a number of times, and returns as a glider, supported by stubby wings. A huge external fuel tank is jettisoned after takeoff and burns up when reentering the atmosphere, but two solid-state booster rockets descend on parachutes and are reusable.

Comparable in size to a DC–9 aircraft, the space shuttle carries payloads of up to 29,500 kg (65,000 lb) in its huge 18.3-m (60-ft) long cargo bay. A 15.25-m (50-ft) mechanical arm is used to manipulate the payload. With this arm, astronauts lift satellites out of the cargo bay and place them into near-Earth orbit.

A total of five space shuttles were built. The first, *Columbia,* began with a two-day mission, making a perfect landing in 1981. This flight marked the start of a series of scientific and military shuttle missions. Shuttle crews placed many satellites in orbit, conducted experiments, and retrieved several satellites for repair. Flights were temporarily halted after *Challenger* exploded shortly after takeoff in 1986, killing the crew of seven people. Although flights resumed on Sep 29 1988, it was apparent that the use of shuttles with human crews was an expensive and dangerous way to place satellites in orbit. NASA once again turned to expendable rockets for routine launches.

Still, there are missions that only humans can do, such as capturing and correcting defects in existing satellites. Space shuttles are also expected to serve in the construction of the US space station *Freedom,* if the station survives budget battles and is ever built.

1981

At Ohio University in Athens OH, scientists transfer genes from other organisms into mice; *See also* 1980 GEN; 1982 GEN

Chinese scientists produce a genetic copy (clone) of a zebra fish; *See also* 1980 GEN; 1982 GEN

MS–DOS (Microsoft Disk Operating System, first called PC–DOS), developed by MicroSoft is adopted by IBM for its PC (personal computer); *See also* 1977 COM

Michell Kapor [b Brooklyn NY, Nov 1 1950] develops Lotus 1–2–3, which adds graphics such as pie charts or bar graphs to a spreadsheet, for use on the IBM PC and compatibles; it is commercially successful; *See also* 1979 COM

A proposed set of technical standards for compact disc recording and playback established in 1980 by Sony, Philips, and Poly-Gram is accepted worldwide, paving the way for introduction of compact discs commercially; *See also* 1972 COM; 1982 COM

The IBM Personal Computer (the PC), using what is to become an industry-standard disk operating system (DOS), is introduced; *See also* 1975 ELE; 1983 ELE

Osborne builds the first portable computer in which disk drives, video monitor, and processor unit are mounted in a single box; it is about the size and weight of a fully packed suitcase

Clive Sinclair develops a minimalist, cheap computer, the ZX81, which has to be connected to a television receiver to be used; more than a million units are sold in ensuing years; *See also* 1976 ELE

Computer chips produced by the Japanese with 64 kilobits (65,536 bits) of memory capture the world market for computer memory; *See also* 1976 ELE; 1984 ELE

Solarplant One in CA, the world's largest solar power station to this time, generating up to 10 megawatts of electricity, is completed; *See also* 1978 ENE; 1988 ENE

A solar furnace, the Eurelios, is completed in Sicily by the European Economic Community (EEC); *See also* 1949 ENE; 1988 ENE

Adam Heller [b Cluj, Romania, Jun 25 1933], Barry Miller [b Passaic NJ, Jan 22 1933], and Ferdinand A. Thiel announce a liquid junction cell that converts 11.5 percent of solar energy into electricity; *See also* 1977 ENE; 1983 ENE

The solar-powered airplane *Solar Challenger* makes its maiden flight in summer; *See also* 1954 ENE; 1979 TRA

1982

Applied Biosystems begins to market a version of an automated gene sequencer (based on the work of Leroy Hood [b Missoula MT, Oct 10 1938]) that can sequence 18,000 uncorrected bases a day, compared with perhaps several hundred bases a year by hand in the 1970s; *See also* 1976 GEN; 1984 GEN

BELLE, Joe Condon and Ken Thomson's dedicated computer for playing chess, contains 1700 chips and evaluates 160,000 positions per second; it rates second in speed chess at the 1982 US National Open; *See also* 1977 GEN; 1989 GEN

In Oct compact disc (CD) players are introduced by CBS/Sony and Philips; the CD is a 120-mm-diameter plastic disk that uses tiny pits read by a laser to reproduce sound or other information; the first CD is *52nd Street*, an album by composer-singer Billy Joel; *See also* 1981 COM; 1984 COM

The Postscript system of desktop publishing is introduced; *See also* 1975 COM; 1985 COM

Martine Kempf develops voice recognition software for use on an Apple computer; this software leads to the Katalvox, a device for operating voice-activated wheelchairs and magnifying devices used in microsurgery; *See also* 1980 COM; 1983 COM

Columbia Data Products announces, only 10 months after the introduction of the IBM PC, the first computer based on the IBM PC that can run programs designed for the IBM machine; such copies are nicknamed clones or PC-compatible computers; *See also* 1981 ELE

Compaq introduces its clone of the IBM Personal Computer (PC); later in the year, Compaq introduces its Portable, the first IBM–PC clone that is portable; *See also* 1981 ELE

Japan starts a nationally funded program to develop fifth-generation computers; they are based on artificial intelligence and use the Prolog language; *See also* 1991 COM

Steve Chen completes the Cray X–MP (extended multiprocessor), which consists of two Cray–1 computers linked in a parallel architecture; it is three times as fast as the Cray–1; *See also* 1976 ELE; 1985 ELE

1981

Aspartame, an artificial sweetener, is introduced in the US; *See also* 1965 FOO; 1983 FOO

Soviet diamond researchers produce both single-crystal diamond films on previously existing diamonds and multiple-crystal diamonds on metal; single-crystal diamonds are much more desirable for many purposes; *See also* 1977 MAT; 1990 MAT

The 3M Company of St. Paul MN introduces a method for creating optical disks that can be written on by lasers; they develop a material that melts at 150 °C (302 ° F) without undergoing chemical changes; *See also* 1979 COM

The genetic code for the hepatitis B surface antigen is found, opening up the possibility of a bioengineered vaccine; *See also* 1980 MED; 1983 MED

Ruth and Victor Nussenzweig [b Sao Paulo, Brazil, Nov 2 1928] of New York University apply for a patent on a malaria vaccine; subsequent trials show that the vaccine is ineffective; *See also* 1934 MED

Gerd Binnig and Heinrich Rohrer develop the scanning tunneling microscope; the surface of a specimen is scanned by measuring a current between a very small tip and the specimen; individual atoms can thus be detected; *See also* 1978 TOO; 1985 TOO

On Apr 12 US astronauts John W. Young and Robert L. Crippen begin the first mission of the first reusable space shuttle, *Columbia*; the mission lasts 2 days, 8 hours and marks the first landing of a US spacecraft on land; *See also* 1975 TRA; 1985 TRA

The world's longest suspension bridge opens over the Humber estuary in Great Britain; it is 1410 m (4626 ft) long; *See also* 1964 TRA

The first regular passenger line of the high-speed Train à grande vitesse (TGV) opens in France; the same year the TGV reaches a record speed of 380 km (236 mi) per hour; *See also* 1964 TRA

1982

A team of doctors led by William DeVries [b Brooklyn NY, Dec 19 1943] implants the first Jarvik 7 artificial heart on Dec 2; the patient, Barney Clark, lives 112 days; *See also* 1969 MED; 1990 MED

The Swedish firm Kabivitrum produces synthetic growth hormone using genetically engineered bacteria; *See also* 1985 MED

The US Food and Drug Administration (FDA) grants approval to Eli Lilly & Company to market human insulin produced by bacteria; *See also* 1980 MED

Swedish physicians attempt to cure Parkinson's disease with dopamine generated by the patient's own adrenal glands; however, implanting in the brain produces temporary or no improvement; *See also* 1987 MED

1983

In Apr Kary Mullis [b Lenoir NC, Dec 28 1944] conceives of the polymerase chain reaction (PCR), a method of multiplying copies of parts of a DNA molecule that will become the basis of genetic fingerprinting and one of the key tools for all sorts of work with genetics; *See also* 1985 GEN

On Oct 20 an international agreement goes into effect that defines the meter as the distance that light travels through a vacuum in 1/299,792,458 second; thus, the speed of light is *exactly* 299,792,458 m per second; *See also* 1967 GEN; 1990 GEN

The Kurzweil company introduces a portable, digital, music keyboard that can store the sounds of 30 different musical instruments; *See also* 1984 GEN

US president Ronald Reagan [b Tampico IL, Feb 6 1911] announces the start of the Strategic Defense Initiative (SDI), soon known as the Star Wars program; it consists of a satellite-based system for intercepting intercontinental ballistic missiles using interceptor missiles and laser beams; program research follows, but not actual defenses; *See also* 1944 TRA

Carl Sagan [b New York City, Nov 9 1934] and an interdisciplinary group of scientists publish a report on the consequences of nuclear war, showing that even a limited nuclear war would have grave consequences; smoke and dust injected into the atmosphere would result in severe climatic changes, termed as nuclear winter, that would destroy most living creatures

The first regular US cellular telephone system goes into operation; by 1987 there are 312 cellular systems operating in 205 cities; *See also* 1979 COM; 1992 COM

Apple's Lisa brings the mouse and pull-down menus to the personal computer; a computer mouse is a device that propels the cursor on the screen by moving it over a hard surface; pressing a button on the mouse sends a command to the computer, depending upon which icon on the screen the cursor is located; although Lisa is too expensive and clumsy to be commercially successful, it leads to the popular Macintosh, which has the same features; *See also* 1968 COM

Intel introduces the Votan 6000 voice recognition unit, consisting of a transaction board, a chip set, and a development system; *See also* 1982 COM; 1988 COM

Fred Cohen coins the term *computer virus* to describe programs that can insert copies of themselves into other programs; he is also the first to write such a program as part of his research in computer security; *See also* 1988 COM

IBM's PC–XT, introduced in Feb, is the first personal computer with a hard disk drive built into the computer; the hard disk is a memory device capable of storing 10 megabytes of information even when the machine is turned off, thus replacing many floppy diskettes; the computer is supplied with an updated disk operating system, DOS 2.0, which allows the creation of subdirectories and the storage of an unrestricted number of files; *See also* 1981 ELE; 1984 ELE

IBM introduces the PC Junior, a scaled-down version of the IBM PC; it is a commercial failure; *See also* 1981 ELE

Immos, a British firm, develops the transputer, a parallel computer in which several processors work simultaneously on a part of a problem, thus speeding up processing considerably; *See also* 1965 ELE; 1985 ELE

Thinking Machines Corporation is founded to build massively parallel computers; *See also* 1965 ELE; 1985 ELE

Intel introduces the 8080 8-bit microprocessor chip; it replaces the 8008 and is more efficient; *See also* 1972 ELE

M. Gurvitch, M.A. Washington, and H.A. Huggins report the development of a Josephson junction suitable for large-scale integrated circuits; it consists of two electrodes of niobium separated by an aluminum-oxide insulating barrier; *See also* 1988 ELE

A team of German and American scientists develops a "wet" solar cell that has an energy conversion efficiency of 9.5 percent; *See also* 1981 ENE; 1989 ENE

The world's largest thin-arch dam—209 m (685 ft) high and 1265 m (4150 ft) long—spans the American River near Auburn CA; *See also* 1979 FOO

Aspartame is approved for use as an artificial sweetener in soft drinks; *See also* 1981 FOO

Kary Mullis

Kary Mullis seems to be one of those California types, Berkeley variety. With a PhD in biochemistry from Berkeley, he drifted for 7 years from job to job, not always directly connected with biochemistry. In 1979, however, he began a close relationship with one of the new California biotechnology giants, the Cetus Corporation. In Mullis's original job, he had to use a small fragment of DNA, called a probe, to locate a matching fragment among millions of DNA pieces in a solution. Although hired by Cetus as a lowly technician, Mullis found ways to automate his work and soon was running the laboratory where he had been assigned. The lab specialized in developing probes to recognize specific parts of a DNA molecule. These probes are used to locate genes, which are, of course, parts of a DNA molecule.

In Apr 1983, while driving up the California coast to his weekend cabin, Kary Mullis thought of a way to make his job much simpler still. Instead of using a probe to pick out a single strand of DNA, he could use the probe to make multiple copies of any part of a DNA molecule. He was so startled by the idea that he had to pull off the road to think about it. When he got back to Cetus to test the idea, it worked on the first try.

Still, it took 4 years after that to work out the details. The new process, called PCR for polymerase chain reaction, quickly became applied to sensitive genetic tests, including the process known as DNA fingerprinting. It also has been used to study ancient forms of life from bits of DNA found in fossils. PCR is essential for the Human Genome Project,

The immunosuppressant cyclosporine is approved by the US Food and Drug Administration, making transplants of organs much safer than they previously had been; *See also* 1953 MED; 1984 MED

Scientists show that a protein produced by genetically engineered yeast can protect chimpanzees against hepatitis B; *See also* 1981 MED;1984 MED

John Buster and Maria Bustillo of the Harbor-UCLA Medical Center in Torrance CA perform the first successful human embryo transfers; *See also* 1950 FOO

Fernand Daffos is the first doctor to use fetal blood taken by a needle through the umbilical cord for diagnosis of disease in the fetus; *See also* 1952 MED; 1984 MED

The US Orphan Drug Act is passed, granting manufacturers of drugs designed to treat diseases with fewer than 200,000 victims an exclusive seven-year marketing period and other incentives

The first five-person crew, consisting of Sally K. Ride, the first US woman in space, and US astronauts Robert L. Crippen, Frederick H. Hauck, John M. Fabian, and Norman E. Thagard, set out on a *Challenger* mission on Jun 18; the Remote Manipulating Structure ("Arm") is first used to deploy and retrieve a satellite; *See also* 1963 TRA 1984 TRA

1983

Kary Mullis (Continued)

the effort to locate all of the parts that make up human DNA. Even more quickly than genetic engineering and monoclonal antibodies have done, PCR has become one of the pillars of the revolution in biotechnology.

PCR works in part by manipulating the copying process built into DNA. The signals for starting and stopping reproduction of a gene or another line of DNA code were discovered while deciphering the entire genetic code in the 1960s. In PCR such a start code is attached to the part of a strand of the DNA double helix for the line of code that is to be duplicated. The modified line is put into a solution where it can react with a good supply of the four bases (chemicals often known by their initials—A, C, G, and T) that make up the code. An enzyme, DNA polymerase, that causes the line of code to be copied is also added to the solution. The start code is copied along with the rest of the line. Thus, as soon as the first copy is made, there are two copies in the solution, each of which has the instructions and ability to copy itself. Thus four copies result, each with the ability to double by copying. After ten such passes, there are over a thousand copies and after twenty passes there are over a million copies. PCR is also known as gene amplification.

The whole process requires warm temperatures to separate the strands of the helix and cooler ones for the actual copying process. In the version of PCR that Mullis developed, the DNA polymerase was degraded by the heat, which meant that more enzyme had to be added after each copying cycle. In 1987 a new version of the DNA polymerase enzyme was found. This enzyme is used by bacteria that normally live in hot springs in Yellowstone National Park. With the bacterial enzyme surviving the process, it was possible to develop a machine that automated the entire chemistry.

1984 Robert Sinsheimer, chancellor of the University of California at Santa Cruz, proposes mapping all of the genes in a human being, a proposal that leads to the Human Genome Project; *See also* 1982 GEN; 1989 GEN

MIDI (the Musical Instrument Digital Interface), a set of standards for interfacing digital musical synthesizers and computers, is introduced; *See also* 1983 GEN

Philips and Sony introduce the CD–ROM (compact disk read-only memory), an optical disk that can store very large amounts of digital data; *See also* 1982 COM

Bill Atkinson of Apple develops MacPaint, a revolutionary graphics package that allows the drawing of pictures on screen; *See also* 1981 COM

A study at the Technical University of Berlin offers the first discussion of computer viruses, including Friday 13, Holland Girl, Trojan Horse, and Christmas Tree; in subsequent years computer viruses become a serious problem, spreading through exchanged diskettes and computer networks; *See also* 1983 COM; 1988 COM

The largest monopoly in the US, the AT&T telephone operation, is broken by the courts into a smaller AT&T (handling only long distance) that retains Bell Labs and seven regional companies that share a new research division, Bellcore; the judge's order permits companies besides AT&T to compete for long-distance service

Digital Productions produces scenes of battles in space created entirely by computer for the film *The Last Starfighter*; to create the 25 minutes of space images, a 400-megaflop supercomputer runs day and night for a month, producing images on a raster of 24 million pixels

Apple introduces the Macintosh, a graphics-based microcomputer that uses icons, a mouse, and an intuitive user interface derived from the Lisa computer; *See also* 1979 ELE; 1987 ELE

In Aug, IBM's PC AT (advanced technology) computer, designed around the 16-bit Intel 80286 processor chip and running at 6 MHz, becomes the first personal computer to use a new chip to expand speed and memory in an existing personal computer architecture; *See also* 1983 ELE; 1987 ELE

Motorola introduces the 68020 version of the 68000 series of microprocessors; it has 32-bit reading and processing capacity; *See also* 1979 ELE; 1990 ELE

Computer chips are manufactured by NEC in Japan with 256 kilobits (262,144 bits) of computer memory; similar chips are manufactured in the US a year later; *See also* 1981 ELE; 1987 ELE

IBM introduces a megabit RAM (random access memory) chip with four times the memory of earlier chips; *See also* 1981 ELE; 1987 ELE

1984

Eric Block and Saleem Ahmad at the State University of NY at Albany synthesize ajoene, a compound found in garlic and believed to function as a blood thinner

More than 2000 die and thousands more are injured by a toxic gas released during an industrial accident at the US-owned Union Carbide plant in Bhopal, India; *See also* 1976 GEN

Surgeon William H. Clewall of the University of Colorado Health Sciences Center at Denver performs the first successful surgery on a fetus before birth; *See also* 1990 MED

The first clinical trials of a vaccine against hepatitis B start on Jun 1; the vaccine is produced by yeast that has been given genes for a surface molecule of the hepatitis virus; *See also* 1983 MED; 1986 MED

Workers at the New York State Department of Health develop a genetically modified vaccinia virus that protects animals against hepatitis B, herpes simplex, and influenza; *See also* 1796 MED; 1983 MED

Japanese scientists discover a immune-system suppresser produced by a fungus, which they name FK506; its results are similar to those of cyclosporine, but it acts by a different mechanism; *See also* 1983 MED

The US government takes steps toward a complete ban on leaded gasoline in order to reduce lead concentrations in the lower atmosphere; *See also* 1978 MAT; 1986 MED

César Milstein of Great Britain, Georges J.F. Köhler of West Germany, and Niels K. Jerne of Denmark share the Nobel Prize in physiology or medicine, Milstein and Köhler for their research on monoclonal antibodies and Jerne for his studies of the immune system; *See also* 1975 MED; 1986 MED

Dennis L. Matthews builds the first successful X-ray laser; the X rays are produced in a plasma created by the vaporization of selenium by a laser; collisions between selenium atoms and free electrons excite the selenium atoms to sufficiently high levels so that they emit soft X rays; *See also* 1980 TOO; 1985 TOO

Jean-Marc Halbout and Daniel Grischkowsky at IBM develop a laser that creates pulses of 12 femtoseconds (12 quadrillionths of a second); *See also* 1980 TOO

Three Soviet cosmonauts begin the Soyuz T 12 mission on Jul 18; Svetlana Savitskaya becomes the first woman to walk in space; *See also* 1983 TRA

Steven P. Jobs

At a time when large computer companies did not believe that there was any future in the few "toy computers," such as the Altair, Steve Jobs had the insight and perseverance to create the first personal computer that conquered a large market. In the early 1970s, Jobs teamed up with Steve Wozniak, an electronics wizard. It was Jobs's salesmanship and vision, combined with Wozniak's talent for achieving wonders with very simple circuits, that sparked the personal computer revolution of the 1980s.

Their first undertaking was an alternative electronics project, a small digital box that could generate the necessary tones for controlling long-distance telephone switching systems without the user being billed. They sold a good number of the boxes, and although the operation was illegal, they probably acquired the experience that led to their computer breakthrough. After a few unsuccessful projects, they built the Apple 1 computer, which consisted of a simple board without power supply or case. But there was a demand for it. So they designed the Apple 2, a complete product housed in a casing designed by Jobs himself. Technically, the Apple 2 was far ahead of its competitors. Although it was capable of displaying colors and disk drives, the price was kept low because of the simplicity of the circuits. By 1977 they had sold 4000 Apple 2s, and a year later they could not keep up with the demand.

Steven P. Jobs (Continued)

The Apple company started its fabulous growth in 1978 and came under the financial leadership of John Scully from Pepsi Cola. Steve Jobs designed both the Apple 3 and the Lisa computer. Lisa was the first computer intended for a large public to incorporate an interface based on icons and the use of a mouse. Both the Apple 3 and Lisa failed to find a market. Lisa, however, was displaced by the success of Jobs's second breakthrough computer, the Macintosh, or Mac. The Mac was a revolutionary design, based on a very fast microprocessor and an operating system based on icons. Initially, sales of the Mac did not take off as fast as Jobs had expected, however, and because of several disagreements between Jobs and the rest of the Apple management, he left Apple in 1985 to form his own company, NeXT.

At NeXT Jobs intended to design a new computer that would outpace all others by superior technology. The NeXT computer became available in 1991. It has the superior graphics and storage capabilities but, because of its high price, has failed to penetrate the personal computer market of the early 1990s and is mainly used by universities.

1985

Alex Jeffreys and coworkers at the University of Leicester in England develop genetic fingerprinting, or identification by genetic matching; the technique becomes vastly more useful after 1987, when it is combined with the use of the polymerase chain reaction (PCR) to amplify genetic material recovered in very small amounts; *See also* 1901 GEN; 1987 GEN

R. Wigginton, Ed Rudder, and Don Breuner develop MacWrite, a word processing program for the Macintosh computer that makes use of icons, graphic illustrations of such items as file cabinets or wastebaskets that can be pointed at (with a mouse) to indicate operations such as saving a file or erasing one; *See also* 1978 COM

Microsoft develops Windows for the IBM PC; this graphics-based user interface allows the operator of an IBM–PC compatible computer to use a mouse and icons to direct the computer, a method that became popular in the Macintosh computer

Paul Brainard writes PageMaker, the first desktop publishing package developed for a microcomputer; it is adopted by the Apple Macintosh computer; later the software becomes available for PC-compatible computers; *See also* 1982 COM

Seiko-Epson in Japan builds a commercial television set with a two-inch screen that is a liquid-crystal display, made with polycrystalline silicon; *See also* 1975 COM

AT&T Bell Laboratories achieves the equivalent of sending 300,000 simultaneous telephone conversations or 200 high-resolution television channels at once over a single optical fiber; *See also* 1978 COM

Thinking Machines introduces the Connection Machine, a computer with 16,384 parallel processors and a speed of 600 million operations per second; at a later stage the number of processors is extended to 65,536 and its speed is 1 billion operations per second; *See also* 1983 ELE; 1991 ELE

Cray Research introduces the Cray 2, a strongly miniaturized supercomputer ten times as powerful as the Cray 1; it contains four processors, has a 2-billion-byte memory and can perform 1 billion floating-point operations per second; later versions have 8-billion-byte memories; *See also* 1976 ELE; 1989 ELE

The US National Science Foundation sets up five national supercomputing centers, in San Diego CA, Pittsburgh PA, Princeton NJ, Ithaca NY, and Champaign-Urbana IL; the last is called the National Center for Supercomputing Applications (NCSA)

Intel introduces the 80386, a 32-bit microprocessor; *See also* 1978 ELE; 1986 ELE

Masaki Togai and Hiroyuki Watanabe, researchers at AT&T Bell Labs, develop a logic chip that operates on fuzzy logic; *See also* 1986 TRA

The Superphénix, the world's largest fast-breeder reactor, producing 1200 megawatts, starts operating in Creys-Malville, France; continuous technical problems cause the reactor to be shut down for most of the time until 1992, when the French government decides to decommission the reactor indefinitely; *See also* 1977 ENE

The "synfuels" concept in the US, designed to develop alternative energy sources based on coal or oil shales, loses nearly all of its funding, reflecting a worldwide petroleum glut; *See also* 1969 ENE; 1979 ENE

The submarine *Saga*, powered by two Stirling motors, is launched by France; *See also* 1817 ENE

1985

Information about lanxides, crosses between ceramics and metals, is released for the first time when the US Defense Department declassifies the three-year-old subject; *See also* 1990 MAT

Richard E. Smalley and Harold W. Kroto discover that some carbon arranges itself in the form of "bucky-balls," molecules shaped like a soccer ball with 60 atoms; they call this form of carbon buckminster-fullerene after Buckminster Fuller, who developed geodesic domes of essentially the same shape as the molecules; *See also* 1926 MAT; 1991 MAT

Lasers are used in the US for the first time to clean out clogged arteries; *See also* 1977 MED

The US Food and Drug Administration approves in Oct the marketing of growth hormone manufactured by bacteria; this is the second drug produced by genetic engineering (after human insulin) to be sold in the US; *See also* 1982 MED

The US Food and Drug Administration approves the implantable cardiac defibrillator, a mechanical device used to prevent the heart from beating too rapidly or from fibrillating (quivering instead of beating); it is implanted in a patient's chest, where it is powered for 5 years by a lithium battery; *See also* 1959 MED

Bert L. Vallee [b Hemer, W Germany, Jun 1 1919] and coworkers find the tumor angiogenesis factor first predicted by Judah Folkman in 1961; it stimulates the growth of new blood vessels; the scientists rename the factor angiogenin

Gerd Binnig, Christoph Gerber, and Calvin Quate develop the atomic force microscope (AFM); a tip made of diamond, silicon, or tungsten is moved over a surface to be investigated; the small movements of the tip caused by bumps of atoms are measured with a scanning tunneling microscope; *See also* 1981 TOO; 1987 TOO

Experimentalists begin to use lasers to handle individual atoms or molecules and tiny clusters of atoms or molecules or even aggregations of some size; light is used to exert a force because photons carry momentum; light pressure works best when particles are already nearly motionless from really low temperatures; *See also* 1975 TOO

A nuclear X-ray laser explodes underground, producing X-ray radiation that is 1,000,000 times brighter that obtained in earlier tests of similar nuclear-powered lasers; this experiment is viewed as an important step toward the "Star Wars" program, or SDI; *See also* 1984 TOO

Atlantis, the fourth US space shuttle, is launched with a five-person crew on Oct 4; this brings the shuttle fleet up to planned size; *See also* 1981 TRA

On Jul 2 Giotto, a joint European mission to study Halley's comet, is launched; it passes closest to the comet at 600 km (375 mi); *See also* 1979 TRA

Deep-sea explorer Robert D. Ballard [b Wichita KA, Jun 30 1942], using Argo, a remote-controlled robot equipped with video cameras, discovers the wreck of the *Titanic,* which sank in 1912; *See also* 1952 TRA; 1989 TRA

1986

A worldwide ban on whaling, with limited exceptions for traditional societies, begins by international agreement; *See also* 1977 GEN

The first DAT (digital audio tape) recorders are demonstrated in Japan; *See also* 1982 COM

In Europe, FM radio stations begin to use the subcarrier signal of FM radio to transmit digital data; the radio data system (RDS) can be used to transmit messages that appear on small display screens attached to suitably equipped radios or for other purposes; *See also* 1992 COM

Terry J. Sejnowski at Johns Hopkins in Baltimore MD develops a neural network computer that can learn to read a text out loud without knowing any pronunciation rules; *See also* 1980 COM

Compaq leaps past IBM by introducing the DeskPro, a computer that uses an advanced 32-bit microprocessor, the Intel 80386; it can run software three times faster than the quickest 16-bit computers; *See also* 1985 ELE

Chernobyl nuclear reactor number 4, near Kiev, USSR, explodes at 1:23 a.m. local time on Apr 26, leading to a catastrophic release of radioactivity that kills dozens of people within a few weeks and forces the mass evacuation of all families within 30 km (18.6 mi) for an indefinite period; *See also* 1979 ENE

The largest dam in the world is built in Venezuela to supply 100,000 megawatts of power to that country; *See also* 1895 ENE

Alcan in Montreal, Canada, develops the aluminum-air battery; aluminum slowly dissolves in salt water, producing aluminum hydroxide and free electrons that move as an electric current through the salt water; *See also* 1965 TRA; 1991 ENE

Nova, an experimental laser fusion device at Lawrence Livermore National Laboratory (LLNL), creates the first fusion reaction induced by a laser; ten laser beams that deliver total energy of 100 trillion watts during one-billionth of a second converge on a hydrogen-filled glass sphere; a small part of the hydrogen nuclei fuse into helium nuclei; *See also* 1958 ENE

Los Alamos National Laboratory in NM uses a two-well system to produce geothermal power; the 4-km (2.5-mi) wells are connected at the bottom; water inserted in one well emerges at a temperature of 190°C (375°F) from the other; the plant produces 4 megawatts; *See also* 1903 ENE

Communicating with light

Light signals are probably the oldest method for transmitting messages over long distances. Ancient Greeks and Romans used torches for communication between hilltops. Coded messages have long been transmitted from ship to ship with lights.

Alexander Bell was the first to experiment with transmitting speech over light beams. The transmitter of his photophone consisted of a mirror reflecting a light beam. The mirror vibrated when sound waves were directed on it via a horn, causing the intensity of the light beam to vary rapidly. The receiver contained a selenium crystal. Its resistance varied with the fluctuations in intensity in the light beam received from the vibrating mirror. The sound waves were reproduced in an ordinary telephone receiver, which Bell had previously invented.

Lasers can create very short light pulses, much shorter than anything that can be produced with conventional light sources, such as electric lamps. Such short light pulses are suitable for transmitting large quantities of information at high speed. Laser beams have been used in the past for transmitting signals by methods like those used in Bell's photophone, but light traveling through the atmosphere suffers from absorption and dispersion.

Physicists discovered in the 19th century that light can travel through curved glass rods. Light travels through such rods because it bounces off the boundary between the glass and the air. At small incident angles, the boundary reflects light just as the flat surface of a pond does, and for the same reason. The index of refraction of air is lower than that of glass, so light striking the boundary is bent.

Glass or quartz rods or silvered tubes are collectively called light guides. In 1880 American engineer William Wheeler proposed piping light into individual rooms in a building using glass rods connected to a central light source. Transmission of telephone signals as modulated light beams through light guides was patented as early as 1934 by Norman French of AT&T. In 1950 Ray D. Kell and George C. Sziklai of RCA proposed transmitting several television channels through light guides.

K. Alex Müller and Georg J. Bednorz discover an oxide combination that is superconducting at 30 K (30°C, or 54°F, above absolute zero), the highest known temperature for superconductivity and a breakthrough that leads to other materials that are superconducting at much higher temperatures; *See also* 1911 MAT; 1987 MAT

Louis Kunkel [b New York City, Oct 13 1949] and coworkers discover the gene that is defective in Duchenne muscular dystrophy, a common, fatal form of muscular dystrophy; *See also* 1978 MED; 1988 MED

The US Food and Drug Administration approves in Jul a hepatitis B vaccine made by yeast, the first vaccine approved for humans that is produced by genetic engineering; *See also* 1984 MED

The US Food and Drug Administration approves OKT3, the first monoclonal antibody to be approved for therapeutic use in humans; it aids in organ transplants; *See also* 1975 MED; 1984 MED

The US government lowers its standards for the permissible amount of lead in air and bans the use of solder containing lead; *See also* 1984 MED; 1991 MED

Tony Hodges patents a split computer keyboard (the two halves can be adjusted to different angles of attack for each hand) to prevent such repetitive stress injuries as carpal tunnel syndrome; *See also* 1993 MED

Gerd Binnig and Heinrich Rohrer of IBM win a Nobel Prize in physics for their 1981 invention of the scanning tunneling microscope (STM), a device used to image surfaces closely, even in terms of individual molecules or atoms; *See also* 1981 TOO

The last mission of the US *Challenger* begins and ends after 73 seconds on Jan 28; O-rings in the solid-fuel boosters wear through and the entire fuel supply explodes, killing schoolteacher Christa McAuliffe and astronauts F. Scobee, M. Smith, R. McNair, E. Onizuka, J. Resnik, and G. Jarvis; *See also* 1985 TRA

Soviet cosmonauts Vladimir Solovyev and Leonid Kizim are launched on the Soyuz T 15 mission on May 5; they are the first cosmonauts to board the *Mir* space station, which was sent up without a crew on Feb 30; *See also* 1973 TRA

The *Wind Star,* a modern wind-powered ship of 134 m (440 ft), is launched; sails on its four 62-m (204-ft) masts are controlled by computer

The European A320 Airbus is the first commercial aircraft to use a "fly-by-wire" system; *See also* 1972 TRA

The Annacis Bridge in Canada is completed using the cable-stayed design, a method involving a deck supported by cables strung from isolated towers at 465 m (1525 ft); its main span is the longest to this date for the cable-stayed design; *See also* 1991 TRA

A subway system controlled by computers using fuzzy logic starts operation in Sendai, Japan, 200 mi (320 km) north of Tokyo; *See also* 1985 ELE

Communicating with light (Continued)

Proposals to use rods or other light guides failed to result in replacement of the use of electricity through wires. Early light guides were more expensive and less effective than wires.

Instead of a rod, one can use a glass fiber for transmitting light. In a fiberscope, a large number of parallel fibers carry a complete image: Each fiber transmits the light from one point of the image. A bronchoscope, used in medicine to investigate the trachea, is such a fiberscope. Besides transmission of the image to the viewer, the fibers also carry the light used for illuminating the field of observation.

In 1966 Charles Kao and George Hockham, two scientists at the Standard Telecommunications Laboratories in England, proposed using optical fibers to replace the traditional cables in telecommunication links. Bell Labs developed an experimental optical-fiber transmission system in 1976 that had the capacity of 45 megabits per second.

The first commercial telephone optical link, 13 km (8 mi) long, was established between Martlesham Heath in Suffolk, England, and Ipswich in 1977. Since then, telephone companies have increasingly replaced coaxial trunk lines with optical-fiber links. The first transatlantic optical fiber, laid in 1988, can carry 37,800 voice channels simultaneously.

1987

For the first time a crime suspect is convicted on the basis of genetic fingerprinting, in Great Britain; later in the year, the same technique is used in resolving a rape case in the US; *See also* 1985 GEN

EuroPACE, a European organization for the training of engineers, scientists, and students via satellite and electronic mail, is founded

Sonic Solutions in San Francisco CA and the National Sound Archive in England develop digital techniques for eliminating unwanted noise from musical records; the American system, Nonoise, replaces noise with sound of the same frequency of adjacent sound, while the English system restores the sound by replacing noisy sections with clear sections from another, identical recording; *See also* 1967 COM

Sega Electronics introduces a three-dimensional video game; images for the left and right eye are displayed in quick alternation (every 1/60 second) on a screen; while the viewer looks at the picture, special liquid-crystal glasses also pass light in quick alternation: When the left image is displayed on screen, the left lens of the glasses is transparent, and vice versa; the alternation of images is so fast that the eye combines it into a three-dimensional image; *See also* 1972 ELE

Telephones become available on Japanese airliners; calls are relayed by satellite; *See also* 1991 COM

Kodak introduces the Fling camera, a disposable camera like the very first Kodak; the entire camera must be returned for film processing; to take some more pictures, one needs a new Fling; *See also* 1888 COM

Apple's Macintosh II and Macintosh SE become the most powerful personal computers available; *See also* 1984 ELE

IBM brings out the Personal System/2 group of personal computers, based on 3½ inch disk drives, hard disks, enhanced graphics, and access to a new operating system that enables interconnections between computers; the system incorporates a bus called Micro Channel Architecture (MCA), which, although incompatible with the normal internal devices for PCs, allows much faster data transfer; *See also* 1984 ELE; 1988 ELE

Computer chips are manufactured with 1 megabit (1000 kilobits or 1,048,576 bits) of computer memory; in Feb, IBM and Nippon Telephone and Telegraph Limited (NTT) of Japan introduce experimental 4- and 16-megabit chips; *See also* 1984 ELE; 1990 ELE

The Numerical Aerodynamic Simulation Facility, an advanced supercomputer devoted to simulation and capable of a top speed of 1,720,000,000 computations a second, starts operations on Mar 9; *See also* 1985 ELE

David Miller [b Hamilton, Scotland, Feb 19 1954] invents at Bell Labs the Symmetric Self-Electro-optic Effect Device (S–SEED), a device that alters its reflectivity to light when irradiated by a laser; S–SEEDs will later be used as components of the first optical computer, built in 1990; *See also* 1990 ELE

1987

A lift-slab building under construction in Bridgeport, CT collapses on Apr 23; the lift-slab technique, in which slabs of flooring are poured at ground level and then raised, comes under suspicion as 28 workers are killed in the disaster; *See also* 1977 FOO

The headquarters of the Internationale Nederlanden Group Bank in Amsterdam, the Netherlands, is completed; its ten skylight-topped towers are contoured to maximize natural lighting, produce solar heating, and deflect the wind, making it one of the best-known examples of the "green architecture" movement for commercial building; *See also* 1979 FOO; 1992 FOO

The French firm Agrotechnic invents a "mechanical cow," a device for producing imitation milk from soybeans; *See also* 1961 FOO; 1988 FOO

Robert Wells of Steamboat Springs CO is granted a patent for a reclosable, tear-out-opening beverage can; *See also* 1980 FOO; 1990 FOO

A team led by Ching-Wu Chu [b Hodnam, China, Dec 2 1941] becomes the first group to make a material that is superconducting at the temperature of liquid nitrogen: 77 K (−196°C or −321°F); *See also* 1986 MAT; 1988 MAT

Herbert Naarmann and N. Theophilou develop a form of polyacetylene that is doped with iodine and that is a better conductor of electricity than copper; *See also* 1977 MAT; 1988 MAT

The Japanese firm Nippon Zeon develops a plastic with memory; on deformation, it bends and keeps its bent shape at low and modest temperatures; it returns to its initial form when heated to a temperature of 37°C (99°F); *See also* 1962 MAT

Georg Bednorz of West Germany and K. Alex Müller of Switzerland win the Nobel Prize in Physics for their discovery of superconductivity in a material at higher temperature than previously known; *See also* 1986 MAT

The antidepressant drug fluoxetine, better known by its trade name of Prozac, is licensed by the US Food and Drug Administration; *See also* 1954 MED

Ignacio Navarro Madrazo announces that implanting cells from a person's adrenal gland in the brain can cure or alleviate Parkinson's disease; earlier experiments along the same lines had been unsuccessful, but Madrazo changes the location of the implant; *See also* 1982 MED

A team led by Hari Reddi at the National Institutes of Health extracts and identifies bone morphogenetic protein (B.M.P.), a substance predicted first by Marshall Urist in the 1960s; B.M.P. encourages new bone to grow and is expected to help cure fractures that are otherwise resistant to healing as well as speed healing in more ordinary fractures and bone damage caused by surgery

Michael Zasloff announces in Aug that he has discovered potent new forms of antibiotics, which he terms magainins, in the skin of the African clawed frog; *See also* 1955 MED

Experimental vaccination of foxes against rabies begins in Belgium, using baits containing a vaccine created by genetic engineering that are dropped from helicopters; the experiment is successful and leads to large-scale vaccination campaigns; *See also* 1885 MED; 1990 MED

The American company 3 D Systems develops a system to produce plastic prototypes of objects designed by CAD techniques; they polymerize a liquid plastic into the computer-designed shape with an ultraviolet laser beam steered by a computer; the newly created object can then be lifted out of the liquid plastic; *See also* 1964 TOO; 1991 TOO

Cornell University in Ithaca NY opens a new laboratory for the development of tiny, submicroscopic tools, which they call the National Nanofabrication Facility; *See also* 1959 TOO; 1988 TOO

Arthur Rich and James Van House develop the positron microscope; the microscope functions in a similar way as an electron microscope, but uses positrons emitted from radioactive source; *See also* 1985 TOO

Soviet cosmonaut Yuri V. Romanenko returns to Earth from the *Mir* space station after 326 days in space, a new record; *See also* 1986 TRA

1988

In Nov representatives of 30 nations meet in Geneva, Switzerland, to form the Intergovernmental Panel on Climate Change; its mission is to consider whether the greenhouse effect will cause global warming; *See also* 1967 GEN

The National Weather Service (NWS) installs the first of 115 Doppler radars for weather forecasting; these radars, using the Doppler effect, can measure the speed and direction of wind and storms by measuring small variations in the wavelength of the reflected radar signals

The US National Academy of Engineering creates the Draper Prize, engineering's equivalent of a Nobel Prize; at $375,000, it is the highest prize in engineering

Yamaha introduces the Disklavier piano, an electronic, digitized version of the automatic piano that uses compact discs (CDs) to direct the motion of the keys instead of punched paper rolls; *See also* 1982 COM; 1983 GEN

Australia introduces the first successful plastic folding money, a $10 bill commemorating the Australian bicentennial; *See also* 1992 MAT

The first transatlantic optical fiber cable is laid; it can carry 37,800 voice channels; *See also* 1956 COM; 1990 COM

The American company Scriptel introduces a method for inputting data into a computer by writing on a screen; *See also* 1960 ELE

Carnegie-Mellon University graduate student Kai-Fu Lee demonstrates that computerized voice recognition systems can be developed that respond correctly most of the time to words spoken by anyone; "training" of the machine to recognize specific voices and intonations is not needed; *See also* 1983 COM

On Nov 2 a computer virus developed by Robert T. Morris, Jr., a graduate student at Cornell University is secretly planted in the Internet computer network; it spreads around the world to more than 60,000 computers, tying up thousands of computers for two days; Morris is eventually fined and expelled from Cornell; *See also* 1983 COM

Arthur Robinson develops a type of map projection that represents countries near the two poles to their true size, and thus deviates considerably from the traditional Mercator projection

Motorola launches its 32-bit 88000 series of RISC (reduced instruction set computing) microprocessors; because they handle fewer different instructions, they can operate much faster than conventional chips, processing as many as 17 million instructions per second; *See also* 1971 ELE; 1990 ELE

In response to the Micro Channel Architecture (MCA) developed by IBM for its PS/2 computers, a group of manufacturers led by Compaq Computer Corporation and Tandy Radio Shack develop the Extended Industry Standard Architecture (EISA), a 32-bit architecture for computers with advanced microprocessors, such as the Intel 80386 and 80486; *See also* 1987 ELE

Steven P. Jobs introduces the NeXT Computer System, a graphics-based system that includes a 256-megabyte optical storage disk and 8 megabytes of RAM; *See also* 1987 ELE

AT&T develops a transistor that responds to the flow of a single electron; *See also* 1947 ELE; 1990 ELE

John L. Gustafson, Gary R. Montry, Robert E. Benner, and coworkers find a way to rewrite problems for computer parallel processing that speeds their solutions by a factor of 1000; previously an increase in speed by a factor of 100 was thought to be the theoretical limit of this method; *See also* 1985 ELE

T. Kotani and coworkers at Fujitsu Laboratories in Japan develop a Josephson microprocessor; it is a 4-bit microprocessor that incorporates Josephson junctions; *See also* 1983 ELE

A vertical-bladed windmill in Hawaii with a single propeller 97.5 m (320 ft) long is installed; it produces 3.2 megawatts of power; *See also* 1941 ENE

Herman Branover completes a magnetohydrodynamic generator, Etgar 5, that converts 46 percent of the energy contained in a liquid mixture of lead and bismuth directly into electricity; it is expected that such generators will eventually have a higher efficiency rate than classical electric generators; *See also* 1971 ENE

Roland Winston [b Moscow, (in Russia), Feb 12 1936] supervises a test of a new mirror system that concentrates sunlight to 60,000 times its normal intensity on Earth; it is believed that the system will have applications in the development of new types of lasers and possibly in developing new materials; *See also* 1981 ENE

1988

Because of concerns about chlorofluorocarbons (CFCs) used in manufacturing foam polystyrene, McDonald's fast food chain replaces its "clamshell" foam package for the Big Mac and other sandwiches with a plastic that is not manufactured with CFCs; *See also* 1975 FOO; 1990 FOO

Australian butcher Dallas Chapman invents a low-fat sausage that also is somewhat lower in cholesterol; *See also* 1987 FOO

Paul May [b New York City, Jul 12 1931] of South Maria CA develops a feed for chickens that results in eggs with less than half the cholesterol content of eggs from chickens fed standard chicken feed; *See also* 1987 FOO

In Jan scientists from Japan's National Research Institute for Metals develop a new high-temperature superconductor based on bismuth, bringing the number of types of high-temperature superconductors to three; In Mar scientists at the University of Arkansas discover a fourth type of high-temperature superconductor based on thallium; *See also* 1987 MAT; 1989 MAT

Researchers at AT&T Bell Labs in Murray Hill NJ report the appearance of magnetic flux lines in ceramic superconducting materials that limit the current density considerably; *See also* 1987 MAT; 1989 MAT

Richard Friend develops a diode made of polyacetylene with an efficiency of 5 percent; *See also* 1987 MAT; 1991 MAT

Frank Filisko [b Lorain OH, Jan 29 1942] develops a fluid that changes its viscosity upon application of an electric field; in a strong field the liquid turns into a gel while it is hyperviscous in a weak field

The Jun 6 issue of *Physical Review Letters* contains the first image of a benzene ring, confirming the ring structure for aromatics first envisioned by Frederick August Kekulé in 1865; the image was produced by scientists at IBM's Almaden Research Center in San Jose CA using the scanning tunneling microscope; *See also* 1980 COM

RU-486, also called the abortion pill, developed by Etienne-Emile Baulieu, is introduced into general use in France; it induces an abortion up to seven weeks after fertilization by blocking receptors for the production of the hormone progesterone; *See also* 1956 MED

Geneticist Philip Leder [b Washington DC, Nov 19 1934] receives a patent for a mouse genetically engineered to be highly susceptible to cancer; the first patented animal in the world (although bacteria had been previously patented) is used in cancer research; *See also* 1980 GEN & ENE; 1992 GEN

The US Food and Drug Administration approves alpha interferon as a treatment for genital warts; *See also* 1980 MED

Italian scientists report that follow-up studies of people exposed to dioxins in a 1976 industrial accident near Seveso, Italy, show no increase in birth defects; *See also* 1957 FOO

Drug Delivery Systems develops the electric skin patch; it contains a battery that passes a tiny current through the skin under the patch, reducing its resistance to the absorption of drugs

Disposable contact lenses go on sale; these can be worn one to seven days without removal or cleaning; *See also* 1965 MED

Francis C. Moon and Rishi Raj use a high-temperature superconductor to build an almost frictionless high-speed bearing; *See also* 1891 TOO; 1988 MAT

Roger Angel in Apr succeeds in casting a 356-cm (140-in.) diameter telescope mirror by a new method in which the molten glass spins in a rotating mold as it cools; centrifugal forces result in a mirror surface that is already close to the desired paraboloid, requiring much less finishing to obtain a useful reflector; *See also* 1948 TOO; 1990 TOO

Long-Sheng Fan and Yu-Chong Tai at the University of California at Berkeley develop an electric micromotor; the motor is built using etching technology used for the manufacture of microchips; *See also* 1987 TOO; 1989 TOO

On Sep 29 the redesigned US space shuttle *Discovery* begins its first flight since the *Challenger* disaster, with a five-person crew; *See also* 1986 TRA

The human-powered aircraft *Daedalus 88*, piloted by Kanellos Kanellopoulos, flies from Crete to the shore of Santorini, where it breaks up just offshore in heavy breezes; the flight of 3 hours 54 minutes covers 119 km (74 mi) and sets new distance and time records for human-powered flight; *See also* 1979 TRA

1989

The Human Genome Project is launched on Jan 3 with James Dewey Watson [b Chicago IL, Apr 6 1928] as its first director; *See also* 1984 GEN

Computer Deep Thought becomes the first computer to beat a master human chess player when it defeats David Levy, who had been winning matches against computers since 1968; however, Gary Kasparov defeats Deep Thought in a two-game match on Oct 22; *See also* 1982 GEN

On Jun 3 Japan initiates daily broadcasts of its analog version of high definition television (HDTV), with a one-hour program featuring the Statue of Liberty and New York Harbor; HDTV pictures are also transmitted between Japan and the US; *See also* 1991 COM

The Japanese Ministry of International Trade and Industry establishes the Laboratory for International Fuzzy Engineering Research; *See also* 1986 TRA

Philips and Sony bring the videodisk to market; *See also* 1927 COM; 1963 COM

Seymour Cray founds the Cray Computer Corporation to develop the Cray 3 supercomputer, using gallium arsenide chips, which are faster than silicon chips; *See also* 1971 ELE; 1991 ELE

The Shippingport Atomic Power Station is decommissioned after 32 years of operation and its still radioactive main reactor taken from PA to Hanford Military Reservation in WA to await approval of a final burial ground; this is the first US nuclear power plant to be dismantled; *See also* 1957 ENE; 1992 ENE

Scientists at the High Technology Center of Boeing develop a stacked photovoltaic cell that converts 37 percent of solar radiation into electricity; the cell consists of two types of photovoltaic materials mounted on top of each other; the upper cell consists of gallium arsenide and captures the energy of blue light, while the second cell is made of gallium antimonide, converting red light into electricity; scientists at Scandia National Laboratories develop a single photovoltaic cell containing silicon that converts 20.3 percent of radiation into electricity; *See also* 1983 ENE; 1992 ENE

High-temperature superconductors

Since the discovery of superconductivity in metals in 1911 by Dutch physicist Heike Kamerlingh Onnes, superconductors have found only a marginal application in technology. The main reason is that metallic superconductors, such as the alloy of niobium and titanium often used in coils for magnets, require cooling to 4 K (−269.15°C, or −452.47°F) for the superconducting effect. Such a temperature is near absolute zero, the theoretical point where there is no heat because molecules have stopped moving entirely. These low temperatures can occur only by bathing the metal in expensive liquid helium. Thus, superconductors have been restricted primarily to applications where the high cost of cooling is justified because the superconducting effect is required, mainly in powerful magnets. These include deflection magnets used in very large particle accelerators and in magnets used for nuclear magnetic resonance imaging (MRI).

In 1986 two scientists working for IBM in Switzerland, Georg Bednorz and Alex Müller, found that certain ceramic materials that normally are electrical insulators become superconducting at low temperatures, but at much higher ones than for metals. The importance of this discovery can be judged by the fact that Bednorz and Müller were awarded the Nobel Prize in physics just a year later.

An impressive series of breakthroughs followed, and for a time the discovery of a new material with a higher critical temperature was announced every few weeks. It looked as if the series of superconducting compounds with increasingly higher critical temperatures would go on indefinitely, and that the creation of room-temperature superconductors would become attainable. However, none achieved a critical temperature higher than the 133 K (−140.15°C or −220.27°F) obtained in 1993 by researchers in Zurich, Switzerland, although there were rumors of 144 K (−129.15°C or −200.47°F) materials soon after.

Oil tanker *Exxon Valdez* is grounded in Prince William Sound in southern AK, leaking 35,000 tons of oil into the sound and damaging wildlife extensively; *See also* 1968 TRA

Y. Tokura, H. Takagi, and Shin-ichi Uchida discover at Tokyo University a high-temperature superconductor in which the superconducting charge carriers are electrons, not holes, as in high-temperature superconductors previously discovered; *See also* 1988 MAT

Several groups of researchers discuss the possibility of flux-line pinning; that is, the immobilization of flux lines by the introduction of impurities or crystal faults to increase the current density of high-temperature superconductors; *See also* 1988 MAT

The Martin Marietta Laboratories in Baltimore MD develop Weldalite 049, an aluminum-lithium alloy that has twice the yield strength (the ability to return to its original shape after being bent) of the leading aluminum-copper alloy used for aerospace

On May 10 Thomas H. Shaffer and coworkers show that an oxygen-suffused liquid called a perfluorocarbon can be breathed by a premature infant to reduce the lung damage that such infants often suffer; in this first test, the infant's lungs had already been damaged beyond repair by conventional treatment, but the fluid treatment is clearly helpful even though it comes too late

At the end of Oct the first-ever conference on nanotechnology, or manufacture of devices on a nanometer (one millionth of a meter), scale takes place in Palo Alto CA, led by Eric Drexler of the Foresight Institute; *See also* 1988 TOO; 1991 TOO

The high-speed Train à grande vitesse (TGV), in France, reaches a record speed of 482 km (300 mi) per hour; *See also* 1981 TRA; 1990 TRA

The US Air Force introduces the Stealth Bomber (B-2), which, because of its shape and of the materials used, is expected to be invisible to radar; in 1991, however, the bomber is observed by British radar during the Persian Gulf War

Robert D. Ballard discovers the wreck of the *Bismarck,* a German battleship sunk by the British in 1941; *See also* 1985 GE

High-temperature superconductors (Continued)

Shortly after the discovery of the high-temperature superconductors, hopes were high that they would soon replace conventional superconductors cooled with liquid helium and be used in various applications. However, a series of difficulties with the ceramic materials—the most notorious one is brittleness that makes them unsuitable for shaping into wires or coils—has delayed practical application.

If the problems of shaping ceramic superconductors in bulk can be solved, important uses will be transformation, transmission, and storage of energy. Magnets made with superconducting coils produce much stronger magnetic fields than those currently used in motors and generators, permitting smaller versions of great power. Superconductors can also make the high-voltage lines that now disfigure the perimeters of large cities a thing of the past. Because a current in a superconducting closed loop flows indefinitely, such a loop can be used for energy storage. The stored energy can be obtained when required by breaking the closed circuit. Plans exist for building huge superconducting coils mounted underground to store thousands of megawatts of energy.

Scientists are also investigating applications of another property of superconducting materials, that of being able to suspend objects in a magnetic field, a phenomenon called magnetic levitation. Maglev trains, moving suspended above their rails, would become more practical. Another application could be magnetic bearings in which the rotating parts do not make mechanical contact with the outer casing.

1990

On Jan 1 the volt (measuring electrical potential) and the ohm (measuring electrical resistance) are defined in atomic terms, replacing the specified 19th-century electrical meters and equipment used in the previous definition; because the volt and ohm are tied to the ampere, which is not redefined, the new definitions—unlike those for the second and meter—will have little application unless they are adopted as fundamental units; *See also* 1967 GEN

Facsimile transmission machines (faxes) that can transmit color become commercially available; *See also* 1980 COM

The state of IA contracts to have a 4500-km (2800-mi) fiber-optic network connecting all 99 counties for broadcasting of college courses, communication among state agencies, distribution of lottery tickets, maintainence of voter registration lists, and linkage of libraries; by 1993 the $200 million project is complete; *See also* 1988 COM

On Jan 29, a team of scientists at Bell Laboratories led by Alan Huang demonstrates the first all-optical processor; the assembly of lasers, lenses, and fast light switches has the same capacity as an electronic chip controlling a dishwasher; although it can perform calculations optically, it is programmed by a separate electronic computer of the ordinary, nonoptical variety; *See also* 1987 ELE; 1993 ELE

Hewlett Packard announces a computer with a RISC (reduced instruction set computing) processor; the RISC processor allows an increase of processing speed since it does not require the parts of the normal instruction set that are seldom used in other microprocessors; later in the year IBM introduces the RS/6000 family of RISC workstations; for many applications, they are as fast as the then current supercomputers; *See also* 1988 ELE

Computer chips are introduced with 4 megabits (4000 kilobits or 4,194,304 bits) of computer memory; in Jun, Hitachi announces a working prototype of a 64-megabit memory chip; *See also* 1987 ELE; 1991 ELE

In Jan, Intel introduces the i486 processor chip, which can operate at a rate of 33 MHz; the first computer equipped with such a chip is the PowerCache 33/4 built by Advanced Logic Research; Intel also launches the iPSC/860 microprocessor chip designed for multiprocessor computers *See also* 1985 ELE; 1993 ELE

Motorola introduces the 68040 version of its 68000 series of microprocessors; the chip incorporates 1.2 million transistors and is adopted by 35 computer manufacturers; *See also* 1984 ELE

IBM develops a transistor that can operate at 75 billion cycles per second; *See also* 1988 ELE; 1992 ELE

Massively parallel computers

Most computers, such as conventional mainframes and personal computers, use von Neumann architecture: A central processor executes instructions one by one. The speed of such a computer is limited by the speed that the data can funnel through the central processor.

During the 1970s engineers developed "pipeline" computers using processors connected in series. For example, the ASC (Advanced Scientific Computer) of 1972, used in processing seismic data, had eight processors controlled by a master processor. The result from one processor was fed to the next one, while the previous processor began to deal the next batch of data.

During the 1980s the first true parallel computers appeared. In such computers the processors are interconnected so that each processor communicates with many others directly. One of the first was the Georgia Cracker, a homemade computer designed to factor large numbers. The Connection Machine, developed at MIT during the late 1980s, contains 64,000 processors, so it and others like it are often called massively parallel. Such computers can calculate, for example, the points of a complex image almost instantly. As a result, massively parallel computers have had their greatest success in simulations of complex phenomena, such as weather patterns.

The Basilica of Our Lady of Peace in Yamoussoukro, Ivory Coast, is completed; it replaces St. Peter's in Rome as the world's largest church; *See also* 1620–29 ARC

The Dutch government approves a plan to repurchase lowland regions reclaimed from the sea and return them to a more natural state of marshland or periodically flooded forest; when complete, the plan affects a tenth of all farmland in the Netherlands, restoring 250,000 hectares (600,000 acres) to an approximation of its original state; this reverses a trend that began about 500 CE; *See also* 1640–49 FOO

Because of concern over plastic's lack of biodegradability in landfills or when otherwise discarded, the fast food chain McDonald's replaces its plastic sandwich packaging with paper; *See also* 1988 FOO

CA inventors Robert DeMars and Spencer Mackay are granted a patent on an easy-opening, reclosable beverage can; *See also* 1987 FOO

On Sep 29, after exactly 83 years of construction, the Washington National Cathedral (officially the Cathedral Church of St. Peter and St. Paul) is completed in Washington DC; the Gothic edifice had been used in incomplete form since 1912

In Jun the Montreal Protocol Amendments, calling for a full worldwide phase-out of the use and manufacture of chlorofluorocarbons (Freons) and other ozone-depleting gases, is passed in London; *See also* 1978 MAT

In Jul workers in the US report that they can grow pure carbon-12 diamond films that conduct heat 50 percent better than natural diamond, which contains 1 percent carbon-13; carbon-12 diamond also can withstand laser radiation better than natural diamond; *See also* 1981 MAT; 1991 MAT

Researchers at Lawrence Livermore National Laboratory develop a silicon dioxide aerogel that weighs only 0.005 gm per cm³ (5 oz per cubic ft)

Fumihiro Wakai and coworkers at Japan's Government Industrial Research Institute in Nagoya develop a stretchable ceramic made from a mixture of silicon nitride, silicon carbide, and other compounds; potential uses of this ceramic material include engine parts that can withstand heat and that can be manufactured into different shapes of high precision; *See also* 1985 MAT

On Jun 7 a genetically engineered live virus is deliberately introduced into the US environment for the first time; the virus is a vaccine against rabies; it is being tested for safety by being mixed with raccoon bait in a northeastern PA wilderness region, where it is expected to immunize raccoons; *See also* 1987 MED

The Jarvik artificial heart is abandoned because of the substandard life quality it imposes on its recipients; *See also* 1982 MED

US medical doctor Ronald Baker patents a noninvasive monitor for fetal heartbeats; the transducer detecting the heartbeats consists of thin sheets of piezopolymer made of polyvinylidene difluoride; *See also* 1984 MED; 1993 MED

In Nov the Keck Telescope on the island of Hawaii sees "first light," becoming the largest optical telescope; its mirror consists of 36 segments, each 1.8 m (6 ft) across; *See also* 1948 TOO; 1988 TOO

In Feb, following a suggestion from astronomer Carl Sagan, the Voyager 1 space probe, near the edge of the solar system, takes a portrait of the system and transmits it back to Earth, presenting the first photograph of the whole solar system from space; *See also* 1975 COM

On Dec 1 at 11:21 a.m. local time Robert Graham Fagg and Philippe Cozette meet and shake hands under the English Channel, signaling the meeting of the two parts of the tunnel under the channel, or Chunnel as it comes to be known; the breakthrough occurs 22.2 km (13.9 mi) from England and 14.5 km (9.7 mi) from France

On Mar 6 Ed Yeilding and J. T. Vida set a record for a flight from CA to MD of 1 hour, 8 minutes, 17 seconds, flying an SR-71 Blackbird spy plane on its way from its CA base, where its use is being discontinued, to the Washington DC area so that the plane can be shown at the Smithsonian; *See also* 1960 GEN

Sweden introduces a new computerized railroad car, the X-2000, on the line between Stockholm and Goteborg, cutting 10 percent off the time possible with even the highest speed trains previously available; the secret of the X-2000 is that special computerized steering and suspension allow it to take curves 30 to 40 percent faster than ordinary railroad cars can do; *See also* 1989 TRA

The space probe Magellan, launched in 1989, reaches Venus and produces a detailed map of its surface using radar and a 3.66-m- (12-ft-) diameter antenna; *See also* 1962 TRA

1991

Philips of the Netherlands develops a light bulb that uses electromagnetic induction to excite a gas to emit light; the bulb has no parts that wear out and lasts for up to 60,000 hours; *See also* 1960 GEN

On a second trip to photograph the wreck of the *Titanic,* Robert D. Ballard introduces the use of superbright mercury-vapor lamps for undersea photography; *See also* 1985 TRA

The formal dissolution of the Soviet Union takes place on Dec 8 as Russia, Belarus, and Ukraine sign an agreement to that effect, establishing instead the Commonwealth of Independent States to encompass former republics within the Soviet Union

Woo Paik and coworkers at General Instrument Corporation's Videocipher division in San Diego CA produce the first working prototype of digital high-definition television (HDTV); *See also* 1989 COM; 1992 COM

On Feb 1 Philips demonstrates at the electronics fair in Las Vegas NV its digital compact cassette; it is of the same size as its analog compact cassette introduced in 1962; both are most familiar for recording music, but also have been used for recording other data; *See also* 1962 COM; 1986 COM

Matsushita in Japan introduces a video recorder that is programmable by voice; the commands have to be given in Japanese, however; *See also* 1979 COM

MITI (the Ministry of Trade and Industry) of Japan launches a $2 billion research program to develop a sixth-generation computer based on neural networks; the program to develop the so-called fifth-generation computer is abandoned; *See also* 1982 ELE; 1986 COM

Skyphone starts equipping airliners with telephones that remain operational anywhere on Earth; only outgoing calls are possible, and the signals are relayed by existing telecommunications satellites; *See also* 1987 COM

After 6 years of testing, *The New York Times* begins to use an ink that does not rub off on readers

Thinking Machines develops the CM-2, a computer with up to 16,000 processors that can reach speeds of 2 teraflops; *See also* 1985 ELE

Cray Research introduces the Cray Y–MP C90 supercomputer; it is equipped with 16 processing chips, and its speed reaches 16 gigaflops (16 billion calculations per second); *See also* 1989 ELE

The Japanese Institute for New Generation Computer Technology develops the Parallel Inference Machine (PIM), a fifth-generation computer that can handle words and images by logical inference, without the requirement that they be represented by numbers; *See also* 1982 ELE

In Feb four Japanese companies—Fujitsu, Matsushita, Mitsubishi, and Toshiba—announce that they have developed an experimental 64-megabyte dynamic random access (D–RAM) memory chip; on Dec 18, IBM and Siemens A.G. of Germany announce that they too have developed jointly a prototype of such a chip; *See also* 1990 ELE

Misha Mahowald and Rodney Douglas build a chip that simulates a neuron; their first chip represents five neurons; *See also* 1986 COM; 1991 COM

Information Storage Devices develops a practical analog chip that can store sound without the requirement that it be converted into digital form; the sound is sampled and stored as charges in the chip, which can distinguish 230 levels in these charges

In Oct computers that use a person's signature for a password come to market; these use pressure and electronic information processing instead of optical processing

A laboratory in France develops paper-thin electrical batteries that are suitable for a wide range of applications, including powering cars; *See also* 1986 ENE

Michael Gratzel and coworkers at the Swiss Federal Institute of Technology in Lausanne patent a type of transparent solar panel that supplies electricity and can be fitted on buildings as ordinary windows; *See also* 1974 ENE; 1992 ENE

The Taliq Corporation of Sunnyside CA begins marketing windows based on liquid crystals that are transparent in the presence of an electric current but opaque when the switch is turned off; at a cost of 20 to 30 percent more than plate glass, such window panels are used primarily as dividers and in a few office buildings; *See also* 1974 FOO

On Apr 19 Jagdish Narayan and Vijay Godbole of NC State University and Carl White of the US Oak Ridge National Laboratory announce that they have grown thin single-crystal diamond films on metal surfaces; their first efforts result in films only 100 sq microns in area; *See also* 1990 MAT; 1992 MAT

Nunez Reguero produces diamond by compressing C_{60} (buckminsterfullerene) molecules at room temperature, in the first production of artificial diamonds without the use of very high temperatures; a high pressure of 150 kilobars, or about 150,000 atmospheres, is still required; *See also* 1985 MAT; 1992 MAT

Scientists in Japan develop a p-n junction diode (transistor) based on diamond; it should resist both ionizing radiation and high temperatures better than conventional diodes made from silicon or germanium; *See also* 1990 MAT

Kenneth Matsumura develops at Berkeley a tiny electrocardiograph, worn like a wristwatch; by continuously monitoring electric signals from the heart, it can give an early warning of heart attack; *See also* 1959 MED

The US government mandates on May 7 tests for lead in tap water; it will take about 20 years for tests to be completed and for any problems found to be corrected; on Oct 7 the US lowers permissible levels of lead in the blood of children; in Nov the US lowers levels of lead that will be allowed to leach from dishes and glassware; *See also* 1979 MED; 1986 MED

Steve Barnard and David Walt develop a fiber-optic sensor for analyzing the chemical composition of blood; it consists of a multistranded optical fiber with several dye probes at its ends; when illuminated with ultraviolet light, specific dye probes glow in the presence of various compounds in the blood sample; *See also* 1951 MED

American physicist William Bennett and his daughter Jean develop the dynamic spectral phonocardiograph (DSP); it contains a sensitive microphone that picks up the sounds of heartbeat; the intensity of the sound is displayed for different frequencies on a monitor; *See also* 1990 MED

James S. Albus [b Louisville KY, May 4 1935], the inventor, and other engineers at the US National Institute of Standards and Technology build the first working model of a Stewart Platform Independent Drive Environmental Robot (SPIDER), a robot crane that is simpler to build and to operate than a standard crane, yet able to lift up to 6 times its own weight and position it accurately; although designated a robot because of its use of sensors and camera "eyes," SPIDER is essentially a crane that uses forces directed against themselves instead of a massive counterweight

Steward Dickson invents a method for producing complex plastic objects by irradiating a liquid polymer with a ultraviolet laser beam; the liquid solidifies when irradiated, and by using a slowly descending platform and a computer-controlled laser beam, the system can create objects that cannot be created using the conventional computer-controlled machine-tool techniques; *See also* 1987 TOO

Donald M. Eigler, Christopher P. Lutz and William E. Rudge at IBM's Almaden Research Center in San Jose CA develop a tunneling device that acts as a switch based on the location of a single atom of xenon; in theory such switches could be used to put the entire collection of the US Library of Congress onto a single disk with a 30-cm (12-in.) diameter; *See also* 1947 ELE; 1989 TOO

The Bendix/King Air Transport Avionics division develops the Traffic Alert and Collision Avoidance System II, which monitors airspace for 65 km (40 mi) around and monitors the position of up to 31 airplanes, ordering evasive action if required

A new record for the length of a cable-stayed main span for a bridge is set at 527 m (1729 ft) by the Skarnsundet Bridge in Norway; *See also* 1986 TRA

Alvon Elrod [b Walhalla SC, Dec 28 1928] and Tim Nelson patent a variable camshaft for car engines; the relative position of the cams is changed electronically to optimize the opening and closing of valves for any engine speed

1992 Nearly 5 years after approving its first patent for a genetically engineered animal, the US Patent and Trademark Office on Dec 29 issues three patents for mice with specific transplanted genes; the patents cover two breeds to mimic human diseases—mice that develop enlarged prostates and mice without immune systems for use in AIDS research—and a strain of mice resistant to viruses for use in developing disease-resistant livestock

The European Patent Office (EPO), 7 years after US Patent and Trademark Office approval, grants a patent to Philip Leder for a mouse, by now christened the oncomouse for its sensitivity to carcinogens; animal rights groups challenge the patent, the first granted by the EPO on an animal

On Jan 6 AT&T announces that it will start selling in May (it turns out to be Aug) a mass-market videophone for about $1500; using compression of signals, the Videophone 2500 can plug into any ordinary telephone jack and transmit and receive a 8.4-cm (3.3-in.) square picture along with sound to any other Videophone 2500 for the same price per call as sound alone; *See also* 1927 COM; 1993 COM

A digital cellular phone system is introduced in the US, tripling user capacity and vastly improving sound quality; *See also* 1983 COM

In Paris an experimental digital FM transmitter starts operation; *See also* 1986 COM

Brian W. Coles, working for Westinghouse Underwater Laser Systems and Applied Remote Technology, develops undersea black-and-white cameras that use laser light to illuminate subjects for undersea photography

IBM develops the silicon-on-insulator (SOI) bipolar transistor; it can operate at 20 GHz, the highest operational frequency ever achieved with a bipolar transistor; *See also* 1990 ELE

The Yankee Rowe nuclear plant in Rowe MA is shut down and retired rather than repaired; *See also* 1955 ENE; 1993 ENE

The Sanyo Electric Company of Japan introduces a kerosene-fueled space heater in which both the speed of the fan and adjustment of heat to match temperature needs are controlled by chaos theory; this permits temperatures to fluctuate more than in conventional feedback systems, which attempt to maintain a steady state by simply turning the heat and fan on or off simultaneously

The biggest array of thin-film photovoltaic modules ever assembled starts operation in Davis CA; built by Advanced Photovoltaic Systems, the 9600 modules convert sunlight into electricity, delivering up to 479 kilowatts, enough for 124 homes; *See also* 1991 ENE

Convergence of modes

A new process in technological development has become apparent during the Information Age; it is termed by some sociologists of technology as the "convergence of modes."

The clearest example of convergence of modes is in the telephone system. Until not so long ago, telephone networks served only for transmitting conversations between two people in a strictly point-to-point fashion. By using wires distinct from those of telegraph and telex systems, the telephone was entirely dedicated to voice communication. Today the telephone network fulfills a number of tasks. It still serves mainly for relaying speech, but now telephone networks also transmit documents via facsimile machines (faxes) and send data between computers. The one-to-one relationship between the communication channel and its use has disappeared.

A specific communications channel, such as an optical fiber, also serves a multiplicity of purposes: It may transmit telephone signals, carry data between computers at high speed, and transmit broadcasting and television programs. Via the same optical cable customers will also have access to interactive on-line services, such as viewing movies on

demand. The opposite is also true: A specific mode of communication now makes use of a variety of physical communications channels. For example, a telephone conversation can travel through ordinary copper wires, coaxial cables, optical fibers, or via microwaves, relayed either by a communications satellite or by a microwave transponder. The erosion of the one-to-one relationship between a specific technology and a specific use has also come to the fore with the recent development of high-definition television (HDTV). Interested parties are not only television broadcasters, television set manufacturers, and video equipment manufacturers, but also the computer industry for the development of advanced graphics and the design of virtual reality systems and the military for the design of high-resolution screens in radar and weapons systems and flight simulators.

The main reason for such a development is technological progress. The application range of digital electronics, embodied in the microchip, has widened enormously. In the past logical circuits served only for number crunching, but now they serve for a multitude of applications. Microchips connect telephone customers with each other, control elec-

In New York City the US National Audubon Society rebuilds the interior of a 19th-century department store as its national headquarters, using concepts that are termed "green architecture"; by reusing an existing structure, the building saves 300 tons of steel, 9000 tons of masonry, and 560 tons of concrete; other savings come from lighting that adjusts to ambient light and the presence or absence of people in the room and an effective ozone-friendly air-conditioning system that needs only half the capacity ordinarily used for a building of its size; all materials used in the rebuilding are renewable or recycled; although costing about 10 percent more to build, the office, by using 61 percent of the heating energy and 68 percent of the electricity of conventional buildings, will repay the extra building costs within 5 years; *See also* 1987 FOO

General Electric reports the synthesis of a three-carat diamond that consists almost entirely of the carbon-13 isotope; the diamond has a higher density and hardness than a carbon-12 diamond found in nature, which contains about 1 percent of the C-13 isotope; *See also* 1991 MAT

Following the success of a special bicentennial plastic bill, Australia begins to produce in Jul plastic $5 bills for circulation as a first step in a three-year plan to offer regular issues of plastic $5, $10, $20, $50, and $100 bills; *See also* 1988 GEN

The US Food and Drug Administration and the Canadian Department of Health and Welfare approve a powerful anticancer drug made from the bark of Pacific yew trees; the drug is taxol, also known as paclitaxel; it is particularly advantageous in treating ovarian cancer; *See also* 1971 MED

The Swedish government announces the first large-scale test of an AIDS vaccine for people already affected with the disease

On Jan 6 the US Food and Drug Administration asks manufacturers to stop making, and doctors to stop installing, silicone breast implants until further studies are done on the safety of the devices; *See also* 1960 MED

On Jun 28 the liver of a baboon is transplanted into a human whose own liver was destroyed by hepatitis; the operation at the University of Pittsburgh Medical Center appears to be a success, but the patient later dies as a result of complications of AIDS; *See also* 1963 MED

The US Food and Drug Administration approves an injectable form of sumatriptan (trade name Imitrex) for the treatment of migraine headaches

A company in France develops an aerosol can in which a small solenoid produces the aerosol electrically, thus eliminating the need of a gas propellant; *See also* 1941 GEN; 1978 MAT

Conductus, Inc. introduces Mr. SQUID, the first high-temperature superconducting device to become available commercially (a SQUID is a superconducting quantum interference device, used for measuring extremely small currents, voltages, and magnetic fields); another high-temperature superconducting device that is also produced (by Illinois Superconductor Corp.) is a dipstick for measuring the level of very cold fluids, such as liquid nitrogen; *See also* 1888 TOO; 1988 MAT

At the First General Conference on Nanotechnology, Ralph Merkle describes computer software developed to design and test machines that have been assembled one molecule at a time; *See also* 1989 TOO

The Main-Danube Canal is complete, providing a water route from the North Sea to the Baltic Sea, when combined with various rivers and lakes, that is only 3500 km (2175 mi) long; although shorter than going around the Strait of Gibraltar and through the Bosporus, the realities of canal and river travel mean that the short way through the canal still takes twice as long as the long way around, although the canal is cheaper, especially for heavy, low-grade materials

---

Convergence of modes (Continued)

tric motors in trains, and generate HDTV images. The electronic camera is another example of blurring between specific technologies and specific applications. While photographic film has only one purpose, the storage of an image, the image on a magnetic diskette in a modern still camera is stored by a technology originally developed for computer data.

Another factor contributing to the convergence of modes is the increasing process of cross-ownership. One company can own television networks, magazines, publishing houses, and cable television systems.

1993

In the US, FM radio stations begin to use a system already in place in Europe to transmit digital data along with the signal used to produce sound; the radio data system (RDS) is used to send messages that appear on a small display screen and for such purposes as replacing checks on the US emergency broadcasting system; *See also* 1986 COM

In a demonstration on Jan 28 of the first system for telephonic simultaneous translation, Japanese researcher Toshiyuki Takezawa speaks the word "moshimoshi" in Kyoto; his word is translated into Japanese text by one computer, which then passes it on to another computer that translates it into English and sends it via a modem and telephone lines to Pittsburgh PA, where yet another computer reads the text and synthesizes the English word "hello"; the process takes 12 seconds

A team led by Harry Jordan and Vincent Heuring at the University of Colorado unveils on Jan 12 the first general-purpose all-optical computer capable of being programmed and manipulating instructions internally; *See also* 1990 ELE

Intel Corp announces on Mar 22 that it is shipping its Pentium microprocessor to computer manufacturers; Pentium is the fifth generation of the basic chip that powers the IBM PC family of computers and their clones; the Pentium chip contains 3.1 million transistors and operates twice as fast as the best fourth-generation Intel chip, the 486DX2; *See also* 1990 ELE

Fujitsu in Japan announces manufacture of a 256-megabit memory chip; *See also* 1991 ELE

The Portland General Electric Company decides to retire its Trojan nuclear plant in Rainier OR rather than repair its corroded heat exchanger; *See also* 1992 ENE

Mazda Motors Corporation announces that it will build a production version of an automobile powered with a Miller-cycle internal combustion engine, which will produce higher power for lower fuel consumption; *See also* 1947 ENE

Magic bullets (part 2)

Since World War II, medical researchers have developed chemotherapeutic agents that can destroy viruses or cancer cells. The strongest magic bullets, however, are made by our own immune system, so long it is not compromised by immune-damaging diseases, such as AIDS.

The human immune system is complex, but one of the most easily accessible parts of it is a group of chemicals called antibodies that circulate in the bloodstream. Antibodies are the key to the success of vaccination, for example. Injection of a vaccine stimulates a kind of white blood cell (lymphocyte) called a B cell to produce antibodies against proteins in the vaccine. Such antibodies home in on specific foreign proteins so that they can be destroyed. Vaccination confers long-lasting immunity because the B cells that have been stimulated to produce the antibodies reproduce and the B-cell line remains in the bloodstream, prepared to produce antibodies against a reinvasion of the specific protein.

In 1975 César Milstein and Georges Köhler developed in England a method of producing large amounts of specific antibodies by a cloning technique. Antigens (chemicals that provoke immune reactions) from human cancer cells are injected into a mouse. The B cells in the mouse start producing antibodies that attack the antigens. These B cells are removed and fused with cancerous mouse B cells, forming hybrid cells called hybridomas that live and reproduce. The hybridomas continue to produce large amounts of antibodies. Probes are used to pick out a hybridoma that produces a specific desired antibody, and this cell is then reproduced in large quantities. Antibodies from such a cell line are called monoclonal because they are produced from the clones of a single hybridoma.

The original mouse-produced antibodies sometimes caused undesired immune reactions when injected into a human. More recently, scientists have used human cells to make the hybridomas, reducing such unwanted reactions.

Just like natural antibodies, monoclonal antibodies track down proteins or other chemicals with exquisite specificity. Thus, they can be used in diagnosis, one of the first applications. By combining a monoclonal antibody with a poison, cells that have a given proteins on their surface can be tracked down by the antibody and destroyed. Scientists are using this approach experimentally against cancer.

Calgene Inc. announces that it will apply for food additive status from the US Food and Drug Administration (FDA) for the marker gene known as *kan'r* in its genetically altered tomato termed the Flavr Savr; the Flavr Savr is scheduled to reach food stores in 1994, with or without FDA approval; *See also* 1980 GEN

In Iraq the Third River Canal between the Tigris and Euphrates is completed, primarily to provide water to wash salt out of the soil; the salt had been introduced by irrigation practices of the past; the canal would also drain large areas of marshland, making them suitable for conventional farming; it runs from Mahmudiya, near Baghdad, to Basra on the Persian Gulf, a distance of 560 km (350 mi); *See also* 1250–59 GEN

A team of scientists from Catalytica Inc. in Mountain View CA discovers a way to make methanol from methane that is cheap and works at a low temperature; the input is methane and air, catalyzed by ionized mercury, and the products are methanol and carbon dioxide; *See also* 1928 MAT

Procter & Gamble Company adds the enzyme cellulase to its popular detergent Cheer; cellulase decomposes cellulose, the main component of plant fibers and many artificial fibers as well; claiming that because cellulase in nature decomposes plant material, this is a natural adjunct to its product, Proctor & Gamble explains that the main reason for the addition is to promote digestion of damaged fibers in cotton fabrics, leaving the undamaged ones intact; *See also* 1957 MAT

Yvonne Bryson, Steven Miles, Erin Balden, and coworkers announce development of a fast and effective way to detect AIDS in newborns and infants; AIDS may be contracted in the womb, a condition that can be detected at birth, or during birth, which cannot be determined until four to six weeks later; *See also* 1990 MED

Susan Perrine and coworkers discover that administration of butyrate, a naturally occurring fatty acid, can successfully treat anemias caused by deformed hemoglobin, such as sickle-cell anemia; butyrate in the bloodstream promotes the development of fetal-type hemoglobin, which, because it is directed by different genes, is not affected by genetic diseases of normal hemoglobin; *See also* 1934 MED

Adam Software Inc. announces on Apr 7 the availability of its CS-ROM called Animated Dissection of Anatomy for Medicine, a virtual reality version of a cadaver, complete with all the organs; it can be dissected by medical students; *See also* 1832 MED

Apple Computer becomes the first company to offer a split computer keyboard for reduction of repetitive stress hand injuries such as carpal tunnel syndrome; the keyboard has been designed without reference to an earlier version patented by Tony Hodges, according to Apple; *See also* 1986 MED

A network of 10 radiotelescopes, called the Very Long Baseline Array (VLBA), uses interferometry to achieve angular resolutions 500 times better than the finest optical telescopes; *See also* 1957 TOO

Early in the year scientists from MIT launch in Antarctic waters the inexpensive tetherless (autonomous) robot submersible Odyssey, designed by James G. Bellingham and coworkers; the 2-m (7-ft) observer can dive to ocean depths as great as 5 km (3 mi); because Odyssey costs about a hundredth as much as the first significant robot submersible, Jason, and a thousandth the cost of the human-piloted *Alvin,* it is thought to inaugurate a new era of almost disposable robots for undersea exploration; *See also* 1962 TOO

In Mar a team from the Woods Hole Oceanographic Institution launches off Bermuda the Autonomous Benthic Explorer (ABE), a tetherless robot submersible designed to sink to ocean depths as great as 6 km (4 mi) and stay there for as long as 1 year if needed, observing for scientists on the ocean's surface; the 2-m (6-ft) ABE was designed by Albert M. Bradley and coworkers; *See also* 1962 TOO

NAME INDEX

Entries are located in two ways. Those in the Timetables are indicated by the year and the first three letters of the subject, such as 1990 TOO (Tools and Devices). Those in overviews or essays are indicated by a page number, preceded by p or by pp. The same system is used in the subject index, which follows. In this index, a middle name in *italics* is the name by which a person is commonly known.

Gaulard, Lucien: 1883 ELE
Gauss, Karl Friedrich: 1833 COM, p 242
Gay-Lussac, Joseph-Louis: 1824 TOO
Ged, William: 1720–29 COM, 1739 COM
Geiger, Hans: 1913 TOO, 1928 TOO
Genghis Kahn (Temujin): 1200–09 GEN
George III (King): 1772 TRA
Gerber, Christoph: 1985 TOO
Gerbert of Aurillac: 950 TOO
Gernsback, Hugo: 1926 GEN, 1928 COM
Gerstner, Franz Anton von: 1832 TRA
Gesner, Conrad von: 1560–69 COM
Ghega, Karl von: 1842 TRA
Giauque, William Francis: 1933 MAT
Gibbon, John H., Jr.: 1935 MED, 1951 MED, 1953 MED
Gibbs, John Dixon: 1883 ELE
Giffard, Henri: 1852 TRA
Gilbert, William: p 246
Gilchrist, Percy C.: 1876 MAT
Gillette, King Camp: 1901 GEN, 1903 TOO
Gillot, Charles: 1872 COM
Gillot, Firmin: 1793 COM, 1823 COM
Gilmore, Jack: 1955 COM
Giocondo, Giovanni (Fra): 1500–09 ARC
Giorgio Martini, Francesco di: 1470–79 ARC
Girard, Pierre-Simon: 1775 ENE, 1798 MAT
Glaser, Peter E.: 1977 ENE
Glass, Louis: 1889 COM
Glauber, Johann R.: 1630–39 FOO
Glenn, John H., Jr.: 1962 TRA, p 395
Glidden, Joseph Farwell: 1874 FOO
Göbel, Heinrich: 1854 GEN
Godbole, Vijay: 1991 MAT
Goddard, James L.: 1968 FOO
Goddard, Robert Hutchings: 1914 TRA, 1919 TRA, 1926 TRA, 1929 TRA
Godfrey, Thomas: 1730–32 TOO
Godley, Paul: 1921 COM
Goethals, George W.: 1914 TRA
Goldberger, Joseph: 1915 MED
Goldmark, Peter Mark: 1940 COM, 1948 COM
Goldschmidt, Johann *Hans*: 1898 MAT
Goldstine, Herman: 1944 ELE, p 376
Gooch, Daniel: 1840 TRA, 1846 TRA
Goodall, Jane: 1960 TOO, pp vi, 3
Goodhue, L.D.: 1941 GEN
Goodpasture, Ernest: 1931 MED
Goodrich, Robert R.: 1950 COM
Goodwin, Hannibal W.: 1887 MAT
Goodyear, Charles: 1839 MAT
Googe, Barnaby: 1570–79 FOO
Gorgas, William C.: 1904 MED
Gossage, William: 1835 MAT
Gould, Charles Henry: 1868 TOO

Gould, Gordon: 1957 TOO
Graham, George: 1710–19 TOO, 1720–29 TOO, 1765 TOO, p 144
Gramme, Zénobe Théophile: 1867 ENE, 1869 ENE, 1871 ENE, 1877 ENE, p 279
Grant, G.B.: 1887 TOO
Grassin-Baledans, Léonce-Eugène: 1860 FOO
Gratzel, Michael: 1991 ENE
Gray, Elisha: 1872 COM, 1876 COM, p 282
Gray, Truman S.: 1931 ELE
Graybeard, David (Chimpanzee): p vi
Green, John Van: 1966 COM
Greene, Catherine: p 216
Greene, Nathaniel: p 216
Greenwood, John: 1790 MED
Gregg, D.: 1963 COM
Grégoire, Jean A.: 1926 TRA
Gregory, James: 1660–69 TOO
Gregory XIII (Pope): 1580–89 GEN
Griffith, A.A.: 1919 MAT, 1920 MAT, 1929 ENE, 1937 ENE
Grijns, Gerrit: 1901 MED
Grimoin-Sanson, Raoul: 1900 COM
Grischkowsky, Daniel: 1984 TOO
Grissom, Virgil I.: 1965 TRA
Grove, William Robert: 1839 ENE, 1840 GEN
Gruentzig, Andreas R.: 1977 MED
Guericke, Otto von: 1640–49 TOO, 1650–59 GEN, 1660–69 ENE, 1660–69 TOO, p 246
Guérin, Camille: 1923 MED
Guibal, Théophile: 1858 TOO
Guillaume, Charles: 1896 MAT
Guillemin, Roger: 1968 MED
Guillet, Leon: 1904 MAT
Gunter, Edmund: 1620–29 TOO
Gurney, Goldsworth: 1825 GEN
Gurvitch, M.: 1983 ELE
Gutenberg, Johann Gensfleisch zum: 1440–49 COM, 1450–59 COM, p 119
Guthrie, Samuel: 1831 MED

Haber, Fritz: 1908 MAT, p 333
Hadfield, Robert Abbott (Sir): 1883 MAT
Hadley, John: 1730–32 TOO
Hadrian (Emperor): 120 ARC, 130 ARC
Haenlein, Paul: 1872 TRA
Hahn: 1774 TOO
Hahnemann, Christian Friedrich *Samuel*: 1810 MED, 1811 MED
Haise, Fred W., Jr.: 1970 TRA
Halbout, Jean-Marc: 1984 TOO
Haldane, John Scott: 1907 MED
Haldane, T.G.N.: 1930 ENE
Hale, Edward Everett: 1869 GEN
Hale, George Ellery: 1936 TOO, 1948 TOO, p 375
Hales, Stephen: 1720–29 MED, 1733 MED, 1752 ARC
Hall, Charles M.: 1886 MAT
Hall, John: 1811 TOO, 1820 TOO
Hall, Marshall: 1830 MED
Hallady, Daniel: 1854 ENE

Haller, Albrecht: 1747 MED
Halley, Edmond: 1710–19 TRA
Hallwach, Wilhelm: 1888 ELE
Halsted, William: 1890 MED
Hammurabi: 1800 BC GEN
Hampson, William: 1895 MAT
Han Chih-Ho: 890 TOO
Hanaman, Franz: 1907 GEN
Hancock, Thomas: 1845 MAT
Hancock, Walter: 1831 TRA
Hardy, James Daniel: 1963 MED
Hare, Robert: 1801 TOO
Hargrave, Lawrence: 1893 GEN
Hargreaves, James: 1764 TOO, p 158
Harington, John (Sir): 1580–89 ARC
Hariot, Thomas: 1580–89 MAT
Harpales: 500 BC ARC
Harrison, C.C.: 1860 COM
Harrison, John: 1720–29 TOO, 1735 TRA, 1759 TRA, 1761 TRA, 1765 TRA, 1772 TRA
Harrison, Joseph: 1859 ENE
Harrison, Ross Granville: 1907 MED
Harrison, William: 1570–79 GEN, 1570–79 FOO
Harsha Vardhana: 650 GEN
Hart, Charles: 1901 FOO
Hartley, Ralph V.L.: 1927 COM, pp 362, 410
Hartmann, Jon: p 30
Harvey, Fred: 1876 FOO
Harvey, James J.: 1893 TOO
Hascall, J.: 1838 FOO
Hata, Sahachiro: 1910 MED, p 356
Hatch, Fred: 1873 FOO
Hauck, Frederick H.: 1983 TRA
Hauksbee, Francis: 1700–09 GEN
Haüy, Valentin: 1783 COM, 1784 COM
Havas, Charles: 1835 COM
Havers, Clopton: 1690–99 MED
Hawkins, John Isaac: 1822 COM
Hayashi: 1939 FOO
Hazen, Elizabeth: 1948 MED
He: p 83
Hecataeus of Miletus: 500 BC COM
Heeger, Alan J.: 1977 MAT
Hele-Shaw, Henry S.: 1924 TRA
Heller, Adam: 1981 ENE
Helmholtz, Hermann: 1851 MED, pp 282, 300, 426
Helmont, Johann Baptista van: 1660–69 MED
Hench, Philip Showalter: 1948 MED
Henle, Friedrich: 1840 MED
Henlein, Peter: 1500–09 TOO
Hennebique, François: 1892 MAT
Henri IV (King): 1600–09 GEN
Henri, Frédéric: 1954 FOO
Henry (Prince, the Navigator): 1420–29 TRA
Henry III (King): 1240–49 ARC
Henry V (King): 1410–19 MAT
Henry VI (King): 1450–59 MAT
Henry VIII (King): 1510–19 COM
Henry, Edward Richard (Sir): 1901 GEN
Henry, Joseph: 1829 TOO, 1831 ENE, 1831 COM, 1832 ENE, 1835 ENE, 1835 COM, 1836 COM, pp 160, 242

Henry, the Boar, of Sens (Archbishop): 1130–39 ARC, p 90
Henson, W.S.: 1842 TRA
Herhan, Louis Etienne: 1793 COM
Hernández, Francisco: 1670–79 FOO
Hero (Heron) of Alexandria: 60 TOO, 100 GEN, pp 61, 101, 156, 304
Herod the Great: 0 BC ARC
Herodotus: p 84
Héroult, Paul-Louis-Toussaint: 1886 MAT
Herrera, Alonso: 1530–39 FOO
Herring, William *Conyers*: 1952 MAT
Herschel, John: 1839 COM
Herschel, William James (Sir): 1860 GEN
Herschel, William: 1781 TOO
Hertz, Heinrich: 1887 ELE, 1888 COM, p 281
Hesiod: p 101
Hess, A.: 1897 MAT
Hess, Walter Rudolf: 1948 MED
Hetherington, John: 1844 ENE
Heuring, Vincent: 1993 ELE
Hewlett, William R.: 1942 TOO
Hickman, Henry: 1824 MED
Hiero II (King): pp 56–57
Hilbert, David: p 352
Hill, John: 1753 GEN
Hill, William: 1911 MED
Hinks, Joseph: 1865 GEN
Hinton, Christopher: 1956 ENE
Hipparchus: 200 BC TOO
Hippocrates of Cos: 400 BC MAT, pp 29, 88, 92
Hire, J.N. de la: 1710–19 TOO
Hock, Julius: 1870 ENE
Hockham, Georges: 1966 COM, p 439
Hodges, Tony: 1986 MED
Hoe, Richard March: 1847 COM, 1853 COM
Hoerni, Jean: 1959 ELE
Hoff, Ted: 1971 ELE
Hoffman, Felix: 1893 MED
Hoffman, Jacob R.: 1869 TOO
Hofmann, August Wilhelm von: 1845 MAT, 1858 MAT
Hofstein, Steven: 1979 ELE
Hog, Ralph: 1540–49 TOO
Hollerith, Herman: 1890 ELE, 1896 COM, 1901 ELE, 1911 ELE
Holly, Alexander L.: 1870 MAT
Holmes, E.L.: 1935 GEN
Holmes, Oliver Wendell: 1843 MED
Holmgren, G.: 1921 MED
Holt, Benjamin: 1904 TRA
Holter, Norman J.: 1959 MED
Holzwarth, Hans: 1903 ENE, 1908 ENE, 1930 ENE
Homer: p 101
Homer, R.F.: 1955 FOO
Honda, Kotaro: 1916 MAT
Honnecourt, Villard de. See Villard de Honnecourt
Honold, G.: 1902 ENE
Hood, Leroy: 1982 GEN
Hooke, Robert: 1650–59 TOO, 1660–69 MAT, 1660–69 TOO, 1670–79 MAT, 1670–79 TOO, 1680–89 COM

Rickover, Hyman George: 1947 TRA, 1954 TRA, 1957 ENE
Ride, Sally K.: 1983 TRA
Righi, Augusto: p 306
Ringer, Sydney: 1883 MED
Rinio, Benedetto: 1410–19 MED
Riquet, Pierre-Paul: 1660–69 TRA, 1680–89 TRA, 1690–99 TRA
Ritchie, Dennis: 1970 COM, 1972 COM
Ritty, James: 1879 TOO
Riva, Carlo: 1710–19 MAT
Robert, Nicolas Louis: 1798 MAT
Roberts, Edward: 1975 ELE
Roberts, Michael: 1958 COM
Roberts, Richard: 1817 TOO
Robins, Benjamin: 1742 TOO
Robinson, Arthur: 1988 COM
Robiquet, Pierre-Jean: 1831 MAT
Rock, John: 1956 MED
Rockefeller, John D.: 1865 MAT
Roebling, Emily Warren: 1883 TRA
Roebling, John Augustus: 1841 MAT, 1845 ARC, 1855 ARC, 1869 ARC, p 160
Roebling, Washington Augustus: 1883 TRA
Roebuck, John: 1746 MAT, 1762 MAT, p 164
Roentgen, Wilhelm Konrad: 1895 MED, p 309
Rogers, Howard G.: 1963 COM
Rohrer, Heinrich: 1981 TOO, 1986 TOO
Romain, P.A.: 1785 TRA
Roman, François: 1680–89 ARC
Romanenko, Yuri V.: 1987 TRA
Romulus: 750 BC GEN
Roosevelt, Franklin D. (President): p 383
Rose, Albert: 1946 COM
Roselius, Ludwig: 1903 FOO
Rosenblatt, Frank: 1960 ELE
Rosetti, Gioaventura: 1540–49 MAT
Rosing, Boris: 1907 COM
Ross, Thomas: 1938 COM
Rous, Francis *Peyton*: p 284
Rowland, F. *Sherwood*: 1974 MAT
Rtcheouloff, Boris: 1927 COM
Rubel, Ira W.: 1904 COM
Rubin (Major): 1880 TOO
Rubin, Benjamin A.: 1965 MED
Rudder, Ed: 1985 COM
Rudge, William E.: 1991 TOO
Rudolph, Paul: 1890 COM
Rudyerd, John: 1700–09 ARC
Ruhmkorff, Heinrich: 1851 ENE
Ruska, Ernst: 1933 TOO
Russell, John Scott: 1834 GEN
Russell, Sidney: 1912 GEN
Rutherford, Daniel: 1794 TOO

Sabin, Albert Bruce: 1957 MED, p 383
Sabine, R.: 1859 COM
Sabine, Wallace Clement Ware: 1895 GEN
Sagan, Carl: 1983 GEN
Saint-Victor, Abel Niepce de: 1847 COM
Saint-Victor, Hugues de: p 89
Sainte-Claire Deville, Henri: 1854 MAT
Sakel, Manfred J.: 1929 MED

Sakharov, Andrei: 1947 ENE
Sakmann, Bert: 1976 MED
Salk, Jonas Edward: 1952 MED, p 383
Salomen, A.: 1913 MED
Salva, Francisco: 1796 COM
Sanctorius, Sanctorius: 1600–09 MED, 1610–19 TOO
Sanger, Frederick: 1953 MED
Sauer, Carl: p 77
Sauerbronn, Drais von: 1818 TRA
Sauria, Charles: 1830 GEN
Saussure, Horace Benedict de: 1766 TRA, 1783 TOO
Savage, John Lucian: 1933 FOO
Savery, Thomas: 1690–99 ENE, 1700–09 ARC, 1700–09 ENE, 1710–19 TRA, pp 156, 159, 182
Savitskaya, Svetlana: 1984 TRA
Sayce, B.J.: 1864 COM
Scaligier, Joseph Justus: 1580–89 GEN
Scaurus, Marcus Aemilius: 100 BC FOO
Schallenberger, Oliver: 1888 ELE
Schawlow, Arthur L.: 1958 TOO, p 421
Scheider, Ralph: 1950 GEN
Scheutz, Edward: 1834 TOO, 1855 TOO
Scheutz, Pehr Georg: 1834 TOO, 1853 GEN, 1855 TOO
Schick, Bela: 1913 MED
Schick, Joseph: 1928 GEN
Schickard, Wilhelm: 1620–29 TOO, 1957 ELE
Schilling, Baron: 1832 COM
Schnitzler, J.: 1860 COM
Schoeffer, Peter: 1450–59 COM, p 119
Schoenbein, Christian: 1846 MAT
Scholl, William: 1904 MED
Schöner, Johannes: 1510–19 COM
Schott, Gaspar: 1660–69 TOO
Schott, Otto: 1888 MAT
Schreyer, Helmuth: p 365
Schrieffer, John R.: 1957 MAT
Schultes, Johann: 1650–59 MED
Schulze, Johann H.: 1720–29 COM
Schurer, Christoph: 1540–49 MAT
Schwarz, Bernard: 1310–19 TOO
Schwarz, David: 1897 TRA
Schweigger, Johann: 1820 TOO
Scobee, Francis: 1983 TRA
Scott de Martinville, Léon: 1856 COM
Scragg, Thomas: 1845 FOO
Scully, John: p 435
Scylax the Younger: 350 BC COM
Seebeck, Thomas *Johann*: 1810 COM, 1821 TOO
Seely, Henry W.: 1882 GEN
Seguin, Marc: 1822 ARC, 1828 TRA
Sejnowski, Terry J.: 1986 COM
Selfridge, Gordon: p 342
Seliger, Robert L.: 1978 TOO
Sellars, William: 1864 GEN
Semmelweiss, Ignaz Philipp: 1847 MED
Semon, Waldo L.: 1930 MAT
Senefelder, Aloys: 1798 COM, 1818 COM
Sennacherib: 700 BC ARC
Senning, Ake: 1959 MED

Senusret I (Pharaoh): 2000 BC COM
Septima Zenobia: 270 COM
Serson: 1744 TRA
Sesotris II: 1900 BC FOO
Shaffer, Thomas H.: 1989 MED
Shannon, Claude Elwood: 1938 COM, 1949 COM, pp 345, 362, 410
Shatalov, Vladimir A.: 1971 TRA
Shaw, J. Cliff: 1957 COM
Shen Kua: 1080–89 TRA
Shepard, Alan B. Jr.: 1961 TRA
Shepard, Edward G.: 1850 GEN
Shi Huangdi (Emperor): 200 BC GEN, 200 BC COM
Shih Tsung: 950 MAT
Shirakawa, Hideki: 1972 MAT, 1977 MAT
Shockley, William Bradford: 1942 ELE, 1948 ELE, 1951 ELE, pp 345, 370, 382
Sholes, Christopher *Latham*: 1867 COM, 1872 COM
Shortlife, Edward: 1970 COM
Shrapnel, Henry: 1784 GEN
Shrayer, Michael: 1975 COM
Shreeve, Herbert: 1903 COM
Shu Jse (King): 500 BC GEN
Shultes, Johann: 1650–59 MED
Siddhartha Gautama (Buddha): 550 BC GEN
Siemens, Ernst *Werner* von: 1842 COM, 1844 TOO, 1846 COM, 1847 GEN, 1857 ENE, 1876 TOO, 1879 TRA
Siemens, Friedrich: 1856 MAT, 1868 MAT
Siemens, William (Sir): 1851 TOO, 1867 ENE, 1868 MAT, 1879 MAT
Siems, Ruth: 1975 FOO
Sigl, Georg: 1851 COM
Sikorsky, Igor: 1913 TRA, 1939 TRA, 1944 TRA
Silliman, Benjamin: 1855 ENE
Simon, Henry: 1881 MAT
Simon, Herbert A.: 1957 COM
Simpson, James (Sir): 1846 MED
Sinclair, Clive (Sir): 1981 ELE
Singer, Isaac Merrit: 1851 TOO, 1857 TOO
Sinsheimer, Robert: 1984 GEN
Sivrac, de (Count): 1790 TRA
Skolnick, Mark H.: 1978 MED
Slater, Mrs. Samuel: 1793 TOO
Sloan, Alfred P.: 1920 GEN
Slotnick, Dan: 1965 ELE
Smalley, Richard E.: 1985 MAT
Smeaton, John: 1752 ENE, 1754 MAT, 1759 ENE, 1759 MAT, 1765 ENE, 1766 TRA, 1772 ENE, pp 95, 103, 160, 164, 182
Smellie, William: 1752 MED
Smirke, Robert: 1823 ARC
Smith, Adam: 1776 GEN
Smith, Beauchamp E.: 1941 ENE
Smith, G.: 1908 COM
Smith, Michael J.: 1983 TRA
Smith, Oberlin: 1888 COM
Snefru (Pharaoh): 2600 BC ARC, 2600 BC COM
Snell, Willebrord: 1610–19 GEN
Snow, C.P.: p 409
Snow, John: 1854 MED
Soane, John: 1788 ARC

Sobrero, Ascanio: 1846 MAT
Soderblom, Olof: 1967 COM
Soemmering, Samuel: 1809 COM, 1812 COM
Solvay, Ernest: 1861 MAT
Sommeiller, Germain: 1858 TOO, 1870 GEN
Sosigenes: 46 BC GEN
Sostratos of Knidos: 300 BC ARC
Southern, John: 1796 ENE
Spallanzani, Lazzaro: 1765 FOO
Sparks, Morgan: 1951 ELE
Spencer, Christopher M.: p 284
Spencer, Percy L.: 1953 FOO
Sperry, Elmer Ambrose: 1908 TRA, 1911 TRA, 1912 TRA, 1914 TRA, p 340
Spill, Daniel: 1868 MAT
Spittler, Adolf: 1897 MAT
Spode II, Josiah: 1797 MAT
Sprague, Frank Julian: 1884 TRA, 1888 TRA
Sprengel, Christian Konrad: 1865 TOO, 1873 TOO
Sprengel, Hermann P.: 1866 TOO, 1871 MAT, 1873 TOO
Squier, George: 1922 GEN
Staite, W.E.: 1845 GEN
Stanley, John: 1822 TOO
Stanley, William: 1885 ELE
Stapp, John Paul: 1949 GEN
Starzl, Thomas: 1963 MED
Staudinger, Hermann: 1926 MAT
Stein, P.: 1956 COM
Steinheil, Adolph: 1866 COM
Steinheil, Karl August: 1837 COM, pp 242, 243
Steinman, David: 1929 TRA
Steinmetz, Charles Proteus: 1910 ENE
Stephen, George A.: 1950 ENE
Stephen, W.E.: 1953 MAT
Stephenson, George: 1814 TRA, 1825 TRA, 1829 TRA, p 160
Stephenson, Joseph: 1841 ENE
Steptoe, Patrick: 1969 MED
Stevens, John C.: 1802 TRA, 1812 TOO, 1836 TRA
Stevinius, Simon: 1580–89 GEN, 1580–89 ENE
Stewart, John: 1767 ENE
Stibitz, Georges: 1937 ELE, 1939 ELE, 1940 COM, p 362
Stirling, Robert: 1817 ENE
Strauss, Joseph: 1937 TRA
Strauss, Lewis: 1955 ENE, p 427
Stringfellow, John: 1868 TRA
Strowger, Almon B.: 1892 COM, p 283
Strutt, Jedediah: 1758 TOO
Stuart, Akroyd: 1890 ENE
Stubbs, J.: 1955 FOO
Stumpf, Johann: 1908 ENE
Sturtevant, Simon: 1610–19 MAT
Su Sung: 1080–89 TOO
Suger (Abbé): 1120–29 ARC, 1140–49 ARC, p 90
Sugg, William: 1879 GEN
Sullivan W.N.: 1941 GEN
Sullivan, Louis: p 298
Sultan (Chimpanzee): p vi
Sulzberger, Jacob: 1834 FOO
Sundback, Gideon: 1913 GEN, 1914 GEN, 1923 GEN
Suomi, Verner E.: 1964 TOO

SUBJECT INDEX

472

477

481

485